MW01517682

Biomanagement of Metal-Contaminated Soils

ENVIRONMENTAL POLLUTION

VOLUME 20

Editors

Brain J. Alloway, *Department of Soil Science, The University of Reading, U.K.*
Jack T. Trevors, *School of Environmental Sciences, University of Guelph,*
Ontario, Canada

Editorial Board

For further volumes:
http://www.springer.com/series/5929

Mohammad Saghir Khan • Almas Zaidi
Reeta Goel • Javed Musarrat
Editors

Biomanagement
of Metal-Contaminated Soils

 Springer

Editors

Mohammad Saghir Khan
Department of Agricultural Microbiology
Faculty of Agricultural Sciences
Aligarh Muslim University
Aligarh, Uttar Pradesh, India
khanms17@rediffmail.com

Reeta Goel
Department of Microbiology
Govind Ballabh Pant University
of Agriculture & Technology
Pantnagar, Uttarakhand, India
rg55@rediffmail.com

Almas Zaidi
Department of Agricultural Microbiology
Faculty of Agricultural Sciences
Aligarh Muslim University
Aligarh, Uttar Pradesh, India
alma29@rediffmail.com

Javed Musarrat
Department of Zoology
King Saud University
Riyadh, Saudi Arabia
musarratj1@yahoo.com

ISSN 1566-0745
ISBN 978-94-007-1913-2   e-ISBN 978-94-007-1914-9
DOI 10.1007/978-94-007-1914-9
Springer Dordrecht Heidelberg London New York

Library of Congress Control Number: 2011935735

Cover image: © 2011 JupiterImages Corporation

Printed on acid-free paper

Springer is part of Springer Science+Business Media (www.springer.com)

Dedicated in memory of Dr. Meena Zaidi

Preface

The soil environment is a major sink for a multitude of chemicals and heavy metals, which inevitably leads to environmental contamination problems. Indeed, a plethora of different types of heavy metals are used and emanated through various human activities including agricultural, urban or industrial. Millions of tonnes of trace elements are produced every year from mines in demand for newer materials. On being discharged into soil, the heavy metals get accumulated and may disturb the soil ecosystem including microbial compositions and biomass, microbial community structure and their biological activities. Moreover, the excessive concentration of metals in soil can also elicit a wide range of visible and physiological symptoms in plants leading to losses in crop productivity. As a result, heavy metal pollution poses a major threat to human health and environment. Unlike many other pollutants, heavy metals cannot be biologically degraded to more or less toxic products and hence persist in the environment. Toxic metal pollution is, therefore, an enigma for scientists how to tackle this issue that has threatened the environment. Management of metal contaminated environment, especially soils, therefore, becomes important, as these soils usually cover large areas that are rendered unsuitable for sustainable agriculture. The remediation of such soils, in turn, could lead to food security across the globe. To address this environmental threat, conventional remediation approaches have been applied, which, however, do not provide acceptable solutions due either to the technological constraints or to the production of large quantities of toxic products. Therefore, the establishment of efficient, inexpensive and safe and environment friendly methodology and techniques for identifying and limiting or preventing metal pollution, causing threats to the agricultural production systems and human health, is earnestly required. In this regard, the management of contaminated soils using microbes or other biological systems to degrade/transform environmental pollutants under controlled conditions to an innocuous state or to levels below concentration limits established by regulatory authorities is required. Bioremediation in this context, applied in the rehabilitation of heavy-metal-contaminated soil, has been found interesting because it provides an ecologically sound and economically viable method for restoration and remediation of derelict soils. In this regard, biological agents including both heterogeneously distributed microbial communities and plants of various origins could play a pivotal role in the management of metal polluted soils. Besides their role in protecting the plants from metal toxicity, the microbes are also well known for their biological activities enhancing the soil fertility and promoting plant growth by providing essential nutrients and growth regulators. Use of such microbes possessing multiple properties of metal resistance/reduction and ability to promote plant growth through different mechanisms in metal contaminated soils make them one of the most suitable choices for bioremediation studies. Advances in understanding the role of microorganisms in such processes, together with the ability to fine-tune their activities using the tools of molecular biology, would, therefore, lead to the development of novel or improved metal bioremediation processes. Both plants and microbial strategies for managing contaminated soils could however be different for different agro-ecosystems. The role of non-living microbial communities in controlling the mobility and bioavailability of metal ions is well known. The use of biomass to extract heavy metals is, therefore, an area of current research.

Biomanagement of Metal Contaminated Soils integrates the frontiers of knowledge on both fundamentals and practical aspects of remediation of metal polluted soils. The book written by experts in the field provide unique, updated and comprehensive information on strategies as to how metal contaminated soils could be remediated, exploited and practiced for increasing the productivity of crops in varied production systems. This book covers a broad area including from sources of heavy-metal pollution to metal toxicity to plants to remediation strategies. Therefore, various bioremediation approaches adopted to remediate contaminated sites and major concerns associated with phytoremediation as a sustainable alternative are reviewed and discussed. Legumes have traditionally been used in soil regeneration, owing to their capacity to increase soil nitrogen due to biological nitrogen fixation. Recently, legumes have also attracted attention for their role in remediation of metal contaminated soils. Given the importance of *Rhizobium*–legume interactions in maintaining soil fertility, attention is paid to explain the role of this symbiosis and approaches employed to genetically engineer legume–*Rhizobium* pairing in order to improve bioremediation. Information relative to the mechanism of metal tolerance and the importance of arbuscular mycorrhizal fungi in the detoxification of metal polluted soils are explored. Research advances in bioremediation of soils and groundwater using plant-based systems, the ecological and evolutionary implications of endophytic bacterial flora of the nickel hyperaccumulator plant *Alyssum bertolonii*, use of biosurfactants of various origins in the removal of heavy metal ions from soils, metal signaling in plants and new possibilities for crop management under cadmium contaminated soils are also broadly covered in this book. Microbial management of highly toxic metals like cadmium and arsenic present in soil, various phytotechnologies employed in remediation of heavy metal contaminated soils and a selective overview of past achievements and current perspective of chromium remediation technologies using promising microorganisms and plants are highlighted separately. Furthermore, the possible genotoxic effects of heavy metals on plants and other organisms and the development and applications of new biomonitoring methodologies for assessment of soil/plant genotoxicity have been sufficiently discussed in this chapter. The application of biomonitoring protocols in conjunction with the genotoxic assessment of contaminated soil will be advantageous in effective management of heavy metal polluted soils. The mobility and availability of toxic metals after soil washing with chelating agents and decontamination of radioactive-contaminated soils are addressed. Removal of heavy metals by microalgal biomass, the intrinsic and extrinsic factors affecting uptake of metals by microalgae and how these microalgae can be helpful in removing metals from polluted environments are reviewed and highlighted. The transgenic approaches centrally important in metal uptake, compartmentalization and/or translocation to organs, improved production of intracellular metal-detoxifying chelators and (over) production of novel enzymes are discussed. Efforts are also directed to obtain better molecular insights into the metallomics and physiology of hyperaccumulating plants, which are likely to provide candidate genes suitable for phytoremediation. The book further describes how the bioremediation potential of heavy-metal resistant novel actinobacteria, like *Amycolatopsis tucumanensis*, and maize plants could be exploited in detoxifying heavy metals in the polluted soil microcosm. The importance of free-living fungi in metal sorption and plant growth promotion in different agro-ecosystems are dealt separately.

This book collectively involves different bioremediation strategies used in metal removal from contaminated environments and crop production in metal stressed agro-ecosystems. This book contains a wealth of information for the person who needs to remove pollutants from soil or water. It describes the degree of success that can be achieved in removing a variety of metals. The knowledge and methodologies described in this book offer invaluable research tools, which may serve as important and updated source material. This edition provides an authoritative overview for individuals interested in bioremediation technologies. This book will, therefore, be of great interest to research scientists, postgraduate students, bioscience professionals, decision makers and farmers who intend to use natural resources for the abatement of metal contamination. It would also serve as a valuable resource for agronomists, environmentalists, soil microbiologists, soil scientists, biologists and biotechnologists involved in the restoration of contaminated lands. Thus, this book will cover the most interesting and applied aspects of phytoremediation and the role of microbial communities in crop productivity in soils contaminated with heavy metals, written by specialists who provide the scientific community with a critical evaluation of the management of metal contaminated soils.

We are highly thankful to our well qualified and proficient colleagues from across the world for providing the state-of-the-art scientific information to make this book a reality. All chapters are well exemplified with appropriate tables and figures, and enriched with extensive and the latest literature. The help and support provided by research scholars in designing and preparing the illustrations presented in this book are greatly acknowledged. We are indeed very grateful to our family members for their untiring and sustained support during the processing of this book. And most of all, we are extremely thankful to our adorable children, Zainab and Butool, for their patient and helpful attitude all through the project. We appreciate the great efforts of book publishing team at Springer-Verlag, the Netherlands, in responding to all our queries very promptly and earnestly. Finally, this book may have some basic mistakes or printing errors that might have occurred inadvertently during compilation, for which we regret in anticipation. If pointed out at any stage they will definitely be corrected and improved in subsequent prints/editions. Suggestions about the text and presentation are most welcome.

Mohammad Saghir Khan
Almas Zaidi
Reeta Goel
Javed Musarrat

Contents

Contributors

Carlos M. Abate Planta Piloto de Procesos Industriales y Microbiológicos (PROIMI-CONICET), Tucumán, Argentina

Universidad Nacional de Tucumán, Avenida Belgrano y Pasaje Caseros, 4000 Tucumán, Argentina

Mamdoh F. Abdel-Sabour Nuclear Research Center, Atomic Energy Authority, P.O. 13759, Cairo, Egypt, freemfs73@yahoo.com

Reda Abd El-Aziz Ibrahim Abou-Shanab Environmental Biotechnology Department, Genetic Engineering and Biotechnology Research Institute (GEBRI), City of Scientific Research and Technological Applications, New Borg El Arab City, Alexandria 21934, Egypt, redaabushanab@yahoo.com

Yeşim Sağ Açıkel Department of Chemical Engineering, Hacettepe University, 06800 Beytepe, Ankara, Turkey, yesims@hacettepe.edu.tr

Ees Ahmad Department of Agricultural Microbiology, Faculty of Agricultural Sciences, Aligarh Muslim University, Aligarh 202002, Uttar Pradesh, India

Virginia H. Albarracín Planta Piloto de Procesos Industriales y Microbiológicos (PROIMI-CONICET), Tucumán, Argentina

Universidad Nacional de Tucumán, Avenida Belgrano y Pasaje Caseros, 4000 Tucumán, Argentina

Sas-Nowosielska Aleksandra Institute for Ecology of Industrial Areas, Kossutha 6 Street, Katowice, Poland, sas@ietu.katowice.pl

Abdulaziz A. Al-Khedhairy Department of Zoology, College of Science, King Saud University, Riyadh 11451, Saudi Arabia

María J. Amoroso Planta Piloto de Procesos Industriales y Microbiológicos (PROIMI-CONICET), Tucumán, Argentina

Universidad del Norte Santo Tomás de Aquino, Tucumán, Argentina

Universidad Nacional de Tucumán, Avenida Belgrano y Pasaje Caseros, 4000 Tucumán, Argentina

K.K.I.U. Aruna Kumara Department of Crop Science, Faculty of Agriculture, University of Ruhuna, Mapalana, Kamburupitiya, Sri Lanka, kkiuaruna@crop.ruh.ac.lk

Marco Bazzicalupo Department of Evolutionary Biology, University of Firenze, via Romana 17, I-50125 Florence, Italy

Claudia S. Benimeli Planta Piloto de Procesos Industriales y Microbiológicos (PROIMI-CONICET), Tucumán, Argentina

Universidad del Norte Santo Tomás de Aquino, Tucumán, Argentina, cbenimeli@yahoo.com.ar

Paula M.L. Castro CBQF/Escola Superior de Biotecnologia, Universidade Católica Portuguesa, Rua Dr. António Bernardino de Almeida, P-4200-072, Porto, Portugal

Miguel Ángel Caviedes Departamento de Microbiología y Parasitología, Facultad de Farmacia, Universidad de Sevilla, Profesor García González, 2, Sevilla 41012, Spain

Nilanjana Das School of Biosciences and Technology, VIT University, Vellore 632014, Tamil Nadu, India, nilanjana00@lycos.com

Andrew Agbontalor Erakhrumen Department of Forest Resources Management, University of Ibadan, Ibadan, Nigeria, erakhrumen@yahoo.com

Etelvina Figueira Centre for Cell Biology, Biology Department, University of Aveiro, Universidade de Aveiro, Aveiro, Portugal, efigueira@ua.pt

Reeta Goel Department of Microbiology, Govind Ballabh Pant University of Agriculture and Technology, Pantnagar 263145, Uttarakhand, India, rg55@rediffmail.com

Abhishek Gupta Department of Microbiology, Govind Ballabh Pant University of Agriculture and Technology, Pantnagar 263145, Uttarakhand, India

Marya Hashmatt Department of Civil and Environmental Engineering, University of Auckland, Auckland, New Zealand

Arshad Javaid Institute of Agricultural Sciences, University of the Punjab, Quai-e-Azam Campus, Lahore, Pakistan, arshadjpk@yahoo.com

Anthea Johnson Department of Civil and Environmental Engineering, University of Auckland, Auckland, New Zealand

Anil Kapri Department of Microbiology, College of Basic Sciences & Humanities, Govind Ballabh Pant University of Agriculture & Technology, Pantnagar 263145, Uttarakhand, India

Mohammad Saghir Khan Department of Agricultural Microbiology, Faculty of Agricultural Sciences, Aligarh Muslim University, Aligarh 202002, Uttar Pradesh, India

Pavel Kotrba Department of Biochemistry and Microbiology, Institute of Chemical Technology, Prague, Technická 5, Prague 166 28, Czech Republic, pavel.kotrba@vscht.cz

Alejandro Lafuente Departamento de Microbiología y Parasitología, Facultad de Farmacia, Universidad de Sevilla, Profesor García González, 2, Sevilla 41012, Spain

Domen Lestan Department of Agronomy, Biotechnical Faculty, Centre for Soil and Environmental Science, University of Ljubljana, Jamnikarjeva 101, Sl-1000 Ljubljana, Slovenia, Domen.Lestan@bf.uni-lj.si

Ana Lima Centre for Cell Biology, Biology Department, University of Aveiro, Universidade de Aveiro, Aveiro, Portugal

Tomas Macek Department of Biochemistry and Microbiology, Institute of Chemical Technology, Prague, Technická 5, Prague 166 28, Czech Republic

IOCB & ICT Joint laboratory, Institute of Organic Chemistry and Biochemistry, Academy of Sciences of the Czech Republic, Flemingovo sq. 2, 166 10 Prague, Czech Republic

Martina Mackova Department of Biochemistry and Microbiology, Institute of Chemical Technology, Prague, Technická 5, Prague 166 28, Czech Republic

F. Xavier Malcata ISMAI – Instituto Superior da Maia, Avenida Carlos Oliveira Campos, Castelo da Maia, Avioso S. Pedro P-4475-690, Portugal

CIMAR/CIIMAR – Centro Interdisciplinar de Investigação Marinha e Ambiental, Rua dos Bragas no 289, P-4050-123, Porto, Portugal, fmalcata@ismai.pt

Lazar Mathew School of Biosciences and Technology, VIT University, Vellore 632014, Tamil Nadu, India, lazarmathew@rediffmail.com

Alessio Mengoni Department of Evolutionary Biology, University of Firenze, via Romana 17, I-50125 Florence, Italy, alessio.mengoni@unifi.it

Ali Seid Mohammed Department of Biology, Bahir Dar University, Bahir Dar, Ethiopia

Cristina M. Monteiro CBQF/Escola Superior de Biotecnologia, Universidade Católica Portuguesa, Rua Dr. António Bernardino de Almeida, P-4200-072 Porto, Portugal

Javed Musarrat Al-Jeraisy Chair for DNA Research, Department of Zoology, College of Science, King Saud University, Riyadh 11451, Saudi Arabia

Department of Agricultural Microbiology, Faculty of Agricultural Sciences, Aligarh Muslim University, Aligarh 202002, Uttar Pradesh, India, musarratj1@yahoo.com

Mohammad Oves Department of Agricultural Microbiology, Faculty of Agricultural Sciences, Aligarh Muslim University, Aligarh 202002, Uttar Pradesh, India

Eloísa Pajuelo Departamento de Microbiología y Parasitología, Facultad de Farmacia, Universidad de Sevilla, Profesor García González, 2, Sevilla 41012, Spain, epajuelo@us.es

Francesco Pini Department of Evolutionary Biology, University of Firenze, via Romana 17, I-50125 Florence, Italy

Marta A. Polti Planta Piloto de Procesos Industriales y Microbiológicos (PROIMI-CONICET), Tucumán, Argentina

Universidad Nacional de Tucumán, Avenida Belgrano y Pasaje Caseros, 4000 Tucumán, Argentina

Jai Prakash Narain Rai Ecotechnology Laboratory, Department of Environmental Sciences, Govind Ballabh Pant University of Agriculture & Technology, Pantnagar 263145, Uttarakhand, India, jamesraionline@gmail.com

Ignacio David Rodríguez-Llorente Departamento de Microbiología y Parasitología, Facultad de Farmacia, Universidad de Sevilla, Profesor García González, 2, Sevilla 41012, Spain

Bhoomika Saluja Department of Microbiology, Govind Ballabh Pant University of Agriculture and Technology, Pantnagar 263145, Uttarakhand, India

Shweta Saraswat Ecotechnology Laboratory, Department of Environmental Sciences, Govind Ballabh Pant University of Agriculture & Technology, Pantnagar 263145, Uttarakhand, India

Maqsood Ahmad Siddiqui Department of Zoology, College of Science, King Saud University, Riyadh 11451, Saudi Arabia

Naresh Singhal Department of Civil and Environmental Engineering, University of Auckland, Auckland, New Zealand, n.singhal@auckland.ac.nz

Metka Udovic Department of Agronomy, Biotechnical Faculty, Centre for Soil and Environmental Science, University of Ljubljana, Jamnikarjeva 101, Sl-1000 Ljubljana, Slovenia

Almas Zaidi Department of Agricultural Microbiology, Faculty of Agricultural Sciences, Aligarh Muslim University, Aligarh 202002, Uttar Pradesh, India, alma29@rediffmail.com

Chapter 1
Heavy Metal Pollution: Source, Impact, and Remedies

Ali Seid Mohammed, Anil Kapri, and Reeta Goel

Abstract Although some heavy metals are essential trace elements, most of them can be toxic to all forms of life at high concentrations due to formation of complex compounds within the cell. Unlike organic pollutants, heavy metals once introduced into the environment cannot be biodegraded. They persist indefinitely and cause pollution of air, water, and soils. Thus, the main strategies of pollution control are to reduce the bioavailability, mobility, and toxicity of metals. Methods for remediation of heavy metal-contaminated environments include physical removal, detoxification, bioleaching, and phytoremediation. Because heavy metals are increasingly found in microbial habitats due to natural and industrial processes, microorganisms have evolved several mechanisms to tolerate their presence by adsorption, complexation, or chemical reduction of metal ions or to use them as terminal electron acceptors in anaerobic respiration. In heavy metals, pollution abatement, microbial sensors, and transformations are getting increased focus because of high efficiency and cost effectiveness. The sources and impacts of heavy metal pollution as well as various remediation techniques are described.

Keywords Bioremediation • Heavy metal • Metal toxicity • Phytoremediation

A.S. Mohammed
Department of Biology, Bahir Dar University, Bahir Dar, Ethiopia

A. Kapri • R. Goel (✉)
Department of Microbiology, College of Basic Sciences & Humanities,
Govind Ballabh Pant University of Agriculture & Technology,
Pantnagar 263145, Uttarakhand, India
e-mail: rg55@rediffmail.com

M.S. Khan et al. (eds.), *Biomanagement of Metal-Contaminated Soils*,
Environmental Pollution 20, DOI 10.1007/978-94-007-1914-9_1,
© Springer Science+Business Media B.V. 2011

1.1   Introduction

Heavy metal is a general collective term that applies to the group of metals and metalloids with density greater than 4 ± 1 g/cm³. Although it is a loosely defined term, it is widely recognized and usually applied to the widespread contaminants of terrestrial and freshwater ecosystems (Duffus 2002). These metals occur naturally in the earth crust and are found in soils, rocks, sediments, waters, and microorganisms with natural background concentrations. Anthropogenic releases of them can give rise to higher concentrations of metals into the environment. Since heavy metals cannot be degraded or destroyed, they persist in the environment. Most of the discussions on heavy metals covered here include cadmium, chromium, copper, mercury, lead, zinc, arsenic, boron, and the platinum group metals.

The global industrial revolution has led to an unprecedented dissemination of toxic substances in the environment. Exposure to these pollutants especially through dietary intake of plant-derived food and beverages, drinking water, or air can have long-term effects on human health (Chaffei et al. 2004; Godt et al. 2006; Järup and Akesson 2009). For example, long-term exposure to lead in humans can cause acute or chronic damage to the nervous system, while long-term exposure of cadmium in humans is associated with renal dysfunction and obstructive lung disease and has also been linked to lung cancer and damage to human's respiratory systems. Due to industrialization, or injudicious applications of agrochemicals like phosphate fertilizers, which show a big load of metals like cadmium, the amounts of heavy metals deposited onto the surface of the Earth are however many times greater than depositions from natural background sources. In Scandinavia, for example, cadmium concentration in agricultural soil increases by 0.2% per year. The presence of heavy metals in waste as a result of their uses in modern society is a matter of ever-growing concern. This article covers information on sources of heavy metals, their harmful effects, problems posed by the disposal and recycling of heavy metal containing products, and to find options for abatement methodologies.

1.2   Sources of Heavy Metals

Heavy metals occur as natural constituents of the earth crust and are also released due to human activities. Then, they become persistent environmental contaminants since they cannot be degraded or destroyed. They enter the body system through food, air, and water and bio-accumulate over a period of time (UNEP/GPA 2004). Heavy metals can be emitted into the environment both by natural and anthropogenic routes.

1.2.1   Natural Sources

Some high heavy metal concentrations in soils could be natural in origin, resulting from weathering of the underlying bedrock. For example, in Great Britain, the

Mendip region soils are found highly enriched in lead, zinc, and cadmium due to the high concentrations of these metals in the bedrock and the presence of mineralized veins (Fuge et al. 1991). This area has a long history of mining and smelting. Soils developed on serpentinite are highly enriched in nickel and chromium and sustain a specialized plant community composed of Ni-tolerant species (Proctor and Baker 1994). However, these soils are restricted to small areas and are easily recognized by their special plant community. In rocks, heavy metals exist as ores in different chemical forms, from which they are recovered as minerals. Heavy metal ores include sulfides, such as iron, arsenic, lead, lead–zinc, cobalt, gold, silver, and nickel sulfides, and oxides, such as aluminum, manganese, gold, selenium, and antimony. Some exist and can be recovered as both sulfide and oxide ores, such as iron, copper, and cobalt. Ore minerals tend to occur in families whereby metals that exist naturally as sulfides would mostly occur together, likewise for oxides. Therefore, sulfides of lead, cadmium, arsenic, and mercury would naturally be found occurring together with sulfides of iron (pyrite, FeS_2) and copper (chalcopyrite, $CuFeS_2$) as minors, which are obtained as by-products of various hydrometallurgical processes or as part of exhaust fumes in pyrometallurgical and other processes that follow after mining to recover them.

Weathering of bedrock with only slightly elevated metal concentrations may result in an enrichment of metals by pedogenic processes (Blaser et al. 2000). Therefore, soils in areas that are not known to have elevated metal concentrations in the bedrock might also show naturally elevated metal concentrations. High HNO_3– extractable lead concentrations of up to 140 mg kg^{-1} were found in the topsoils of a remote site (Mount la Schera) in the Swiss National Park at an altitude of 2,400 m above sea level, far away from industry and major traffic routes (Bernd et al. 2001). However, the speculation is that long-distance aerosol could be another reason. Sulfur-bearing compounds when combined with oxygen in water vapor form sulfuric acid, and hence the phenomenon is known as acid drainage. Generally, strong acid forming processes in nature involve exposure of metal sulfides enriched with heavy metals or metalloids to atmospheric air, which leads to oxidation and the production of acid and/or heavy metal-rich water (Evangelou 1998). In addition, weathering of coalmine waste can produce alkaline compounds, heavy metals, and sediments. Acid drainage, alkaline compounds, heavy metals, and sediment leached from mine waste into groundwater or washed away by rainwater can pollute streams, rivers, and lakes. Acid drainage has various anthropogenic and natural sources, but the most extensive and widely known source is one related to mining coal and various metal ores. During mining processes, some metals are left behind as tailings scattered in open and partially covered pits, while some are transported through wind and flood, creating various environmental problems (Habashi 1992). Heavy metals are basically recovered from their ores by mineral processing operations (UNEP/GPA 2004).

1.2.2 Anthropogenic Sources

Generally, metals are emitted during mining and processing activities. In some cases, even long after mining activities have ceased, the emitted metals continue to

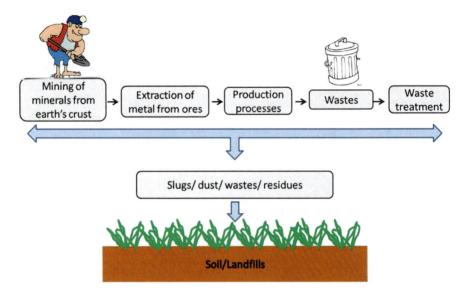

Fig. 1.1 Schematic representation depicting flow of heavy metals into waste

persist in the environment. For example, Peplow (1999) reported that hard rock mines operate from 5 to 15 years until the minerals are depleted, but metal contamination that occurs as a consequence of hard rock mining persists for hundreds of years after the cessation of mining operations. Apart from mining operations, mercury is introduced into the environment through cosmetic products as well as manufacturing processes like making of sodium hydroxide. Heavy metals are emitted both in elemental and compound (organic and inorganic) forms. Anthropogenic sources of emission are the various industrial point sources including former and present mining sites, foundries and smelters, combustion by-products and traffics (UNEP/GPA 2004). Cadmium is released as a by-product of zinc (and occasionally lead) refining; lead is emitted during its mining and smelting activities, from automobile exhausts (by combustion of petroleum fuels treated with tetraethyl lead antiknock) and from old lead paints; mercury is emitted by the degassing of the earth's crust. According to Ross (1994), the anthropogenic sources of metal contamination can be divided to five main groups:

1. Metalliferous mining and smelting (As, Cd, Pb, Hg)
2. Industry (As, Cd, Cr, Co, Cu, Hg, Ni, Zn)
3. Atmospheric deposition (As, Cd, Cr, Cu, Pb, Hg, U)
4. Agriculture (As, Cd, Cu, Pb, Si, U, Zn)
5. Waste disposal (As, Cd, Cr, Cu, Pb, Hg, Zn)

From all the above-mentioned sources, heavy metals may end up in solid/liquid waste during all life cycle phases of the products (Fig. 1.1). The figure actually depicts the overall flow of heavy metal contamination in practice and each step in the figure consists of several minor steps not indicated in the figure.

1.2.3 Activities Generating Heavy Metals

Sources of heavy metal vary from place to places. Generally, these include sewage and stormwater discharges; landfills and cemeteries; incinerators and crematoria and motor vehicles, while a range of human activities like electroplating, smelting, dentists, laboratories, timber preservation, drum reconditioning, waste storage and treatment, metal treatment, sheep and cattle dips, scrap metal yards, tanning, chemical manufacturers, production and use of accumulators, mercury lamps, thermometers, utensils, batteries, etc., also adds substantial amounts of metals to the environment.

1.3 Nature of Heavy Metal Pollution

A pollutant is any substance in the environment, which causes objectionable effects, impairing the environment, reducing the quality of life, and eventually causing death. Such a substance has to be present in the environment beyond a set or tolerance limit, which could be either a desirable or acceptable limit. Hence, environmental pollution is the presence of a pollutant in the environment such as air, water, and soil, which may be poisonous or toxic and cause harm to living forms in the polluted environment. Toxic heavy metals in air, soil, and water are global problems that are a growing threat to the environment.

1.3.1 Heavy Metals and Air Pollution

Both natural and manmade sources are responsible for increasing heavy metals in the air. Natural emissions come from wind-born soil particles, volcanoes, forest fires, sea-salt sprays, and biogenic sources. However, the anthropogenic sources of atmospheric emissions, through diverse human activities, exceed the natural fluxes for most metals. Even the metals emitted naturally in wind-blown dusts are often of industrial origin. Some of the prominent sources of atmospheric pollution are: burning fossil fuel to generate energy (V, Ni, Hg, Se, Sn), automobile exhaust (Pb), insecticides (As), manufacturing of steel (Mn, Cr), smelting (As, Cu, Zn), etc. Lead is the most pervasive environmental pollutant. Despite reduction of its content in petrol/diesel by many countries, it still accounts for roughly two third of global lead emission. One third of the world's urban population is estimated to be exposed to marginal or unacceptable lead concentrations (Athar and Vohora 1995). Thus, emission of metals into the air is probably the greatest source of heavy metal pollution, which in turn contaminates aquatic ecosystem and soils through atmospheric fallout.

1.3.2   Water Pollution

Contaminants behave in different ways when added to water. Nonconservative materials including organics, some inorganics, and many microorganisms are degraded by natural self-purification processes so that their concentration is reduced with time. The rate of decay of these materials, however, depends on the type of pollutant, the receiving water quality, temperature, and other environmental factors. Many inorganic substances are not affected by natural processes, and hence their concentration is reduced only by dilution. Conservative pollutants are often unaffected by normal water and wastewater treatment processes so that their presence in a particular water source may limit its use. Most of these materials originate from industrial discharges and would include heavy metals from metal finishing and plating operations, insect repellents from textile manufacture, herbicides and pesticides, etc.

Rapid industrialization and accelerating global development over the past two centuries have greatly increased the rate at which trace metals are released into the environment. As a result, many of the freshwater bodies are becoming greatly altered. According to Balogh et al. (2009), Lake Pepin, a natural lake on the upper Mississippi River (USA), reveals the historical trends in trace metal use and discharge in the watershed. Both diffuse and point sources have contributed to trace metal loadings in the river and accumulation in the lake. Prior to European settlement, trace metals accumulating in Lake Pepin came primarily from diffuse, natural sediment sources throughout the watershed. Later, with increasing human development in the watershed, municipal and industrial wastes added trace metal to the river and lake. The great Ganges river of India has also been found to have high heavy metals in sediments and fish (Gupta et al. 2008). Substantial changes in waste generation and treatment practices after the 1960s have, however, reduced trace metal inputs.

Arsenic is toxic (see Sect. 1.4.2) and a known carcinogen, whose safe limit in drinking water in most countries is 10–50 µg/l. Despite similar mandates found in many countries, arsenic contamination remains a worldwide threat. Arsenic concentrations are higher in groundwater than in surface water where the presence of arsenic is mainly due to dissolved minerals from weathered rocks and soils. Additionally, in groundwater from the area surrounding and including Hanoi, Vietnam, arsenic concentrations have been found to range from 1 to 3,050 µg/l with an average concentration of 159 µg/l. In highly affected areas, arsenic concentrations averaged over 400 µg/l. Water analyzed after treatment had concentrations ranging from 25 to 91 µg/l but with 50% of wells tested still had over 50 µg/l arsenic (Berg et al. 2001). High arsenic concentrations pose a chronic health threat to millions drinking contaminated water.

1.3.3   Soil Pollution

Rapid industrialization and subsequent expansion of the population have considerably increased industrial and municipal wastewater discharge and other pollutants in many countries. Soil is a major sink for heavy metals released into the environment.

Many soils in industrialized countries are affected by acid deposition, mine waste and organic refuses, such as sewage sludge that introduce pollutants to the soil. According to Moral et al. (2005), The level of pollution of soils by heavy metals depends on the retention capacity of soil, especially on physicochemical properties (mineralogy, grain size, organic matter) affecting soil particle surfaces and also on the chemical properties of the metal. These metals may be retained by soil components in the near surface soil horizons or may precipitate or co-precipitate as sulfides, carbonates, oxides or hydroxides with Fe, Mn, Ca, etc. In arid zones, carbonate effectively immobilizes heavy metals by providing an adsorbing or nucleating surface and by buffering pH at values where metals hydrolyze and precipitate. The mobility of trace metals reflects their capacity to pass from one soil compartment to another where the element is bound less energetically, the ultimate compartment being soil solution, which determines the bioavailability. The distribution of metals among various compartments or chemical forms can be measured by sequential extraction procedures. Knowledge of how contaminants are partitioned among various chemical forms allows a better insight into degradation of soil and water quality following the input of metals around mining and metallurgical plants. Therefore, soils pollution by heavy metals occurring both on surface and in deeper layers of soil is of great concern for environmental quality control. The pattern of pollutant content is the synergistic result of mixed processes, including diffusion of deposited airborne particulate matter, fluvial deposition of contaminated sediments and irregular leaching of soil layers, assisted by rainwater, even down to groundwater (McLean and Bledsoe 1992). The impact of heavy metals resulting from mining and ore roasting on soil is attenuated by several processes such as adsorption, precipitation, and complex formation with soil compounds.

Soil pollution with heavy metals is multidimensional. Upon entering the soil in large amounts, heavy metals primarily affect biological characteristics: the total content of microorganism changes, their species diversity reduced, and the intensity of basic microbiological processes and the activity of soil enzymes decreases. In addition, heavy metals also changes humus content, structure, and pH of soils (Levin et al. 1989). These processes ultimately lead to the partial or, in some cases, complete loss of soil fertility. Any increase in contamination emission may also affect the crop productivity adversely. There are a number of factors influencing the concentration of heavy metals in plants and soils. These factors include climate, irrigation, atmospheric deposition, the nature of the soils on which the plant is grown and time of harvesting. Heavy metal contamination derived from anthropogenic sources is one of the severest; this can strongly influence their speciation and hence bioavailability (Lester 1987).

1.4 Impacts of Heavy Metals

Metals play an integral role in the life processes of organisms. Some metals, such as cobalt, chromium, copper, iron, potassium, magnesium, manganese, sodium, nickel, and zinc, are essential, serving as micronutrients used for (1) redox-processes,

(2) to stabilize molecules through electrostatic interactions, (3) acting as components of various enzymes, and (4) regulation of osmotic pressure (Bruins et al. 2000). Other metals, like silver, cadmium, gold, lead, and mercury, have no biological function and are nonessential and potentially toxic to organisms. Toxicity of nonessential metals occurs through the displacement of essential metals from their native binding sites or through ligand interactions (Nies 1999; Bruins et al. 2000). For example, Hg^{2+}, Cd^{2+}, and Ag^{2+} can bind to SH group of proteins, and thus inhibit the activity of enzymes (Nies 1999). Moreover, both essential and nonessential metals can (1) damage cell membranes, (2) alter enzyme specificity, (3) disrupt cellular functions, and (4) damage the structure of DNA at high concentration (Bruins et al. 2000). To have a physiological or toxic effect, most metal ions have to enter into the cell. Many divalent metal cations (e.g., Mn^{2+}, Fe^{2+}, Co^{2+}, Ni^{2+}, Cu^{2+}, and Zn^{2+}) are structurally very similar. Also, the structure of oxy-anions such as chromate resembles that of sulfate, and the same is true for arsenate and phosphate. Thus, to be able to differentiate between structurally very similar metal ions, the organism's uptake systems have to be tightly regulated.

1.4.1 Impact on the Environment

Some soil types have great resistance to pollution in general and to heavy metals in particular. However, even those soils are not always capable of resisting the effects of pollutants on their properties. Hence, the assessment of heavy metal toxicity to biological and ecological properties and the extent of decrease in soil biological activity may be used as a parameter characterizing the effects of heavy metals on the soil. It is expedient to use indices of biological activity for monitoring and diagnosis of soil pollution. The results of pollution with heavy metal are not always unequivocal. In most cases, a decrease in soil biological activity is observed (Levin et al. 1989). In some cases, however, researchers noted an increase in the content of microorganisms, soil enzymatic activity, and other parameters. Thus, it is important to take into account a significant spatial and temporal variation in biological characteristics of the soils.

Bacteria (especially spore-forming) are relatively sensitive to pollution, which is followed by actinomyces and microscopic fungi. The maximum toxic effect of heavy metals on soil microorganisms was observed in the initial period after pollution: in most cases, the numbers of microorganisms decrease significantly. The structure of the soil microflora changed significantly after pollution with heavy metals. The proportion of microscopic fungi and, sometimes, actinomyces will increase at high concentrations of heavy metals in the soil (Kolesnikov et al. 2000). Heavy metals also impact communities and cause food chain contamination. Food chain contamination refers to the potential for the soil metals to cause harm to animals that feed on the plants and soil mesofauna (animals living among the litter and inside the microscopic crevices of the site soil). According to US-EPA (2007), soil particles on the plants or the soil mesofauna may result in high enough levels of contaminants that are toxic to animals that consume them. For example, if *shrews* at a contaminated site feed on *earthworms*, the *shrews* will be exposed to high concentrations of contaminants in the soils. This is the case because earthworms generally consist of over 50% soil by weight. Consumption of soil through earthworm ingestion results in high body burdens for *shrews*. This then could lead to an increase in body burden for birds that prey on the shrews. Soil extractions, such as dilute Ca $(NO_3^{-1})_2$, have been shown to be related to earthworm available metals and offer one way to evaluate this risk. According to Kolesnikov et al. (2000), changes can also occur in communities of different soil microorganisms. When heavy metal content increased, their species diversity became significantly lower, especially in the case of microscopic fungi. Besides, heavy metal pollution has a strong effect on the qualitative composition of humus; they have also indicated that all the high values of microbial abundance were accounted for the development of a small group of microorganisms that are tolerant to heavy metals.

1.4.2   Heavy Metal Toxicity

The Dangerous Substances Directive (76/464/EEC) of the European Commission defines dangerous chemicals as those that are toxic, persistent, and/or bioaccumulative. Unlike many organic pollutants, which eventually degrade to CO_2 and water, heavy metals are nondestructible and hence accumulate in the environment, especially in lakes, estuaries or marine sediments, and soils. Metals can be transported from one environmental compartment to another.

Many of the heavy metals are toxic to organisms at low concentrations. However, some heavy metals, such as copper and zinc, are also essential elements required in minute quantities. Concentrations of essential elements in organisms are normally homeostatically controlled, with uptake from the environment regulated according to nutritional demands. Their effects on the organisms are manifested when the regulation mechanism breaks down as a result of either insufficient (deficiency) or excess (toxicity) metals. Whether the source of heavy metals is natural or anthropogenic, the concentrations in terrestrial and aquatic organisms are determined by the size of the source and adsorption/precipitation in soils and sediments. The extent of adsorption depends on nature of the metals, the absorbent, the physicochemical characteristics of the environment (e.g., pH, water hardness, and redox potential) and the concentration of other metals as well as complex chemicals present in the soil or water (river or lake). Thus, concentration of metal in bioavailable form is not necessarily proportional to the total concentration of the metal.

The accumulation of heavy metals in plant tissues eventually leads to toxicity and change in plant community (Gimmler et al. 2002; Kim and McBride 2009; John et al. 2009). The toxic metals in soils are reported to inhibit root and shoot growth, affect nutrient uptake and homeostasis, and are frequently accumulated by agriculturally important crops. Thereafter, they enter the food chain with a significant amount of potential to impair animal and/or human health. The reduction in biomass of plants growing on metal-contaminated soil has been found to be due to the direct consequence on the chlorophyll synthesis and photosynthesis inhibition (Dong et al. 2005; Shamsi et al. 2007), carotenoids inhibition (John et al. 2009), inhibition of various enzyme activities, and induction of oxidative stress including alterations of enzymes in the antioxidant defense system (Kachout et al. 2009; Dazy et al. 2009). Since an increased metal concentration in soil is reported to affect soil microbial properties, such as respiration rate and enzyme activity, it is considered as a very useful indicator of soil pollutions (Brookes 1995; Szili-Kovács et al. 1999). However, the short-term and long-term effects of metals depend on the type of metals and soil characteristics (Németh and Kádár 2005). The free ions are generally the most bioavailable forms of metals and are often considered as the best indicator of toxicity. Metals exert toxic effects after they enter into biochemical reactions of the organism and typical responses are inhibition of growth, suppression of oxygen consumption, and impairment of reproduction and tissue repair (Duruibe et al. 2007). However, there are exceptions such as for mercury, whose organic form (methylmercury) is more toxic than the inorganic ion. The biotoxic effects of heavy metals refer to the harmful effects of heavy metals to the body when consumed above the biological (recommended) limits. Although individual metals exhibit specific signs of toxicity, general signs associated with cadmium, lead, arsenic, mercury, zinc, copper, and aluminum poisoning include gastrointestinal (GI) disorders, diarrhea, stomatitis, tremor, hemoglobinuria causing a rust-red color to stool, ataxia, paralysis, vomiting and convulsion, depression, and pneumonia when volatile vapors and fumes are inhaled (McCluggage 1991).

The nature of effects could be toxic (acute, chronic, or sub-chronic), neurotoxic, carcinogenic, mutagenic, or teratogenic. Among metals, cadmium is toxic at extremely low levels. In humans, long-term exposure results in renal dysfunction, characterized by tubular proteinuria. High exposure can lead to obstructive lung disease, cadmium pneumonitis, resulting from inhaled dusts and fumes. It is characterized by chest pain, cough with foamy and bloody sputum, and death of the lining of the lung tissues because of excessive accumulation of watery fluids. Cadmium is also associated with bone defects, namely, osteomalacia, osteoporosis and spontaneous fractures, increased blood pressure, and myocardic dysfunctions. Depending on the severity of exposure, the symptoms of effects include nausea, vomiting, abdominal cramps, dyspnea, and muscular weakness. Severe exposure may result in pulmonary edema and death. Pulmonary effects (emphysema, bronchiolitis, and alveolitis) and renal effects may occur following subchronic inhalation exposure to cadmium and its compounds (European Commission 2002a).

Lead is the other most significant toxin of the heavy metals, and the inorganic forms are absorbed through ingestion by food and water, and inhalation (Ferner 2001). A notably serious effect of lead toxicity is its steratogenic effect. Lead poisoning also causes inhibition of the synthesis of hemoglobin; dysfunctions in the kidneys, joints and reproductive systems, cardiovascular system, and acute and chronic damage to the central nervous system (CNS) and peripheral nervous system (PNS). Other effects include damage to the gastrointestinal tract (GIT) and urinary tract resulting in bloody urine, neurological disorder, and severe and permanent brain damage. While inorganic forms of lead typically affect the CNS, PNS, GIT, and other biosystems, organic forms predominantly affect the CNS (Lenntech Water Treatment and Air Purification 2004). Lead affects children leading to the poor development of the grey matter of the brain and consequently poor intelligence quotient (IQ) (Udedi 2003). Its absorption in the body is enhanced by Ca and Zn deficiencies. Acute and chronic effects of lead result in psychosis.

Zinc has been reported to cause the same signs of illness as does lead and can easily be mistakenly diagnosed as lead poisoning (McCluggage 1991). Zinc is considered to be relatively nontoxic, especially if taken orally. However, excess amount can cause system dysfunctions that result in impairment of growth and reproduction (Nolan 2003). The clinical signs of zinc toxicosis have been reported as vomiting, diarrhea, bloody urine, icterus (yellow mucus membrane), liver failure, kidney failure, and anemia. According to Rai and Pal (2002), inhalation of dust containing Cr in high oxidation states (IV) and (VI) is associated with malignant growth in the respiratory tract and painless perforation in nasal septum. Among these, trivalent and hexavalent states are the most stable and common in terrestrial environments. Hexavalent chromium is the form considered to be the greatest threat because of its high solubility, its ability to penetrate cell membranes, and its strong oxidizing ability. Hence, Cr (+6) is more toxic than Cr (+3) because of its high rate of absorption on living surface. Cr (+6) exists only as oxy-species such as CrO_3, CrO_4, and Cr_2O_7, and is a strong oxidizing agent.

Cadmium (Cd) causes chronic poisoning. The incubation period for chronic cadmium intoxication varies considerably usually between 5 and 10 years but in some cases upto 30 years. During the first phase of poisoning, a yellow discoloration of teeth, "cadmium ring," is formed, the sense of smell is lost, and mouth becomes dry; subsequently, the number of red blood cells is diminished, which results in impairment of bone marrow. The most characteristic feature of diseases is lumbar pains and leg myalgia. These conditions continue for several years until the patient becomes bed ridden and clinical conditions progress rapidly. Urinary excretion of albuminous substances results from the severe kidney damage. Cadmium induced disturbances in calcium metabolism accompanied by softening of bones; fractures and skeletal deformations take place with a marked decrease in body height up to 30 cm (Rai and Pal 2002).

Mercury is toxic and has no known function in human biochemistry and physiology. Inorganic forms of mercury cause spontaneous abortion, congenital malformation, and GI disorders (like corrosive esophagitis and hematochezia). Poisoning by its organic forms, which include monomethyl and dimenthylmercury, presents with erethism (an abnormal irritation or sensitivity of an organ or body part to stimulation), acrodynia (Pink disease, which is characterized by rash and desquamation of the hands and feet), gingivitis, stomatitis, neurological disorders, total damage to the brain and CNS, and is also associated with congenital malformation (Lenntech Water Treatment and Air Purification 2004).

As with lead and mercury, arsenic toxicity symptoms are dependent on the chemical form ingested (Ferner 2001). Arsenic acts to coagulate protein, forms complexes with coenzymes, and inhibits the production of adenosine triphosphate (ATP) during respiration. It is possibly carcinogenic in compounds of all its oxidation states and high-level exposure can cause death (USDOL 2004). Arsenic toxicity also presents a disorder, which is similar to, and often confused with Guillain-Barre syndrome, an anti-immune disorder that occurs when the body's immune system mistakenly attacks part of the PNS, resulting in nerve inflammation that causes muscle weakness (Kantor 2006; NINDS 2007).

1.4.3   Health Hazards

There are many chemical compounds whose presence in water and food could be harmful or fatal to human life and it is necessary to consider two aspects of the problem in assessing potential hazards. An acute effect could be produced by the accidental discharge of sufficient toxic matter into a water source to produce more or less immediate symptoms in consumers. This form of contamination is fortunately rare and usually the contaminant would produce obvious effects in the water source such as fish kills, strong tastes and odors, etc., which would provide a warning even if the accident had not been reported to the authorities. A more insidious type of chemical contamination occurs when the contaminant produces a long-term hazard due to exposure to minute concentrations, perhaps over many years. In this situation, the determination of allowable levels for the particular contaminants is extremely difficult since scientific evidence is very limited and difficult to interpret. Probably, one of the earliest chemical contamination problems arose from the use of lead piping and tanks in domestic plumbing. Soft acidic waters from upland catchments tend to be plumbosolvent so that significant amounts of lead can be dissolved in the water, particularly when standing overnight in service connections. Lead is a cumulative poison and current concern about lead in the environment requires that allowable levels of lead are kept as low as possible.

Sewage containing mercury was released by Chisso's chemicals works into Minimata Bay in Japan. The mercury accumulates in sea creatures, leading eventually to mercury poisoning in the population. In the 1950s, residents of Minamata, began experiencing unusual symptoms, including numbness, vision problems, and convulsions. Several hundred people died. The cause was discovered to be mercury ingestion by poisoned fish and thousands of people. Since then, Japan has had one of the strictest environmental laws in the industrialized world. In 1997, after a massive cleanup, Japan announced that the bay had been cleared of the contaminant (Microsoft Encarta 2008). In 1947, an unusual and painful disease of rheumatic nature was recorded in the case of 44 patients from a village on the banks of *Jintsu* River, Toyama prefecture, Japan. During subsequent years, it became known as "itai-itai" disease (meaning "ouch-ouch") in accordance with the patient shrieks resulting from painful skeletal deformities (Rai and Pal 2002).

In 1960, fatal incidents of lung cancer were reported from the Kiryama factory of Nippon-Denki concern on the Islands of Hokkaido; medical warnings were issued that inhalation of dust containing Cr in high oxidation states (IV) and

(VI) was associated with malignant growth in the respiratory tract and painless perforation in nasal septum among trivalent and hexavalent states being the most stable and common in terrestrial environments. In 1986, a massive chemical spill from a plant in Basel reversed 10 years of progress. Nearly 30 t of toxic waste, including fungicides and mercury, entered the Rhine. The spill, called the greatest nonnuclear disaster in Europe in a decade, killed 500,000 fish and forced the closing of water systems in West Germany, France, and the Netherlands (Rai and Pal 2002). Toxic chemicals in water from a burst dam belonging to a mine contaminate the Coto de Donana nature reserve in southern Spain. Spanish nature reserve was also contaminated after the environmental disaster. About five million m^3 of mud, containing sulfur, lead, copper, zinc, and cadmium, had flown down into the Rio Guadimar. Experts have estimated that the Europe's largest bird sanctuary, as well as Spain's agriculture and fisheries, suffered permanent damage from the pollution. Arsenic is a hazard over vast areas of West Bengal (India) and Bangladesh and has contaminated the groundwater in the Ganges, Mekong, and Red River (Sen and Chakrabarti 2009).

1.5   Abatement of Heavy Metals Pollution

The bioavailability of contaminants poses a health risk to animals and humans who may be exposed to contaminated sites. Possible exposure pathways include ingestion of contaminated soil or water from the site, direct contact with contaminated soil or water, inhalation of contaminants adhered to dust in the air, and ingestion of food items (i.e., plants or animals) that have accumulated from exposure to contaminated soil or water. Managing the risks posed by contaminants at a site involves understanding the possible pathways and applying appropriate remedial measures to mitigate, treat, or remove sources.

1.5.1   Physicochemical Methods

The most common physical methods used for remediation of heavy metal pollution include excavation and land fill thermal treatment, acid leaching, and electro-reclamation. Further, chemical precipitation, chemical oxidation and reduction, ion exchange, filtration, electrochemical treatment, reverse osmosis, freeze crystallization, electrodialysis, cementation, starch xanthate adsorption, and solvent extraction can be used for removing of heavy metal ions from diluted solutions. Owing to their higher operational cost or difficulty to treat solid and liquid wastes, most of these methods are expensive and ineffective, especially when the metal ions are dissolved in large volume of solutions. In response to the growing problems, federal and state governments in different countries have instituted environmental regulations to protect the quality of surface and ground water from heavy metal pollutants, such as

Cd, Cu, Pb, Hg, Cr, and Fe. To meet the state guidelines for heavy metal discharge, companies in the western countries often use chemical precipitation or chelating agents. For acid mine drainage and wastewater treatment plants, the typical means of removing heavy metals is usually accomplished through pH neutralization and precipitation with lime, peroxide addition, reverse osmosis, and ion exchange. A major disadvantage of the liming process, however, is the need for large doses of alkaline materials to increase and maintain pH values of 4 to above 6.5 for optimal metal removal (McDonald and Grandt 1981). Additionally, pH neutralization typically requires that the materials be appreciably fine-grained to provide the necessary reactive surface area. Furthermore, liming produces secondary wastes, such as metal hydroxide sludges, that necessitate highly regulated and costly disposal (Wang et al. 1996).

As an alternative to the liming process many companies have developed chelating ligands to precipitate heavy metals from aqueous systems. Precipitation is the most common and widely used method in wastewater treatment. Undesired metal ions from effluents can be precipitated as insoluble metal hydroxides by adding calcium or sodium hydroxide. They can also be precipitated as more insoluble metal sulfides by adding sodium sulfide, sodium hydrosulfide, or ferrous sulfide. The insoluble precipitate formed is then allowed to settle in the sedimentation tanks by addition of coagulating agents. In order to be competitive economically, many of these chelating ligands are simple, easy to obtain, and, in general, offer minimal bonding for heavy metals. Ion exchange is a stoichiometric and reversible chemical reaction in which a metal ion (e.g., Fe^{2+}) is exchanged for a similarly charged ion (e.g., H^+) attached to an embolized particle. Some of the methods for removal of heavy metals make use of naturally existing inorganic zeolites, synthetic resin materials, clay minerals, activated carbon, and synthetic organic ion exchangers contained in a series of columns in ion exchanging system. While, regeneration of columns can be achieved by adding acidic solution and metal ions may be recovered from eluent by evaporation.

Two basic approaches are commonly employed to treat contaminated soils. The first involves a phase transfer in which the contaminants are moved from one phase into another. The second approach involves destruction or transformation of the contaminants into harmless products (Evangelou 1998). These two approaches can be utilized by incorporating a number of technologies, including high–low temperature thermal treatments, radiofrequency heating, steam stripping and vacuum extraction, aeration, soil flashing, and others. High-temperature treatment systems involve destruction of contaminants through complex oxidation, whereas low-temperature systems increase the rate of phase transfer (solid to liquid and then to gaseous phase), and thus encourage contaminant partitioning from the soil. Large and small mining sites, landfills, and industrial sites such as refineries, smelters, foundries, milling and plating facilities, and other sites with contaminated or disturbed soils exhibit a variety of problems that often can be addressed effectively and directly through the use of soil amendments (Fig. 1.2). With the addition of appropriate soil amendments, metals in the amended area are chemically precipitated and/or sequestered by complexation and sorption mechanisms within the contaminated substrate.

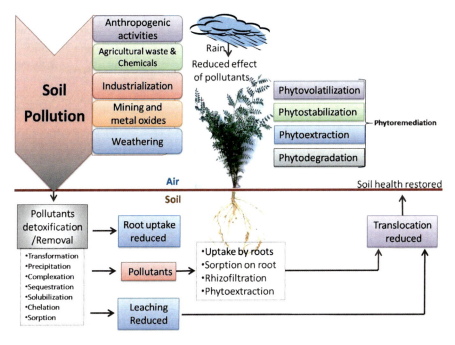

Fig. 1.2 Influence of soil amendments and plants in the remediation of metal-contaminated soils

Metal availability to plants is minimized, and metal leaching into groundwater can be reduced. In certain cases, metal availability below the treated area is also reduced.

When applied properly, soil amendments reduce exposure by limiting many of the exposure pathways and immobilizing contaminants to limit their bioavailability. The addition of amendments restores soil quality by balancing pH, adding organic matter, increasing water holding capacity, re-establishing microbial communities, and alleviating compaction. As such, the use of soil amendments enables site remediation, revegetation and revitalization, and reuse. There is a definite need for new and more effective physicochemical methods and reagents to meet the growing environmental problem. Many reagents in the market today either lack the necessary binding criteria or pose too many environmental risks. For this reason, ligands utilizing multiple binding sites for heavy metals and mimicking biological systems look to be a possible answer to heavy metal remediation.

1.5.2   Biochemical Methods

It has been known for a long time that various living and dead microorganisms can remove heavy toxic ions from solutions (Sterritt and Lester 1996). In addition, their applications are important in the general environment and in areas where potential exists for both clean wastewater and heavy metal recovery. The methods are now recognized not only as viable alternatives but a desirable alternative and/or addition to the traditional remediation technologies. First, the unique interaction between microorganisms and heavy metals and then their use in pollution abatement will be discussed.

1.5.2.1 Heavy Metals and Microorganisms

As mentioned earlier, the major problem of heavy metal concentration is ion imbalance. Microorganisms have evolved mechanisms to solve this problem by using two types of uptake systems for metal ions. One is fast, unspecific, and driven by the chemiosmotic gradient across the cytoplasmic membrane of bacteria. Since this mechanism is used by a variety of substrates, it is constitutively expressed (Nies 1999). The second type of uptake system has high substrate specificity. It is slower, often uses ATP hydrolysis as the energy source, and is only produced by the cell in times of need, starvation, or a special metabolic situation (Nies and Silver 1995). Even though microorganisms have specific uptake systems, high concentrations of nonessential metals may be transported across the cell by a constitutively expressed unspecific system. This "open gate" is the one reason why metal ions are toxic to microorganisms. As a consequence, microorganisms have been forced to develop metal-ion homeostasis factors and metal-resistance determinants (Bruins et al. 2000). Because metal ions cannot be degraded or modified like toxic organic compounds, there are six possible mechanisms for a metal resistance system: (1) exclusion by permeability barrier, (2) intra- and extra-cellular sequestration, (3) active efflux pumps, (4) enzymatic reduction, (5) and (6) reduction in the sensitivity of cellular targets to metal ions. One or more of these resistance mechanisms allow microorganisms to function better in metal-contaminated environments.

1.5.2.2 Biosorption and Floatation

Sorption refers to the taking in or holding of something, either by absorption or adsorption. The ability of microorganisms to bind metals from aqueous solution in some cases selectively is known as biosorption and microorganisms responsible for this process are called biosorbents. The biosorption of heavy metals ions by microorganisms is a promising property with a great potential for industrial applications. The use of biological substrates as a metal concentrator from dissolved metal ions, applying for example marine algae, as well as the ability of several microorganisms to remove metal ions from aqueous solutions is well studied (Zouboulis and Matis 1998).

Biosorption reactions are metabolism independent and proceed rapidly by any one or a combination of the metal binding mechanisms like coordination, complexation,

ion-exchange, physical adsorption (e.g., electrostatic), or inorganic microprecipitation. These, processes can occur, whether the organism is living or dead; and may be facilitated by microbial viability. Nonliving microbial biomass is usually preferable in the biosorption technology as this precludes the necessity of adding nutrients required for microbial development, maintaining sterility of the process and adjusting parameters for favorable microbial growth. Various kinds of bacteria, fungi, and algae have accordingly been identified with biosorption ability. Mechanisms of biosorption are complex. Generally, the biomass contact with a solution of metals is realized by different techniques – in flasks on shakers, in columns, tanks, reactors, fermenters, and other vessels. The modes by which the microorganisms remove metal ions from solution can be extracellular accumulation/precipitation, cell surface sorption or complexation, and intracellular accumulation. Industrial applications of biosorption often make use of dead biomass, which does not require the supplementary addition of nutrients and it can be also exposed to environment of high toxicity. Experimental (mostly laboratory batch) results have been previously presented, applying with actinomycetes, fungi, yeasts, etc., as the respecting biosorbent materials for metals (Kefala et al. 1999). Nonliving biomass showed greater binding capacities for cadmium than living biomass; this observation was accounted to an ion exchange mechanism (Kefala et al. 1999). Metabolism-independent binding of metal ions to fungal and yeast cell walls is usually a rapid process and large amounts of metals may be bound and removed.

Centrifugation, being a conventional separation method in microbiology and biochemistry, is relatively expensive, considering the power demand per unit of microbial cells recovered; therefore, alternative biomass separation methods, such as flotation, are being examined. Flotation nowadays is considered as a well-established unit operation in the field of mineral and environmental technology. Flotation, following metal biosorption, was proved to be a useful and effective separation method of metal-loaded biomass, producing efficient removals, usually over 95%. The two processes can effectively operate in combination, in what was termed *biosorptive flotation*. The main critical parameters affecting both of them, which need careful control, are solution pH and ionic strength (Zouboulis and Matis 1997). As an example, flotation, which includes different techniques, such as foam or bubble fractionation, foam separation, or froth flotation, was examined for the separation of metal-loaded baker's yeast, *Saccharomyces cerevisiae* (Zouboulis et al. 2001). Among biosorbents, fungi were found to bio-accumulate metal and radionuclide species by several physicochemical and biological mechanisms, including extra-cellular binding by metabolites and biopolymers, binding to specific polypeptides, as well as metabolism-dependent accumulation (Zouboulis and Matis 1998). However, to date, the most promising approach for metal removal by fungi is biosorption. The fungal cell wall is considered to contain two main components: interwoven skeletal framework micro-fibrils, usually of chitin, embedded in an amorphous layer of proteins and various polysaccharides (Zouboulis et al. 1999).

1.5.2.3   Biotransformation

Some bacteria have evolved mechanisms to detoxify heavy metals, and some even use them for respiration. Microbial interactions with metals may have several implications for the environment. They play a large role in the biogeochemical cycling of toxic heavy metals and also in cleaning up metal-contaminated environments. Microbial transformations of metals serve various functions. Generally, microbial transformations of metals can be divided into two broad categories: redox conversions of inorganic forms; and conversions from inorganic to organic form and vice versa, typically methylation and demethylation. Through oxidation of iron, sulfur, manganese, and arsenic, microbes can obtain energy (Tebo et al. 1997; Santini et al. 2000). On the other hand, reduction of metals can occur through dissimilatory reduction where microorganisms utilize metals as a terminal electron acceptor for anaerobic respiration. In addition, microorganisms may possess reduction mechanisms that are not coupled to respiration, but instead are thought to impart metal resistance. For example, aerobic and anaerobic reduction of heavy metals is widespread detoxification mechanisms among microorganisms.

Microbial methylation plays an important role in the biogeochemical cycle of metals, because methylated compounds are often volatile. For example, mercury (Hg^{2+}/Hg II) can be biomethylated by a number of different bacterial species (e.g., *Pseudomonas* sp., *Escherichia* sp., *Bacillus* sp., and *Clostridium* sp.) to gaseous methylmercury, which is the most toxic and most readily accumulated form of mercury. Also, biomethylation of arsenic to gaseous arsines; selenium to volatile dimethyl selenide; and lead to dimethyl lead (Pongratz and Heumann 1999) has been observed in various soil environments. In addition to redox conversions and methylation reactions, acidophilic iron and sulfur-oxidizing bacteria are able to leach high concentrations of arsenic, cadmium, copper, cobalt, nickel, and zinc from contaminated soils. On the other hand, metals can be precipitated as insoluble sulfides indirectly by the metabolic activity of sulfate-reducing bacteria (White et al. 1997; Lloyd and Lovley 2001). Sulfate-reducing bacteria are anaerobic heterotrophs utilizing a range of organic substrates with $SO4^{2-}$ as the terminal electron acceptor. In general, microbiological processes can either solubilize metals (White et al. 1997), thereby increasing their bioavailability and potential toxicity, or immobilize them, and thereby reduce the bioavailability of metals.

1.5.2.4   Pollution Monitoring Biosensors

The genes responsible for microbial metal resistance mechanism are organized in operons and are usually found in plasmids of resistant bacteria (Ramanathan et al. 1997). The expression of the resistance genes is tightly regulated and induced by the presence of specific metals in the cellular environment (Ramanathan et al. 1997). Because of the specificity of this regulation, the promoters and regulatory genes from these resistance operons can be used to construct metal-specific biosensors (promoter-reporter gene fusions). By using metal-specific bacterial sensors in

Table 1.1 Metal-specific sensor strains used in different laboratories

Metal	Reporter	Host strain
Antimony (Sb^{3+})	*Luc*	*Staphylococcus aureus, Bacillus subtilis, E. coli*
Arsenic (As^{3+})	*Luc*	*S. aureus, B. subtilis, E. coli*
Arsenic (As^{3+})	*Lux*	*E. coli*
Arsenic (As^{3+})	*Luc*	*E. coli, Pseudomonas fluorescens*
Arsenic (As^{5+})	*Lux*	*E. coli*
Arsenic (As^{5+})	*Luc*	*E. coli*
Cadmium (Cd^{2+})	*Luc*	*S. aureus, B. subtilis*
Cobalt (Co^{2+})	*Lux*	*Ralstonia eutropha*
Copper (Cu^{2+})	*Lux*	*P. fluorescens*
Lead (Pb^{2+})	*Luc*	*S. aureus, B. subtilis*
Mercury (Hg^{2+})	*Lux*	*E. coli*
Mercury (Hg^{2+})	*Luc*	*E. coli*
Mercury (Hg^{2+})	*Lux*	*E. coli, Pseudomonas putida*
Mercury (Hg^{2+})	*Lac*	*E. coli*
Mercury (Hg^{2+})	*Gfp*	*E. coli*
Mercury (Hg^{2+})	*Luc*	*E. coli, P. fluorescens*
Nickel (Ni^{2+})	*Lux*	*R. eutropha*

Adapted from Turpeinen (2002)

addition to chemical analyses, it is possible to distinguish the bioavailable metal concentration from the total metal concentration of the samples. Various metal-specific sensor strains have been developed and applied in different laboratories (Table 1.1) and were collected by Turpeinen (2002). These sensor strains are all based on the concept of metal responsive regulation unit, which regulates expression of a sensitive reporter gene. Reporter genes include those that code for bioluminescent proteins, such as bacterial luciferase (*luxAB*) and firefly luciferase (*lucFF*) or for β-galactosidase, which can be detected electrochemically or by using chemiluminescent substrates (Bontidean et al. 2000). The light produced can be measured by a variety of instruments, including luminometers, photometers, and liquid-scintillation counters.

1.5.2.5 Bioremediation

Many organisms have developed chromosomally or extrachromosomally controlled detoxification mechanisms to overcome the detrimental effects of heavy metals (Silver and Phung 1996). These resistance mechanisms take several forms, such as extracellular precipitation and exclusion, binding to the cell surface, and intracellular sequestration. Binding of metal cations on the outer surface of bacterial cells has become one of the most attractive means for bioremediation of industrial wastes and other metal-polluted environments. Valuable metals can be entrapped and recovered from negatively charged microbial surfaces (McLean and Beveridge 1990). The interest in bioremediation processes has greatly increased by the turn of the twentieth century. Some microorganisms act in the biosphere as geochemical agents promoting precipitations, transformations, or dissolutions of minerals. The use of these microorganisms could offer new tools to degrade or to transform toxic contaminants. Sulfate-reducing bacteria (SRB) constitute a group of anaerobic prokaryotes, commonly found in contaminated environments by heavy metals, metalloids, or other pollutants that are lethal to other microorganisms. SRB are able to couple the oxidation of organic compounds or hydrogen with the reduction of sulfate. It has been proposed that SRB are able to detoxify contaminated environments by an indirect chemical reduction of heavy metal via the production of H_2S, which is the end product of the dissimilatory sulfate reduction.

Arsenic biogeochemistry is a good example of microbe–metal interactions in the environment. Many studies have been done on microbial metabolism of arsenic in aquatic environments and the effects microbes have on the speciation and mobilization of arsenic. Since aquatic sediments can be anaerobic, and because arsenic concentrations in sediments can range from 100 to 300 µg/l, microbe-mediated arsenic reduction may be common. Earlier, some scientists have found that the addition of arsenate to an anaerobic sediment resulted in the accumulation of arsenite, indicating the reduction of arsenate to arsenite by microbes. Ahmann et al. (1997) further showed that native microorganisms from the Aberjona watershed were responsible for the arsenic flux in the anoxic contaminated sediments. In reducing conditions, it was found that arsenate was the dominant form of arsenic. They also found that dissimilatory iron-reducing bacteria (DIRB) and sulfate-reducing bacteria (SRB) are capable of both arsenic reduction and oxidation and thus may contribute to the cycling of arsenic in sediments. Microbial reduction of arsenate is important because arsenite (the reduced form) is more toxic and more soluble (and thus, more mobile) than arsenate, which forms relatively insoluble, non-bioavailable compounds with ferrous oxides and manganese oxides. Speciation of arsenic is affected or controlled by not only oxidation and reduction processes by microbes, but also by methylation by microbes, and adsorption to other particles. It was found that DIRB responsible for the dissolution of iron oxides bound to arsenic can also free soluble arsenic into the sediment (Cummings et al. 1999). Another study done on arsenic biogeochemistry in Lake Biwa Japan showed that arsenic concentration and speciation may also depend on eutrophication (Sohrin et al. 1997). They also found an interesting cycling of arsenic in the presence of nitrate, rapidly re-oxidizing any arsenate that had been produced. Thus, in some environments, both oxidation and reduction of arsenic may occur. Several other potential bacterial strains for heavy metal bioremediation have also been reported by our group (Table 1.2). Rani et al. (2008) reported a *Proteus vulgaris* strain KNP3 that reduced copper concentration in the soil as well as in the pigeon pea (*Cajanus cajan* var. UPS-120), and promote plant growth. The data also suggested that dual role of bacteria can decrease both the level of copper in soil and the metal load in plants. Further, Rani et al. (2009) also reported in-situ studies whereby upon seed bacterization, cadmium-resistant acidophilic *Pseudomonas putida* 62BN and alkalophilic *Pseudomonas monteilli* 97AN strains were able to enhance agronomical parameters of soybean (*Glycine max* var. PS-1347), in the presence of cadmium in acidic and alkaline soils, respectively. Similarly, Tripathi et al. (2005) reported that *Pseudomonas putida* KNP9 reduced

Table 1.2 Bacterial strains used for bioremediation as reported by our group

Accession no.	Bacteria	Strain	Property	References
AY970345	*Bacillus cereus* 16S r RNA	AG27	Arsenic resistant	Satlewal et al. (2010)
AY970346	*B. cereus* 16S rRNA	AG24	Arsenic resistant	Satlewal et al. (2010)
DQ205432	*Proteus vulgaris* 16S r RNA	KNP3	Pb and Cd resistant	Rani et al. (2008)
DQ517938	*B.cereus* Arsenate reductase gene	AG27	As resistant	Gupta (2006)
DQ517939	*B.cereus* Arsenate reductase like gene	AG24	As resistant	Gupta (2006)
EU512943	*Pseudomonas monteilli*	97AN	Cd resistant	Rani et al. (2009)
EU512944	*Pseudomonas putida*	62BN	Cd resistant	Rani et al. (2009)
DQ205427	*P. putida* 16S r RNA	KNP9	Siderophore producing	Rani and Goel (2009a, b) and Tripathi et al. (2005)
EU512945	*P. veronii*	GCP1	P- solubilizing	Rani (2009)
EU512946	*Enterobacter amnigenus*	GRS3	P- solubilizing	Rani (2009)
EF207715	*P. putida*	710A	Cd resistant	Rani (2009)
EF207716	*Comamonas aquatica*	710B	Cd resistant	Rani (2009)

the cadmium accumulation in mung bean plants. Further, Gupta et al. 2005 reported mercury-resistant strains of *Pseudomonas fluorescens viz.* PRS$_9$Hg^r and GRS$_1$Hg^r, which simultaneously had plant growth-promoting characters like indoleacetic acid (IAA) production, P-solubilization, and siderophore production. Therefore, these potential isolates can be ideal candidates for bioremediation strategies. Laverman et al. (1995) showed that the bacterial strain SES- 3 could grow using a diversity of electron acceptors, including Fe (III), thiosulfate, and arsenate coupled to the oxidation of lactate to acetate.

1.5.3 Phytoremediation

The term phytoremediation refers to the use of plants to extract, sequester, and/or detoxify pollutants and has been an effective, nonintrusive, inexpensive, aesthetically pleasing, socially accepted technology to remediate polluted soils (Alkorta et al. 2004; Garbisu et al. 2002). Phytoremediation is widely viewed as ecologically responsible alternative to the environmentally destructive physical remediation methods currently practiced.

The survey of hyperaccumulating plants was started by the early 1990s. They are often small plants, such as *Alyssum murale*, which grows on metamorphic rocks, *Brassica juncea*, the Indian mustard, which extracts lead, or Thlaspi, which accumulates zinc and nickel. About 400 species have been identified, including 300 that accumulate only nickel. An endemic tree in New Caledonia, *Sebertia acuminata*, contains up to 20% of nickel in its sap and is green (nickel is generally toxic to plants at a concentration of 0.005%). In the Democratic Republic of Congo, the number of plants accumulating copper and cobalt is highest: 24 and 26 species,

Table 1.3 Some hyperaccumulator plants with respective metal species

Metal	Species
Zinc	*Typha caerulescens*
Cadmium	*T. caerulescens*
Nickel	*Berkheya coddii*
Selenium	*Astragalus racemosa*
Thallium	*Iberis intermedia*
Copper	*Ipomoea alpina*
Cobalt	*Haumaniastrum*
Arsenic	*P. vittata*
Zinc, nickel, cadmium	*Thlaspi caerulescens*
Zn/Cd	*T. caerulescens, Arabidopsis halleri*
Ni	*Hybanthus floribundus* subsp. *adpressus, H. floribundus* subsp. *Floribundus, Pimelea leptospermoides*

respectively. The accumulation efficiency is not generally very high. For some metals like silver, mercury, and arsenium, there are yet no plants known to accumulate them. However, in 2000, the team of Lena Ma of the University of Florida, Gainesville, identified a fern, *Pteris vittata*, which tolerates and accumulates arsenium, while conserving a very rapid growth and a high biomass. Edenspace, a company from Virginia specialized in phytoremediation and acquired the rights to commercialize the fern (now called Edenfern™) by signing an exclusive license agreement in 2000 with the University of Florida, which patented the use of the fern in phytoremediation. In the USA, seven or eight similar companies were already in existence in 2002, where the value of the potential market for phytoremediation was estimated at $100 million (Tastemain 2002 cited in Sasson 2004). The US phytoremediation market was expected to expand more than tenfold between 1998 and 2005, to over $214 million (Evans and Furlong 2003). Salt et al. (1995) divided plant-based heavy metals remedies into following areas: (1) Phytoextraction: the use of pollutant-accumulating plants to remove metals or organics from soil by concentrating them in the harvestable parts, (2) Rhizofiltration: the use of plant roots to absorb and absorb pollutants, mainly metals, from water and aqueous waste streams, (3) Phytostabilization: the use of plants to reduce the bioavailability of pollutants in the environment, and (4) Phytovolatilization: the use of plants to volatilize pollutants (Sasson 2000). The idea of using plants to remediate metal-polluted soils has emerged from the discovery of "hyperaccumulators" defined as plants, often endemic to naturally mineralized soils, that accumulate high concentrations of metals in their foliage (Brooks 1998). According to Baker (1981), plants growing on metalliferous soils can be grouped into three categories: (1) Excluders, where metal concentrations in the shoot are maintained, up to a critical value, at a low level across a wide range of soil concentration. (2) Accumulators, where metals are concentrated in above-ground plant parts from low to high soil concentrations. (3) Indicators, where internal concentration reflects external levels. Some of the examples of metal hyperaccumulators compiled from different sources are presented in Table 1.3 (Alkorta et al. 2004; Liang et al. 2009; Kachenko et al. 2009).

1.6   Genetic Engineering: The Way Forward

The molecular basis of heavy metal accumulation is being studied with a view to transferring the relevant genes to plant species having a wider geographic and ecological distribution. Transgenesis applied to phytoremediation is certainly incipient. Its application on a large scale is questioned largely due to risks associated with the transfer of the bacterial transgenes to plants that when consumed by herbivorous animals or humans may lead to metal toxicity. However, genetic transformation of the microorganisms involved in bioremediation could enhance the process through the introduction of genes controlling specific degradation pathways; it can be also aimed at degrading recalcitrant compounds such as pesticides and other xeno-substances.

Researchers discovered many bacteria that had developed high tolerance to heavy metals, related to the binding of these metals to their proteins, for example, metallothionein that binds mercury. As naturally thriving mercury-tolerant bacteria are rare and cannot be grown easily in culture, researchers at Cornell University, Ithaca, New York have successfully inserted the metallo-thionein gene into *E. coli*. Thus, a sufficiently large number of genetically engineered bacteria could treat mercury-polluted water inside a bioreactor. The efficiency of this procedure was high, as mercury was removed from polluted water down to a few nanograms per liter. Once the bacteria died, they were incinerated to recuperate the accumulated pure mercury (European Commission 2002b). In another study, researchers at the University of Georgia, Athens introduced two foreign genes from *E. coli* for the synthesis of two enzymes: one that catalyzes the transformation of arsenate into arsenite, the other that induces the formation of a complex with arsenite, into the genome of *Arabidopsis thaliana* (Dhankher et al. 2002). Recently, Chen et al. (2008) reported Hg^{2+} removal by *E. coli* cells engineered to express a Hg^{2+} transport system and metallothionein accumulated Hg^{2+} effectively over a concentration range of 0.2–4 mg/l in batch systems. Guo et al. (2008) also developed transgenic plants with increased tolerance for and accumulation of heavy metals and metalloids from soil by simultaneous overexpression of phytochelatins (PCs) and glutathione (GSH) genes in *Arabidopsis thaliana*.

1.7   Conclusion

Heavy metals occur as natural constituents of the earth crust and are released in the environment due to human activities and natural phenomena. Due to industrialization, the amounts of heavy metals deposited onto the surface of the earth are many times greater than arising from natural background sources. In rocks, heavy metals exist as ores in different chemical forms, from where they are recovered as minerals. Acid drainage has various anthropogenic and natural sources, but the most widely known source is one related to mining coal and various metal ores. Both natural and manmade sources contaminate the air, water, and soil ecosystems. The accelerating global development over the past two centuries have greatly increased the rate at which trace metals are released to the global environment and many of the famous freshwater bodies are becoming greatly altered. Soil is a major sink for heavy metals released into the environment. Many soils in industrialized countries are affected by acid deposition, mine waste, and organic refuses, such as sewage sludge, that introduce the pollutants to soils. The pattern of soil pollutant content is the synergistic result of mixed processes, including diffusion of deposited airborne particulate matter, fluvial deposition of contaminated sediments, and irregular leaching of soil layers, assisted by rainwater, even down to groundwater.

While some metals are essential with integral role in life processes, others are nonessential and potentially toxic. Both essential and nonessential metals can damage cell membranes; alter enzyme specificity; disrupt cellular functions; and alter DNA at high concentration. Besides, unlike many organic pollutants, heavy metals tend to accumulate in the environment. Though there are exceptions, free ions are generally the most bioavailable form of a metal, and the free ion concentration is often the best indicator of toxicity. Biochemical methods that use various living and dead microorganisms have been known for a long time and can remove toxic ions from solutions with or without heavy metal recovery. The methods are now recognized as viable and desirable alternative technologies. Some microorganisms act in the biosphere as geochemical agents promoting precipitations, transformations, or dissolutions of minerals. The use of these microorganisms for bioremediation offers new tools to degrade or transform toxic contaminants. Phytoremediation, the use of plants to extract, sequester, and/or detoxify pollutants, has been reported to be an effective, nonintrusive, inexpensive, aesthetically pleasing, and socially acceptable technology for polluted soil amendment. While different environmental protection agencies are appreciating the role of environmental biotechnology, particularly for small-scale industries in developing countries, enhanced research efforts in the area are a renowned global interest.

References

Ahmann AH, Krumholz LR, Hemond HF, Lovley DR, Morel FMM (1997) Microbial mobilization of arsenic from sediments of the Aberjona watershed. Environ Sci Technol 3:2923–2930

Alkorta I, Hernández-Allica J, Becerril JM, Amezaga I, Albizu I, Garbisu C (2004) Recent findings on the phytoremediation of soils contaminated with environmentally toxic heavy metals and metalloids such as zinc, cadmium, lead, and arsenic. Rev Environ Sci Biotechnol 3:71–90

Athar M, Vohora SB (1995) Heavy metals and environment. New Age International (P) Ltd., Publishers Reprint 2001, Google Book cited 28 Mar 2010

Baker AJM (1981) Accumulators and excluders-strategies in the response of plants to heavy metals. J Plant Nutr 3:643–654

Balogh SJ, Engstrom DR, Almendinger JE, McDermott C, Hu J, Nollet YH, Meyer ML, Johnson DK (2009) A sediment record of trace metal loadings in the upper Mississippi River. J Paleolimnol 41:623–639

Berg M, Tran CH, Nguyen TC, Pham HV, Schertenleib R, Giger W (2001) Arsenic contamination of groundwater and drinking water in Vietnam: a human health drinking threat. Environ Sci Technol 35:2621–2626

Bernd N, Jean-Marc O, Mathias S, Rainer S, Werner H, Victor K (2001) Elevated lead and zinc contents in remote alpine soils of the Swiss national park. J Environ Qual 30:919–926

Blaser P, Zimmermann S, Luster J, Shotyk W (2000) Critical examination of trace element enrichments and depletions in soils: As, Cr, Cu, Ni, Pb, and Zn in Swiss forest soils. Sci Total Environ 249:257–280

Bontidean I, Lloyd JR, Hobman JL, Wilson JR, Csöregi E, Mattiasson B, Brown NL (2000) Bacterial metal-resistance proteins and their use in biosensors for the detection of bioavailable heavy metals. J Inorg Biochem 79:225–229

Brookes PC (1995) The use of microbial parameters in monitoring soil pollution by heavy metals. Biol Fertil Soils 19:269–279

Brooks RR (1998) Phytoarcheology and hyperaccumulators. In: Brooks RR (ed.) Plants that hyperaccumulate heavy metals. CAB International, Oxon, pp 153–180

Bruins MR, Kapil S, Oehme FW (2000) Microbial resistance to metals in the environment. Ecotoxicol Environ Safe 45:198–207

Chaffei C, Pageau K, Suzuki A, Gouia H, Ghorbal MH, Daubresse CM (2004) Cadmium toxicity induced changes in nitrogen management in *Lycopersicon esculentum* leading to a metabolic safeguard through an amino acid storage strategy. Plant Cell Physiol 45:1681–1693

Chen S, EunKi Kim E, Shuler ML, Wilson DB (2008) Hg^{2+} removal by genetically engineered *Escherichia coli* in a hollow fiber bioreactor. Biotechnol Prog 14:667–671

Cummings DE, JrF C, Fendorf S, Rosenzweig RF (1999) Arsenic mobilization by the dissimilatory Fe(III) reducing bacterium *Shewanella alga* Br Y. Environ Sci Technol 33:723–729

Dazy M, Masfaraud JF, Férard JF (2009) Induction of oxidative stress biomarkers associated with heavy metal stress in *Fontinalis antipyretica* Hedw. Chemosphere 75:297–302

Dhankher OP, Li Y, Rosen BP, Shi J, Salt D, Senecoff JF, Sashti NA, Meagher RB (2002) Engineering tolerance and hyperaccumulation of arsenic in plants by combining arsenate reductase and γ-glutamylcysteine synthetase expression. Nat Biotechnol 20:1140–1145

Dong J, Fei-bo W, Guo-ping Z (2005) Effect of cadmium on growth and photosynthesis of tomato seedlings. J Zhejiang Univ Sci B 6:974–980

Duffus JH (2002) "Heavy Metals" – a meaningless term. Pure Appl Chem 74:793–807

Duruibe JO, Ogwuegbu MOC, Egwurugwu JN (2007) Heavy metal pollution and human biotoxic effects. Int J Phys Sci 2(5):112–118

European Commission (2002) Wonders of life. Stories from life sciences research (from the Fourth and Fifth Framework Programmes). Office of Official Publications of the European Communities, Luxembourg, p 27

European Commission (2002a) Heavy metals in wastes, European Commission on Environment, Denmark. http://ec.europa.eu/environment/waste/studies/pdf/heavy_metalsreport.pdf

Evangelou VP (1998) Environmental soil and water chemistry: principles and applications. Wiley, New York

Evans GM, Furlong JC (2003) Environmental biotechnology theory and application. Wiley, West Sussex, pp 143–170

Ferner DJ (2001) Toxicity, heavy metals. eMed J 2:1

Fuge R, Glover SP, Pearce NJG, Perkins WT (1991) Some observations on heavy metal concentrations in soils of the Mendip region of north Somerset. Environ Geochem Health 13:193–196

Garbisu C, Hernández-Allica J, Barrutia O, Alkorta I, Becerril JM (2002) Phytoremediation: a technology using green plants to remove contaminants from polluted areas. Rev Environ Health 17:75–90

Gimmler H, Carandang J, Boots A, Reisberg E, Woitke M (2002) Heavy metal content and distribution within a woody plant during and after seven years continuous growth on municipal solid waste (MSW) bottom slag rich in heavy metals. J Appl Bot Food Qual 76:203–217

Godt J, Franziska S, Christian GS, Vera E, Paul B, Andrea R, David AG (2006) The toxicity of cadmium and resulting hazards for human health. J Occup Med Toxicol 1:22

Guo J, Dai X, Xu W, Ma M (2008) Over-expressing GSH1 and AsPCS1 simultaneously increases the tolerance and accumulation of cadmium and arsenic in *Arabidopsis thaliana*. Chemosphere 72:1020–1026

Gupta A (2006) Diversified arsenic resistant microbial population from industrial and ground water sources and their molecular characterization. PhD thesis, G. B. Pant University of Agriculture & Technology, Pantnagar

Gupta A, Rai V, Bagdwal N, Goel R (2005) In situ characterization of mercury resistant growth promoting fluorescent pseudomonads. Microbiol Res 160:385–388

Gupta A, Rai DK, Pandey RS, Sharma B (2008) Analysis of some heavy metals in the riverine water, sediments and fish from river Ganges at Allahabad. Environ Monit Assess 157:449–458

Habashi F (1992) Environmental issues in the metallurgical industry: progress and problems. In: Singhal RK et al (eds.) Environmental issues and waste management in energy and mineral production. Balkema, Rotherdam, pp 1143–1153

Järup L, Akesson A (2009) Current status of cadmium as an environmental health problem. Toxicol Appl Pharmacol 238:201–208

John RP, Ahmad P, Gadgil K, Sharma S (2009) Heavy metal toxicity: effect on plant growth, biochemical parameters and metal accumulation by *Brassica juncea* L. Int J Plant Prod 3:65–76

Kachenko AG, Bhatia NP, Siegele R, Walsh KB, Singh B (2009) Nickel, Zn and Cd localisation in seeds of metal hyperaccumulators using μ-PIXE spectroscopy. Nucl Instrum Methods Phys Res B Beam Interact Mater Atoms 267:2176–2180

Kachout SS, Mansoura AB, Leclerc JC, Mechergui R, Rejeb MN, Ouerghi Z (2009) Effects of heavy metals on antioxidant activities of *Atriplex hortensis* and *A. rosea*. J Food Agr Environ 7(938):945

Kantor D (2006) Guillain-Barre syndrome, the medical encyclopedia, National Library of Medicine and National Institute of Health. http://www.nlm.nih.gov/medlineplus/

Kefala MI, Zouboulis AI, Matis KA (1999) Biosorption of cadmium ions by *Actinomycetes* and separation by flotation. Environ Pollut 104:283–293

Kim B, McBride MB (2009) Phytotoxic Effects of Cu and Zn on soybeans grown in field-aged soils: their additive and interactive actions. J Environ Qual 38:2253–2259

Kolesnikov SI, Kazeev KS, Varkov VF (2000) Effects of heavy metal pollution on the ecological and biological characteristics of common Chernozem. Russ J Ecol 31:174–181

Laverman AM, Blum JS, Schaefer JK, Phillips EP, Lovley DR, Oremland RS (1995) Growth of strain SES-3 with arsenate and other diverse electron acceptors. Appl Environ Microbiol 61:3556–3561

Lenntech Water Treatment and Air Purification (2004) Water treatment. Lenntech, Rotterdamseweg. www.excelwater.com/thp/filters/Water-Purification.htm

Lester JN (1987) Heavy metals in wastewater and sludge treatment processes. CRC, Boca Raton, p 208

Levin SV, Guzev VS, Aseeva IV, Bab'eva IP, Marfenina OE, Umarov MM (1989) Heavy metals as a factor of anthropogenic impact on the soil microbiota. Mosk Gos Univ, pp 5–46

Liang MH, Lin TH, Chiou JM, Yeh KC (2009) Model evaluation of the phytoextraction potential of heavy metal hyperaccumulators and non hyperaccumulators. Environ Pollut 157:1945–1952

Lloyd JR, Lovley DR (2001) Microbial detoxification of metals and radionuclides. Curr Opin Biotechnol 12:248–253

McCluggage D (1991) Heavy metal poisoning, NCS Magazine, Bird Hospital, CO. http//:www.cockatiels.org/articles/Diseases/metals.html

McDonald DG, Grandt AF (1981) Limestone- lime treatment of acid mine drainage-full scale. EPA project summary, US-EPA-600/S7-81-033

McLean RJC, Beveridge TJ (1990) Metal binding capacity of bacterial surfaces and their ability to form mineralized aggregates. In: Ehrlich HL, Brierly CL (eds.) Microbial mineral recovery. Mc Graw-Hill, New York, pp 185–222

McLean JE, Bledsoe BE (1992) Behavior of metals in soils. In Ground water issue EPA/540/S-92/018, Ground Water and Ecosystems Restoration Division, US-EPA, Ada

Microsoft Encarta (2008) Encarta encyclopedia. Encyclopedia DVD, Microsoft Corporation, Redmond

Moral R, Gilkes RJ, Jordán MM (2005) Distribution of heavy metals in calcareous and non-calcareous soils in Spain. Water Air Soil Pollut 162:127–142

National Institute of Neurological Disorders and Stroke, NINDS (2007) Guillain-Barre syndrome, Guillain-Barre syndrome fact sheet. http://www.ninds.nih.gov/disorders/gbs/details_gbs.htm

Németh T, Kádár I (2005) Leaching of microelement contaminants: a long term field study. Z Naturforsch 60:260–264

Nies DH (1999) Microbial heavy-metal resistance. Appl Microbiol Biotechnol 51:730–750

Nies DH, Silver S (1995) Ion efflux systems involved in bacterial metal resistances. Ind J Microbiol 14:186–199

Nolan K (2003) Copper toxicity syndrome. J Orthomol Psychiatry 12:270–282

Peplow D (1999) Environmental impacts of mining in eastern Washington. Center for Water and Watershed Studies Fact Sheet, University of Washington, Seattle

Pongratz R, Heumann KG (1999) Production of methylated mercury, lead and cadmium by marine bacteria as a significant natural source for atmospheric heavy metals in polar regions. Chemosphere 39:89–102

Proctor J, Baker AJM (1994) The importance of nickel for plant growth in ultramafic (serpentine) soils. In: Ross SM (ed.) Toxic metals in soil–plant systems. Wiley, Chichester, pp 417–432

Rai UN, Pal A (2002) Health hazards of heavy metals. International Society of Environmental Botanist 8(1)

Ramanathan S, Ensor M, Daunert S (1997) Bacterial biosensors for monitoring toxic metals. Trends Biotechnol 15:500–506

Rani A (2009) Proteomic studies and functional characterization of cadmium resistant bacteria. PhD thesis, GB Pant University of Agriculture & Technology, Pantnagar

Rani A, Goel R (2009a) In-situ bioremediation effect of Pseudomonas putida KNP9 strain on cadmium and lead toxicity. In: Sayyed RZ, Patil AS (eds.) Biotechnology emerging trends. Scientific Publisher, Jodhpur, pp 379–384

Rani A, Goel R (2009b) Strategies for crop improvement in contaminated soils using metal tolerant bioinoculants. In: Khan MS, Zaidi A, Musarrat J (eds.) Microbial strategies for crop improvement. Springer, Berlin, pp 85–104

Rani A, Shouche YS, Goel R (2008) Declination of copper toxicity in pigeon pea and soil system by growth promoting proteus vulgaris KNP3 strain. Curr Microbiol 57:78–82

Rani A, Shouche YS, Goel R (2009) Comparative assessment of in situ bioremediation potential of cadmium resistant acidophilic Pseudomonas putida 62BN and alkalophilic Pseudomonas monteilli 97AN strains on soybean. Int Biodet Biodegrad 63:62–66

Ross S (1994) Toxic metals in soil-plant systems. Wiley, Chichester

Salt DE, Blaylock M, Kumar NPBA, Dushenkov V, Ensley BD, Chet I, Raskin I (1995) Phytoremediation: a novel strategy for the removal of toxic metals from the environment using plants. Biotechnology 13:468–474

Santini JM, Sly LI, Schnagl RD, Macy JM (2000) A new chemolitoautotrophic arsenite-oxidizing bacterium isolated from a gold-mine: phylogenetic, physiological, and preliminary biochemical studies. Appl Environ Microbiol 66:92–97

Sasson A (2000) Biotechnologies in developing countries: present and future. Regional and sub regional co-operation, and joint ventures, vol 3. UNESCO, Paris, Future-oriented studies, pp 103

Sasson A (2004) Industrial and environmental biotechnology achievements, prospects, and perceptions. UNU-IAS Report. Center Pacifico-Yokohama, Japan

Satlewal A, Goel R, Garg GK (2010) Industrially useful microbial bioresources. In: Maheshwari DK, Dubey RC, Saravanamuthu R (eds.) Industrial exploitation of microorganisms. IK International Publication Ltd, New Delhi, pp 390–405

Sen R, Chakrabarti S (2009) Biotechnology applications to environmental remediation in resource exploitation. Curr Sci 97:25

Shamsi IH, Wei K, Jilani G, Zhang GP (2007) Interactions of cadmium and aluminum toxicity in their effect on growth and physiological parameters in soybean. J Zhejiang Univ Sci B 8:181–188

Silver S, Phung LT (1996) Bacterial heavy metal resistance: new surprises. Annu Rev Microbiol 50:753–789

Sohrin Y, Matsui M, Kawashima M, Hojo M, Hasegawa H (1997) Arsenic biogeochemistry affected by eutrophication in Lake Biwa, Japan. Environ Sci Technol 31:2712–2720

Sterritt RM, Lester JM (1996) Heavy metals immobilization by bacterial extracellular polymers. In: Eccles H, Hunt J (eds.) Immobilization of ions by bio-sorption. Chemistry Society, London, pp 121–134

Szili-Kovács T, Anton A, Gulyás F (1999) Effect of Cd, Ni and Cu on some microbial properties of a calcareous chernozem soil. In: Kubát J, Prague (ed.) Proceedings of the 2nd Symposium on the pathways and consequences of the dissemination of pollutants in the biosphere, Prague, pp 88–102

Tebo BM, Ghiorse WC, van Waasbergen LG, Siering PL, Caspi R (1997) Bacterially-mediated mineral formation: insights into manganese (II) oxidation from molecular genetics and biochemical studies. Rev Miner Geochem 35:225–266

Tripathi M, Munot HP, Souche Y, Meyer JM, Goel R (2005) Isolation and functional characterization of siderophore producing lead and cadmium resistant *Pseudomonas putida* KNP9. Curr Microbiol 50:233–237

Turpeinen R (2002) Interactions between metals, microbes and plants–Bioremediation of arsenic and lead contaminated soils. Academic dissertation in environmental ecology, University of Helsinki, Helsinki

Udedi SS (2003) From guinea worm scourge to metal toxicity in Ebonyi State. Chem Niger as New Millennium Unfolds 2:13–14

United Nations Environmental Protection/Global Program of Action (UNEP/GPA) (2004) Why the marine environment needs protection from heavy Metals, heavy metals. UNEP/GPA Coordination Office. http://www.oceansatlas.org/unatlas/uses/uneptextsph/wastesph/2602gpa

United States Department of Labor (USDOL) (2004) Occupational Safety and Health Administration (OSHA); Safety and health topics: heavy metals. USDOL Publication, Washington, DC. http://www.osha.gov/SLTC/metalsheavy/index.html

US EPA (2007) The use of soil amendments for remediation, revitalization and reuse. http://www.epa.org. Cited 29 Mar 2010

Wang W, Zhenghe X, Finck J (1996) Fundamental study of an ambient temperature ferrite process in the treatment of acid mine drainage. Environ Sci Technol 30:2604–2608

White C, Sayer JA, Gadd GM (1997) Microbial solubilization and immobilization of toxic metals: key biochemical processes for treatment of contamination. FEMS Microbiol Rev 20:503–516

Zouboulis AI, Matis KA (1997) Removal of metal ions from dilute solutions by sorptive flotation. Crit Rev Environ Sci Tech 27:195–235

Zouboulis AI, Matis KA (1998) The biosorption process: an innovation in reclamation of toxic metals. In: Gallios GP, Matis KA (eds.) Mineral processing and the environment. Kluwer, Dordrecht, pp 361–385

Zouboulis AI, Rousou EG, Matis KA, Hancock IC (1999) Removal of toxic metals from aqueous mixtures: part 1. Biosorption. J Chem Technol Biotechnol 74:429–436

Zouboulis AI, Matis KA, Lazaridis NK (2001) Removal of metal ions from simulated wastewater by *Saccharomyces* yeast biomass; combining biosorption and flotation processes. Sep Sci Technol 36:349–365

Chapter 2
Metal–Plant Interactions: Toxicity and Tolerance

Anthea Johnson, Naresh Singhal, and Marya Hashmatt

Abstract Plants are primarily exposed to metals through the soil from where they may be absorbed by root tissues and transported into the shoots. The presence of metals at toxic levels can elicit a wide range of visible and physiological symptoms in plants. In addition to deformation and discoloration of tissues, effects include inhibition of seed germination, decreased root and shoot growth, decreased rates of photosynthesis and transpiration, damage to proteins and membranes, nutrient imbalances, and altered enzyme activity. Some metals cause oxidative stress through their participation in reactions that produce reactive oxygen species. Oxidative stress results in a range of general effects including damage to membranes and a range of biomolecules. Other effects of metals include direct substitution in biomolecules and conformational changes in proteins and enzymes. Plants respond to toxicity by either producing metal-binding compounds such as phytochelatins, sequestering metals into specific tissues, or by addressing oxidative damage via the antioxidant system. Metal tolerance may be enhanced through systems already utilized by plants, including chelators, phytohormones, and relationships with soil microorganisms. This chapter outlines the plant uptake of metals and their effects, in addition to mechanisms by which plants tolerate high metal levels.

Keywords Metal bioavailability • Phytotoxicity • Metal tolerance

A. Johnson • N. Singhal (✉) • M. Hashmatt
Department of Civil and Environmental Engineering, University of Auckland,
Auckland, New Zealand
e-mail: n.singhal@auckland.ac.nz

M.S. Khan et al. (eds.), *Biomanagement of Metal-Contaminated Soils*,
Environmental Pollution 20, DOI 10.1007/978-94-007-1914-9_2,
© Springer Science+Business Media B.V. 2011

2.1   Introduction

Potentially toxic metals are widely spread in urban (Wong et al. 2006; Revitt and Ellis 1980), contaminated (Pichtel et al. 2000; Berti and Jacobs 1998; Martin et al. 1982), and natural environments (Yang et al. 2002a). Although metals are found naturally in minerals and soils, anthropogenic activities such as mining (Boularbah et al. 2006; Colbourn and Thornton 1978), wastewater disposal (Cajuste et al. 1991), and fertilizer applications (Raven and Loeppert 1997) have increased the levels of many metals in soils. Furthermore, certain metals are known to have adverse effects on the health of plants, humans, and other organisms (White and Brown 2010; Clemens 2006; Alloway 1995; Domingo 1994; Kusaka 1993; Wagner 1993). Although many metals are required for plant nutrition, higher levels of these metals are potentially toxic to plants when adequately bioavailable (Adriano 2001). Metal phytotoxicity depends on soil composition, plant genotypes, and the type and concentration of metals (Orcutt et al. 2000). Symptoms of phytotoxicity often include reduced germination and growth, with visible discoloration and deformation of leaf and root tissues (Kabata-Pendias and Pendias 2001). Other symptoms are less visible. Furthermore, metals may bind to proteins and inactivate enzymes (Van Assche and Clijsters 1990). Metal exposure can also result in oxidative stress, both directly and indirectly, due to the production of reactive oxygen species (ROS) that damage various cell components (Dietz et al. 1999). The inhibition of physiological processes such as photosynthesis and transpiration are other common consequences of excess metal levels (Walley and Huerta 2010; Benzarti et al. 2008).

To overcome metal stress, plants have evolved mechanisms to tolerate and detoxify high levels of metals. For example, metals may be bound extracellularly to organic ligands and cell walls, or be detoxified within the cell (Hall 2002). Enzymes produce antioxidants that reduce the damaging effects of ROS. Plant metal tolerance may also be assisted by externally applied compounds analogous to those produced naturally by plants. Some of the potential methods for enhancing plant metal tolerance include the application of metal-binding chelators and growth-promoting phytohormones (Lopez et al. 2005, 2009; Israr and Sahi 2008).

The expression "heavy metal" is ill-defined and may be misleading, sometimes even used to describe metalloids (semi-metals) and nonmetals. However, the term is pervasive, persistent, and ubiquitous, and can be found widely distributed in scientific literature, governmental regulations, and popular culture (Duffus 2002). Despite attempts to displace the term with something more meaningful, the common usage of this expression has ensured its continued existence (Nieboer and Richardson 1980). Although heavy metals have often been defined based on density, with a lower limit ranging from 3.5 to 7 g/cm^3 (Duffus 2002), this property has no physiological significance to plants. Terms more relevant to phytotoxicity should therefore be based on biological effects and chemical properties (Appenroth 2010). The term "toxic metal" may suffice, although some metals that are toxic at high concentrations are essential at low concentrations. Chemical properties, such as those based on the periodic table ("s-group," "d-group," and "p-group"), may also be used, as these are frequently relevant to biological effects (Duffus 2002).

2.2 Plant Structure

Roots anchor and support the plants and also play an important role in the transfer of water and nutrients from soil to aerial plant parts. Key features of root structure are schematically illustrated in Fig. 2.1. The outer cells of the root form the epidermis, which depending on the plant species and growth conditions, may have waxy layers, forming the exodermis (Nobel 2005). Fine root hairs may project radially from epidermal cells, increasing the effective root surface area available for water and nutrient absorption. A layer of cells at the inner boundary of the cortex is known as the endodermis, which is typically only one cell thick in angiosperms. Spanning the radial walls of endodermal cells is the casparian band, which is highly lignified and coated with suberin, a waxy substance. The endodermis is therefore capable of preventing extracellular water and solute transport between the cortex and the stele. In some plant species, casparian bands can also be present as lamellae in the cortex and exodermis (Enstone et al. 2003; Hose et al. 2001; Peterson 1987). The endodermis binds the predominantly vascular tissue known as the stele. The xylem is continuous from roots to leaves and conducts water and nutrients required for growth. Carbon compounds (photosynthate) are transported to growing regions of the plant in the phloem, generally in the opposite direction to the flow of the relatively dilute xylem sap.

In the plants, the primary photosynthetic organs are leaves that serve as the principle areas for conversion of light into chemical energy. Important structural components of the leaf are illustrated in Fig. 2.2. A thin waxy layer known as the cuticle covers the outer surface of the leaf, reducing water loss from the plant. Gas exchange within the leaf occurs in the large air spaces present between the spongy cells, which are packed with chloroplasts for photosynthesis. Like the cuticle, the epidermis lines the lower area of the leaf. The stomata are small holes in the epidermis bounded by specialized concave-shaped cells known as guard cells, which regulate opening and closing of stomata (Cutler et al. 2008).

2.3 Metal Uptake and Transport

Water and mineral elements are absorbed from soil by plant roots. The soil solution contains dissolved mineral elements that are available for uptake by roots. The dissolved nutrients can be transported radially across the root via apoplastic and symplastic pathways, or combinations of both. The apoplastic pathway involves solute transfer through extracellular fluid and gas spaces between and within cell walls. In the symplastic pathway, water and solutes are transferred intracellularly, passing from cell to cell through tubular channels known as plasmodesmata that connect the cytoplasm of adjacent cells (Steudle and Ranathunge 2007; Steudle and Peterson 1998). In this way, solutes absorbed at the root surface by root hairs and epidermal cells can be transported across the root cortex, through the endodermis

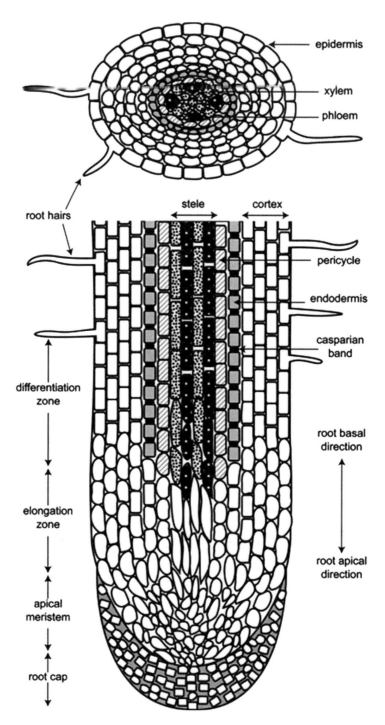

Fig. 2.1 Schematic cross section and longitudinal section of a young root

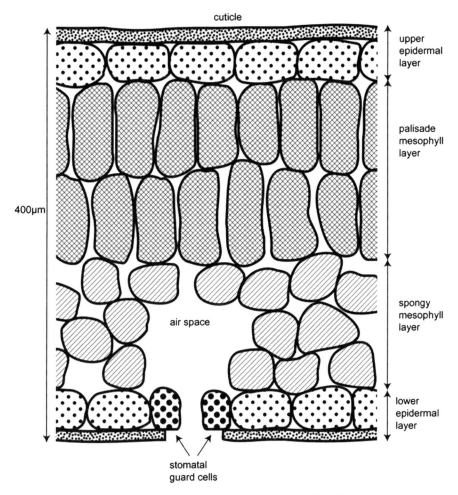

Fig. 2.2 Schematic diagram of a leaf cross section, showing the major cell types

and into the parenchymal cells where they may then enter the conductive vessels of the xylem. Solutes may also be exchanged between the apoplastic and symplastic pathways by crossing the plasma membrane in the transcellular pathway. This step allows the plant a certain degree of selectivity and control over metal uptake. Pathways of radial solute transport in root tissue are illustrated in Fig. 2.3. As the radial walls of the endodermis are blocked by hydrophobic casparian strips, water must enter the cytoplasm of endodermal cells to continue moving into the root. In the entire pathway of water movement from soil to air via the plant, the endodermal cells are the only place where water is forced into the cell cytoplasm (Salisbury and Ross 1992).

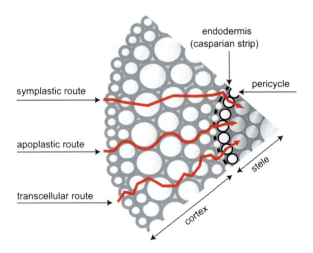

Fig. 2.3 Apoplastic, symplastic, and transcellular routes of radial solute transport in the root

2.3.1   Metal Bioavailability

Bioavailability of metals to plants refers to the amount of metal accessible for uptake by plants in the environment (Rand 1995). The bioavailable fraction of metal thus often refers to the proportion of total metal available in the free ionic form; however, this may not always be an accurate measure (Nolan et al. 2003). It is a function of the total metal concentration and physiochemical and biological factors (Berthelin et al. 1995). Metal bioavailability is influenced by physical and chemical factors such as the soil pH, redox potential and the proportions of clays, mineral components (such carbonates and oxides), and soil organic matter (Plassard et al. 2000). It is also adjusted by biological processes such as biosorption, bioaccumulation, and solubilization (Wu et al. 2006; Ernst 1996). Plants are exposed to metals principally through the aqueous phase of the soil, with soil-bound metals unavailable for direct uptake by roots (Plette et al. 1999). Metal availability for plants is generally greater at lower soil pH (Alloway et al. 1988). For example, cadmium uptake from soil decreases as the pH increases from acidic to neutral values. However, in alkaline conditions, metal uptake may begin to increase (Podar and Ramsey 2005).

Plant access to metals can be facilitated by root exudates such as organic and amino acids, which bind to soil bound metal ions and convert them to a more soluble form. Under nutrient deficient conditions, plants increase the fraction of metal-binding ligands known as phytometallophores in root exudates (Fan et al. 1997). This enables plants to enhance the release of essential cations from the soil matrix. Metal binding siderophores are also produced by rhizosphere bacteria (Whiting et al. 2001). Plants also produce metal binding ligands in response to excess levels of toxic metals (Mohanpuria et al. 2007). In many plants, metals absorbed from soil are accumulated principally in root tissues (Andrade et al. 2008; Kadukova et al. 2008; Chaignon et al. 2002). Metal binding phytochelatins (PC) may be used for long distance transport of metals from the root to the shoot tissue in some plants (Polette et al. 2000). Processes influencing metal uptake and transport within the soil and plants are schematically illustrated in Fig. 2.4.

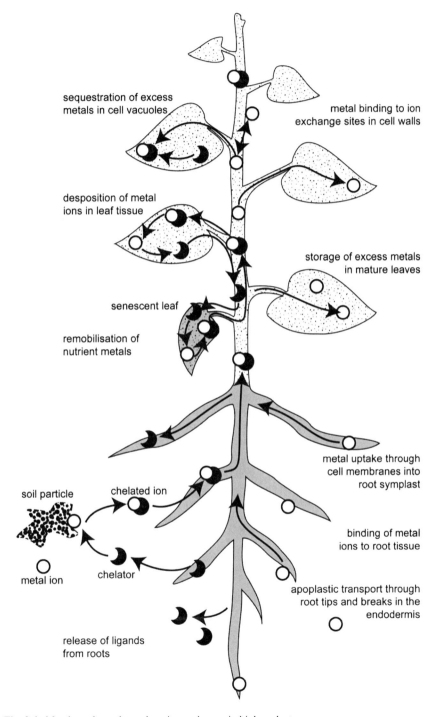

Fig. 2.4 Metal uptake and translocation pathways in higher plants

2.3.2   Root Uptake

Plants are capable of adjusting the pH of soil in the root zone to facilitate access to nutrients (Crowley et al. 1991). A decrease in the pH of the root zone is usually achieved through the release of protons at the root surface (Chaignon et al. 2002; Muranyi et al. 1994); however, root respiration and acid exudation may also acidify the rhizosphere (Hinsinger 2001). The spatial distribution of metals in soil has also been shown to influence plant health and metal uptake. An experiment comparing homogeneous and heterogeneous distributions of Cd and Zn in soils found that *Brassica juncea* tolerated heterogeneously distributed Zn with no growth reduction at levels equivalent to those causing severe phytotoxicity when applied homogeneously (705 mg/kg soil) (Podar et al. 2004). Shoot yield in heterogeneous treatments was up to 24 times greater, and the masses of Cd and Zn extracted from soil were both sixfold higher than in homogeneous treatment.

2.3.3   Foliar Uptake

The leaves have the ability to uptake both essential and nonessential metals. Although the aqueous phase uptake of metals is restricted by epicuticular waxes on the leaf surface, hydrophilic cuticular pores provide routes for the exchange of water, gases, and mineral elements through the leaf surface. Metals such as Mn and Cu are often applied in foliar sprays, in areas where nutrient acquisition from the soil may be limiting (Marschner 1995; Karhadkar and Kannan 1984). Foliar uptake may also contribute to the uptake of metals in the solid phase. For example, in creosote bush (*Larrea tridentate*), windblown particulates less than 10 μm in diameter have been shown to enter the plant through leaf stomata (Polette et al. 2000).

2.4   Metal Phytomtoxicity

Metals have been found to influence plants by three main mechanisms including (1) the production of reactive oxygen species (caused by redox active transition metals), (2) binding to functional groups in biomolecules (non-redox active metals), and (3) the displacement of other metals from biomolecules (Schutzendubel and Polle 2002). These mechanisms can act independently or simultaneously, leading to a range of toxic effects. Some effects may be manifest as visible symptoms, while others cause changes in plant cell structures and interference with normal physiological processes. Roots are the primary plant organs that are in direct contact with metal contaminated soil, and are generally more sensitive to metal toxicity (Seregin and Ivanov 2001). For this reason, root elongation is often used as an indicator of plant sensitivity to metals (Di Salvatore et al. 2008; Karataglis 1987) although care needs to be taken in the interpretation and comparability of root elongation results, especially between species with differing root morphology (Baker 1987).

2.4.1 Visible Symptoms

Common metal phytotoxicity symptoms include growth inhibition, chlorosis, and deformed roots (Kabata-Pendias and Pendias 2001). Although plant roots are valuable indicators of metal phytotoxicity, they are generally not readily visible, and shoot symptoms are often the first toxicity indications. Visible phytotoxicity symptoms for different metals are presented in Table 2.1.

2.4.2 Physiological Changes

High concentrations of metals can weaken the growth and development of plants by limiting seed germination, damaging photosynthetic apparatus and cell membranes, reducing transpiration, altering enzyme activity, and causing peroxidation of lipids (Monni et al. 2001). Major components of plant cells that may be affected by the presence of excess metals are illustrated in Fig. 2.5, and a range of physiological effects observed in plants for several relevant metals are presented in Table 2.2.

2.4.2.1 Inhibition of Germination

The effect of metals on germination of seeds has been well documented, with metals capable of altering the expression of key enzymes within the seed that initiate the germination process (Ahsan et al. 2007; Peralta-Videa et al. 2002; Wierzbicka and Obidzinska 1998). For example, germination success of chickpea (*Cicer arietinum*) was significantly reduced by 54% with Zn (10 mM) and by 73% with Pb (5 mM) (Atici et al. 2005), with similar results observed for Cd (applied at 5 mM) (Atici et al. 2003). Germination of rice (*Oryza sativa*) decreased to 75% with the application of 1.5 mM Cu, and was completely inhibited by 2 mM Cu (Ahsan et al. 2007). In another germination study, metal phytotoxicity followed the order: Cd > Cu > Ni > Pb for lettuce (*Lactuca sativa*), broccoli (*Brassica oleracea*), and tomato (*Lycopersicon esculentum*) seedlings, while for radish (*Raphanus sativus*), the order was Cd = Ni > Pb > Cu, with lettuce found to be the most metal-sensitive species (Di Salvatore et al. 2008). Reductions in *Hypericum perforatum* germination success of 21% and 28% were observed due to 25 and 50 mM Ni, respectively (Murch et al. 2003).

Changes in phytohormone levels due to metal exposure during germination have also been observed. As an example, in germinating chickpea seeds, Pb reduced the levels of gibberellic acid, while increasing abscisic acid and zeatin. High concentrations (1–10 mM) of Zn were found to decrease zeatin, zeatin riboside,

Table 2.1 Visible symptoms of metal phytotoxicity

Metal	Symptoms		References
	Shoots	Roots	
Cadmium	Progressive chlorosis (young to old leaves). Red/purple leaf veins and petioles. Browning of leaf margins. Wilting stems. Stunted shoot growth. Necrotic leaf tips. Decreased leaf area.	Roots stunted, brown, and coralloid.	McKenna et al. (1993), Pandey and Sharma (2002), Sayed (1997), and Van Engelen et al. (2007)
Copper	Interveinal chlorosis beginning in young leaves. Necrosis of leaf tips. Brown spots on leaf lamina. Stunted shoot growth.	Brown root tips. Stunted root growth.	Chatterjee et al. (2006) and Gunawardana et al. (2010)
Cobalt	Chlorosis of young leaves. Reddish-purple leaf margins. Leaves become brittle.	Poor root development.	Keeling et al. (2003) and Pandey and Sharma (2002)
Lead	Stunted leaf growth. Dark green leaves.	Reduced root elongation.	Gunawardana et al. (2010) and Lopez et al. (2009)
Manganese	Chlorosis of leaf margins. Puckering and crinkling of leaves. Necrotic spots and margins. Leaf abscission. Loss of apical dominance.	Root growth inhibition. Progressive browning of roots from tips.	El-Jaoual and Cox (1998) and Sirkar and Amin (1974)
Nickel	Progressive chlorosis beginning on margins of young leaves. Necrosis of young leaves and shoot apex. Stunted growth. Increase in red pigmentation.	Root growth inhibition.	Murch et al. (2003) and Pandey and Sharma (2002)
Zinc	Progressive chlorosis starting in young leaves. Purple-red color in leaves. Limited shoot growth.	Limited root elongation.	Ebbs and Kochian (1997) and Lee et al. (1996)

Including data from Kabata-Pendias and Pendias (2001) and Shaw et al. (2004)

and gibberellic acid and increase abscisic acid levels (Atici et al. 2005). Abscisic acid is produced in response to plant stress (Monni et al. 2001) while gibberellic acid affects processes such as germination, leaf expansion, fruiting, and flowering (Salisbury and Ross 1992). Zeatin and zeatin riboside are cytokinins, hormones that influence plant growth through their effects on cell division.

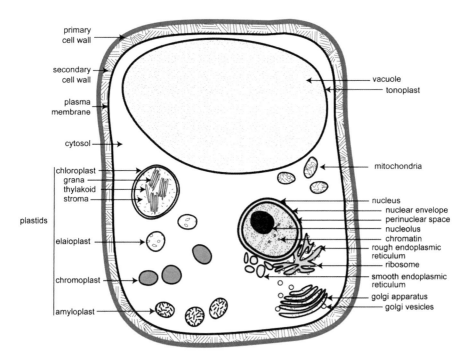

Fig. 2.5 Components of plant cells

2.4.2.2   Decreased Growth

Elevated metal levels can change the normal patterns of plant growth and development, both directly and indirectly. Nickel for example has been found to reduce the growth rate of *Hypericum perforatum* seedlings by 30% and 38%, when applied at 25 and 50 mM respectively (Murch et al. 2003). Stunted growth has been observed as a toxicity symptom for many metals. Effects of heavy metals are not limited to terrestrial plants. For example, in aquatic plants, accumulation of Cr in plant tissues was found to reduce plant biomass (Vajpayee et al. 2001; Gupta et al. 1994; Sen et al. 1987). Growth is driven by cell elongation. Decreased cell elongation can be a result of lowered turgor, reduced synthesis of cell components, or lack of hormonal stimulus. Cell division and root elongation are inhibited by excess levels of Cu and Zn (Karataglis 1980). Metals can also influence plant behavior and alter the timing of growth stages. High levels of Cu, Pb, and Zn in soil caused both later bud break in oak and maple trees and premature senescence in autumn, resulting in a reduced growing season (Bell et al. 1985). Metals may interfere with the utilization of energy in plants, through altering enzyme activity, respiration, and ATP levels (El-Jaoual and Cox 1998).

Indirect effects of metals include those due to metal influence on insects and microorganisms. Plants can benefit from mycorrhizal fungi, which retrieve nutrients from the soil, increasing the effective surface area of the root tissue (Allen 1991).

Table 2.2 Physiological effects of selected metals

Metal	Exposure level	Plant species	Toxicity symptoms	References
Al	0.2 mM	Roman nettle (*Urtica pilulifera*)	>75% decrease in stomatal area.	Özyiğit and Akinci (2009)
Cd	0.2 mM	Roman nettle	>70% decrease in stomatal area.	Özyiğit and Akinci (2009)
	0.03 mM	Safflower (*Carthamus tinctorius*)	Reduced leaf and stomatal area. Transpiration rate decreased by 50%.	Sayed (1997)
	42 mg/kg-soil	Barley (*Hordeum vulgare*)	Reduced gas exchange in leaves. Decreased electron transport activity. Reduced photosynthetic pigments.	Vassilev et al. (2004)
	0.3 mM	Algae (*Chlamydomonas reinhardtii*)	Nitrate uptake rate inhibited by 20%. Sulfate uptake rate increased by 40%	Mosulén et al. (2003)
	0.5 mM	Cabbage (*Brassica oleracea*)	Decreased Fe uptake, chlorophyll content, water potential, and transpiration rate. Increased accumulation of proline in leaves.	Pandey and Sharma (2002)
	186 mg/kg soil	Indian mustard (*Brassica juncea*)	Leaf chlorosis, stunted growth, necrosis of 25% of plants.	Van Engelen et al. (2007)
	0.025 mM	Tomato (*Solanum lycopersicon*)	Induction of oxidative stress, lipid peroxidation.	Chamseddine et al. (2009)
	0.316 µM	Indian mustard	Root distortions.	McKenna et al. (1993)
	20 µM	Sunflower (*Helianthus annuus*)	No visible toxicity symptoms. Significant decrease in chlorophyll content.	Andrade et al. (2008)
	0.05 mM	Soybean (*Glycine max*)	Transpiration and stomatal conductance decreased by 40%. Decrease in relative leaf water content.	Leita et al. (1995)
Co	20 mg/kg soil	*Berkheya coddii*	Significantly reduced biomass in hyperaccumulating species.	Keeling et al. (2003)
	0.5 mM	Cabbage	Decreased Fe uptake, chlorophyll content, water potential, and transpiration rate. Increased accumulation of proline in leaves.	Pandey and Sharma (2002)
Cr(VI)	50 mg/kg soil	Alfalfa (*Medicago sativa*)	Inhibition of germination.	Peralta-Videa et al. (2002)

Metal	Concentration	Plant	Effect	Reference
Cu	0.3 mM	Algae (*C. reinhardtii*)	Nitrate uptake rate inhibited by 20%.	Mosulén et al. (2003)
	0.1–0.2 mM	Radish (*Raphanus sativus*)	Reduced biomass and Fe levels. Decreased chlorophyll and antioxidant enzyme activities.	Chatterjee et al. (2006)
	0.025 mM	Tomato	Induction of oxidative stress, lipid peroxidation.	Chamseddine et al. (2009)
	1.5 mM	Rice (*Oryza sativa*)	Inhibition of germination, decreased root, and shoot growth.	Ahsan et al. (2007)
	2 mM	Rice	Complete inhibition of germination.	Ahsan et al. (2007)
	0.1–1 mM	Oat (*Avena sativa*)	Increasing Cu levels decreased chlorophyll and carotenoid content in leaves, increased membrane permeability and malondialdehyde content.	Luna et al. (1994)
	1.5 mM	Barley	Decreased levels of photosynthesis and ammonium assimilation proteins.	Demirevska-Kepova et al. (2004)
Hg	0.8 mM	Broad bean (*Vicia faba*)	Inhibited stomatal response due to blockage of water and ion channels.	Yang et al. (2004)
Mn	18.3 mM	Barley	Decreased levels of ammonium assimilation proteins, ascorbate and ascorbate peroxidase.	Demirevska-Kepova et al. (2004)
Ni	0.5 mM	Cabbage	Decreased Fe uptake, chlorophyll content, water potential, and transpiration rate. Increased accumulation of proline in leaves.	Pandey and Sharma (2002)
	25 mM	St John's wort (*Hypericum perforatum*)	Reduced germination. Growth rate reduced by 30%. Overall 40% reduction in shoot length.	Murch et al. (2003)
Pb	0.8 mM	Broad bean	Inhibited stomatal response.	Yang et al. (2004)
Zn	1 g/kg soil	Lettuce (*Lactuca sativa*)	Significant decrease in biomass at pH 6.3.	Podar and Ramsey (2005)
	0.5 g/kg soil	Lettuce	Significant decrease in biomass at pH 7.0.	Podar and Ramsey (2005)
	0.8 mM	Broad bean	Inhibited stomatal response.	Yang et al. (2004)
	0.05 g/kg-soil	Lettuce	Significant decrease in biomass at pH 7.7.	Podar and Ramsey (2005)
	0.7 g/kg soil	Indian mustard	Significantly reduced biomass and plant survival.	Podar et al. (2004)

In metal-contaminated soils, mycorrhizal fungi that are tolerant to high metal levels can benefit host plants by providing nutrients and reducing metal transfer to roots (Hildebrandt et al. 2007; Schutzendubel and Polle 2002; Dueck et al. 1986). Scots pine (*Pinus sylvestris*) exposed to high levels of soil Cd were mycorrhizally supplied with higher levels of nutrients when inoculated with a Cd-tolerant isolate of the ectomycorrhizal fungus *Suillus luteus*, than with Cd-sensitive isolates (Krznaric et al. 2009). Reductions in tree growth have been attributed to the possible failure of mycorrhizal colonization and establishment due to the presence of high levels of metals in soil (Burton et al. 1984).

2.4.2.3 Membrane Damage

Plant membranes are one of the first structures to suffer from the effects of toxic metal stress (Foyer et al. 1997), with effects on both structure and function. Biological membranes consist of a lipid bilayer and generally contain embedded proteins with various functions. Metal-induced lipid peroxidation damages the structure of membranes, and hence their ability to maintain ion homeostasis in the cytoplasm. Metal binding to sulfhydryl groups and active sites of proteins and enzymes results in deactivation of membrane-bound proteins. High Cu levels were found to increase lipid peroxidation and the permeability of membranes in oat (*Avena sativa*) leaves, resulting in senescence (Luna et al. 1994). In wheat (*Triticum aestivum*) leaves, lipid peroxidation, indicated by malondialdehyde levels, increased proportionally with Cr and Zn exposure (Panda et al. 2003). Ion leakage occurs across membranes damaged by the presence of metals. In *Amaranthus* seedlings, Cd and Pb were found to cause membrane injury with corresponding leakage of solutes (Bhattacharjee 1997). Membrane damage due to Cu exposure increased the leakiness of cell membranes in *Mimulus guttatus*, and allowed higher rates of diffusive K efflux and Cu influx from the cell (Strange and Macnair 1991). The effect of Cu was rapidly observed, and found to be greater in non-tolerant genotypes, suggesting that Cu tolerance is provided by some constitutive feature of the cell membrane in Cu tolerant genotypes, and not due to a protective metal-induced response such as phytochelatin synthesis (Strange and Macnair 1991). Membrane-bound enzymes are also influenced by high metal levels. Activities of H^+-ATPase in *Cucumis sativus* root cell plasma membranes were reduced after exposure to 10 or 100 μM Cd, Cu, or Ni (Janicka-Russak et al. 2008). Exposure to Cu and Cd has previously been found to decrease plasma membrane H^+-ATPase in sunflower (*Helianthus annuus*) and wheat (Fodor et al. 1995). In addition, metal contamination can alter plant biosynthesis of secondary metabolites. In a study using the medicinal plant St. John's wort, 25 mM Ni was found to decrease by more than 15-fold the production of the medicinal components hypericin and pseudohypericin, and completely prevented the production of hyperforin (Murch et al. 2003).

2.5 Inhibition of Physiological Processes

2.5.1 Oxidative Damage in Plants

Oxidative damage is one of the key consequences of excess metal stress in plants and is due to the effects of reactive oxygen species (ROS) in plant tissues (Apel and Hirt 2004). These are produced in the presence of metals, but also by normal plant processes and as a result of biotic and abiotic stressors (Rio et al. 2009; Aroca et al. 2005). Reactive oxygen species can damage cells and metabolic processes and are countered by the antioxidant system. Such ROS include hydrogen peroxide (H_2O_2), and superoxide $(O_2^{\cdot-})$ and hydroxyl $(^{\cdot}OH)$ radicals. These are generated naturally in plants as part of redox reactions and mitochondrial and photosynthetic processes, but are accumulated in greater quantities due to the presence of metal ions. Redox-active metals such as Fe, Cr, Cu, and Mn can catalyze the formation of hydroxyl radicals. In the presence of Fe and the enzymatically generated ROS superoxide and hydrogen peroxide, the Fenton reactions (Eqs. 2.1 and 2.2) produce the highly reactive (and consequently more toxic) hydroxyl radical (Kehrer 2000).

$$Fe^{3+} + O_2^{\cdot-} \rightarrow Fe^{2+} + O_2 \tag{2.1}$$

$$H_2O_2 + Fe^{2+} \rightarrow {}^{\cdot}OH + OH^- + Fe^{3+} \tag{2.2}$$

The net reaction (Eq. 2.3) is widely known as the Häber-Weiss reaction, which uses hydrogen peroxide and the superoxide radical to generate hydroxyl radicals (Kehrer 2000).

$$H_2O_2 + O_2^{\cdot-} \rightarrow {}^{\cdot}OH + OH^- + O_2 \tag{2.3}$$

However, the Häber-Weiss reaction does not occur directly in vivo and in aqueous systems has a near-zero rate constant ($k=0.13\pm0.07$ $M^{-1}s^{-1}$) (Weinstein and Bielski 1979). It has only been directly observed in vitro in the gas phase (Blanksby et al. 2007). Thus, the Häber-Weiss reaction can only occur when catalyzed by transition metals (such as in the Fenton reactions) and the physiological significance of Eq. 2.3 has been fervently debated (Koppenol 2002; Liochev and Fridovich 2002; Koppenol 2001). Irrespective of the equations used, generation of the hydroxyl radical in plants proceeds in the presence of transition metal catalysts, with Fe being the most important of these in biological systems (Liochev 1999). Other redox-inactive metals (such as Cd, Ni, and Zn) do not participate in these reactions, but increase oxidative stress by interfering with the biological processes that produce

ROS (such as photosynthesis) and disrupting the mechanisms employed by organisms to neutralize ROS (Dietz et al. 1999).

ROS are produced during mitochondrial respiration and photosynthesis, and in response to pathogens, environmental stress (such as excess metals, drought, UV radiation, extreme temperatures, and air pollutants), and mechanical stress (Aroca et al. 2005; Foyer et al. 1997). ROS such as superoxide are sometimes utilized for their toxicity and can be produced by plants in the immune response to pathogens (Merzlyak et al. 1990). Although superoxide is not a particularly toxic ROS (Jabs et al. 1996), it has a role in immune signaling and is generated in the initial stage of the hypersensitive response (Delledonne et al. 2001). The hypersensitive response is a method of apoptosis (programmed cell death) and is a key part of the plant immune system, the molecular basis of which has been reviewed by Nimchuk et al. (2003). Superoxide is produced at the plasma membrane by the enzyme NAD(P)H oxidase in response to triggers by pathogens (Foyer et al. 1997) and is unable to traverse biological membranes. Superoxide has been found to trigger runaway cell death in *Arabidopsis thaliana* mutant lacking a necessary negative regulatory control (Jabs et al. 1996). This limits some of the damaging effects until oxidative burst is initiated (Delledonne et al. 2001). Hydrogen peroxide can readily cross membranes, and therefore can cause more widespread damage in cells and organelles (Boominathan and Doran 2003). The accumulation of H_2O_2 is another step in the hypersensitive response, but is also triggered in response to excess levels of metals such as Cd (Boominathan and Doran 2003) and Ni (Hao et al. 2006).

ROS are reasonably nonspecific, however, and can injure structures and processes in healthy plant tissue. Exposure to ROS causes damage to biomolecules such as phospholipids, DNA, proteins, and enzymes (Hao et al. 2006; Aust et al. 1985). The oxidation of proteins and enzymes can result in their deactivation (Boominathan and Doran 2003). ROS-induced lipid peroxidation leads to deterioration of membrane integrity, resulting in leakage of ions. This type of acute injury can also result in cellular necrosis. For example, the exposure of *Nicotiana tabaccum* to 0.178 mM Cd resulted in a fivefold increase in levels of H_2O_2 in roots, increased lipid peroxidation, and caused cessation of root growth (Boominathan and Doran 2003). Additionally, redox-active metals such as Cu can participate directly in the oxidative damage of polyunsaturated lipids (De Vos et al. 1993). The accumulation of malondialdehyde, a decomposition product of polyunsaturated fatty acids, is commonly used as an indicator of oxidative stress in plants (Tripathi et al. 2006; Demiral and Türkan 2005). Thus, as ROS are produced naturally by plants, and to a greater degree when under stress, plants have a need for effective scavenging mechanisms in order to prevent uncontrolled damage. The antioxidant system in plants provides plants with some defense against ROS and is comprised of enzymatic and nonenzymatic components. Important antioxidants in plants include the molecules glutathione and ascorbate, and enzymes such catalase, superoxide dismutase, glutathione peroxidase, and ascorbate peroxidase (Boominathan and Doran 2003). The enzyme superoxide dismutase has an important role in all aerobic organisms but is found in particularly high levels in the chloroplasts of green leaves (Marschner 1995). It is responsible for the conversion of superoxide into H_2O_2 as shown in Eq. 2.4 (Marschner 1995).

$$O_2^{\cdot-} + O_2^{\cdot-} + 2H^+ \xrightarrow{\text{superoxide dismutase}} H_2O_2 + O_2 \qquad (2.4)$$

Hydrogen peroxide can then be dismutated to H_2O and O_2 by catalases and peroxidases, rendering it nontoxic. The production of antioxidants by plants can be triggered by the presence of ROS, metals, and certain stress-related molecules including salicylic acid and glutathione (Boominathan and Doran 2003; Foyer et al. 1997).

2.5.2 Photosynthesis

Photosynthesis is the process by which atmospheric CO_2 is converted to carbohydrates in plants. Interference of metals with photosynthetic processes has been observed for many metals and plant species (Kupper et al. 2009; Benzarti et al. 2008). Metal toxicity is often observed as leaf chlorosis, the visible symptom of reduced photosynthetic pigments such as chlorophyll and carotenoids (MacFarlane and Burchett 2001). Metals such as Cr and Pb can impair photosynthesis (Panda et al. 2003) by reducing the levels of these pigments (Choudhury and Panda 2005; Rai et al. 1992). Chlorosis has been observed as a primary toxicity symptom for many metals (Table 2.1). This can be the consequence of inhibited synthesis of chlorophyll, increased chlorophyll degradation in the presence of metals (Patsikka et al. 2002), or a reduction in chloroplast density (Baryla et al. 2001). In a novel approach, the unicellular protist, *Euglena gracilis*, both with and without chloroplasts, was used as model system to test the effects of Cu and Cr(VI). The light-dependent reactions of photosynthesis were found to be particularly sensitive to these metals, with the light-independent reactions and respiration less sensitive. Increased production of ROS from the disturbed photosynthetic reactions resulted in greater degradation of carotenoids (Rocchetta and Küpper 2009). Chlorophyll production is inhibited in the presence of Cr and Pb due to their influence on key enzymes controlling chlorophyll biosynthesis (Geebelen et al. 2002; Vajpayee et al. 2000). High levels of Cu damage chloroplast proteins, including RuBisCO, a key enzyme catalyzing photosynthesis and the most abundant leaf protein (Demirevska-Kepova et al. 2004). Metals applied at concentrations from 1 to 10 mg/L were found to affect the photosynthetic activity of seagrass (*Halophila ovalis*). Photosynthetic processes were more sensitive to the presence of Cu and Zn than to Pb and Cd (Ralph and Burchett 1998). Metals such as Cd can also damage the structure of chloroplasts and thylakoids (Baszynski et al. 1980), but the effects differ between and within species (Vassilev et al. 2004). Experiments with tomato cultivars found that the application of Cd at levels above 1 μmol/L decreased the leaf chlorophyll content, net photosynthetic rate, and intracellular CO_2 levels (Dong et al. 2005).

2.5.3 Water Relations

Metal toxicity can also alter the water balance in plants. A wide range of responses have been observed due to stress caused by excess levels of metals. These include reductions in water loss, size of stomata, area of leaves and diameter of xylem vessels, along with increased stomatal resistance, leaf abscission, and suberization of roots (Barcelo and Poschenrieder 1990). In Roman nettle, exposures to Al and Cd decreased stomatal diameter, and thus the area available for gas exchange (Özyi it and Akinci 2009). The application of Cd was found to decrease transpiration of saf-flower (*Carthamus tinctorius*), by reducing leaf area, relative water content, and stomatal opening (Sayed 1997). Stomatal conductance and transpiration rate in soy-bean (*Glycine max*) were found to decrease due to Cd exposure (Leita et al. 1995). Metal exposure has been found to inhibit stomatal responses to light (opening) and darkness (closing). In an experiment using leaf tissue of broad bean (*Vicia faba*), stomatal opening and closing were inhibited by the application of Hg, Pb, and Zn at 0.8 mM, but remained unchanged at the same levels of Ca, K, and Mg (Yang et al. 2004). The effects of Hg, Pb, and Zn were thought to be due to the ability of these metals to block water and ion channels in membranes of stomatal guard cells (Yang et al. 2004). In wheat, Hg has previously been found to block water channels (Zhang and Tyerman 1999).

2.5.4 Nutrient Deficiency

High levels of metals may also impede plant growth indirectly by depriving plants of nutrients required for growth. This may be due to the inhibition of root growth and transpiration, or due to competition by the metal for uptake carriers. The reduc-tions in root growth due to toxic metal exposure can therefore limit nutrient uptake, due to reduced root area available for mineral absorption. For example, exposure of *Brassica oleracea* to the transition metals Cd, Co, and Ni decreased both root bio-mass and transpiration, with reduced Fe uptake (Pandey and Sharma 2002). Another impact of metal toxicity on plant nutrition is the potential loss of essential cations due to competition or substitution by other metal ions. When grown in a mixture of metals (50 mg/kg-soil of each of Cd, Cu, Ni, and Zn), P and S levels increased in shoot tissue of *Medicago sativa*, but Fe and Ca levels were significantly reduced (Peralta-Videa et al. 2002). These effects were thought to be caused by the presence of Cu, which had previously been shown to reduce Ca uptake in *Zea mays* (Ouzounidou et al. 1995) and *Lotus purshianus* (Lin and Wu 1994). When Fe uptake is reduced, deficiency symptoms such as chlorosis may be visible. Excess levels of Cd in soil were suspected to have reduced the uptake of nutrient cations such as Zn, Ca, and Fe by *Brassica juncea*, resulting in leaf chlorosis (Van Engelen et al. 2007). Exposure of radish (*Raphanus sativus*) to high Cu levels (0.2 mM) had visible toxicity symptoms similar to Fe deficiency (Chatterjee et al. 2006). In other studies, elevated concentrations of Cr(VI) have been found to inhibit uptake of nearly all the essential nutrients (Shanker et al. 2005). Potential reasons for this include the binding of toxic metals to nonspecific carriers for ion uptake or the inhibition of ATPase proton pumps in the plasma membrane (Obata et al. 1996; Zaccheo et al. 1982). Additionally, metal interference with enzymes (such as nitrate reductase and ferric reductase) that facilitate nutrient acquisition can also result in deficiency of specific nutrients (Shanker et al. 2005).

2.6 Metal Essentiality and Toxicity

Metals disturb plant development and metabolism at different stages of growth. An essential metal can affect plants when present in either deficient or excess amounts. With those elements, plant growth and yields are enhanced as the metal uptake is increased; metal uptake beyond a certain level however may either have no effect or induce toxic effects that lead to decreased growth and yields. In the case of nonessential elements, deficiency-toxicity symptoms are not observed. The initial supply of these metals does not affect plant growth, but above a certain concentration, the plant can exhibit toxic symptoms (Shaw et al. 2004).

Among metals, manganese is required at low concentrations for plant nutrition and tends to accumulate in plant roots (Wheeler and Power 1995). The threshold for toxicity of Mn is however reasonably high (White and Brown 2010). Nickel at low concentration (0.1 µg/g dry weight) is essential for plant growth, but is toxic at levels greater than 20–301 µg/g dry weight (White and Brown 2010). Excess Ni has been found to disturb water balance, plant development (including cell division and root branching), metal detoxification responses, and photosynthesis (Seregin and Kozhevnikova 2006). Copper is a key component of many metalloenzymes, including superoxide dismutase and ascorbate oxidase (Maksymiec 1997) and is essential for photosynthesis (Fernandes and Henriques 1991). Although Cu is an essential micronutrient, required at levels of 1–5 µg/g dry weight, high concentrations of Cu induce toxicity in plant tissues (White and Brown 2010). Other common soil contaminants that are nonessential metals for plants include Cd, Cr, and Pb, with toxicity thresholds in dry leaf tissue of 5–10 µg/g, 1–2 µg/g, and 10–20 µg/g, respectively (White and Brown 2010; Lombi et al. 2000). High concentrations of Cd can inhibit plant growth, root development, and photosynthesis (Sanita di Toppi and Gabbrielli 1999; Barcelo and Poschenrieder 1990). The presence of Cr can inhibit seed germination, degrade pigmentation, cause nutrient deficiencies, damage antioxidant enzymes, and can induce oxidative stress in plants (Panda et al. 2003; Panda and Choudhury 2005). The mechanism of Pb toxicity on the other hand is thought to be due to its ability to bind to nucleic acids and alter the arrangement of chromatin, which inhibits DNA replication and transcription, leading to reduced cell division and plant growth (Johnson 1998). Seed germination is also affected by the presence of Pb (Atici et al. 2005).

2.7 Toxicity Sequence

The toxicity of metals to plants varies between, and even within, plant species (Wang et al. 2004). Additionally, factors such as plant age, developmental stage, nutrient status, and stress levels also influence metal toxicity (Shaw and Rout 1998;

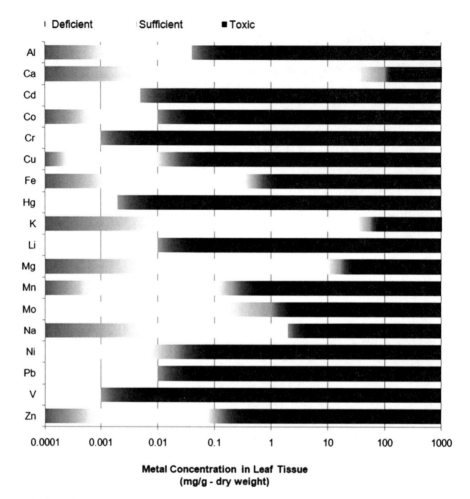

Fig. 2.6 Metal levels required in plant leaf tissues for beneficial and toxic effects (Data from White and Brown 2010)

Maksymiec et al. 1995). It is therefore difficult to ascertain the relative toxicity of different metals for any given plant. A broad indication of metal essentiality and toxicity is illustrated in Fig. 2.6. A general sequence of toxicity for biologically relevant metals, which decreases approximately in the order are: Cr > V > Hg > Cd > Pb = Co > Cu > Ni > Li > Al > Zn > Fe > Mo > Mn > Na > Mg > K > Ca. Toxicity sequences for a range of organisms have also been described (Nieboer and Richardson 1980). For flowering plants, the sequence was: Hg > Cd > Ti > Te > Pb > Bi ~ Sb (Fergusson 1990). The toxicity sequence obtained for the freshwater plant duckweed (*Lemna minor*) was found to be: Ag > Cd > Hg > Tl > Cu > Ni > Zn > Co > Cr(VI) (Appenroth 2010). Toxicity sequence also varies within the plant, depending on the type of tissue affected. For example, in the shoot tissue of wheat, the sequence was

Cu > La > Ga = Fe > Zn > Mn while in root tissue, the order was Cu > La = Fe > Zn > Ga > Mn (Wheeler and Power 1995). The variability in toxicity sequences between different plants illustrates the inherent difficulties in predicting metal toxicities based on the properties of the metal only.

To further confound toxicity predictions, the simultaneous presence of different metals can alter their toxic effects. Metal co-contamination has been found to increase overall toxicity symptoms (Karataglis 1980). The presence of a metal can have synergistic or antagonistic effects on the uptake of other metals. In *Brassica juncea*, the application of Zn at low to moderate levels (113–340 mg/kg soil) was found to suppress Cd uptake by 40%. High levels of Zn (705 mg/kg soil), however, had no such antagonistic effect, possibly due to Zn phytotoxicity (Podar et al. 2004). Similar results have previously been found in lettuce and spinach (*Spinacia oleracea*) (McKenna et al. 1993).

2.8   Metal Tolerance Mechanisms

Two main strategies for plants to circumvent metal toxicity were suggested by Baker (1981). These are exclusion, where the plant limits the entry and movement of metals into cells, and accumulation, where metals taken into plant cells are detoxified through mechanisms such as chelation, active efflux, sequestration to vacuoles (compartmentalization), and binding to the cell wall matrix (Prasad 1995; Baker 1987). Other mechanisms, such as the increased production of antioxidants, are also involved (Sharma and Dietz 2006; Ruley et al. 2004). Metal availability may be reduced through alterations in soil pH (Hinsinger 2001). Within plants, metal ions can be bound to the cell walls (Poulter et al. 1985) or precipitated ("phytoexcreted") from the leaves (Manousaki et al. 2008). To reduce the toxicity of metals such as Pb and Cd, plants may deposit them inside vacuoles (Conn and Gilliham 2010) or bind them to sulfur compounds (Sharma et al. 2005).

2.8.1   Metal Accumulation

Plants capable of accumulating extraordinarily high levels of metals are referred to as hyperaccumulators (Brooks et al. 1977). Hyperaccumulation thresholds of different plants for different metals have been developed over the years (Reeves and Baker 2000; Watanabe 1997; Reeves and Brooks 1983) and are presented in Table 2.3.

Among various plants, *Alyssum*, *Brassica*, and *Thlaspi* have been found to be the most effective and dominant metal-tolerant and accumulating plants (Prasad and Freitas 2003), although only a few species of each genus have been confirmed as hyperaccumulators. Hyperaccumulators tend to accumulate either Ni, Cu/Co, or Cd/Pb/Zn (Raskin et al. 1994). Some metals are more amenable to hyperaccumulation, due to their availability in soil. For example, several hundred Ni hyperaccumulators exist, while only a few plants are able to naturally hyperaccumulate Pb or Cr,

Table 2.3 Metal thresholds levels of certain hyperaccumulating plant species

Metal	Threshold leaf concentration (mg/g dry weight)	Examples of hyperaccumulating plant species
Cd	0.1	*Thlaspi caerulescens*
		Sedum alfredii
Co	1	*Aeolanthus biformifolius*
		Berkheya coddii
Cr	1	*Sutera fodina*
		Leptospermum scoparium
Cu	1	*Aeolanthus biformifolius*
		Commelina communis
		Crassula helmsii
Mn	10	*Macadamia neurophylla*
Ni	1	*Alyssum bertolonii*
		Alyssum murale
		Sebertia acuminata
		Berkheya coddii
Pb	1	*Hemidesmus indicus*
Zn	10	*Thlaspi caerulescens*
		Sedum alfredii

Compiled from Baker and Brooks (1989), Deng et al. (2008), Kupper et al. (2009), Lyon et al. (1968), Malaisse et al. (1979), Robinson et al. (1997), Wang et al. (2004), Whiting et al. (2001), and Yang et al. (2002b)

due to the relative insolubility of these metals in soil solution (Hossner et al. 1998). Hyperaccumulators may play some roles in the phytoremediation of metals by phytoextraction (Lasat 2002; Cunningham and Ow 1996; Chaney 1983). Hyperaccumulating plant species are often capable of survival in soils where other crops may fail, as they can tolerate abnormally high levels of metal in soils and tissue. Comparison of metal uptake factors (concentration in shoots/concentration in soil) of 100–2,200 in radish (Lorenz et al. 1997), a nonaccumulator, and 3,500–85,000 in the hyperaccumulator *Thlaspi caerulescens* grown in the same soils (Knight et al. 1997) demonstrates the impressive accumulation capacity of hyperaccumulators. However, hyperaccumulating plants often have low biomass (Zhao et al. 2001), grow slowly, and may be relatively difficult to establish due to lengthy germination periods (Knight et al. 1997).

Metal hyperaccumulators have been investigated to determine the mechanisms responsible for their metal tolerance and accumulation abilities (Clemens et al. 2002). This is generally with a view to incorporating these properties into high biomass plants for enhancing the effectiveness of metal phytoextraction technologies (Kramer and Chardonnens 2001). Potential reasons for hyperaccumulating plants to cope with accelerated metal levels may be due to improved versions of tolerance mechanisms that are present in other plants. These may include systems such as cytoplasmic chelation, vacuolar sequestration, cell wall binding, and antioxidative stress responses (Hall 2002). Metal hyperaccumulators have been found to tolerate excess metal stress using an enhanced antioxidative response. High levels of important antioxidant enzymes were observed in the hyperaccumulator *T. caerulescens* even in the absence of metals. When 0.178 mM Cd was applied (a level preventing root growth in *Nicotiana tabacum*), root growth of *T. caerulescens* continued, the production of catalase was increased, and low levels of H_2O_2 were maintained, although lipid peroxidation occurred (Boominathan and Doran 2003). Metal uptake may also be facilitated by increased levels of metal binding root exudates. This has been considered as a possible mechanism used by metal hyperaccumulators (Knight et al. 1997). In a study comparing the metal complexing abilities of exudates from hyperaccumulating plants (*T. caerulescens*) to those from the nonaccumulators wheat and canola (*Brassica napus*), exudates from the nonaccumulators were actually found to be more effective at mobilizing metals from resins (Zhao et al. 2001). Only in recent times have some of the genetic mechanisms for metal hyperaccumulation become clearer. These are outside the scope of this chapter, but have been recently reviewed elsewhere (Kramer 2010).

2.8.2   Metal Detoxification

Plant sensitivity to toxic metals can be partially countered by the endogenous synthesis of metal-binding proteins including metallothioneins (MTs) and phytochelatins (PCs) (Mejare and Bulow 2001). PCs can be synthesized in response to high metal levels (Rauser 1995, 1999) and are functionally similar to MTs, the metal binding proteins found in most living organisms (Grill et al. 1987). Glutathione, an important component of the plant antioxidant system, also has an essential role as a PC precursor. PCs are synthesized in plants by the enzyme phytochelatin synthase in response to metal ions, Cd in particular, followed in enzyme activating ability by Ag, Bi, Pb, Zn, Cu, Hg, and Au (Grill et al. 1989). In tomato and *Silene vulgaris*, PCs were produced in response to metals, but were found to bind only to Cu and Cd ions, with Pb and Zn bound to lower molecular weight ligands (Leopold et al. 1999). However, PCs were not present in Cu-tolerant *S. vulgaris*, *Armeria maritima*, or *Minuartia verna* populations growing naturally in Cu-contaminated mine spoil, suggesting that PC synthesis may be a transient response to high levels of metals, and not the sole basis for metal tolerance in these species.

Cadmium-sensitive *Arabidopsis thaliana* mutants were found to lack the ability to synthesize glutathione and thus scavenging the toxic metal (Howden et al. 1995a, b). The hyperaccumulator *T. caerulescens* has been found to naturally produce very high levels of glutathione, which may confer some resilience against metal toxicity (Boominathan and Doran 2003). Exposure to metals increases plant production of phytochelatins. For example, cadmium applied to *Camellia sinensis* induced increased transcription of the genes responsible for glutathione biosynthetic enzymes (Mohanpuria et al. 2007). These mechanisms provide plants some protection against excess metal stress (Yadav 2010). Phytosiderophores, an iron chelating compounds, are another metabolite produced by many plants including members of graminaceous monocotyledons, as part of their strategy for mobilizing Fe from soil. In addition to mobilizing iron, phytosiderophores can also bind to other metals such as Zn

(Hopkins et al. 1998; Zhang et al. 1991). Wheat and barley released phytosiderophores from root tissues, particularly when under Fe-deficient conditions. Phytosiderophores were able to mobilize Fe and Cd from the solid phase. Although plant uptake of Cu, Fe, Mn, and Zn increased under Fe stress due to phytosiderophore release, Cd–phytosiderophore complexes were not taken up by the plants (Shenker et al. 2001). Plants may also change the oxidation state of redox active metals in order to reduce their toxicity. The creosote bush (*Larrea tridentata*) as an example has been found to internally reduce copper from Cu^{2+} to Cu^+ as it is translocated with the help of a PC from roots to leaves (Polette et al. 2000).

2.8.3   *Antioxidant Response*

In order to reduce the damaging effects of high metal levels, plants produce antioxidants. Antioxidant synthesis is an indicator of metal toxicity and the synthesis of antioxidant enzymes increases under metal stress (Chamseddine et al. 2009). Hyperaccumulating populations of *Commelina communis* were found to tolerate high levels of Cu (>1 mg/g dry weight) in leaf tissue without toxicity; however, in nonaccumulating populations, Cu toxicity induced the activity of the antioxidant enzymes superoxide dismutase, guaiacol peroxidase, and ascorbate peroxidase (Wang et al. 2004). In cell cultures of the Pb-tolerant shrub *Sesbania drummondii*, exposure to Cu and Pb induced the activity of superoxide dismutase and catalase, but decreased guaiacol peroxidase activity (Sharma et al. 2005). Plants may also be able to tolerate metals through the rapid repair of damaged membranes. This response has been thought to utilize metallothioniens and certain proteins involved in lipid metabolism (Salt et al. 1998).

2.9   Enhancing Metal Tolerance

Metal tolerance may be acquired in plants, when plants are repeatedly exposed to low level of toxic metals. For example, in bean plants, oxidative stress induced by Cu exposure doubled the activity of the antioxidative enzymes ascorbate peroxidase and catalase, but reduced glutathione reductase activity. This stress pretreatment conferred increased tolerance to subsequent oxidative stress induced by exposure to methyl viologen and SO_2 (Shainberg et al. 2001). Some compounds that could enhance the metal tolerance ability of plants are briefly discussed in the following section.

2.9.1   *Chelators*

The application of metal chelators can help plants to tolerate high metal concentrations. The chelators like nitrilotriacetic acid and ethylenediaminedisuccinic acid have been found to improve the copper tolerance of perennial ryegrass, *Lolium perenne* (Johnson et al. 2009). The presence of strong aminopolycarboxylic acid chelating agents reduced Pb toxicity in *Sesbania drummondii* (Ruley et al. 2006). Plant growth was significantly higher in treatments with chelators and Pb, compared to treatments with Pb alone. Accumulation of Pb in shoot tissue increased up to 40-folds in the presence of chelators, without adversely affecting photosynthesis (Ruley et al. 2006). However, strong chelating agents can be toxic and are therefore often not appropriate for application to soil.

2.9.2   Phytohormones

Certain plant hormones can help alleviate some of the detrimental effects of toxic metal exposure (Gadallah and El-Enany 1999). Nearly all plant processes are regulated by very low concentrations of plant hormones. These include substances such as auxins, cytokinins, gibberellins, abscisic acid, ethylene, brassinosteroids, and jasmonic acid. These are used by the plant to regulate processes such as germination, cell expansion, cell division, root development, stem elongation, pathogen response, flowering, fruit ripening, and senescence (Gray 2004). Indole-3-acetic acid (IAA) is the primary auxin, and regulates cell division, expansion and differentiation, controlling apical dominance and the development of lateral roots (Gray 2004; Kende and Zeevaart 1997). Besides acting as a growth regulator, IAA has also been found to increase metal uptake, which could indeed be of benefit in phytoextraction applications. For example, application of 100 μM IAA increased Pb accumulation in shoot tissues of *S. drummondii* by a factor of more than 6.5, although growth and photosynthetic activity were not affected. After scanning electron microscope observations, Pb was found primarily in vascular tissues of stems and leaves (Israr and Sahi 2008).

The application of kinetin, a cytokinin (cell division-promoting hormone) reduced and reversed the toxic effects of Cd. Kinetin at 15 mg/L, applied as a foliar spray, significantly increased leaf area, transpiration rate, leaf turgidity and stomatal area in safflower (*Carthamus tinctorius*) plants exposed to Cd (Sayed 1997). In alfalfa, the combined application of 100 μM IAA and 100 μM kinetin increased Pb accumulation in stems and leaves (Lopez et al. 2009). There are reports that suggest that the combined application of chelating amendments and phytohormones may play an important role in enhancing the efficiency of phytoremediation technology. To substantiate this hypothesis, Lopez et al. (2005) in a study observed that the application of 0.2 mM ethylenediaminetetraacetic acid (EDTA) increased Pb accumulation in alfalfa plants by three times compared to those observed for sole application of Pb. The mixture of EDTA and 100 μM IAA, however, increased the Pb accumulation in plants by 28-folds (Lopez et al. 2005). This dramatic increase in metal accumulation in alfalfa plants was thought to be due to the ability of IAA to activate plasma membrane ATPases and increase carrier-mediated ion transport through the membrane.

2.9.3   Importance of Microorganisms in Metal Removal

The rhizosphere bacteria able to colonize plant roots and facilitate plant growth are generally called as plant growth-promoting rhizobacteria (PGPR). The PGPR assist plant growth through their abilities to act as biofertilizers, rhizoremediators, phyto-stimulators, and stress controllers (Lugtenberg and Kamilova 2009). These PGPR may help plants overcome metal toxicity (Khan et al. 2009). The potential for using microbially synthesized compounds for improving plant growth and metal tolerance has, however, recently been investigated. The production of growth-promoting phytohormones and metal-mobilizing chelators by bacteria may be valuable in phytoremediation. Other symbiotic fungal associations, for example, arbuscular mycorrhizal (AM) fungi in the rhizosphere, may also assist plants by supplying nutrients essential to plants, when toxic levels of metals reduce root growth (Krznaric et al. 2009). Mycorrhiza have been found to limit Cd toxicity of their host plant, possibly by releasing metal-binding chelators, or by supplying glutathione to plant roots (Schutzendubel and Polle 2002). Besides PGPR and AM-fungi, strains of actinomycetes such as *Streptomyces* have also been found capable of simultane-ously producing auxins and siderophores and thus reducing metal toxicity to plants (Dimkpa et al. 2009; Dimkpa et al. 2008). The presence of metals inhibited auxin production by *Streptomyces*, particularly in the absence of siderophores. However, siderophore production was stimulated by the presence of metals such as Al, Cd, Cu, and Ni, enabling the binding of these metals and reducing their inhibitory effects on auxin synthesis (Dimkpa et al. 2008, 2009).

2.10   Conclusion

Metals are ubiquitous in the environment. Certain metals are necessary as nutri-tional elements, while others are not beneficial to plants at any level. In some areas, human activities have increased the distribution and concentration of metals in such a way as to increase the risk of toxic effects in plants. Toxicity can occur either directly through plant contact with metals in contaminated soils, or indirectly via effects on rhizosphere microorganisms. Due to the enormous range of potential biomolecular targets in plants, there is a great deal of difficulty in predicting metal toxicity. Although some correlations have been developed based on chemical prop-erties, there are still metals and plants that interact unpredictably. Hence, many different toxicity responses are observed in plants, with even different populations of the same species expressing different tolerance to metals. A reason for this is that plants can develop adaptations and tolerance mechanisms that are induced in the presence of stressors such as metals.

In many situations, plant establishment and survival will be adversely affected by the presence of excess metals. Many plant species are important agronomically, while others may be utilized for soil remediation applications such as phytostabilization and phytoextraction. Metal tolerance and concentrations in plant tissues will be of great importance in these fields. Enhanced tolerance and controlled accumulation of metals may be possible through similar mechanisms to those already developed by plants. These include metal-binding chelators, phytohormones, and associations with microbes such as bacteria and fungi. The exploration and utilization of these natural mechanisms for metal detoxification and accumulation may provide potential opportunities for enhanced phytoremediation technologies.

References

Adriano DC (ed.) (2001) Trace elements in terrestrial environments: biogeochemistry, bioavailability, and risks of metals. Springer, New York, p 867

Ahsan N, Lee D-G, Lee S-H, Kang KY, Lee JJ, Kim PJ, Yoon H-S, Kim J-S, Lee B-H (2007) Excess copper induced physiological and proteomic changes in germinating rice seeds. Chemosphere 67:1182–1193

Allen MF (1991) The ecology of mycorrhizae. Cambridge University Press, Cambridge

Alloway BJ (ed.) (1995) Heavy metals in soils. Blackie Academic & Professional, Glasgow, p 390

Alloway BJ, Thornton I, Smart GA, Sherlock JC, Quinn MJ (1988) Metal availability. Sci Total Environ 75:41–69

Andrade SAL, Silveira A, Jorge R, de Abreu M (2008) Cadmium accumulation in sunflower plants influenced by arbuscular mycorrhiza. Int J Phytoremediation 10:1–13

Apel K, Hirt H (2004) Reactive oxygen species: metabolism, oxidative stress, and signal transduction. Annu Rev Plant Biol 55:373–399

Appenroth K-J (2010) Definition of "heavy metals" and their role in biological systems. In: Sherameti I, Varma A (eds.) Soil heavy metals, vol 19. Springer, Berlin, pp 19–29

Aroca R, Amodeo G, Fernandez-Illescas S, Herman EM, Chaumont F, Chrispeels MJ (2005) The role of aquaporins and membrane damage in chilling and hydrogen peroxide induced changes in the hydraulic conductance of maize roots. Plant Physiol 137:341–353

Atici O, Agar G, Battal P (2003) Interaction between endogenous plant hormones and alpha-amylase in germinating chickpea seeds under cadmium exposure. Fresenius Environ Bull 12:781–785

Atici O, Agar G, Battal P (2005) Changes in phytohormone contents in chickpea seeds germinating under lead or zinc stress. Biol Plant 49:215–222

Aust SD, Morehouse LA, Thomas CE (1985) Role of metals in oxygen radical reactions. J Free Radic Biol Med 1:3–25

Baker A (1981) Accumulators and excluders-strategies in the response of plants to heavy metals. J Plant Nutr 3:643–654

Baker AJM (1987) Metal tolerance. New Phytol 106:93–111

Baker AJM, Brooks RR (1989) Terrestrial higher plants which hyperaccumulate metallic elements – a review of their distribution, ecology and phytochemistry. Biorecovery 1:81–126

Barcelo J, Poschenrieder C (1990) Plant water relations as affected by heavy metal stress: a review. J Plant Nutr 13:1–37

Baryla A, Carrier P, Franck F, Coulomb C, Sahut C, Havaux M (2001) Leaf chlorosis in oilseed rape plants (*Brassica napus*) grown on cadmium-polluted soil: causes and consequences for photosynthesis and growth. Planta 212:696–709

Baszynski T, Wajda L, Krol M, Wolinska D, Krupa Z, Tukendorf A (1980) Photosynthetic activities of cadmium-treated tomato plants. Physiol Plant 48:365–370

Bell R, Labovitz ML, Sullivan DP (1985) Delay in leaf flush associated with a heavy metal-enriched soil. Econ Geol 80:1407

Benzarti S, Mohri S, Ono Y (2008) Plant response to heavy metal toxicity: comparative study between the hyperaccumulator *Thlaspi caerulescens* (ecotype Ganges) and nonaccumulator plants: lettuce, radish, and alfalfa. Environ Toxicol 23:607–616

Berthelin J, Munier-Lamy C, Leyval C (1995) Effect of microorganisms on mobility of heavy metals in soils. In: Huang PM, Berthelin J, Bollag JM, McGill WD, Page AL (eds.) Environmental impact of soil component interactions, vol 2. CRC Press, Boca Raton, pp 3–17

Berti WR, Jacobs LW (1998) Distribution of trace elements in soil from repeated sewage sludge applications. J Environ Qual 27:1280

Bhattacharjee S (1997) Membrane lipid peroxidation, free radical scavangers and ethylene evolution in *Amaranthus* as affected by lead and cadmium. Biol Plant 40:131–135

Blanksby SJ, Bierbaum Veronica M, Ellison GB, Kato S (2007) Superoxide does react with peroxides: direct observation of the Haber-Weiss reaction in the gas phase. Angew Chem Int Ed 46:4948–4950

Boominathan R, Doran PM (2003) Cadmium tolerance and antioxidative defenses in hairy roots of the cadmium hyperaccumulator, *Thlaspi caerulescens*. Biotechnol Bioeng 83:158–167

Boularbah A, Schwartz C, Bitton G, Aboudrar W, Ouhammou A, Morel J (2006) Heavy metal contamination from mining sites in South Morocco: 2. Assessment of metal accumulation and toxicity in plants. Chemosphere 63:811–817

Brooks RR, Lee J, Reeves RD, Jaffre T (1977) Detection of nickeliferous rocks by analysis of herbarium specimens of indicator plants. J Geochem Explor 7:49–57

Burton KW, Morgan E, Roig A (1984) The influence of heavy metals upon the growth of sitka-spruce in South Wales forests. Plant Soil 78:271–282

Cajuste LJ, Carrillo RG, Cota EG, Laird RJ (1991) The distribution of metals from wastewater in the mexican valley of Mezquital. Water Air Soil Pollut 57:763–771

Chaignon V, Bedin F, Hinsinger P (2002) Copper bioavailability and rhizosphere pH changes as affected by nitrogen supply for tomato and oilseed rape cropped on an acidic and a calcareous soil. Plant Soil 243:219–228

Chamseddine M, Wided BA, Guy H, Marie-Edith C, Fatma J (2009) Cadmium and copper induction of oxidative stress and antioxidative response in tomato (*Solanum lycopersicon*) leaves. Plant Growth Regul 57:89–99

Chaney RL (1983) Plant uptake of inorganic waste constituents. In: Parr JF, Marsh PB, Kla JM (eds.) Land treatment of hazard wastes. Noyes Data Corp, Park Ridge, pp 50–76

Chatterjee C, Sinha P, Dube BK, Gopal R (2006) Excess copper-induced oxidative damages and changes in radish physiology. Commun Soil Sci Plant Anal 37:2069–2076

Choudhury S, Panda S (2005) Toxic effects, oxidative stress and ultrastructural changes in moss *Taxithelium Nepalense* (Schwaegr) Broth under chromium and lead phytotoxicity. Water Air Soil Pollut 167:73–90

Clemens S (2006) Toxic metal accumulation, responses to exposure and mechanisms of tolerance in plants. Biochimie 88:1707–1719

Clemens S, Palmgren MG, Kramer U (2002) A long way ahead: understanding and engineering plant metal accumulation. Trends Plant Sci 7:309–315

Colbourn P, Thornton I (1978) Lead pollution in agricultural soils. J Soil Sci 29(4):513–526

Conn S, Gilliham M (2010) Comparative physiology of elemental distributions in plants. Ann Bot 105:1081–1102

Crowley D, Wang Y, Reid C, Szaniszlo P (1991) Mechanisms of iron acquisition from siderophores by microorganisms and plants. Plant Soil 130:179–198

Cunningham SD, Ow DW (1996) Promises and prospects of phytoremediation. Plant Physiol 110:715–719

Cutler DF, Botha T, Stevenson DW (2008) Plant anatomy: an applied approach. Blackwell Publishing, Oxford, p 312

De Vos C, Bookum V, Vooijs R, Schat H, Dekok L (1993) Effect of copper on fatty acid composition and peroxidation of lipids in roots of copper tolerant and sensitive *Silene cucubalus*. Plant Physiol Biochem 31:151–158

Delledonne M, Zeier J, Marocco A, Lamb C (2001) Signal interactions between nitric oxide and reactive oxygen intermediates in the plant hypersensitive disease resistance response. Proc Natl Acad Sci USA 98:13454–13459

Demiral T, Türkan I (2005) Comparative lipid peroxidation, antioxidant defense systems and proline content in roots of two rice cultivars differing in salt tolerance. Environ Exp Bot 53:247–257

Demirevska-Kepova K, Simova-Stoilova L, Stoyanova Z, Hölzer R, Feller U (2004) Biochemical changes in barley plants after excessive supply of copper and manganese. Environ Exp Bot 52:253–266

Deng D, Deng J, Li J, Zhang J, Hu M, Lin Z, Liao B (2008) Accumulation of zinc, cadmium, and lead in four populations of *Sedum alfredii* growing on lead/zinc mine spoils. J Integr Plant Biol 50:691–698

Di Salvatore M, Carafa AM, Carratù G (2008) Assessment of heavy metals phytotoxicity using seed germination and root elongation tests: a comparison of two growth substrates. Chemosphere 73:1461–1464

Dietz K, Baier M, Kramer U (1999) Free radicals and reactive oxygen species as mediators of heavy metal toxicity in plants. In: Prasad MNV (ed.) Heavy metal stress in plants: from molecules to ecosystems. Springer, Berlin, p 73

Dimkpa CO, Svatos A, Dabrowska P, Schmidt A, Boland W, Kothe E (2008) Involvement of siderophores in the reduction of metal-induced inhibition of auxin synthesis in *Streptomyces* spp. Chemosphere 74:19–25

Dimkpa CO, Merten D, Svatoš A, Büchel G, Kothe E (2009) Metal-induced oxidative stress impacting plant growth in contaminated soil is alleviated by microbial siderophores. Soil Biol Biochem 41:154–162

Domingo JL (1994) Metal-induced developmental toxicity in mammals. J Toxicol Environ Health 42:123–141

Dong J, FB Wu, GP Zhang (2005) Effect of cadmium on growth and photosynthesis of tomato seedlings. J Zhejiang Univ Sci 6B:974–980

Dueck TA, Visser P, Ernst WHO, Schat H (1986) Vesicular-arbuscular mycorrhizae decrease zinc-toxicity to grasses growing in zinc-polluted soil. Soil Biol Biochem 18:331–333

Duffus JH (2002) "Heavy metals" – a meaningless term? Pure Appl Chem 74:793–807

Ebbs SD, Kochian LV (1997) Toxicity of zinc and copper to Brassica species: implications for phytoremediation. J Environ Qual 26:776–781

El-Jaoual T, Cox DA (1998) Manganese toxicity in plants. J Plant Nutr 21:353–386

Enstone DE, Peterson CA, Ma F (2003) Root endodermis and exodermis: structure, function, and responses to the environment. J Plant Growth Regul 21(4):335–351

Ernst WHO (1996) Bioavailability of heavy metals and decontamination of soils by plants. Appl Geochem 11:163–167

Fan TW, Lane AN, Pedler J, Crowley D, Higashi RM (1997) Comprehensive analysis of organic ligands in whole root exudates using nuclear magnetic resonance and gas chromatography-mass spectrometry. Anal Biochem 251:57–68

Fergusson JE (1990) Heavy elements: chemistry, environmental impact and health effects. Pergamon Press, Oxford

Fernandes JC, Henriques FS (1991) Biochemical, physiological, and structural effects of excess copper in plants. Bot Rev 57:246–273

Fodor E, Szabo-Nagy A, Erdei L (1995) The effects of cadmium on the fluidity and H^+-ATPase activity of plasma membrane from sunflower and wheat roots. J Plant Physiol 147:87–92

Foyer CH, Lopez-Delgado H, Dat JF, Scott IM (1997) Hydrogen peroxide- and glutathione-associated mechanisms of acclimatory stress tolerance and signalling. Physiol Plantarum 100:241–254

Gadallah MAA, El-Enany AE (1999) Role of kinetin in alleviation of copper and zinc toxicity in *Lupinus termis* plants. Plant Growth Regul 29:151–160

Geebelen W, Vangrosveld J, Adriano DC, Van Poucke LC, Clijsters H (2002) Effects of Pb–EDTA and EDTA on oxidative stress reactions and mineral uptake in *Phaseolus vulgaris*. Physiol Plant 115:377–384

Gray WM (2004) Hormonal regulation of plant growth and development. PLoS Biol 2:e311

Grill E, Winnacker EL, Zenk MH (1987) Phytochelatins, a class of heavy-metal-binding peptides from plants, are functionally analogous to metallothioneins. Proc Natl Acad Sci USA 84:439–443

Grill E, Loffler S, Winnacker EL, Zenk MH (1989) Phytochelatins, the heavy-metal binding peptides of plants, are synthesized from glutathione by a specific g-glutamylcysteine dipeptidyl transpeptidase (phytochelatin synthase). Proc Natl Acad Sci USA 86:6838–6842

Gunawardana WB, Singhal N, Johnson A (2010) Amendments and their combined application for enhanced copper, cadmium, lead uptake by *Lolium perenne*. Plant Soil 329:283–294

Gupta M, Sinha S, Chandra P (1994) Uptake and toxicity of metals in *Scirpus lacustris* L. and *Bacopa monnieri* L. J Environ Sci Health A 29:2185–2202

Hall JL (2002) Cellular mechanisms for heavy metal detoxification and tolerance. J Exp Bot 53:1–11

Hao F, Wang X, Chen J (2006) Involvement of plasma-membrane NADPH oxidase in nickel-induced oxidative stress in roots of wheat seedlings. Plant Sci 170:151–158

Hildebrandt U, Regvar M, Bothe H (2007) Arbuscular mycorrhiza and heavy metal tolerance. Phytochemistry 68:139–146

Hinsinger P (2001) Bioavailability of soil inorganic P in the rhizosphere as affected by root-induced chemical changes: a review. Plant Soil 237:173–195

Hopkins BG, Whitney DA, Lamond RE, Jolley VD (1998) Phytosiderophore release by sorghum, wheat, and corn under zinc deficiency. J Plant Nutr 21:2623–2637

Hose E, Clarkson DT, Steudle E, Schreiber L, Hartung W (2001) The exodermis: a variable apoplastic barrier. J Exp Bot 52(365):2245–2264

Hossner LR, Loeppert RH, Newton RJ, Szaniszlo PJ, Attrep MJ (1998) Literature review: phytoaccumulation of chromium, uranium, and plutonium in plant systems. Amarillo National Resource Center for Plutonium, Amarillo

Howden R, Goldsbrough PB, Andersen CR, Cobbett CS (1995a) Cadmium-sensitive, cad1 mutants of *Arabidopsis thaliana* are phytochelatin deficient. Plant Physiol 107:1059–1066

Howden R, Andersen CR, Goldsbrough PB, Cobbett CC (1995b) A cadmium-sensitive, glutathione-deficient mutant of *Arabidopsis thaliana*. Plant Physiol 107:1067–1073

Israr M, Sahi SV (2008) Promising role of plant hormones in translocation of lead in *Sesbania drummondii* shoots. Environ Pollut 153:29–36

Jabs T, Dietrich RA, Dang JL (1996) Initiation of runaway cell death in an *Arabidopsis* mutant by extracellular superoxide. Science 273:1853–1856

Janicka-Russak M, Kabala K, Burzynski M, Klobus G (2008) Response of plasma membrane H^+-ATPase to heavy metal stress in *Cucumis sativus* roots. J Exp Bot 59:3721–3728

Johnson FM (1998) The genetic effects of environmental lead. Mutat Res Rev Mut Res 410:123–140

Johnson AC, Gunawardana WB, Singhal N (2009) Amendments for enhancing copper uptake by *Brassica juncea* and *Lolium perenne* from solution. Int J Phytoremediation 11:215–234

Kabata-Pendias A, Pendias H (2001) Trace elements in soils and plants. CRC Press, Boca Raton

Kadukova J, Manousaki E, Kalogerakis N (2008) Pb and Cd accumulation and phyto-excretion by salt cedar (*Tamarix smyrnensis* Bunge). Int J Phytoremediation 10:31–46

Karataglis SS (1980) Zinc and copper effects on metal-tolerant and non-tolerant clones of *Agrostis tenuis* (Poaceae). Plant Syst Evol 134:173–182

Karataglis S (1987) Estimation of the toxicity of different metals, using as criterion the degree of root elongation in *Triticum aestivum* seedlings. Phyton 26:209–217

Karhadkar AD, Kannan S (1984) Transport patterns of foliar and root absorbed copper in bean seedlings. J Plant Nutr Soil Sci 7:1443–1452

Keeling SM, Stewart RB, Anderson CWN, Robinson BH (2003) Nickel and cobalt phytoextraction by the hyperaccumulator *Berkheya coddii*: implications for polymetallic phytomining and phytoremediation. Int J Phytoremediation 5:235–244

Kehrer JP (2000) The Haber-Weiss reaction and mechanisms of toxicity. Toxicology 149:43–50

Kende H, Zeevaart JAD (1997) The five "classical" plant hormones. Plant Cell 9:1197–1210

Khan MS, Zaidi A, Wani PA, Oves M (2009) Role of plant growth promoting rhizobacteria in the remediation of metal contaminated soils. Environ Chem Lett 7:1–19

Knight B, Zhao FJ, McGrath SP, Shen ZG (1997) Zinc and cadmium uptake by the hyperaccumulator *Thlaspi caerulescens* in contaminated soils and its effects on the concentration and chemical speciation of metals in soil solution. Plant Soil 197:71–78

Koppenol WH (2001) The Haber-Weiss cycle – 70 years later. Redox Rep 6:229–234

Koppenol WH (2002) The Haber-Weiss cycle – 71 years later. Redox Rep 7:59–60

Kramer U (2010) Metal hyperaccumulation in plants. Annu Rev Plant Biol 61:517–534

Kramer U, Chardonnens AN (2001) The use of transgenic plants in the bioremediation of soils contaminated with trace elements. Appl Microbiol Biotechnol 55:661–672

Krznaric E, Verbruggen N, Wevers JHL, Carleer R, Vangronsveld J, Colpaert JV (2009) Cd-tolerant *Suillus luteus*: a fungal insurance for pines exposed to Cd. Environ Pollut 157:1581–1588

Kupper H, Gotz B, Mijovilovich A, Kupper FC, Meyer-Klaucke W (2009) Complexation and toxicity of copper in higher plants. I. Characterization of copper accumulation, speciation, and toxicity in *Crassula helmsii* as a new copper accumulator. Plant Physiol 151:702–714

Kusaka Y (1993) Occupational diseases caused by exposure to sensitizing metals. Sangyo Igaku 35:75

Lasat MM (2002) Phytoextraction of toxic metals: a review of biological mechanisms. J Environ Qual 31:109–120

Lee C, Choi J, Pak C (1996) Micronutrient toxicity in seed geranium (*Pelargonium* x *hortorum* Baley). J Am Soc Hort Sci 121:77–82

Leita L, Marchiol L, Martin M, Peressotti A, Vedove GD, Zerbi G (1995) Transpiration dynamics in cadmium-treated soybean (*Glycine max* L.) plants. J Agron Crop Sci 175:153–156

Leopold I, Gunther D, Schmidt J, Neumann D (1999) Phytochelatins and heavy metal tolerance. Phytochemistry 50:1323–1328

Lin SL, Wu L (1994) Effects of copper concentration on mineral nutrient uptake and copper accumulation in protein of copper tolerant and non-tolerant *Lotus purshianus* L. Ecotoxicol Environ Saf 29:214–228

Liochev SI (1999) The mechanism of "Fenton-like" reactions and their importance for biological systems. A biologist's view. Met Ions Biol Syst 36:1–39

Liochev SI, Fridovich I (2002) The Haber-Weiss cycle – 70 years later: an alternative view. Redox Rep 7:55–57

Lombi E, Zhao FJ, Dunham SJ, McGrath SP (2000) Cadmium accumulation in populations of *Thlaspi caerulescens* and *Thlaspi goesingense*. New Phytol 145:11–20

Lopez ML, Peralta-Videa JR, Benitez T, Gardea-Torresdey JL (2005) Enhancement of lead uptake by alfalfa (*Medicago sativa*) using EDTA and a plant growth promoter. Chemosphere 61:595–598

Lopez ML, Peralta-Videa JR, Parsons JG, Gardea-Torresdey JL, Duarte-Gardea M (2009) Effect of indole-3-acetic acid, kinetin, and ethylenediaminetetraacetic acid on plant growth and uptake and translocation of lead, micronutrients, and macronutrients in alfalfa plants. Int J Phytoremediation 11:131–149

Lorenz SE, Hamon RE, Holm PE, Domingues HC, Sequiera E, Christensen TH, McGrath SP (1997) Cadmium and zinc in plants and soil solutions from contaminated soils. Plant Soil 189:21–31

Lugtenberg B, Kamilova F (2009) Plant-growth-promoting rhizobacteria. Annu Rev Microbiol 63:541–556

Luna CM, Gonzalez CA, Trippi VS (1994) Oxidative damage caused by an excess of copper in oat leaves. Plant Cell Physiol 35:11–15

Lyon GL, Brooks RR, Peterson PJ, Butler GW (1968) Trace elements in a New Zealand serpentine flora. Plant Soil 29:225–240

MacFarlane GR, Burchett MD (2001) Photosynthetic pigments and peroxidase activity as indicators of heavy metal stress in the grey mangrove, *Avicennia marina* (Forsk.) Vierh. Mar Pollut Bull 42:233–240

Maksymiec W (1997) Effect of copper on cellular processes in higher plants. Photosynthetica 34:321–342

Maksymiec W, Bednara J, Baszynski T (1995) Responses of runner bean plants to excess copper as a function of plant growth stages: effects on morphology and structure of primary leaves and their chloroplast ultrastructure. Photosynthetica 31:427–436

Malaisse F, Grégoire J, Morrison RS, Brooks RR, Reeves RD (1979) Copper and cobalt in vegetation of Fungurume, Shaba Province, Zaïre. Oikos 33:472–478

Manousaki E, Kadukova J, Papadantonakis N, Kalogerakis N (2008) Phytoextraction and phyto-excretion of Cd by the leaves of *Tamarix smyrnensis* growing on contaminated non-saline and saline soils. Environ Res 106:326–332

Marschner H (1995) Mineral nutrition of higher plants. Academic, London, p 889

Martin MH, Duncan EM, Coughtrey PJ (1982) The distribution of heavy metals in a contaminated woodland ecosystem. Environ Pollut B 3:147–157

McKenna IM, Chaney RL, Williams FM (1993) The effects of cadmium and zinc interactions on the accumulation and tissue distribution of zinc and cadmium in lettuce and spinach. Environ Pollut 79:113–120

Mejare M, Bulow L (2001) Metal-binding proteins and peptides in bioremediation and phytoremediation of heavy metals. Trends Biotechnol 19:67–73

Merzlyak MN, Reshetnikova IV, Chivkunova OB, Ivanova DG, Maximova NI (1990) Hydrogen peroxide- and superoxide-dependent fatty acid breakdown in *Phytophthora infestans* zoospores. Plant Sci 72:207–212

Mohanpuria P, Rana NK, Yadav SK (2007) Cadmium induced oxidative stress influence on glutathione metabolic genes of *Camellia sinensis* (L.) O. Kuntze. Environ Toxicol 22:368–374

Monni S, Uhlig C, Hansen E, Magel E (2001) Ecophysiological responses of *Empetrum nigrum* to heavy metal pollution. Environ Pollut 112:121–129

Mosulén S, Domínguez MJ, Vigara J, Vílchez C, Guiraum A, Vega JM (2003) Metal toxicity in *Chlamydomonas reinhardtii*. Effect on sulfate and nitrate assimilation. Biomol Eng 20:199–203

Muranyi A, Seeling B, Ladewig E, Jungk A (1994) Acidification in the rhizosphere of rape seedlings and in bulk soil by nitrification and ammonium uptake. Z Pflanzenernaehr Bodenkd 157:61–65

Murch SJ, Haq K, Rupasinghe HPV, Saxena PK (2003) Nickel contamination affects growth and secondary metabolite composition of St. John's wort (*Hypericum perforatum* L.). Environ Exp Bot 49:251–257

Nieboer E, Richardson DHS (1980) The replacement of the nondescript term heavy metals by a biologically and chemically significant classification of metal ions. Environ Pollut Ser B 1(1):3–26

Nimchuk Z, Eulgem T, Holt BFI, Dangl JL (2003) Recognition and response in the plant immune system. Annu Rev Genet 37:579–609

Nobel PS (2005) Physicochemical and environmental plant physiology. Elsevier Academic, San Diego, p 567

Nolan AL, Lombi E, McLaughlin MJ (2003) Metal bioaccumulation and toxicity in soils-why bother with speciation? Aust J Chem 56:77–92

Obata H, Inoue N, Umebayashi M (1996) Effect of Cd on plasma membrane ATPase from plant roots differing in tolerance to Cd. Soil Sci Plant Nutr 42:361–366

Orcutt D, Nilsen E, Hale M (2000) The physiology of plants under stress: soil and biotic factors. Wiley, New York

Ouzounidou G, Ciamporova M, Moustakas M, Karataglis S (1995) Responses of maize (*Zea mays* L.) plants to copper stress -I. Growth, mineral content and ultrastructure of roots. Environ Exp Bot 35:167–176

Özyiğit İ, Akinci Ş (2009) Effects of some stress factors (aluminum, cadmium and drought) on stomata of Roman nettle (*Urtica pilulifera* L.). Not Bot Hort Agrobot Cluj 37:108–115

Panda S, Choudhury S (2005) Chromium stress in plants. Braz J Plant Physiol 17:95–102

Panda S, Chaudhury I, Khan M (2003) Heavy metals induce lipid peroxidation and affect antioxidants in wheat leaves. Biol Plant 46:289–294

Pandey N, Sharma CP (2002) Effect of heavy metals Co^{2+}, Ni^{2+} and Cd^{2+} on growth and metabolism of cabbage. Plant Sci 163:753–758

Patsikka E, Kairavuo M, Sersen F, Aro E-M, Tyystjarvi E (2002) Excess copper predisposes photosystem II to photoinhibition in vivo by outcompeting iron and causing decrease in leaf chlorophyll. Plant Physiol 129:1359–1367

Peralta-Videa JR, Gardea-Torresdey JL, Gomez E, Tiemann KJ, Parsons JG, Carrillo G (2002) Effect of mixed cadmium, copper, nickel and zinc at different pHs upon alfalfa growth and heavy metal uptake. Environ Pollut 119:291–301

Peterson CA (1987) The exodermal Casparian band of onion roots blocks the apoplastic movement of sulphate ions. J Exp Bot 38(197):2068–2081

Pichtel J, Kuroiwa K, Sawyerr HT (2000) Distribution of Pb, Cd and Ba in soils and plants of two contaminated sites. Environ Pollut 110:171–178

Plassard F, Winiarski T, Petit-Ramel M (2000) Retention and distribution of three heavy metals in a carbonated soil: comparison between batch and unsaturated column studies. J Contam Hydrol 42:99–111

Plette ACC, Nederlof MM, Temminghoff EJM, Van Riemsdijk WH (1999) Bioavailability of heavy metals in terrestrial and aquatic systems: a quantitative approach. Environ Toxicol Chem 18:1882–1890

Podar D, Ramsey MH (2005) Effect of alkaline pH and associated Zn on the concentration and total uptake of Cd by lettuce: comparison with predictions from the CLEA model. Sci Total Environ 347:53–63

Podar D, Ramsey MH, Hutchings MJ (2004) Effect of cadmium, zinc and substrate heterogeneity on yield, shoot metal concentration and metal uptake by *Brassica juncea*: Implications for human health risk assessment and phytoremediation. New Phytol 163:313–324

Polette LA, Gardea-Torresdey JL, Chianelli RR, George GN, Pickering IJ, Arenas J (2000) XAS and microscopy studies of the uptake and bio-transformation of copper in *Larrea tridentata* (creosote bush). Microchem J 65:227–236

Poulter A, Collin HA, Thurman DA, Hardwick K (1985) The role of the cell wall in the mechanism of lead and zinc tolerance in *Anthoxanthum odoratum* L. Plant Sci 42:61–66

Prasad MNV (1995) Cadmium toxicity and tolerance in vascular plants. Environ Exp Bot 35:525–545

Prasad MNV, Freitas H (2003) Metal hyperaccumulation in plants – biodiversity prospecting for phytoremediation technology. Electron J Biotechnol 6:285–321

Rai U, Tripathi R, Kumar N (1992) Bioaccumulation of chromium and toxicity on growth, photosynthetic pigments, photosynthesis, nitrate reductase activity and protein content in a chlorococcalean green alga *Gaucocystis nostochinearum* Itzigsohn. Chemosphere 25:1721–1732

Ralph PJ, Burchett MD (1998) Photosynthetic response of *Halophila ovalis* to heavy metal stress. Environ Pollut 103:91–101

Rand GM (1995) Fundamentals of aquatic toxicology: effects, environmental fate, and risk assessment. CRC Press, Boca Raton, p 1125

Raskin I, Kumar PBAN, Dushenkov S, Salt DE (1994) Bioconcentration of heavy metals by plants. Curr Opin Biotechnol 5:285–290

Rauser W (1995) Phytochelatins and related peptides structure, biosynthesis, and function. Plant Physiol 109:1141

Rauser WE (1999) Structure and function of metal chelators produced by plants; the case for organic acids, amino acids, phytin and metallothioneins. Cell Biochem Biophys 31:19–48

Raven KP, Loeppert RH (1997) Trace element composition of fertilizers and soil amendments. J Environ Qual 26:551–557

Reeves RD, Baker AJM (2000) Metal-accumulating plants. In: Raskin I, Ensley BD (eds.) Phytoremediation of toxic metals: using plants to clean up the environment. Wiley, New York, pp 193–229

Reeves R, Brooks RR (1983) European species of *Thlaspi* L. (Cruciferae) as indicators of nickel and zinc. J Geochem Explor 18:275–283

Revitt DM, Ellis JB (1980) Rain water leachates of heavy metals in road surface sediments. Water Res 14:1403–1407

Rio LA, Puppo A, Bolwell GP, Daudi A (2009) Reactive oxygen species in plant–pathogen interactions. In: Río LAd, Puppo A (eds.) Reactive oxygen species in plant signaling. Springer, Berlin, pp 113–133

Robinson BH, Chiarucci A, Brooks RR, Petit D, Kirkman JH, Gregg PEH, De Dominicis V (1997) The nickel hyperaccumulator plant *Alyssum bertolonii* as a potential agent for phytoremediation and phytomining of nickel. J Geochem Explor 59:75–86

Rocchetta I, Küpper H (2009) Chromium- and copper-induced inhibition of photosynthesis in *Euglena gracilis* analysed on the single-cell level by fluorescence kinetic microscopy. New Phytol 182:405–420

Ruley AT, Sharma NC, Sahi SV (2004) Antioxidant defense in a lead accumulating plant, *Sesbania drummondii*. Plant Physiol Biochem 42:899–906

Ruley AT, Sharma NC, Sahi SV, Singh SR, Sajwan KS (2006) Effects of lead and chelators on growth, photosynthetic activity and Pb uptake in *Sesbania drummondii* grown in soil. Environ Pollut 144:11–18

Salisbury FB, Ross CW (1992) Plant physiology. Wadsworth, Belmont, p 682

Salt DE, Smith RD, Raskin I (1998) Phytoremediation. Annu Rev Plant Physiol Plant Mol Biol 49:643–668

Sanita di Toppi L, Gabbrielli R (1999) Response to cadmium in higher plants. Environ Exp Bot 41:105–130

Sayed S (1997) Effect of cadmium and kinetin on transpiration rate, stomatal opening and leaf relative water content in safflower plants. J Islam Acad Sci 10:73–80

Schutzendubel A, Polle A (2002) Plant responses to abiotic stresses: heavy metal-induced oxidative stress and protection by mycorrhization. J Exp Bot 53:1351–1365

Sen A, Mondal N, Mandal S (1987) Studies of uptake and toxic effects of Cr(VI) on *Pistia stratiotes*. Water Sci Technol 19:119–127

Seregin IV, Ivanov VB (2001) Physiological aspects of cadmium and lead toxic effects on higher plants. Russ J Plant Physiol 48:523–544

Seregin IV, Kozhevnikova AD (2006) Physiological role of nickel and its toxic effects on higher plants. Russ J Plant Physiol 53:257–277

Shainberg O, Rubin B, Rabinowitch HD, Tel-Or E (2001) Loading beans with sublethal levels of copper enhances conditioning to oxidative stress. J Plant Physiol 158:1415–1421

Shanker AK, Cervantes C, Loza-Tavera H, Avudainayagam S (2005) Chromium toxicity in plants. Environ Int 31:739–753

Sharma SS, Dietz K-J (2006) The significance of amino acids and amino acid-derived molecules in plant responses and adaptation to heavy metal stress. J Exp Bot 57:711–726

Sharma NC, Sahi SV, Jain JC (2005) *Sesbania drummondii* cell cultures: ICP-MS determination of the accumulation of Pb and Cu. Microchem J 81:163–169

Shaw BP, Rout NP (1998) Age-dependent responses of *Phaseolus aureus* Roxb. to inorganic salts of mercury and cadmium. Acta Physiol Plant 20:85–90

Shaw BP, Sahu SK, Mishra RK (2004) Heavy metal induced oxidative damage in terrestrial plants. In: Prasad M (ed.) Heavy metal stress in plants: from biomolecules to ecosystems. Springer, Berlin, pp 84–126

Shenker M, Fan TWM, Crowley DE (2001) Phytosiderophores influence on cadmium mobilization and uptake by wheat and barley plants. J Environ Qual 30:2091–2098

Sirkar S, Amin JV (1974) The manganese toxicity of cotton. Plant Physiol 54(4):539–543

Steudle E, Peterson CA (1998) How does water get through roots? J Exp Bot 49:775–788

Steudle E, Ranathunge K (2007) Apoplastic water transport in roots. In: Sattelmacher B, Horst WJ (eds.) The apoplast of higher plants: compartment of storage, transport and reactions. Springer, Berlin, pp 119–130

Strange J, Macnair MR (1991) Evidence for a role for the cell membrane in copper tolerance of *Mimulus guttatus* Fischer ex DC. New Phytol 119:383–388

Tripathi B, Mehta S, Amar A, Gaur J (2006) Oxidative stress in *Scenedesmus sp.* during short- and long-term exposure to Cu^{2+} and Zn^{2+}. Chemosphere 62:538–544

Vajpayee P, Tripathi RD, Rai UN, Ali MB, Singh SN (2000) Chromium (VI) accumulation reduces chlorophyll biosynthesis, nitrate reductase activity and protein content in *Nymphaea alba* L. Chemosphere 41:1075–1082

Vajpayee P, Rai U, Ali M, Tripathi R, Yadav V, Sinha S, Singh S (2001) Chromium-induced physiologic changes in *Vallisneria spiralis* L. and its role in phytoremediation of tannery effluent. Bull Environ Contam Toxicol 67:246–256

Van Assche F, Clijsters H (1990) Effects of metals on enzyme activity in plants. Plant Cell Environ 13:195–206

Van Engelen DL, Sharpe-Pedler RC, Moorhead KK (2007) Effect of chelating agents and solubility of cadmium complexes on uptake from soil by *Brassica juncea*. Chemosphere 68:401–408

Vassilev A, Lidon F, Scotti P, Da Graca M, Yordanov I (2004) Cadmium-induced changes in chloroplast lipids and photosystem activities in barley plants. Biol Plant 48:153–156

Wagner GJ (1993) Accumulation of cadmium in crop plants and its consequences to human health. Adv Agron 51:173–212

Walley JW, Huerta AJ (2010) Exposure to environmentally relevant levels of cadmium primarily impacts transpiration in field-grown soybean. J Plant Nutr 33:1519–1530

Wang H, Shan X-Q, Wen B, Zhang S, Wang Z-J (2004) Responses of antioxidative enzymes to accumulation of copper in a copper hyperaccumulator of *Commoelina communis*. Arch Environ Contam Toxicol 47:185–192

Watanabe ME (1997) Phytoremediation on the brink of commercialisation. Environ Sci Technol 31:182–186

Weinstein J, Bielski BH (1979) Kinetics of the interaction of HO_2 and O_2-radicals with hydrogen peroxide. The Haber-Weiss reaction. J Am Chem Soc 101:58–62

Wheeler DM, Power IL (1995) Comparison of plant uptake and plant toxicity of various ions in wheat. Plant Soil 172:167–173

White PJ, Brown PH (2010) Plant nutrition for sustainable development and global health. Ann Bot 105:1073–1080

Whiting SN, De Souza MP, Terry N (2001) Rhizosphere bacteria mobilize Zn for hyperaccumulation by *Thlaspi caerulescens*. Environ Sci Technol 35:3144

Wierzbicka M, Obidzinska J (1998) The effect of lead on seed imbition and germination in different plant species. Plant Sci 137:155–171

Wong CSC, Li X, Thornton I (2006) Urban environmental geochemistry of trace metals. Environ Pollut 142:1–16

Wu SC, Luo YM, Cheung KC, Wong MH (2006) Influence of bacteria on Pb and Zn speciation, mobility and bioavailability in soil: a laboratory study. Environ Pollut 144:765–773

Yadav SK (2010) Heavy metals toxicity in plants: an overview on the role of glutathione and phytochelatins in heavy metal stress tolerance of plants. S Afr J Bot 76:167–179

Yang H, Rose NL, Battarbee RW, Monteith D (2002a) Trace metal distribution in the sediments of the whole lake basin for Lochnagar, Scotland: a palaeolimnological assessment. Hydrobiologia 479:51–61

Yang X, Long X, Ni W, Fu C (2002b) *Sedum alfredii* H: a new Zn hyperaccumulating plant first found in China. Chin Sci Bull 47:1634–1637

Yang HM, Zhang XY, Wang GX (2004) Effects of heavy metals on stomatal movements in broad bean leaves. Russ J Plant Physiol 51:464–468

Zaccheo P, Genevini P, Cocucci S (1982) Chromium ions toxicity on the membrane transport mechanism in segments of maize seedling roots. J Plant Nutr 5:1217–1227

Zhang WH, Tyerman SD (1999) Inhibition of water channels by $HgCl_2$ in intact wheat root cells. Plant Physiol 120:849

Zhang F, Römheld V, Marschner H (1991) Diurnal rhythm of release of phytosiderophores and uptake rate of zinc in iron-deficient wheat. Soil Sci Plant Nutr 37:671–678

Zhao FJ, Hamon RE, McLaughlin MJ (2001) Root exudates of the hyperaccumulator *Thlaspi caerulescens* do not enhance metal mobilization. New Phytol 151:613–620

Chapter 3
Bioremediation: New Approaches and Trends

Reda Abd El-Aziz Ibrahim Abou-Shanab

Abstract Ever increasing human activities including agricultural, urban, or industrial are a major source of environmental pollution. Toxic metal pollution of waters, air, and soils is one of the potential problems, which is an enigma for scientists how to tackle this problem that has threatened the environment. To solve this, conventional remediation approaches have been used, which, however, do not provide acceptable solutions. The development of an alternative remediation strategy for the abatement of a contaminated medium is important for environmental conservation and human health. Bioremediation, an attractive and novel technology, is a multidisciplinary approach that uses biological systems to degrade/transform and/or to rid the soil and water of pollutants. This technology involves the use of plants (phytoremediation), plant–microbe interactions (rhizoremediation), and microbial communities involving stimulation of viable native microbial population (biostimulation), artificial introduction of viable population (bioaugmentation), bioaccumulation (live cells), and use of dead microbial biomass (biosorption) to clean up the contaminated sites. Bioremediation is simple, can be applied over large areas, environmentally friendly, and inexpensive. The use of genetic engineering to further modify plants for uptake, transport, and sequester metal opens up new avenues for enhancing efficiency of phytoremediation. Various bioremediation approaches adopted to remediate contaminated sites and major concerns associated with phytoremediation as a sustainable alternative are reviewed and discussed.

Keywords Genetic engineering • Heavy metals • Organic pollutants • Phytoremediation

R.A.I. Abou-Shanab (✉)
Environmental Biotechnology Department, Genetic Engineering and Biotechnology
Research Institute (GEBRI), City of Scientific Research and Technological Applications,
New Borg El Arab City, Alexandria 21934, Egypt
e-mail: redaabushanab@yahoo.com

M.S. Khan et al. (eds.), *Biomanagement of Metal-Contaminated Soils*,
Environmental Pollution 20, DOI 10.1007/978-94-007-1914-9_3,
© Springer Science+Business Media B.V. 2011

3.1   Introduction

Pollution of the surrounding environment has accelerated dramatically after the industrial revolution, leaving a legacy faced by modern society. The primary source of such pollutions is the burning of fossil fuels, mining and smelting of metalliferous ores, municipal wastes, agro-chemicals, and sewage. Migration of contaminants from neighboring contaminated land to a non-contaminated sites as vapors and leachate through the soil, or as dust or, spread of sewage sludge, further contributes to the contamination of natural ecosystems (Lopez-Errasquin and Vazquez 2003; Khan 2005). A wide range of materials causing contamination includes heavy metals, inorganic and organic compounds, oil and tars, toxic and explosive gases, combustible and putrescible substances, hazardous wastes, and explosives (Cheng 2003). Even where contamination levels are relatively low, such sites present major clean-up challenges as the levels may still be above regulatory limits. The excessive deposits of heavy metals thus pose a critical threat to human health and the environment due to their non-degrading ability, low solubility, and carcinogenic and mutagenic activity (Diels et al. 2002).

   In addition to the inorganic compounds, soils and water systems may also be contaminated by organic compounds including chlorinated solvents like, trichloroethylene; explosives such as trinitrotoluene (TNT) and 1, 3, 5-trinitro-1, 3, 5-hexahydrotriazine (RDX); petroleum hydrocarbons including benzene, toluene, and xylene (BTX); polyaromatic hydrocarbons (PAHs) and pesticides. While many of these compounds can be metabolized by soil bacteria, this process is usually slow and inefficient, in part due to occurrence of relatively low numbers of degradative microorganisms in soil (Glick 2003, 2010; Karamalidis et al. 2010). In some cases, soils, however, could be contaminated to such an extent that it may be classified as a hazardous waste (Berti and Jacob 1996). Soils polluted with single or mixture of heavy metals is thus receiving increasing attentions from the public as well as governmental bodies, particularly in developing countries (Yanez et al. 2002). The remediation of sites contaminated with heavy metals is hence a major environmental concern because these usually cover large areas that are rendered unsuitable for agricultural and other human use. There is therefore, an urgent need to develop suitable onsite remediation technologies in order to reduce the use of offsite contaminants for remediation. To address these problems, numerous techniques have been designed and developed that could be adopted to clean up contaminated soils (Ellis 1992; McEldowney et al. 1993).

3.2   Remediation Technologies

Remediation of metal-contaminated soils is particularly challenging. Unlike organic compounds, metals cannot be degraded, and the cleanup usually requires their physical or chemical removal (Lasat 2002). Various physical, chemical and biological processes are already in use to remediate contaminated soils (Smith et al. 1995;

Mulligan et al. 2001). These processes either "decontaminate" the soil, or "stabilize" the pollutant within it. Decontamination reduces the amount of pollutants from soil by removing them; stabilization, however, does not reduce the quantity of pollutant at a site, but makes use of soil amendments to alter the soil chemistry and sequester or absorb the pollutant into the matrix so as to reduce or eliminate environmental risks (Burns et al. 1996). The choice of remediation strategy however depends on the nature of the contaminant(s). The most commonly used method for cleaning up metal polluted sites globally includes bulk excavation and land-filling of contaminated materials (Huang and Cunningham 1996; Begonia et al. 1998). This process is extremely expensive and very disruptive to the site/ecosystem (Gardea-Torresdey et al. 2004). Other commonly available remediation technologies include acid washing and solidification/stabilization (US-EPA 1997).

Soil washing, like excavation and land-filling, is an *ex-situ* technique. It does not detoxify or significantly alter the contaminant, but transfers it from soil matrix to washing fluid. For metals, this technique often involves solubilization and suspension of metal ions through spraying or immersion in acid (HCl and/or acetic acid) solution. Following the washing step, the solution is moved to a "clarifies" tank where metal salts are separated via precipitation or flocculating agents. This process like excavation and land-filling is mechanical in nature, and relatively expensive (Alexander et al. 1997). Large volumes of acid by-product must be dealt with as hazardous wastes, and, in most cases, the physical structure of the soil is damaged by the acids to an extent where soil's biological activity is lost (Baker et al. 1994). Solidification/stabilization is classified as an immobilization technique, in which treatment agents are mixed or injected into contaminant materials (Wills 1988; Mench et al. 1994). Stabilization involves the addition of binding or buffering agents such as calcium carbonate ($CaCO_3$) or clay, while solidification involves encapsulation of contaminated soil in a solid (often cement) matrix. The final product may range from a crumbly, solid-like mixture to a monolithic block (US-EPA 1997).

The costs associated with soil remediation are highly variable and depend on (1) the nature of contaminant, (2) soil properties like structure, pH, moisture contents, temperature, and redox potential, (3) nutritional state of soils, (4) microbial composition, and (5) the volume of material to be remediated. Techniques that remediate soils in situ are generally less expensive than those that require excavation. On an average, remediation costs are US$10–100/m³ of soil for volatile or water-soluble pollutants remediated *in-situ*, US$60–300/m³ by compounds handled by land-filling or low-temperature thermal treatment, and US$200–700/m³ for materials requiring special landfill arrangements or high-temperature thermal treatments. The incineration of contaminated soil can cost up to US$1000/m³. Soil removal and replacement with clean soil is even more expensive, coasting between $8 and $24 million per hectare per meter of soil depth removed (Cunningham et al. 1995; Glass 2000). Certain materials like, radionuclides require even more intensive management techniques that can cost well beyond US$1,000–3,000/m³ of soil (Cunningham and Berti 1993). These practices usually generate secondary waste, destroy soil fertility and adversely affect its physical structure, remove biological activity from the treated soil, and are very expensive (Saxena et al. 1999;

Pulford and Watson 2003). Physical and chemical methods of remediation of contaminated soils are mainly applicable to relatively small areas and are unsuitable for very large areas such as a typical mining site or industrially/agrochemically contaminated soils.

3.2.1   Bioremediation

Bioremediation uses plants and/or microorganisms, such as bacteria, protozoa, and fungi; to degrade contaminants into a less toxic or non-toxic compounds (Pierzynski et al. 1994; US-EPA 1996). The three basic components of any bioremediation process include (1) microorganisms or plant, (2) a potentially biodegradable contaminant, and (3) a bioreactor in which the process can take place. Proper temperature, oxygen, and sufficient nutrient levels are required for functional bioreactor. The microbes in the bioreactor use carbon of the organic contaminants as a source of energy, and in doing so, degrade the contaminant. Biochemical processes such as bioleaching involving bacteria like *Thiobacillus* spp. and *Aspergillus niger* fungus, biosorption of low concentrations of metals in water by algal or bacterial cells, bio-oxidation or bioreduction of metal contaminants by *Bacillus subtilis* and sulfate reducing bacteria (SRB), and biomethylation of metals such as As, Cd, Hg, or Pb have shown promises and could be used for soil sediment treatments (Mulligan et al. 2001; Yoshida et al. 2006; Abou-Shanab et al. 2007).

Bioremediation can be applied both *ex-situ* and *in-situ*. With *ex-situ* bioremediation, the contaminated soil is excavated or the groundwater is extracted prior to treatment while *in-situ* remediation does not require excavation or extraction. As a result, the contaminated soil or groundwater serves as the bioreactor. The microorganisms may occur at the site naturally or be introduced from other locations. The conditions required by microbes for soil remediation are outlined in Table 3.1.

3.2.1.1   *Ex-Situ* Bioremediation

The primary methods used in *ex-situ* bioremediation are slurry-phase and solid-phase treatment. In slurry-phase treatment, contaminated soil is combined with H_2O_2 and other additives in a bioreactor. The resultant slurry is then mixed continuously to keep the microorganisms in contact with the contaminants. Upon completion of the treatment, H_2O_2 is removed from the solids, which are either disposed of or treated further if still contaminated. With solid-phase treatment, soils are remediated in above ground treatment areas equipped with collection systems to prevent contaminants from escaping (Pierzynski et al. 1994; US-EPA 1996). Land farming, soil bio-piles, and composting are three types of solid-phase treatment. Land farming involves spreading the contaminated soil thinly over land or a pad with a leachate-collection system. In some land-farming cases, reduction of contaminant concentrations may actually be due more to volatilization, leaching, or dilution through

Table 3.1 Environmental variables and optimum condition for microbial activity for soil bioremediation

Environmental factors	Optimum conditions	Conditions for microbial activity
Available soil moisture	25–85% water holding capacity	25–28% of water holding capacity
Oxygen	>0.2 mg/L DO, >10% air-filled pore space for aerobic degradation	Aerobic, minimum air-filled pore space of 10%
Redox potential	Eh > 50 mV	–
Nutrients	C:N:P = 120:10:1 M ratio	N and P for microbial growth
pH	6.5–8.0	5.5–8.5
Temperature	20–30°C	15–45°C
Contaminants	Hydrocarbon 5–10% of dry weight of soil	Not too toxic
Heavy metals	700 ppm	Total content 2,000 ppm
Soil type	–	Low clay or silt content

Adapted from Vidali (2001)

mixing with uncontaminated soil than from actual degradation by microorganisms. In a study, Genouw et al. (1994) demonstrated that land farming can be used effectively to clean up oil sludge applied to soil, but only if appropriate technical measures (e.g., nutrient and organic amendment, inoculation, and tillage) are employed and sufficient time (at least 15 years) is allowed for bioremediation to take place as also reported by Loehr and Webster (1996) who found similar results for creosote-contaminated soils. With soil bio-piles, contaminated soil is piled in heaps several meters high over an air distribution system (US-EPA 1996). A vacuum pump is used to pull air through the bio-pile. As a result, volatile contaminants are easily controlled since they are usually the part of the air stream pulled through the pile. In composting, biodegradable contaminants are mixed with straw, hay, or corn cobs to make it easier to achieve optimum levels of air and water (Bollag 1992). The compost can be (1) formed into piles and aerated with blowers or vacuum pumps, (2) placed in a treatment vessel where it is mixed and aerated, or (3) placed in long piles known as windrows and periodically mixed using tractors or similar equipment. Compost piles typically have elevated temperatures due to microbial activity, which sets them apart from bio-piles.

3.2.1.2 *In-Situ* Bioremediation

In-situ bioremediation is similar to phytoremediation as it uses microorganisms on-site to degrade contaminants. However, *in-situ* bioremediation does not involve the use of plants and generally employs more invasive engineering techniques than phytoremediation. For example, the oxygen required by aerobic microorganisms during *in-situ* bioremediation may be provided by pumping air into the soil above the water table, in a process known as *bioventing,* or by delivering the oxygen in liquid form as H_2O_2 (US-EPA 1996). *In-situ* bioremediation has been successful in remediating groundwater as well as surface soils and sub soils contaminated with petroleum hydrocarbons (Pierzynski et al. 1994). As an alternative, scientists have begun to develop technological approaches involving plants to remove organic and inorganic contaminants from the soil (Wild et al. 2005; Glick 2010).

3.3 Phytoremediation

Recognition of the ecological and human health hazards of the pollutants has led to the development of several technologies for remediation. However, due to the excessive cost of some of these technologies, attention has been diverted toward developing alternate/complementary technologies such as, the use of plants and microorganisms as bioremediators (Schneegurt et al. 2001; Sar and D'Souza 2002; Melo and D'Souza 2004; Abou-Shanab et al. 2003a, 2006, 2007, 2008). Compared to the conventional methods, the biomass-based systems are more acceptable due to low cost coupled with high efficiency of detoxification of even very dilute effluents and minimizing the disposable sludge volume. This technology also offers the flexibility for developing nondestructive desorption techniques for biomass regeneration and/or quantitative metal recovery. Among these techniques, phytoremediation involves the use of metabolically viable green plants and their associated microorganisms for *in-situ* risk reduction and/or removal of contaminants from contaminated soil, water, sediments, and air. Specially selected or engineered plants are used in the process (Abd El-Rahman et al. 2008). Risk reduction can be through a process of removal, degradation of a contaminant or a combination of any of these factors. Phytoremediation is energy efficient and aseptically pleasing method of remediating sites with low to moderate levels of contamination and can be used in conjunction with other more traditional remedial methods as a finishing step to the remedial process.

3.3.1 General Advantages of Phytoremediation

To facilitate comparison with physical remediation systems, we have redefined plants as "solar-driven pumping and filtering systems" that have "measurable loading, degrading, and fouling" capacities. Roots may similarly be described as "exploratory, liquid-phase extractors" that can find, alter, and/or translocate elements and compounds against large chemical gradients (Cunningham and Berti 1993). Plants can also be a cost-effective alternative to physical remediation systems (Glass 2000). In many cases, phytoremediation has been found to be less than half the price of alternative methods. Phytoremediation is less disruptive to the environment and does not involve waiting for new plant communities to recolonize the site. It also has the potential to treat sites polluted with multiple pollutants.

Thus, the use of phytoremediation for metal removal/detoxification from metal-poisoned soils offers advantages, like (1) low cost, (2) plants can be easily grown and monitored, (3) the recovery and reuse of valuable products is easy, (4) since it uses naturally occurring organisms, the natural state of the environment can be preserved, and (5) and plants can be engineered for desired traits.

3.3.2 General Limitations of Phytoremediation

Like other remediation technologies, phytoremediation has also certain disadvantages. For example, it is a process that is dependent on the depth of the roots and the tolerance of the plant to the contaminant. Plants are alive; their roots require O_2, H_2O_2, and nutrients. Soil texture, pH, salinity, pollutant concentrations, and the presence of other toxins must be within the limits of plant tolerance. Contaminants that are highly water soluble may leach outside the root zone and require containment (Cunningham et al. 1996; McIntyre and Lewis 1997). Phytoremediation is also frequently slower than physicochemical processes, and requires longer periods for remediating contaminated sites. Exposure of animals to hyperaccumulator plants can also be a concern to environmentalists as herbivorous animals may accumulate contaminants in their tissues, which in turn could affect the whole food web (Boyd et al. 2007). Despite these limitations, phytoremediation can be effective in cases where large surface areas of relatively immobile contaminants exist in the surface soils.

3.4 Phytoremediation Strategies

Phytoremediation includes (1) phytoextraction (phytoaccumulation), (2) rhizofiltration, (3) phytostabilization, (4) phytodegradation (phytotransformation), (5) rhizodegradation, and (6) phytovolatilization.

3.4.1 Phytoextraction

Phytoextraction is the process by which plant roots take up metals from the soil and translocate them to above soil tissues. Different plants have different abilities to uptake and withstand high levels of pollutants, and hence, many plants may be used. This is of particular interest for sites that are polluted with more than one metal. Interest in phytoremediation has grown significantly following the identification of metal hyperaccumulator plants. Hyperaccumulators are conventionally defined as species capable of accumulating metals at levels 100-fold greater than those typically measured in common non-accumulator plants. Thus, a hyperaccumulator will concentrate more than 10 ppm Hg; 100 ppm Cd; 1,000 ppm Co, Cr, Cu, or Pb; 10,000 ppm Ni or Zn. Once the plants have grown and absorbed the metals, they are harvested and disposed of safely (Baker and Brooks 1989; Baker et al. 2000). There are approximately 400 known metal hyperaccumulators (Reeves and Baker 2000) whose number is increasing. However, the remediation potential of many of these plants is limited because of their slow growth and low biomass yielding ability. While, the ideal plant species for phytoremediation should have high biomass with high metal accumulation in the shoot tissues (Chaney et al. 2000; Lasat 2002; McGrath et al. 2002). This process is repeated several times to reduce contamination to acceptable levels. In some cases, it is possible to recycle the metals through a process known as phytomining, though this is usually reserved for use with precious metals. Metal compounds that have been successfully phytoextracted include Zn, Cu, and Ni, but there is promising research being completed on Pb and Cr absorbing plants (Luo et al. 2005; Hsiao et al. 2007; Braud et al. 2009).

3.4.2 Rhizofiltration

Rhizofiltration is similar in concept to phytoextraction but is concerned with the remediation of contaminated groundwater rather than the remediation of polluted soils. The contaminants are either adsorbed onto the root surface or are absorbed by the plant roots. Plants used for rhizofiltration are not planted directly *in-situ* but are acclimated to the pollutant first. Plants are hydroponically grown in clean water rather than soil, until a large root system has developed. Once a large root system is in place, the water supply is substituted for a polluted water supply to acclimatize the plant. After the plants become acclimatized, they are planted in the polluted area where the roots uptake the polluted water and the contaminants along with it. As the roots become saturated, they are harvested and disposed of safely. Repeated treatments of the site can reduce pollution to suitable levels as was exemplified in Chernobyl where sunflower *(Helianthus annuus)* was grown in radioactively contaminated pools (Mahesh et al. 2008; Vera Tomé et al. 2008).

3.4.3 Phytostabilization

Phytostabilization involves the use of certain plants to immobilize soil and water contaminants. Contaminants are absorbed and accumulated by roots, adsorbed onto the roots, or precipitated in the rhizosphere. This reduces or even prevents the mobility of the contaminants into the groundwater or air, and also reduces the bioavailability of the contaminant, thus preventing the spread of metals through the food chain. This technique can also be used to reestablish a plant community on sites that have been denuded due to the high levels of metal contamination. Once a community of tolerant species has been established, the potential for wind erosion (and thus spread of the pollutant) is reduced and leaching of the soil contaminants is also reduced (Claudia Santibáñez et al. 2008; Ivano et al. 2008).

3.4.4 Phytodegradation

Phytodegradation is the degradation or breakdown of organic contaminants by internal and external metabolic processes driven by the plant. Ex-planta metabolic processes hydrolyze organic compounds into smaller units that can be absorbed by the plant (Suresh and Ravishankar 2004). Some contaminants can be absorbed by the plant and are then broken down by plant enzymes. These smaller pollutant molecules may then be used as metabolites by the plant as it grows and thus are incorporated into the plant tissues (Xiaoxue et al. 2008).

3.4.5 Rhizodegradation

Rhizodegradation (also called enhanced rhizosphere biodegradation, phytostimulation, and plant-assisted bioremediation) is the breakdown of organic contaminants in the soil by soil dwelling microbes, which is enhanced by the rhizosphere's presence. Certain soil dwelling microbes digest organic pollutants such as fuels and solvents, producing harmless products through a process known as bioremediation. The types of plants growing in the contaminated area influence the amount, diversity, and activity of microbial populations (Kirk et al. 2005). Plant root exudates such as sugars, alcohols, and organic acids are used as C source for the soil microflora and enhance microbial growth and activity. Some of these compounds may also act as chemotactic signals for certain microbes. The plant roots also loosen the soil and transport water to the rhizosphere, thus additionally enhancing microbial activity (Arshad et al. 2008; Gerhardt et al. 2009).

3.4.6 Phytovolatilization

Phytovolatilization is the process where plants uptake contaminants that are water soluble and release them into the atmosphere as they transpire the water. The contaminant may become modified along the way, as the water travels along the plant's vascular system from the roots to the leaves, whereby the contaminants evaporate or *volatilize* into the air surrounding the plant (Abd El-Rahman et al. 2008; Zhu and Rosen 2009). The major advantage of this method is that the contaminant (for example mercuric ion) may be transformed into a less toxic substance (elemental Hg).

However, mercury released into the atmosphere may again be recycled by precipitation and are redeposited back into lakes and oceans and thereby may cause problems. Mercury volatization by genetically modified tobacco (*N. tabacum*) and *Arabidopsis thaliana* (Meagher et al. 2000) and yellow poplar (*Liriodendron tulipifera*) (Rugh et al. 1998) and selenium volatization by Indian mustard and canola (*Brassica napus*) (Bañuelos et al. 1997) is reported.

3.5   Environmental Factors Affecting Phytoremediation

3.5.1   Soil Types and Organic Matter Contents

A variety of environmental factors affect or alter the performance of phytoremediation technology. Of these, soil types and organic matter (OM) content can limit the bioavailability of organic and inorganic contaminants. In terms of the influence of soil structure, Alexander et al. (1997) identified that phenanthrene may be trapped within and sorbed to the surfaces of nanopores (soil pores with diameters <100 nm) that are inaccessible to organisms (i.e., not bioavailable). Soil texture can also affect phytoremediation efforts by influencing the bioavailability of the contaminant (Brady and Weil 1996). Soil organic matter binds to lipophilic compounds and reduces their bioavailability (Cunningham et al. 1996). A high organic carbon content (>5%) in soil usually leads to strong adsorption and, therefore, low availability, while a moderate organic carbon content (1–5%) may lead to limited availability (Otten et al. 1997). Soil type may influence the quality or quantity of root exudates, which may influence phytoremediation (Bachmann and Kinzel 1992; Siciliano and Germida 1997).

3.5.2   Soil Water/Moisture

Water content in soil and wetlands affects plant/microbial growth and the availability of oxygen required for aerobic respiration (Eweis et al. 1998). Water is not only a major component of living organisms; it also serves as a transport medium to carry nutrients to biota and carry wastes away.

3.5.3   Temperature

Temperature affects the rates at which the various mechanisms of phytoremediation take place. In general, the rate of microbial degradation or transformation doubles for every 10°C increase in temperature (Eweis et al. 1998). Simonich and Hites (1994) reported that concentrations of PAHs in plants were higher during spring and autumn when ambient temperatures were relatively low compared to summer. Conversely, during the summer, when ambient temperatures were higher, lower concentrations of PAHs were found in the plants.

3.5.4 Light

Sunlight can transform parent compounds into other compounds, which may have different toxicities and bioavailability than the original compounds. Photomodifications of PAHs by ultraviolet light can occur in contaminated water or on the surface of soil increasing the polarity, water solubility, and toxicity of the compounds prior to uptake by the plant (McConkey et al. 1997).

3.5.5 Weathering Process

Weathering processes include volatilization, evapotranspiration, photomodification, hydrolysis, leaching, and biotransformation of the contaminant. These processes selectively reduce the concentration of easily degradable contaminants, with the more recalcitrant compounds remaining in the soil. These various environmental factors cause weathering, the loss of certain fractions of the contaminant mixture, with the end result being that only the more resistant compounds remain in the soil (Cunningham et al. 1996).

3.6 Techniques Used to Enhance Phytoremediation Process

Since phytoremediation is a relatively slow process, it may require years to reduce metal contents in soil to a safe and acceptable level. To make phytoremediation a viable technology, it is required to identify plants with fast growing ability and producing massive root system and excessively accumulating metal-tolerant capabilities. Alternatively, common plants can be, engineered with as yet unidentified hyperaccumulation genes. However, many fast growing and high biomass-producing plants such as vetiver grass (*Chrysopogon zizanioides*) and hemp (*Cannabis sativa*) may not be defined as metal hyperaccumulators, but are metal tolerant allowing them to grow in soil with high metal concentrations. The possibilities of using such plant species, which are easily growing in different climates, and using their biomass in non-food industries, can make them ideal plants for phytoremediation purposes (Linger et al. 2002; Khan 2003). Despite certain limitations, phytoremediation is considered a safe and long-term strategy for removing/reducing the toxicity of metal-contaminated soils (Cunningham et al. 1995). In order to promote the efficiency of this novel technique, various strategies have been tested in recent times. Some of the methods adopted to enhance the efficiency of this technique are discussed in the following section.

3.6.1 Chelator-Induced Phytoextraction

The uptake of metals by plants is frequently restricted by limitations of contaminant bioavailability. And hence, in order to enhance the metal uptake by plants, soil amendments with metal chelating agents such as EDTA, HEDTA, DTPA, EGTA, NTA, citrate, and hydroxylamine to make metals bioavailable and absorbable by plant roots have shown promises (Evangelou et al. 2007; Leštan et al. 2008). The type of chelate and its time of application are however important. It has also been suggested that by increasing plant biomass, phytoextraction can be increased (Ebbs and Kochian 1997). So far, researches in chelate-assisted phytoremediation have focused mainly on searching high efficiency chelates (Kos and Lestan 2003; Chen et al. 2010). Among plants, *Brassica juncea* initially had very little ability to absorb Pb from contaminated soils. However, Blaylock et al. (1997) and Huang et al. (1997) identified methods to aid or "induce" Pb phytoextraction from soils and reported that by adding EDTA, Pb could be desorbed from soil so it could move to the roots, and secondly, the Pb-EDTA chelate could leak through root membranes and be transported to shoots with transpiration. Consequently, *B. juncea* has become one of the first identified Pb accumulators (Kumar et al. 1995), as it accumulates high concentrations of Pb when EDTA is applied to soils—much more than average plant species. Part of the success was however, suggested later on due to the injury of the root membranes caused by EDTA and not to Pb chelation (Vassil et al. 1998). In a series of experiments thereafter, numerous groups attempted to find ways to make "chelator-induced *in-situ* phytoextraction" effective and safe in the environment, but the added chelating agents caused unavoidable leaching of chelated metals (e.g., Pb) down the soil profile (Romkens et al. 2001; Madrid et al. 2003; Wu et al. 2004). And in several field tests of EDTA on firing range soils in the USA, rapid leaching of Pb to groundwater proved an unacceptable side effect of chelator-induced Pb phytoextraction.

3.6.2 Plant Growth Regulators

Use of plant growth regulators (PGR) such as auxins and cytokinins has shown to enhance phytoremediation abilities of non-hyperaccumulating plants by increasing their growth and biomass (Fuentes et al. 2000; Pe et al. 2000). For example, Patten and Glick (1996) reported enhanced bioavailability of iron by applying plant hormone indol-acetic acid (IAA) via a mechanism different from that involving siderophores. IAA is also produced by many plant growth-promoting rhizobacteria

(PGPR) such as *Pseudomonad* and *Acinetobacter* strains (Reed and Glick 2005; Trotel-Aziz et al. 2008), which result in enhanced uptake of Fe, Zn, Mg, Ca, K, and P by crop plants (Lippmann et al. 1995). Usefulness of PGPR is, however, limited under nutrient-deficient conditions. Fertilizers have been used to help plants to increase their biomass and to extract more metals (Shetty et al. 1995). Further research needs to be carried out to find suitable combination of plant, PGPR, and soil types in order to investigate their potential(s) in increasing metal uptake by hyperaccumulator plants and improving the process of phytoextraction (Abou-Shanab et al. 2010).

3.6.3   Plant Growth-Promoting Rhizobacteria

Beneficial free-living soil bacteria are generally referred to as plant growth-promoting rhizobacteria and are found in association with the roots of many different plants (Glick et al. 1999). The high concentration of bacteria around the roots (rhizosphere) presumably occurs because of the presence of high levels of nutrients (especially small molecules such as amino acids, sugars, and organic acids) that are exuded from the roots of most plants, and can then be used to support bacterial growth and metabolism (Penrose and Glick 2001). The PGPR enhance plant growth by atmospheric N_2-fixation, phytohormone production, specific enzymatic activity, and plant protection from diseases by producing anti-biotic and other pathogen-depressing substances such as siderophores and chelating agents (Kamnev and van der Lelie 2000; Recep et al. 2009; Vleesschauwer and Höfte 2009). Microbial cells can produce and sense signal molecules, allowing the whole population to spread as a biofilm over the root surface and initiating a concerted action when a particular population density is achieved (Daniels et al. 2004). Free-living as well as symbiotic PGPR can enhance plant growth directly by providing bioavailable P to plants, fixing N for plant use, sequestering trace elements like iron for plants by siderophores, producing plant hormones and lowering of plant ethylene levels (Ahmad et al. 2008; Marques et al. 2010). The use of PGPR in phytoremediation technologies is relatively new, which can aid plant growth on contaminated sites (Burd et al. 2000; Gerhardt et al. 2009) and enhance detoxification of soil (Mayak et al. 2004; Dary et al. 2010). The properties of plants like high biomass production, low-level contaminant uptake, plant nutrition, and health, used for phytoremediation, can be improved by PGPR but it is important to choose PGPR that can survive and colonize, when used in phytoremediation practices.

3.6.3.1   Remediation of Heavy Metals by PGPR

While growing in metal contaminated soils, plants might be able to withstand some of the inhibitory effects of high metal concentrations; two features of most plants could result in a decrease in growth and viability. That is, in the presence of high levels of metals, most plants (1) synthesize stress ethylene and (2) become severely iron deficient. However, PGPR may be used to relieve some of the toxicity of metals to plants (Khan et al. 2009). This could occur in two different ways. As indicated earlier, the use of (1-aminocyclopropane-1-carboxylic acid) ACC deaminase-containing PGPR is reported to decrease the level of stress ethylene in plants growing in soil that contained high levels of metal (Belimov et al. 2005; Safronova et al. 2006). In addition, plants are able to take up and utilize bacterial siderophores. Plant siderophores bind to iron with a much lower affinity than bacterial siderophores so that in metal-contaminated soils, a plant is unable to accumulate a sufficient amount of iron unless bacterial siderophores are present (Glick 2003).

Soil microorganisms are known to affect the metal mobility and availability to the plants, through acidification, and redox changes or by producing iron chelators and siderophores for ensuring the iron availability, and/or mobilizing the metal phosphates (Burd et al. 2000; Guan et al. 2001; Abou-Shanab et al. 2003b). A large proportion of metal contaminants are unavailable for the root uptake, because heavy metals in soils are generally bound to organic and inorganic soil constituents, or alternatively, present as insoluble precipitates. Hence, how to increase the availability of metals to plants in soils is critical for the success of phytoremediation (Ernst 1996; Kukier et al. 2004). In a study, Abou-Shanab et al. (2006) reported the effect of certain rhizobacteria on nickel uptake. They indicated that rhizobacteria facilitated the release of Ni from the non-soluble phases in the soil, thus enhancing the availability of Ni to *Alyssum murale*. There is however, a need to improve our understanding of the mechanisms involved in transfer and mobilization of heavy metals by the rhizosphere microbes. A possible explanation might be acid and siderophore production and P-solubilization.

3.6.3.2 Remediation of Organic Contaminants by PGPR

Although PGPR was first used for prompting the plant growth and for the bio-control of plant diseases, much attention has recently been paid on bioremediation with PGPR (Narasimhan et al. 2003; Huang et al. 2005). In contrast with inorganic compounds, microorganisms can degrade and even mineralize organic compounds in association with plants (Saleh et al. 2004; Glick 2010). Hence discovery of effective pathways for degradation and mineralization of organic compounds may play an important role in the future. So far, bacteria capable of degrading certain kind of organic pollutant, such as polychlorinated biphenyls (PCBs) have been isolated from a range of sites and the pathways and encoding genes have also been well studied (Brazil et al. 1995). But most of these bacteria cannot survive in the near-starvation conditions found in soils, including the rhizosphere (Normander et al. 1999). Several effective methods have been developed to improve the degradation efficiency and the tolerance of bacteria to contaminants in soils. According to Huang et al. (2004a, b), the addition of PGPR increased the organic pollutant (polycyclic aromatic hydrocarbon and creosote) removal probably by enhancing plants germination and survival in soils that were heavily contaminated and by stimulating the plants to grow faster and accumulate more root biomass.

Facing a variety of environmental contaminants such as total petroleum hydrocarbons (TPHs), remediation technology even with both PGPR and plants may still be low in efficiency. The combination of PGPR and specific contaminant-degrading bacteria was found to be effective (Ajithkumar et al. 1998). Huang et al. (2005) thus developed a multi-process phytoremediation system (MPPS) where they used both PGPR and specific contaminant-degrading bacteria to treat TPHs. In this system, specific contaminant-degrading bacteria can be selected according to the properties of contaminants. They can rapidly metabolize some readily available compounds while the role of PGPR is still to facilitate plant growth and increase the plant tolerance to pollutants.

3.7 Breakthroughs in Phytoremediation: Novel Transgenic Approaches

Many genes are involved in metal uptake, translocation and sequestration and transfer of any of these genes into candidate plants is a possible strategy for genetic engineering of plants for improved phytoremediation traits. By using the tools of molecular biology/genetic engineering, transgenic plants with higher metal accumulating ability can be developed. In this context, transfer or over-expression of genes is likely to lead to enhanced metal uptake, translocation, sequestration, or intracellular targeting. For developing efficient transgenic plants for phytoremediation purposes, genes can be transferred from hyperaccumulators or from other sources. Different genes, which have been used for the development of transgenic plants, are listed in Table 3.2. Some of the areas that have been attempted for genetic manipulation include (1) metallothioneins, phytochelatins, and metal chelators, (2) metal transporters, (3) metabolic pathways, (4) oxidative stress mechanisms, (5) roots of plants, and (6) biomass.

3.7.1 Metallothioneins, Phytochelatins, and Metal Chelators

Metallothionein genes have been cloned and introduced into several plant species. Transfer of human MT-2 gene in tobacco (*Nicotiana tabacum*) or oil seed rape (*Brassica napus*) resulted in plants with enhanced Cd tolerance (Misra and Gedamu 1989) and pea (*Pisum sativum*) MT gene in *Arabidopsis thaliana* enhanced Cu accumulation (Evans et al. 1992). Transgenic plants with increased phytochelatin (PC) levels through over-expression of cysteine synthase resulted in enhanced Cd tolerance (Harada et al. 2001). In other study, yeast CUP-1 gene transferred to cauliflower (*Brassica oleracea* L. *botrytis*) resulted in 16-fold higher Cd tolerance and accumulation (Hasegawa et al. 1997). Various MT genes-mouse MTI, human MTIA, human MT II, Chinese hamster MT II, yeast CUP I, and pea ps MT A- have

Table 3.2 Selected examples of transgenic plants for metal tolerance/phytoremediation

Target plant	Gene transferred	Origin	Transgene effects	Reference
Tobacco, oil seed rape	MT-2 gene	Human	Cd tolerance	Misra and Gedamu (1989)
Tobacco	MT 1 gene	Mouse	Cd tolerance	Pan et al. (1994)
Arabidopsis	MTA gene	Pea	Cu accumulation	Evans et al. (1992)
Tobacco	CUP-1 gene	Yeast	Cu accumulation	Thomas et al. (2003)
Cauliflower	CUP-1 gene	Yeast	Cd accumulation	Hasegawa et al. (1997)
Indian mustard	Glutathione synthetase	Rice	Cd tolerance	Zhu et al. (1999a)
Indian mustard	g-Glutamylcysteine synthetase	E. coli	Cd tolerance	Zhu et al. (1999b)
Tobacco	CAX-2 (Vacuolar transporters)	A. thaliana	Accumulation of Cd, Ca, and Mn	Hirschi et al. (2000)
Tobacco	Nt CBP4	Tobacco	Ni tolerance and Pb accumulation	Arazi et al. (1999)
Tobacco	FRE-1 and FRE-2	Yeast	More Fe content	Samuelsen et al. (1998)
Arabidopsis	Glutathione-s-Transferase	Tobacco	Al, Cu, Na tolerant	Ezaki et al. (2000)
Tobacco, Rice	Ferretin	Soybean	Increased iron accumulation	Goto et al. (1998, 1999)
Arabidopsis	Zn transporters ZAT(At MTPI)	Arabidopsis	Zn accumulation	Van der Zaal et al. (1999)
Indian mustard	Arsenate reductase g-glutamylcysteine synthetase	Bacteria	As tolerance	Dhankher et al. (2002)
Arabidopsis	Znt A-heavy metal transporters	E. coli	Cd and Pb resistance	Lee et al. (2003)
A. thaliana	Selenocysteine methyl transferase	A. bisculatus	Resistance to selenite	Ellis et al. (2004)
Arabidopsis	YCF1	Yeast	Cd and Pb tolerance	Song et al. (2003)
Arabidopsis	Se-cys lyase	Mouse	Se tolerance and accumulation	Pilon et al. (2003)
A. thaliana	merP	B. megaterium	Hg²⁺ tolerance and accumulation	Hsieh et al. (2009)
Nicotiana glauca	Phytochelatin synthase (Ta PCS)	Wheat	Pb accumulation	Gisbert et al. (2003)

been transferred to *N. tabacum*, *Brassica* species and *A. thaliana* (Misra and Gedamu 1989; Maiti et al. 1991; Evans et al. 1992; Brandle et al. 1993; Elmayan and Tepfer 1994; Hasegawa et al. 1997), resulting in constitutively enhanced Cd tolerance in these plants. In most cases, metal uptake was not markedly altered in transgenic plants. However, when MT was of plant origin as in the case of Ps MTA from

P. sativum expressed in *A. thaliana*, more Cu accumulated in the roots of the transformed plants than control plants (Evans et al. 1992). Similarly, transgenic *B. juncea* over-expressing different enzymes involved in PC synthesis were shown to extract more Cd, Cr, Cu, Pb, and Zn than wild plants (Zhu et al. 1999a, b). Transgenic Indian mustard with higher levels of glutathione and PC were developed by over expression of two enzymes-g-glutamylcysteine synthetase (g-ECS) or glutathione synthetase (GS), which showed enhanced Cd tolerance and accumulation (Zhu et al. 1999a, b).

3.7.2   Metal Transporters

Genetic manipulation of metal transporters is known to alter metal tolerance/accumulation in plants (Pedas et al. 2009). Presently, over 100 ZIP family members have been identified. The ZIP family is represented in all the eukaryotic kingdoms, including animals, plants, protists, and fungi, but members are also found in archaea and bacteria. ZIP proteins from plants are capable of transporting Cd^{2+}, Fe^{3+}/Fe^{2+}, Mn^{2+}, Ni^{2+}, Co^{2+}, Cu^{2+}, and Zn^{2+} (Eckhardt et al. 2001; Grotz and Guerinot 2006; Pedas et al. 2009). Transfer of Zn transporter-ZAT gene from *Thalspi goesingense* to *A. thaliana* resulted in twofold higher Zn accumulation in roots (Van der Zaal et al. 1999). Introduction of calcium vacuolar transporter CAX-2 from *A. thaliana* to tobacco resulted in enhanced accumulation of Ca, Cd, and Mn (Hirschi et al. 2000). Enhanced Ni tolerance was obtained by transfer of another transporter gene-NtCBP4, which encodes for a calmodulin binding protein (Arazi et al. 1999). Transfer of yeast protein (YCF 1), a member of ABC transporter family involved in transfer of Cd into vacuoles by conjugation with glutathione, when transferred to *A. thaliana* was shown to over-express and resulted in transgenic plants with enhanced lead and cadmium tolerance (Song et al. 2003). A variety of ferric reductases have been shown to aid the acquisition of iron, for example the FRE family of metalloreductases in yeast (Dancis et al. 1992) and the FRO protein in plants (Robinson et al. 1999). In yeast, both FRE-1 and FRE-2 have been shown to reduce copper as well as iron and to increase copper uptake (Georgatsou et al. 1997; Wyman et al. 2008). Transfer of yeast FRE-1 and FRE-2 genes encoding ferric reductase when transferred to tobacco was shown to enhance the iron content of the plants 1.5-fold (Samuelsen et al. 1998).

3.7.3   Alteration of Metabolic Pathways

New metabolic pathways can be introduced into plants for hyperaccumulation or phytovolatilization as in the case of *Mer*A and *Mer*B genes that were introduced into plants likes *A. thaliana* and *N. tabacum*, which resulted in plants being several fold tolerant to Hg and volatilized elemental mercury (Bizily et al. 2000;

Abd El-Rahman et al. 2008; Ruiz and Daniell 2009). Dhankher et al. (2002) developed transgenic *Arabidopsis* plants that could transport oxyanion arsenate to aboveground, reduce to arsenite, and sequester it to thiol peptide complexes by transfer of *Escherichia coli* ars C and g-ECS genes. Transgenic plants with enhanced potential for detoxification of xenobiotics such as trichloro ethylene, pentachlorophenol, trinitro toluene, glycerol trinitrate, atrazine, ethylene dibromide, metolachlor, and hexahydro-1,3,5-trinitro-1,3,5-triazine are a few successful examples of utilization of transgenic technology (Table 3.3).

3.7.4   Alteration of Oxidative Stress Mechanisms

Alteration of oxidative stress–related enzymes may also result in altered metal tolerance as reported in the case of enhanced Al tolerance by over-expression of glutathione-S-transferase and peroxidase (Ezaki et al. 2000). Over-expression of 1-aminocyclopropane-1-carboxylic acid (ACC) deaminase led to an enhanced accumulation of a variety of metals (Grichko et al. 2000).

3.7.5   Alteration in Roots

It is essential to have plants with highly branched root systems with large surface area for efficient uptake of toxic metals. The hairy roots induced in some of the hyperaccumulators (*B. juncea, Chenopodium amaranticolor, A. bertolonii*, and *T. caerulescens*) were shown to have high efficiency for rhizofiltration of radionuclides (Eapen et al. 2003) and heavy metals (Nedelkoska and Doran 2000).

3.7.6   Alteration in Biomass

Biomass of known hyperaccumulators can be altered by introduction of genes, which affect phytohormone synthesis resulting in enhanced biomass. Recently, biosynthetic pathways have been elucidated for most of the plant hormone classes and genes encoding many of the enzymes have been cloned. These advances offer new opportunities to manipulate hormone content and regulate their biosynthesis (Hedden and Phillips 2000; Grattapaglia et al. 2009). Increased gibberellins biosynthesis in engineered trees for example, *Populus and Eucalyptus*, promoted the growth and biomass production (Erikson et al. 2000). However, little work has been done in this area for improving biomass of plants for phytoremediation. Since each metal has a specific mechanism for uptake, translocation, and sequestration, it becomes essential to design suitable strategies for developing transgenic plants specific for each characteristic.

Table 3.3 Selected examples of transgenic plants for enhanced phytoremediation of organic pollutants

Target plant	Gene transferred	Origin	Transgene effects	Reference
Arabidopsis thaliana	NfsA	E. coli	The plants showed higher nitroreductase activity and 7–8 times higher uptake compared with wild plants	Kurumata et al. (2005)
A. thaliana	XplA and XplB	Rhodococcus Rhodochorus	Enhanced degradation of RDX	Jackson et al. (2007)
A. thaliana	743B4, 73C1	A. thaliana	Overexpression of UGTs genes resulted in the enhanced detoxification of TNT and enhanced root growth	Gandia-Herrero et al. (2008)
Brassica juncea	γ-ECS, GS	Brassica juncea	Overexpression of ECS and GS resulted in enhanced tolerance to atrazine, 1-chloro-2, 4-dinitrobenzene, phenanthrene, metolachlor	Flocco et al. (2004)
Hybrid aspen (Populus tremuloides)	pnrA	Pseudomonas putida	The transgenic aspen (hybrid) was shown to tolerate and take up greater amounts of TNT from contaminated water and soil	Van Dillewijn et al. (2008)
Hybrid poplar (P. tremula x p. alba)	CYP450 2E1	Rabbit	Increased removal of TCE, vinyl chloride, carbon tetrachloride, benzene, and chloroform from hydroponic solution and air	Doty et al. (2007)
Nicotiana tabaccum	NfsI	Enterobacter. Cloaceae	The transgenic plants removed high amount of TNT from the test solution and reduction of TNT to 4-hydroxylamino-2, 6-dinitrotoluene	Hannink et al. (2007)
N. tabaccum	CYP450 2E1	Human	Oxidation of TCE and ethylene dibromide	Doty et al. (2000)
N. tabaccum	tpx1 and tpx2	Lycopersicon Esculentum	Hairy cultures of transgenic tobacco showed enhance removal of phenol	Alderete et al. (2009)
N. tabaccum	Bphc	PCB-degrading bacteria	Enhanced degradation of PCBs	Chrastilova et al. (2007)
N. tabaccum	CYP450E1	Human	Enhanced degradation of anthracene and chloropyriphos	Dixit et al. (2008)
Oryza sativa	CYP1A1	Human	Enhanced metabolism of chlorotoluron, norflurazon	Kawahigashi et al. (2008)
O. sativa	Protox	Bacillus subtilis	Tolerance to diphenyl ether herbicide oxyflufen	Jung et al. (2008)

3.8 Example of Genetically Engineered Plant

3.8.1 Mercury Detoxification Using Transgenic Plants

Mercury and mercurial compounds are hazardous to all biological organisms. Bacteria have evolved mechanisms for colonizing mercury-contaminated environments, and an operon of mercury resistance (*mer*) genes encoding for transporters and enzymes for biochemical detoxification (Summers 1986) has been identified. Mercury-resistant bacteria (*B. megaterium* MB1) convert organic and ionic mercury compounds to the volatile and less toxic elemental form Hg (O), which rapidly evaporates through cell surface (Huang et al. 1999). Genetically engineered plants with *mer* A and *mer* B genes were produced in three plant species *A. thaliana* (Bizily et al. 2000), *N. tabacum,* and *Liriodendron tulipifera* L. (Rugh et al. 2000; Abd El-Rahman et al. 2008) and have demonstrated that transgenic plants could grow in the presence of toxic levels of organic and inorganic mercury (Fig. 3.1).

To improve the expression of *mer* genes in plants, the bacterial *merA* DNA sequence was modified by reducing the GC content in a 9% block of the protein coding region and adding plant regulatory elements (Rugh et al. 1996). When transferred to *A. thaliana* and tobacco, the new gene construct (*mer A*) conferred resistance to 50 mm Hg (II) suggesting that *merA* plants enzymatically reduce Hg (II)

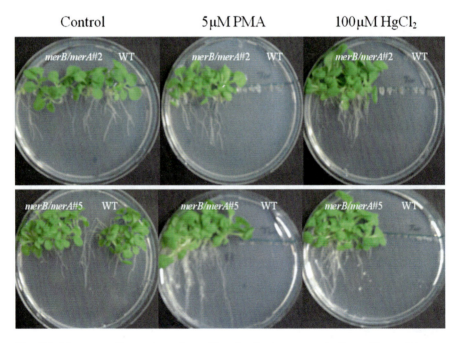

Fig. 3.1 Mercury resistance test of *merB/merA* tobacco transgenic lines. *Wt* = wild type, *PMA* = phenylmercuric acetate

and evaporate away Hg (0). Three modified *merA* constructs were used for transformation of yellow poplar proembryogenic masses, each having different amounts of altered coding sequences. Each of these constructs was shown to confer Hg (II) resistance (Rugh et al. 2000). Transgenic *Populous deltoids* over-expressing *merA9* and *merA*18 gene when exposed to Hg (II) evolved two- to fourfold Hg (0) relative to wild plant (Che et al. 2003). These transgenic trees when grown in soil with 40 ppm of Hg (II) developed higher biomass. Subcellular targeting of methyl-mercury lyase was shown to enhance its specific activity for organic mercury detox-ification in plants (Bizily et al. 2003). If phytovolatilization is unsuitable due to the hazards of releasing Hg (0), alternate strategies should be explored. One option is to develop plants that sequester high mercury loads in harvestable tissues. The strategy for mercury sequestration may be further enhanced by root-specific expression of *mer A* and *mer B* genes to detoxify charged mercurials prior to transport to shoots. Plant expression of modified mercury transport genes, *mer P* and *mer T*, may pro-vide a means of improving mercury uptake and organelle and tissue-specific targeting.

3.9 Conclusion

The present review provides the scientific understanding required to harness natural processes and to develop and design methods to accelerate these processes for the bioremediation of contaminated soil environments. Despite certain limiting factors, the bioremediation technology including phytoremediation is considered a promis-ing option for decontaminating metal polluted soils. Moreover, the rapid advances in science, has led us to better understand and apply these technology more effec-tively on sites contaminated with heavy metals. The use of culture-independent techniques and other molecular tools has obviously assisted us to better explain the microbial community dynamics, structure, and composition, which in turn has pro-vided insight into the finer details of bioremediation, a safer and reliable technology. The optimization of different processes of bioremediation or engineering plants/ microbes has now started showing impact on the ecosystem. However, still there are problems that need to be urgently addressed by the scientists before this technology is realized at commercial scale, for example, how abilities like fast growth, high biomass, extensive root systems, metal tolerance, and harvesting of plants could be improved. Even though, no such plant has been described so far, high biomass non-accumulators that are fast growing can be engineered to achieve some of the prop-erties of the hyperaccumulators. Furthermore, to allow remediation within a reasonable period, metal uptake and plant yields have to be enhanced dramatically. For this, a continuous search for metal hyperaccumulators, as well as engineering common plants with suitable functioning genes are required. To achieve this, a multidisciplinary strategy by involving plant biologists, soil chemists, microbiolo-gists, and environmental engineers is required for greater success of bioremediation/ phytoremediation, which could serve as a viable soil cleanup technique.

References

Abd El-Rahman RA, Abou-Shanab RA, Moawad H (2008) Mercury detoxification using genetic engineered *Nicotiana tabacum*. Global NEST J 10:432–438

Abou Shanab RI, Angle JS, Delorme TA, Chaney RL, van Berkum P, Moawad H, Ghanem K, Ghozlan HA (2003a) Rhizobacterial effects on nickel extraction from soil and uptake by *Alyssum murale*. New Phytol 158:219–224

Abou-Shanab RI, Delorme TA, Angle JS, Chaney RL, Ghanem K, Moawad H, Ghozlan HA (2003b) Phenotypic characterization of microbes in the rhizosphere of *Alyssum murale*. Int J Phytoremediation 5:367–379

Abou-Shanab RAI, Angle JS, Chaney RL (2006) Bacterial inoculants affecting nickel uptake by *Alyssum murale* from low, moderate and high Ni soils. Soil Biol Biochem 38:2882–2889

Abou-Shanab RAI, Angle JS, van Berkum P (2007) Chromate-tolerant bacteria for enhanced metal uptake by *Eichhornia crassipes* (Mart.). Int J Phytoremediation 9:91–105

Abou-Shanab RAI, Ghanem KM, Ghanem NB, Al-Kolaibe AM (2008) The role of bacteria on heavy-metals extraction and uptake by plants growing on multi-metal contaminated soils. World J Microbiol Biotechnol 24:253–262

Abou-Shanab RAI, Angle JS, Delorme TA, Chaney RL, van Berkum P, Ghozlan HA, Ghanem K, Moawad H (2010) Characterization of Ni-resistant bacteria in the rhizosphere of the hyperaccumulator *Alyssum murale* by 16 S rRNA gene sequence analysis. World J Microbiol Biotechnol 26:101–108

Ahmad F, Ahmad I, Khan MS (2008) Screening of free-living rhizospheric bacteria for their multiple plant growth promoting activities. Microbiol Res 163:173–181

Ajithkumar PV, Gangadhara KP, Manilal P, Kunhi AAM (1998) Soil inoculation with *Pseudomonas aeruginosa* 3MT eliminates the inhibitory effect of 3-chloroand 4-chlorobenzoate on tomato seed germination. Soil Biol Biochem 30:1053–1059

Alderete LGS, Talano MA, Ibannez SG, Purro S, Agostini E (2009) Establishment of transgenic tobacco hairy roots expressing basic peroxidases and its application for phenol removal. J Biotechnol 139:273–279

Alexander M, Hatzinger PB, Kelsey JW, Kottler BD, Nam K (1997) Sequestration and realistic risk from toxic chemicals remaining after bioremediation. Ann NY Acad Sci 829:1–5

Arazi T, Sunkar R, Kaplan B, Fromm HA (1999) Tobacco plasma membrane calmodulin binding transporter confers Ni^+ tolerance and Pb^{2+} hypersensitivity in transgenic plants. Plant J 20:71–82

Arshad M, Saleem M, Hussain S (2008) Perspectives of bacterial ACC deaminase in phytoremediation. Trends Biotechnol 8:356–362

Bachmann G, Kinzel H (1992) Physiological and ecological aspects of the interactions between plant roots and rhizosphere soil. Soil Biol Biochem 24:543–552

Baker AJM, Brooks RR (1989) Terrestrial higher plants which hyperaccumulate metallic elements. A review of their distribution ecology and phytochemistry. Biorecovery 1:81–126

Baker AJM, McGrath SP, Sidoli CMD, Reeves RD (1994) The possibility of in situ heavy metal decontamination of polluted soils using crops of metal-accumulating plants. Res Conserv Recycl 11:41–49

Baker AJM, McGrath SP, Reeves RD, Smith JAC (2000) Metal hyperaccumulator plants: A review of the ecology and physiology of a biological resource for phytoremediation of metal polluted soils. In: Terry N, Banuelos G (eds.) Phytoremediation of contaminated soil and water. Lewis Publishers, Boca Raton, pp 85–107

Bañuelos GS, Ajwa HA, Mackey B, Wu LL, Cook C, Akohoue S, Zambrzuski S (1997) Evaluation of different plant species used for phytoremediation of high soil selenium. J Environ Qual 26:639–646

Begonia GB, Davis CD, Begonia MFT, Gray CN (1998) Growth response of Indian mustard (*Brassica juncea* (L.) czern.) and its phytoextraction of lead contaminated soil. Bull Environ Contam Toxicol 61:38–43

Belimov AA, Hontzeas N, Safronova VI, Demchinskaya SV, Piluzza G, Bullitta S, Glick BR (2005) Cadmium-tolerant plant growth-promoting bacteria associated with the roots of Indian mustard (*Brassica juncea* L.Czern.). Soil Biol Biochem 37:241–250

Berti WR, Jacob LW (1996) Chemistry and phytotoxicity of soil trace elements from repeated sewage sludge application. J Environ Qual 25:1025–1032

Bizily SP, Rugh CL, Meagher RB (2000) Phytodetoxification of hazardous organomercurials by genetically engineered plants. Nat Biotechnol 18:213–217

Bizily SP, Kim T, Kandasamy MK, Meagher RB (2003) Subcellular targeting of methylmercury lyase enhances its specific activity for organic mercury detoxification in plants. Plant Physiol 131:463–471

Blaylock MJ, Salt DE, Dushenkhov S, Zakharova O, Gussman C, Kapulnik Y, Ensley BD, Raskin I (1997) Enhanced accumulation of Pb in Indian mustard by soil-applied chelating agents. Environ Sci Technol 31:860–865

Bollag JM (1992) Decontaminating soil with enzymes: an *in situ* method using phenolic and anilinic compounds. Environ Sci Technol 26:1876–1881

Boyd RS, Davis MA, Wall MA, Balkwill K (2007) Host-herbivore studies of Stenoscepa sp. (Orthoptera:Pyrgomorphidae), a high-Ni herbivore of the South African Ni hyperaccumulator *Berkheya coddii* (Asteraceae). Insect Sci 14:133–143

Brady NC, Weil RR (1996) The nature and properties of soils. Prentice Hall, Upper Saddle River

Brandle JE, Labbe H, Hattori J, Miki BL (1993) Field performance and heavy metal concentrations of transgenic flue cured tobacco expressing a mammalian metallothionein-h-glucuronidase gene fusion. Genome 36:255–260

Braud A, Jézéquel K, Bazot S, Lebeau T (2009) Enhanced phytoextraction of an agricultural Cr- and Pb-contaminated soil by bioaugmentation with siderophore-producing bacteria. Chemosphere 74:280–286

Brazil GM, Kenefick L, Callanan M, Haro A, de Lorenzo V, Dowling DN (1995) Construction of a rhizosphere pseudomonad with potential to degrade polychlorinated biphenyls and detection of bph gene expression in the rhizosphere. Appl Environ Microbiol 61:1946–1952

Burd GI, Dixon DG, Glick RR (2000) Plant growth promoting bacteria that decrease heavy metal toxicity in plants. Can J Microbiol 46:237–245

Burns RG, Rogers S, McGhee I (1996) Remediation of inorganics and organics in industrial and urban contaminated soil. In: Naidu R, Kookana RS, Oliver DP, Rogers S, McLaughlin MJ (eds.) Contaminants and the soil environment in the Australasia-Pacific region. Kluwer Academic Publishers, London, pp 125–181

Chaney RL, Li YM, Brown SL, Homer FA, Malik M, Angle JS (2000) Improving metal hyperaccumulator wild plants to develop phytoextraction systems: approaches and progress. In: Terry N, Banuelos G (eds.) Phytoremediation of contaminated soil and water. Lewis Publishers, Boca Raton, pp 129–158

Che D, Meagher RB, Heaton ACP, Lima A, Rugh CL, Merkle SA (2003) Expression of mercuric ion reductase in Eastern cottonwood (*Populus deltoides*) confers mercuric ion reduction and resistance. Plant Biotechnol J 1:311

Chen JC, Wang KS, Chen H, Lu CY, Li HC, Peng TH, Chang SH (2010) Phytoremediation of Cr(III) by *Ipomonea aquatica* (water spinach) from water in the presence of EDTA and chloride: Effects of Cr speciation. Bioresour Technol 101:3033–3039

Cheng S (2003) Heavy metal pollution in China: origin, pattern and control. Environ Sci Pollut Res Int 10:192–198

Chrastilova Z, Mackova M, Novakova M, Macek T, Szekeres M (2007) Transgenic plants for effective phytoremediation of persistent toxic organic pollutants present in the environment (Abstracts). J Biotechnol 131S:S38

Claudia S, Cesar V, Rosanna G (2008) Phytostabilization of copper mine tailings with biosolids: Implications for metal uptake and productivity of *Lolium perenne*. Sci Total Environ 395:1–10

Cunningham SD, Berti WR (1993) Remediation of contaminated soils with green plants: an overview. In Vitro Cell Dev Biol 29:207–212

Cunningham SD, Berti WR, Huang JW (1995) Phytoremediation of contaminated soils. Trends Biotechnol 13:393–397

Cunningham SD, Anderson TA, Schwab AP, Hsu FC (1996) Phytoremediation of soils contaminated with organic pollutants. Adv Agron 56:55–114

Dancis A, Roman DG, Anderson GJ, Hinnebusch AG, Klausner RD (1992) Ferric reductase of *Saccharomyces cerevisiae*: molecular characterization, role in iron uptake, and transcriptional control by iron. Proc Natl Acad Sci USA 89:3869–3873

Daniels R, Vanderleyden J, Michiels J (2004) Quorum sensing and swarming migration in bacteria. FEMS Microbiol Rev 28:261–289

Dary M, Chamber-Pérez MA, Palomares AJ, Pajuelo E (2010) "*In situ*" phytostabilization of heavy metal polluted soils using *Lupinus luteus* inoculated with metal resistant plant-growth promoting rhizobacteria. J Hazard Mater 177:323–330

Dhankher OP, Li Y, Rosen BP, Shi J, Salt D, Senecoff JF (2002) Engineering tolerance and hyperaccumulation of arsenic in plants by combining arsenate reductase and g-glutamylcysteine synthetase expression. Nat Biotechnol 20:1140–1145

Diels L, van der Lelie N, Bastiaens L (2002) New development in treatment of heavy metal contaminated soils. Rev Environ Sci Biotechnol 1:75–82

Dixit P, Singh S, Mukherjee PK, Eapen S (2008) Development of transgenic plants with cytochrome P450E1 gene and glutathione-S-transferase gene for degradation of organic pollutants (Abstracts). J Biotechnol 136S:S692–S693

Doty SL, Shang QT, Wilson AM, Moore AL, Newman LA, Strand SE, Gordon MP (2000) Enhanced metabolism of halogenated hydrocarbons in transgenic plants contain mammalian P450 2E1. Proc Natl Acad Sci USA 97:6287–6291

Doty SL, Shang QT, Wilson AM, Moore AL, Newman LA, Strand SE (2007) Enhanced metabolism of halogenated hydrocarbons in transgenic plants contain mammalian P450 2E1. Proc Natl Acad Sci USA 97:6287–6291

Eapen S, Suseelan K, Tivarekar S, Kotwal S, Mitra R (2003) Potential for rhizofiltration of uranium using hairy root cultures of *Brassica juncea* and *Chenopodium amaranticolor*. Environ Res 91:127–133

Ebbs DS, Kochian LV (1997) Toxicity of zinc and copper to *Brassica* species: implications for phytoremediation. J Environ Qual 26:776–781

Eckhardt U, Marques AM, Buckhout TJ (2001) Two iron-regulated cation transporters from tomato complement metal uptake-deficient yeast mutants. Plant Mol Biol 45:437–448

Ellis B (1992) On site and in situ treatment of contaminated sites. In: Rees FJ (ed.) Contaminated land treatment technologies. Society of Chemical Industry. Elsevier Applied Science, London, pp 30–46

Ellis DR, Sors TG, Brunk DG, Albrecht C, Orser C, Lahner B (2004) Production of S methyl selenocysteine in transgenic plants expressing selenocysteine methyltransferase. BMC Plant Biol 28:4

Elmayan T, Tepfer M (1994) Synthesis of a bifunctional metallothionein h glucuronidase fusion protein in transgenic tobacco plants as a means of reducing leaf cadmium levels. Plant J 6:433–440

Erikson ME, Israelsson M, Olsson O, Moritz T (2000) Increased giberellin biosynthesis in transgenic trees promotes growth, biomass production and xylem fiber length. Nat Biotechnol 18:784–788

Ernst WHO (1996) Bioavailability of heavy metals and decontamination of soil by plants. Appl Geochem 11:163–167

Evangelou MWH, Bauer U, Ebel M, Schaeffer A (2007) The influence of EDDS and EDTA on the uptake of heavy metals of Cd and Cu from soil with tobacco *Nicotiana tabacum*. Chemosphere 68:345–353

Evans KM, Gatehouse JA, Lindsay WP, Shi J, Tommey AM, Robinson NJ (1992) Expression of the pea metallothionein like gene Ps MTA in *Escherichia coli* and *Arabidopsis thaliana* and analysis of trace metal ion accumulation:implications of Ps MTA function. Plant Mol Biol 20:1019–1028

Eweis JB, Ergas SJ, Chang DPY, Schroeder ED (1998) Bioremediation principles. McGraw-Hill, Toronto

Ezaki B, Gardner RC, Ezaki Y, Matsumoto H (2000) Expression of aluminium induced genes in transgenic *Arabidopsis* plants can ameliorate aluminium stress and/or oxidative stress. Plant Physiol 122:657–665

Flocco CG, Lindblom SD, Smits EAHP (2004) Overexpression of enzymes involved in glutathione synthesis enhances tolerance to organic pollutants in Brassica juncea. Int J Phytoremediation 6:289–304

Fuentes HD, Khoo CS, Pe T, Muir S, Khan AG (2000) Phytoremediation of a contaminated mine site using plant growth regulators to increase heavy metal uptake. In: Sanches MA, Vergara F, Castro SH (eds.) Waste treatment and environmental impact in the mining industry. University of Concepcion Press, Victor Lamas, Concepcion, pp 427–435

Gandia-Herrero F, Lorenz A, Larson T, Graham IA, Bowles J, Rylott EL (2008) Detoxification of the explosive 2,4,6- trinitrotoluene in *Arabidopsis*: discovery of bifunctional O and C-glucosyltransferases. Plant J 56:963–974

Gardea-Torresdey JL, Peralta-Videa JR, Montes M, de la Rosa G, Corral-Diaz B (2004) Bioaccumulation of cadmium, chromium and copper by *Convolvulus arvensis* L.: impact on plant growth and uptake of nutritional elements. Bioresour Technol 92:229–235

Genouw G, de Naeyer F, van Meenen P, vam de Werf H, de Nijs W, Verstraete W (1994) Degradation of oil sludge by land farming - a case study at the Ghent harbour. Biodegradation 5:37–46

Georgatsou E, Mavrogiannis LA, Fragiadakis GS, Alexandraki D (1997) The yeast Fre1p/Fre2p cupric reductases facilitate copper uptake and are regulated by the copper-modulated Mac1p activator. J Biol Chem 272:13786–13792

Gerhardt KE, Huang X, Glick BR, Greenberg BM (2009) Phytoremediation and rhizoremediation of organic soil contaminants: potential and challenges. Plant Sci 176:20–30

Gisbert C, Ros R, De Haro A, Walker DJ, Pilar Bernal M, Serrano R (2003) A plant genetically modified that accumulates Pb is especially promising for phytoremediation. Biochem Biophys Res Commun 303:440–445

Glass DJ (2000) Economic potential of phytoremediation. In: Raskin I, Ensley BD (eds.) Phytoremediation of toxic metals – using plants to clean up the environment. Wiley, New York, pp 15–31

Glick BR (2003) Phytoremediation: synergistic use of plants and bacteria to clean up the environment. Biotechnol Adv 21:383–393

Glick BR (2010) Using soil bacteria to facilitate phytoremediation. Biotechnol Adv 28:367–374

Glick BR, Patten CL, Holguin G, Penrose DM (1999) Biochemical and genetic mechanisms used by plant growth-promoting bacteria. Imperial College Press, London

Goto F, Yoshihara T, Saiki H (1998) Iron accumulation in tobacco plants expressing soybean ferritin gene. Transgenic Res 7:173–180

Goto F, Yoshihara T, Shigemoto N, Toki S, Takaiwa F (1999) Iron accumulation in rice seed by soya bean ferritin gene. Nat Biotechnol 17:282–286

Grattapaglia D, Plomion C, Kirst M, Sederoff RR (2009) Genomics of growth traits in forest trees. Curr Opin Plant Biol 12:148–156

Grichko VP, Filby B, Glick BR (2000) Increased ability of transgenic plants expressing the bacterial enzyme ACC deaminase to accumulate Cd, Co, Cu, Ni, Pb and zinc. J Biotechnol 81:45–53

Grotz N, Guerinot ML (2006) Molecular aspects of Cu, Fe and Zn homeostasis in plants. Biochim Biophys Acta 1763:595–608

Guan LL, Kanoh K, Kamino K (2001) Effect of exogenous siderophores on iron uptake activity of marine bacteria under iron limited conditions. Appl Environ Microbiol 67:1710–1717

Hannink NK, Subramanian M, Rosser SJ, Basran A, Murray JAH, Shanks JV (2007) Enhanced transformation of TNT by tobacco plants expressing a bacterial nitroreductase. Int J Phytoremediation 9:385–401

Harada E, Choi YE, Tsuchisaka A, Obata H, Sano H (2001) Transgenic tobacco plants expressing a rice cysteine synthase gene are tolerant to toxic levels of cadmium. Plant Physiol 158:655–661

Hasegawa I, Terada E, Sunairi M, Wakita H, Shinmachi F, Noguchi A (1997) Genetic improvement of heavy metal tolerance in plants by transfer of the yeast metallothionein gene (CUPI). Plant Soil 196:277–281

Hedden P, Phillips AL (2000) Manipulation of hormone biosynthetic genes in transgenic plants. Curr Opin Biotechnol 1:130–137

Hirschi KD, Korenkov VD, Wilganowski NL, Wagner GJ (2000) Expression of *Arabidopsis* CAX2 in tobacco altered metal accumulation and increased manganese tolerance. Plant Physiol 124:125–133

Hsiao KH, Kao P, Hseu ZY (2007) Effects of chelators on chromium and nickel uptake by *Brassica juncea* on serpentine-mine tailings for phytoextraction. J Hazard Mater 148:366–376

Hsieh J, Chen CY, Chiu M, Chein MJC, Endo G, Huang CC (2009) Expressing a bacterial mercuric ion binding protein in plant for phytoremediation of heavy metals. J Hazard Mater 161:920–925

Huang JW, Cunningham JD (1996) Lead phytoextraction: species variation in lead uptake and translocation. New Phytol 134:75–84

Huang JW, Chen J, Berti WR, Cunningham SD (1997) Phytoremediation of lead-contaminated soils: role of synthetic chelates in lead phytoextraction. Environ Sci Technol 31:800–805

Huang CC, Narita M, Yamagata T, Itoh Y, Endo G (1999) Structure analysis of a class II transposon encoding the mercury resistance of the Gram-positive bacterium, *Bacillus megaterium* MB1, a strain isolated from Minamata Bay, Japan. Gene 234:361–369

Huang XD, El-Alawi Y, Penrose DM, Glick BR, Greenberg BM (2004a) Responses of three grass species to creosote during phytoremediation. Environ Pollut 130:453–463

Huang XD, El-Alawi Y, Penrose DM, Glick BR, Greenberg BM (2004b) Multi-process phytoremediation system for removal of polycyclic aromatic hydrocarbons from contaminated soils. Environ Pollut 130:465–476

Huang XD, El-Alawi Y, Gurska J, Glick BR, Greenberg BM (2005) A multi-process phytoremediation system for decontamination of persistent total petroleum hydrocarbons (TPHs) from soils. Microchem J 81:139–147

Ivano B, Jo''rg L, Madeleine S, Gu''nthardt G, Beat F (2008) Heavy metal accumulation and phytostabilisation potential of tree fine roots in a contaminated soil. Environ Pollut 152:559–568

Jackson EG, Rylott EL, Fournier D, Hawari J, Bruce NC (2007) Exploring the biochemical properties and remediation applications of the unusual explosive-degrading P450 system XplA/B. Proc Natl Acad Sci USA 104:16822–16827

Jung S, Lee HJ, Lee Y, Kang K, Kim YS, Grimm B (2008) Toxic tetrapyrrole accumulation in protoporphyrinogrn IX oxidase overexpressing transgenic rice plants. Plant Mol Biol 67:535–546

Kamnev AA, van der Lelie N (2000) Chemical and biological parameters as tools to evaluate and improve heavy metal phytoremediation. Biosci Rep 20:239–258

Karamalidis AK, Evangelou AC, Karabika E, Koukkou AI, Drainas C, Voudrias EA (2010) Laboratory scale bioremediation of petroleum-contaminated soil by indigenous microorganisms and added *Pseudomonas aeruginosa* strain Spet. Bioresour Technol 101:6545–6552

Kawahigashi H, Hirose S, Ohkawa H, Ohkawa Y (2008) Transgenic rice plants expressing human P450 genes involved in xenobiotic metabolism for phytoremediation. J Mol Microbiol Biotechnol 15:212–219

Khan AG (2003) Vetiver grass as an ideal phytosymbiont for Glomalian fungi for ecological restoration of derelict land. In: Truong P, Hanping X (eds.) Proceedings of the third international conference on vetiver and exhibition: vetiver and water. China Agricultural Press, Guangzou, pp 466–474

Khan AG (2005) Role of soil microbes in the rhizospheres of plants growing on tracemetal contaminated soils in phytoremediation. J Trace Elem Med Biol 18:355–364

Khan MS, Zaidi A, Wani PA, Oves M (2009) Role of plant growth promoting rhizobacteria in the remediation of metal contaminated soils. Environ Chem Lett 7:1–19

Kirk J, Klironomos J, Lee H, Trevors JT (2005) The effects of perennial ryegrass and alfalfa on microbial abundance and diversity in petroleum contaminated soil. Environ Pollut 133:455–465

Kos B, Lestan D (2003) Induced phytoextraction/soil washing of lead using biodegradable chelate and permeable barriers. Environ Sci Technol 37:624–629

Kukier U, Peters CA, Chaney RL, Angle JS, Roseberg RJ (2004) The effect of pH on metal accumulation in two *Alyssum* species. J Environ Qual 32:2090–2102

Kumar PBAN, Dushenkov V, Motto H, Raskin I (1995) Phytoextraction: the use of plants to remove heavy metals from soils. Environ Sci Technol 29:1232–1238

Kurumata M, Takahashi M, Sakamotoa A, Ramos JL, Nepovim A, Vanek T (2005) Tolerance to and uptake and degradation of 2, 4, 6-trinitrotoluene (TNT) are enhanced by the expression of a bacterial nitroreductase gene in *Arabidopsis thaliana*. Z Naturforsch C 60:272–278

Lasat MM (2002) Phytoextraction of toxic metals: a review of biological mechanisms. J Environ Qual 31:109–120

Lee J, Bae H, Jeong J, Lee JY, Yang YY, Hwang I (2003) Functional expression of heavy metal transporter in Arabidopsis enhances resistance to and decreases uptake of heavy metals. Plant Physiol 133:589–596

Leštan D, Luo C, Li X (2008) The use of chelating agents in the remediation of metal-contaminated soils: a review. Environ Pollut 153:3–13

Linger P, Mussing J, Fischer H, Kobert J (2002) Industrial hemp (*Cannabis sativa* L.) growing on heavy metal contaminated soil: fibre quality and phytoremediation potential. Ind Crops Prod 16:33–42

Lippmann B, Leinhos V, Bergmann H (1995) Influence of auxin producing rhizobacteria on root morphology and nutrient accumulation of crops. 1. Changes in root morphology and nutrient accumulation in maize (*Zea mays* L.) caused by inoculation with indol-3 acetic acid (IAA) producing Pseudomonas and *Acinetobacter* strains of IAA applied exogenously. Angew Bot 69:31–36

Loehr RC, Webster MT (1996) Performance of long-term, field-scale bioremediation processes. J Hazard Mater 50:105–128

Lopez-Errasquin E, Vazquez CV (2003) Tolerance and uptake of heavy metals by *Trichoderma atroviride* isolated from sludge. Chemosphere 50:137–143

Luo C, Shen Z, Li X (2005) Enhanced phytoextraction of Cu, Pb, Zn and Cd with EDTA and EDDS. Chemosphere 59:1–11

Madrid F, Liphadzi MS, Kirkham MB (2003) Heavy metal displacement in chelate-irrigated soil during phytoremediation. J Hydrol 272:107–119

Mahesh WJ, Jagath C, Kasturiarachchi R, Kularatne KA, Suren LJW (2008) Contribution of water hyacinth (*Eichhornia crassipes* (Mart.) Solms) grown under different nutrient conditions to Fe-removal mechanisms in constructed wetlands. J Environ Manage 87:450–460

Maiti IB, Hunt AG, Wagner GJ, Yeargan R, Hunt AG (1991) Light inducible and tissue specific expression of a chimeric mouse metallothionein cDNA gene in tobacco. Plant Sci 76:99–107

Marques APGC, Pires C, Moreira H, Rangel AOSS, Castro PML (2010) Assessment of the plant growth promotion abilities of six bacterial isolates using *Zea mays* as indicator plant. Soil Biol Biochem 42:1229–1235

Mayak S, Tirosh S, Glick BR (2004) Plant growth promoting bacteria that confer resistance to water stress in tomatoes and peppers. Plant Physiol 166:525–530

McConkey BJ, Duxbury CL, Dixon DG, Greenberg BM (1997) Toxicity of a PAH photooxidation product to the bacteria *Photobacterium phosphoreum* and the duckweed *Lemna gibba*: effects of phenanthrene and its primary photoproduct, phenanthrenequinone. Environ Toxicol Chem 16:892–899

McEldowney S, Hardman DJ, Waite S (1993) Treatment technologies. In: McEldowney S, Hardman J, Waite S (eds.) Pollution, ecology and biotreatment. Longman Singapore Publishers Pte. Ltd, Singapore, pp 48–58

McGrath SP, Zhao FJ, Lombi E (2002) Phytoremediation of metals, metalloids and radionuclides. Adv Agron 75:1–56

McIntyre T, Lewis GM (1997) The advancement of phytoremediation as an innovative environmental technology for stabilization, remediation, or restoration of contaminated sites in Canada: a discussion paper. J Soil Contam 6:227–241

Meagher RB, Rugh CL, Kandasamy MK, Gragson G, Wang NJ (2000) Engineered phytoremediation of mercury pollution in soil and water using bacterial genes. In: Terry N, Bañuelos G (eds.) Phytoremediation of contaminated soil and water. Lewis Publishers, Boca Raton, pp 201–219

Melo JS, D'Souza SF (2004) Removal of chromium by mucilaginous seeds of Ocimum basilicum. Bioresour Technol 92:51–155

Mench MJ, Didier VL, Loffer M, Gomez A, Masson P (1994) A mimicked in situ remediation study of metal-contaminated soils with emphasis on cadmium and lead. J Environ Qual 23:58–63

Misra S, Gedamu L (1989) Heavy metal tolerant transgenic *Brassica napus* L and Nicotiana tabacum L plants. Theor Appl Genet 78:16–18

Mulligan CN, Young RN, Gibbs BF (2001) Remediation technologies for metal-contaminated soils and groundwater: an evaluation. Eng Geol 60:193–207

Narasimhan K, Basheer C, Bajic VB, Swarup S (2003) Enhancement of plant-microbe interactions using a rhizosphere metabolomics-driven approach and its application in the removal of polychlorinated biphenyls. Plant Physiol 132:146–153

Nedelkoska TJ, Doran PM (2000) Hyperaccumulation of cadmium by hairy roots of Thlaspi caerulescens. Biotechnol Bioeng 67:607–615

Normander B, Hendriksen NB, Nybroe O (1999) Green fluorescent protein-marked *Pseudomonas fluorescens*: localization, viability, and activity in the natural barley rhizosphere. Appl Environ Microbiol 65:4646–4651

Otten A, Alphenaar A, Pijls C, Spuij F, de Wit H (1997) *In situ* soil remediation. Kluwer Academic Publishers, Boston

Pan A, Yang M, Tie F, Li L, Chen Z, Ru B (1994) Expression of mouse metallothionein-1-gene confers cadmium resistance in transgenic tobacco plants. Plant Mol Biol 24:341–351

Patten CL, Glick BR (1996) Bacterial biosynthesis of indol-3 acetic acid. Can J Microbiol 42:207–220

Pe T, Fuentes HD, Khoo CS, Muir S, Khan AG (2000) Preliminary experimental results in phytoremediation of a contaminated mine site using plant growth regulators to increase heavy metal uptake. In: Handbook and abstracts 15th Australian statistical conference, Adelaid Hilton International, Adelaide, South Australia, pp 143–144

Pedas P, Schjoerring JK, Husted S (2009) Identification and characterization of zinc-starvation-induced ZIP transporters from barley roots. Plant Physiol Biochem 47:377–383

Penrose DM, Glick BR (2001) Levels of 1-aminocyclopropane-1-carboxylic acid (ACC) in exudates and extracts of canola seeds treated with plant growth promoting bacteria. Can J Microbiol 47:368–372

Pierzynski GM, Sims JT, Vance GF (1994) Soils and environmental quality. Lewis Publishers, Ann Arbor

Pilon M, Owen JD, Garifullina GF, Kurihara T, Mihara H, Esaki N (2003) Enhanced selenium tolerance and accumulation in transgenic Arabidopsis expressing a mouse selenocysteine lyase. Plant Physiol 131:1250–1257

Pulford ID, Watson C (2003) Phytoremediation of heavy metal contaminated land by trees e a review. Environ Int 29:529–540

Recep K, Fikrettin S, Erkol D, Cafer E (2009) Biological control of the potato dry rot caused by *Fusarium* species using PGPR strains. Biol Control 50:194–198

Reed MLE, Glick BR (2005) Growth of canola (*Brassica napus*) in the presence of plant growth promoting bacteria and either copper or polycyclic aromatic hydrocarbons. Can J Microbiol 51:1061–1069

Reeves RD, Baker AJH (2000) Metal accumulating plants. In: Raskin I, Ensley BD (eds.) Phytoremediation of toxic metals: using plants to clean up the environment. Wiley, New York, pp 193–229

Robinson NJ, Procter CM, Connolly EL, Guerinot ML (1999) A ferric-chelate reductase for iron uptake from soils. Nature 397:694–697

Romkens P, Bouwman L, Japenga J, Draaisma C (2001) Potentials and drawbacks of chelate-enhanced phytoremediation of soils. Environ Pollut 116:109–121

Rugh CL, Wilde D, Stack NM, Thompson DM, Summers AO, Meagher RB (1996) Mercuric ion reduction and resistance in transgenic *Arabidopsis thaliana* plants expressing a modified bacterial merA gene. Proc Natl Acad Sci USA 93:3182–3187

Rugh CL, Senecoff JF, Meagher RB, Merkle SA (1998) Development of transgenic yellow poplar for mercury phytoremediation. Nat Biotechnol 16:925–928

Rugh CL, Bizily SP, Meagher RB (2000) Phytoremediation of environmental mercury pollution. In: Raskin I, Ensley BD (eds.) Phytoremediation of toxic metals using plants to clean up the environment. Wiley, New York, pp 151–171

Ruiz ON, Daniell H (2009) Genetic engineering to enhance mercury phytoremediation. Curr Opin Biotechnol 20:213–219

Safronova VI, Stepanok VV, Engqvist GL, Alekseyev YV, Belimov AA (2006) Rootassociated bacteria containing 1-aminocyclopropane-1-carboxylate deaminase improve growth and nutrient uptake by pea genotypes cultivated in cadmium supplemented soil. Biol Fertil Soils 42:267–272

Saleh S, Huang XD, Greenberg BM, Glick BR (2004) Phytoremediation of persistent organic contaminants in the environment. In: Singh A, Ward O (eds.) Soil biology, vol 1, Applied bioremediation and phytoremediation. Springer, Berlin, pp 115–134

Samuelsen AI, Martin RC, Mok DWS, Machteld CM (1998) Expression of the yeast FRE genes in transgenic tobacco. Plant Physiol 118:51–58

Sar P, D'Souza SF (2002) Biosorption of thorium (IV) by a *Pseudomonas* strain. Biotechnol Lett 24:239–243

Saxena PK, Krishnaraj S, Dan T, Perras MR, Vettaakkorumakankav NN (1999) Phytoremediation of heavy metal contaminated and polluted soils. In: Prasad MNV, Hagemeyer J (eds.) Heavy metal stress in plants: from molecules to ecosystems. Springer, Berlin, pp 305–329

Schneegurt MA, Jain JC, Menicucci FR, Brown SA, Kemner KM, Garofalo DF (2001) Biomass byproducts for the remediation of waste waters contaminated with toxic metals. Environ Sci Technol 35:3786

Shetty KG, Hetrick BAD, Schwab AP (1995) Effects of mycorrhizae and fertilizer amendments on zinc tolerance of plants. Environ Pollut 88:307–314

Siciliano SD, Germida JJ (1997) Bacterial inoculants of forage grasses that enhance degradation of 2-chlorobenzoic acid in soil. Environ Toxicol Chem 16:1098–1104

Simonich SL, Hites RA (1994) Importance of vegetation in removing polycyclic aromatic hydrocarbons from the atmosphere. Nature 370:49–51

Smith LA, Means JL, Chen A, Alleman B, Chapma CC, Tixier JR, Brauning SE, Gavaskar AR, Royer MD (1995) Remedial options for metal contaminated sites. Lewis, Boca Raton

Song WY, Sohn EJ, Martinoia E, Lee YJ, Yang YY, Jasinski M (2003) Engineering tolerance and accumulation of lead and cadmium in transgenic plants. Nat Biotechnol 21:914–919

Summers AO (1986) Organization, expression and evolution of genes for mercury resistance. Annu Rev Microbiol 40:607–634

Suresh B, Ravishankar G (2004) Phytoremediation - a novel and promising approach for environmental clean-up. Crit Rev Biotechnol 24:97–124

Thomas JC, Davies EC, Malick FK, Endreszi C, Williams CR, Abbas M (2003) Yeast metallothionein in transgenic tobacco promotes copper uptake from contaminated soils. Biotechnol Prog 19:273–280

Trotel-Aziz P, Couderchet M, Biagianti S, Aziz A (2008) Characterization of new bacterial biocontrol agents *Acinetobacter*, *Bacillus*, *Pantoea* and *Pseudomonas* spp. mediating grapevine resistance against *Botrytis cinerea*. Environ Exp Bot 64:21–32

United States Environmental Protection Agency (US-EPA) (1996) A citizen's guide to bioremediation-technology fact sheet. Office of Solid Waste and Emergency Response. EPA 542-F-96-007

United States Environmental Protection Agency (US-EPA) (1997) Technology alternatives for the remediation of soils contaminated with As, Cd, Cr, Hg, and Pb. (Report EPA/540/S-97/500). United States Environmental Protection Agency, Washington, DC, pp 1–21

Van der Zaal BJ, Neuteboom LW, Pinas JE, Chardonnen AN, Schat H, Verkleij JAC (1999) Overexpression of a novel Arabidopsis gene related to putative zinc transporter genes from animals can lead to enhanced zinc resistance and accumulation. Plant Physiol 119:1047–1055

Van Dillewijn P, Couselo JL, Corredoira E, Delgado E, Wittich RM, Ballester A (2008) Bioremediation of 2, 4, 6-trinitrotoluene by bacterial nitroreductase expressing transgenic aspen. Environ Sci Technol 42:7405–7410

Vassil AD, Kapulnik Y, Raskin I, Salt DE (1998) The role of EDTA in lead transport and accumulation by Indian mustard. Plant Physiol 117:447–453

Vera Tomé F, Blanco Rodríguezb P, Lozano JC (2008) Elimination of natural uranium and 226Ra from contaminated waters by rhizofiltration using *Helianthus annuus* L. Sci Total Environ 393:51–357

Vidali M (2001) Bioremediation: an overview. Pure Appl Chem 73:1163–1172

Vleesschauwer D, Höfte M (2009) Rhizobacteria-induced systemic resistance. Adv Bot Res 51:223–281

Wild E, Dent J, Thomas GO, Jones KC (2005) Direct observation of organic contaminant uptake, storage, and metabolism within plant roots. Environ Sci Technol 39:3695–3702

Wills B (1988) Mineral processing technology, 4th edn. Pergamon Press, Oxford

Wu LH, Luo YM, Xing XR, Christie P (2004) EDTA-enhanced phytoremediation of heavy metal-contaminated soil with Indian mustard and associated potential leaching risk. Agric Ecosyst Environ 102:307–318

Wyman S, Simpson RJ, McKie AT, Sharp PA (2008) Dcytb (Cybrd1) functions as both a ferric and a cupric reductase *in vitro*. FEBS Lett 582:1901–1906

Xiaoxue W, Ningfeng W, Guo J, Xiaoyu C, Jian T, Bin Y, Yunliu F (2008) Phytodegradation of organophosphorus compounds by transgenic plants expressing a bacterial organophosphorus hydrolase. Biochem Biophys Res Comm 365:453–458

Yanez L, Ortiz D, Calderon J, Batres L, Carrizales L, Mejia J (2002) Overview of human health and chemical mixtures: problems facing developing countries. Environ Health Perspect 10:901–909

Yoshida N, Ikeda R, Okuno T (2006) Identification and characterization of heavy metal-resistant unicellular alga isolated from soil and its potential for phytoremediation. Bioresour Technol 97:1843–1849

Zhu Y, Rosen BP (2009) Perspectives for genetic engineering for the phytoremediation of arsenic-contaminated environments: from imagination to reality? Curr Opin Biotechnol 20:220–224

Zhu Y, Pilon-Smits EAH, Jouanin L, Terry N (1999a) Overexpression of glutathione synthetase in *Brassica juncea* enhances cadmium tolerance and accumulation. Plant Physiol 119:73–79

Zhu Y, Pilon-Smits EA, Tarun AS, Weber SU, Jouanin L, Terry N (1999b) Cadmium tolerance and accumulation in Indian mustard is enhanced by overexpressing g-glutamylcysteine synthetase. Plant Physiol 121:1169–1177

Chapter 4
Legume–*Rhizobium* Symbioses as a Tool for Bioremediation of Heavy Metal Polluted Soils

Eloísa Pajuelo, Ignacio David Rodríguez-Llorente, Alejandro Lafuente, and Miguel Ángel Caviedes

Abstract Legumes have traditionally been used in soil regeneration, owing to their capacity to increase soil nitrogen due to biological nitrogen fixation. Recently, legumes have attracted attention for their role in remediation of metal-contaminated soils. Legumes accumulate heavy metals mainly in roots and show a low level of metal translocation to the shoot. The main application of these plants is thus in metal phytostabilization. However, high concentrations of heavy metals in soil lead to a decrease in the symbiotic properties of legumes, which could be due to a decrease in the number of rhizobial infections. In order to identify a best legume–*Rhizobium* partnership for bioremediation purposes, selection of plant varieties and rhizobia resistant to heavy metal is required. Different approaches directed to improve metal bioremediation potential of legumes have been undertaken; from inoculation with rhizosphere bacterial consortia resistant to heavy metals to genetic engineering. Inoculation of legume plants with appropriate inocula containing rhizobia and heavy metal-resistant plant growth-promoting rhizobacteria (PGPR) and/or mycorrhiza has been found as an interesting option to improve plant performance under stressed conditions. The role of *Rhizobium*–legume symbiosis and approaches employed to genetically engineer legume–*Rhizobium* interactions in order to improve bioremediation are reviewed and discussed.

Keywords Bioremediation • Legumes • Rhizobia • Symbiosis

E. Pajuelo (✉) • I.D. Rodríguez-Llorente • A. Lafuente • M.Á. Caviedes
Departamento de Microbiología y Parasitología, Facultad de Farmacia,
Universidad de Sevilla, Profesor García González, 2, Sevilla 41012, Spain
e-mail: epajuelo@us.es

M.S. Khan et al. (eds.), *Biomanagement of Metal-Contaminated Soils*,
Environmental Pollution 20, DOI 10.1007/978-94-007-1914-9_4,
© Springer Science+Business Media B.V. 2011

4.1 Introduction

Global industrialization is the main source of releasing toxic compounds into the biosphere, which poses a greater risk for human health, wildlife, and environment (http://www.epa.org). Heavy metals, metalloids, and radionuclides are some of the most toxic and persistent pollutants. Unlike organic compounds, metals cannot be degraded although toxicity can be minimized by altering metal speciation or bioavailability. Biologically based technologies, collectively known as bioremediation, a powerful alternative to most traditional physicochemical remediation techniques, have become the preferred choice that can be integrated with other technologies for effective remediation of metal polluted sites (Van Aken 2008; Wood 2008; Ghosh and Singh 2005; Jorgensen 2007). Within bioremediation, rhizoremediation is a combination of two methodologies, phytoremediation and bioaugmentation. The term rhizoremediation refers to the combined use of plants and rhizosphere microorganisms in order to improve the bioremediation capacity of plants (Khan et al. 2009; Khan 2005; Kuiper et al. 2004; Glick 2003). The term rhizosphere was first introduced by Hiltner in 1904 and refers to the portion of soil under the direct influence of plant roots; it is the interface between plant root, soil, and the community of rhizosphere microorganisms associated with plant roots. The rhizosphere shows a higher microbial density (10^2–10^4 fold) compared to the microorganisms inhabiting bulk soil (Hinsinger et al. 2005), as plant exudates favors microbial growth. Plants release a variety of organic compounds in the rhizosphere that serve as carbon (C) sources for heterogeneously distributed microbial communities. As much as 20% of C fixed by a plant may be released from its roots. Besides increasing microbial populations, the organic compounds increase the metabolic activity of microbes. The colonization of the plant root allows the microorganisms to move deeper into soil layers, and thereafter, increases the contact of detoxifying microorganisms and soil contaminants (Kidd et al. 2009; Gerhardt et al. 2009). The presence and survival of rhizosphere microorganisms have important consequences for plants, such as providing plants the defense against pathogens (biocontrol), can be used as biofertilizers (for example, phosphate solubilizers, nitrogen fixers), and stimulate plant growth by secreting phytohormones, like auxins, by inhibiting ethylene accumulation via the expression of aminocyclopropane deaminase activity, etc. The utilization of selected inoculants possessing multiple properties improves plant yields, both in contaminated and non-contaminated soils.

Rhizoremediation is an attractive process since plant roots provide a large surface area for a large population of bacteria and transport the colonizing bacteria to deeper soil layers (Anderson et al. 1993). To achieve optimum inoculation effects, the microbial populations, therefore, have to be carefully selected with multiple growth-promoting activities, like the ability to resist/tolerate soil contaminants, ability to survive and colonize, and even to compete with native rhizosphere microbial populations. Therefore, while choosing inoculants for rhizoremediation, one should focus on the selection of native soil microorganisms for remediation of metal-contaminated soils. In this context, rhizobacteria have been used in phytoremediation of polluted soils (Doty 2008; Zhuang et al. 2007).

In the rhizosphere, metal mobilization as a consequence of plant root growth and to the metabolic activity of rhizosphere microorganisms increases the mobility and bioavailability of metals. Processes such as the secretion of protons, organic acids, and chelating agents increase the transport of metals to the plant root and thereafter increase plant phytoextraction (Ma et al. 2009; Abou-Shanab et al. 2008; Sheng and Xia 2006). On the contrary, the mobility and bioavailability of metals in soils can be reduced by microorganisms through processes like bioprecipitation or biosorption or by plants through phytostabilization. All these mechanisms can act together helping metal immobilization (Méndez and Maier 2008). Besides the effect on metal mobility and bioavailability, or on the degradation of organics (Zhuang et al. 2007), the PGPR such as *Pseudomonas*, *Acinetobacter*, *Achromobacter*, *Flavobacterium*, *Bacillus*, *Nocardia*, and *Rhizobium* increases plant yield and biomass and improves soil quality, and the content of organic matter or the amount of N. Mycorrhizal fungi is also used to improve the phytoremediation capacity of many plants. Since this falls out of the scope of this chapter, only an example is given on the inoculation of legume plants with AM-fungi for improving phytoremediation. The inoculation of pea (*Pissum sativum*) plants with *Glomus intraradices* attenuated cadmium stress, when pea was grown in Cd-polluted soils. The AM-fungus increased plant biomass and photosynthetic activity, and protected the plant against metal stress. Furthermore, the concentration of cadmium in tissues of AM-inoculated pea plants tends to decrease compared to non-inoculated plants (Rivera-Becerril et al. 2002). Recently, Stephanie et al. (2011) determined the impact of rhizobial inoculation on legume while growing in metal-enriched soils. In this study, the legume plant *Anthyllis vulneraria* subsp. carpatica from a mine site and of a non-metallicolous subsp. praeopera from nonpolluted soil were bacterized with a metallicolous or a non-metallicolous compatible *Mesorhizobium* spp. and grown on low and high heavy metal (like, Zn, Pb, and Cd) contaminated soils. The *M. metallidurans*–inoculated *A. vulneraria* plants had many nodules even when grown in metal-contaminated soils while the non-metallicolous *A. vulneraria* died after a few weeks despite *Rhizobium* inoculation. In addition, in metal-polluted soils, 80% of the total N was derived from BNF resulting between metallicolous *A. vulneraria* and the *Mesorhizobium*. This finding thus suggests that the legumes like *A. vulneraria* expressing a high N_2-fixing ability could be used to facilitate a low-maintenance plant cover and for stabilizing the vegetation in soils contaminated with heavy metals.

4.2   The Legume–*Rhizobium* Symbiotic Interaction

One of the major factors that limit plant growth is the deficiency of certain nutrients in the soil, especially nitrogen (N) and phosphorus (P). Different genera of plants have solved this problem via beneficial interactions with microbes inhabiting rhizosphere. One of the widely studied beneficial plant–microbe interaction is that between Gram-negative soil bacteria, collectively known as rhizobia, and legumes. The use of legume plants for soil N enrichment is a very old agricultural practice (De Hoff and Hirsch 2003; Graham and Vance 2003) and has been employed for regeneration of arid and degraded lands (Méndez and Maier 2008; Requena et al. 2002; Piha et al. 1995). Recently, there has been an increasing interest in the use of legume–*Rhizobium* symbiosis as a tool for bioremediation of both heavy metals (Dary et al. 2010; Pastor et al. 2003; Sriprang et al. 2002, 2003), and some organic compounds (Doty et al. 2003). Legume plants can interact with different rhizosphere microorganisms, including bacteria and mycorrhizal fungi. Besides its importance in agriculture, the legume–*Rhizobium* symbiotic interaction is also an important model for plant-bacteria signaling and for plant organogenesis, since the N fixing nodule is a model of a *de novo* formed plant organ.

Rhizobia currently includes 13 genera with 76 species of α- and β-proteobacteria (Velázquez et al. 2010; Weir 2009), while new rhizobia are being discovered. On the other hand, all members of the leguminosae family are not able to nodulate, and few other related higher plants are known to establish this symbiosis (Sprent 2007). The rhizobia–legume symbiotic interaction results in the formation of nodule, a highly organized structure in which atmospheric N is converted into ammonia by the bacteria, allowing the plant to grow without an external supply of reduced N. The formation of the root nodules involves a complex molecular dialogue between both symbiotic partners; an exchange of signals induced by the spatially and temporally regulated expression of specific genes. Legumes excrete secondary metabolites, mainly (iso) flavonoids, that induce the expression of bacterial nodulation (*nod*) genes: involved in the production of lipochitooligosaccharide Nod factors (Cooper 2007; Gibson et al. 2008). The Nod factors elicit several plant responses, such as root hair curling and the induction of cortical cell division leading to nodule formation (Oldroyd and Downie 2008). They are also essential for the expression of nodule-specific plant genes, the nodulins.

4.2.1 Promotion of the Bacterial Infection

The root epidermis is the first point of contact for rhizobia and determines where, when, and how many nodules will be formed. Bacterial infection can occur either through root hairs (intracellular infection) or by crack invasion (intercellular infection), usually at points of epidermal damage, generally caused by the emergence of lateral roots (Den Herder et al. 2006). The specificity of the rhizobia–legume interaction is expressed mainly at the first steps of the symbiosis and is controlled by the Nod factors via interaction with specific plant receptors. Initial Nod factor perception occurs in the epidermis, but Nod factor signaling is also important during cortex invasion and may be important for bacterial release into nodule cells (Oldroyd and Downie 2008). Kinases with N-acetylglucosamine-binding lysin motifs (LysM) in the extracellular domain have been described as Nod factor receptors (Radutoiu et al. 2003). Nod factor perception induces calcium spiking in the nucleus and changes in root hair growth linked to the induction of calcium gradients at the tip of root hair cells (Oldroyd and Downie 2006). A calcium and calmodulin-dependent protein kinase and at least three transcriptional regulators have been involved in

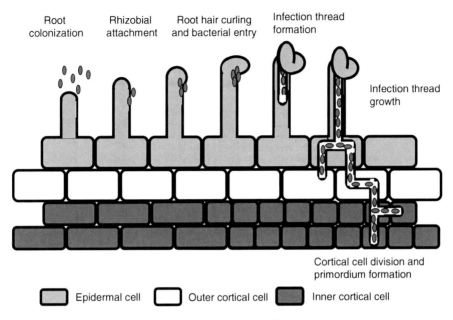

Fig. 4.1 Scheme highlighting the early events of legume–*Rhizobium* interaction and nodulation

perception and transduction of the calcium-spiking signal, but the exact mechanism of such perception and how this activates transcription remains unclear. In addition to root hair deformation, calcium spiking, and early nodulation genes (*ENOD*) induction, Nod factors also induce pre-infection thread structures in epidermal cells (van Brussel et al. 1992) and cortical cell divisions leading to the formation of a nodule meristem (Geurts et al. 2005).

4.2.2 Bacterial Infection

The different steps of the bacterial infection are schematized in Fig. 4.1. The process begins with the attachment of rhizobia to the root epidermis, mainly root hair cells. Rhizobia are able to attach the legume roots and root hairs in preference to other bacteria and hence, rhizobial numbers in the rhizosphere and the probability of specific strains being infective are enhanced (Downie 2010). Rhizobia have multiple mechanisms like they secrete polysaccharides and/or surface proteins that enable them to attach to roots (Rodríguez-Navarro et al. 2007). The symbiosis ontogeny in common legumes starts from penetration of rhizobia into the root hairs and the formation of a specific tubular structures called infection threads (Gage 2004). This process is initiated by the inhibition of root hair growth and the initiation of a new growth axis, such that root hairs curl around the attached bacteria (van Batenburg et al. 1986). Rhizobia proliferate in the root hair curl and an inversion of root hair tip growth takes place, leading to the infection thread formation. The lumen of this structure is similar to an intracellular space (Brewin 2004). The rhizobia divide at the growing infection thread tip and a column of bacteria is formed. Rhizobial surface polysaccharides play a critical role during this stage of the infection, probably acting as specific signals. Different evidence indicates that they could also prevent plant defense reactions. The role of these compounds has been widely studied in the *Sinorhizobium meliloti–Medicago truncatula* model interaction (Jones et al. 2007). The infection thread elongates and ramifies through the root cortical cell layers, following the direction indicated by the Nod factor–induced pre-infection thread structures. Finally, the infection thread penetrates inside the emerging nodule and bacteria are internalized in nodule cells.

4.2.3   Nodule Formation

Epidermal and cortical responses occur in parallel during the symbiotic interaction. While the epidermis regulates bacterial infection, the root cortex controls the formation of a nodule. An important step during the symbiotic interaction is the activation of the mitotic cell cycle in cortical cells, and regulators of this process play an important role during nodule primordium formation (Cebolla et al. 1999). Although epidermal and cortical responses can be separated, so that bacterial infection can occur in the absence of nodule organogenesis and vice versa (Oldroyd and Downie 2008), to produce a bacterially infected nodule, both processes should be coordinated. A nodule primordium develops close to the place of bacterial infection, such that the growing infection thread invades the dividing nodule cells. At this point, rhizobia are released into the plant cell cytoplasm (via an endocytosis-like process) to form the so-called symbiosomes, in which rhizobia are surrounded by a plant-derived plasma membrane (Roth and Stacey 1989). Inside symbiosome, the rhizobia differentiate into nitrogen-fixing bacteroids.

4.3   The Legume–*Rhizobium* Symbiosis as a Tool for Bioremediation

4.3.1   The Microsymbiont and Heavy Metals

Some light metals, such as Ca, Na, K, Mg, and other heavy metals, like Co, Cr, Cu, Fe, Mn, Ni, or Zn, play fundamental roles in the living process of the microorganisms. Some of these metals are essential micronutrients acting as cofactors of enzymes, or they act in redox process or participate in osmoregulation (Silver and Phung 1996, 2005). While, others, such as Cd, Hg, and Pb have, no known biological function and adversely affect microbial cells through process like oxidative stress, binding to enzymes and other proteins, and damage membranes and DNA (Nies and Silver 2007). Anyway, metals with important biological activity are also toxic at elevated concentrations. In order to exhibit toxicity, metals must first enter the microbial cells. The problem of the uptake of metals with biological functions has been solved by microorganisms by duplicating different transport systems. Cytoplasmic membrane uptake systems can be grouped into high rate and rather unspecific secondary transport systems, which supply the basic need for a range of metal ions; and into highly substrate-specific, inducible, primary transport systems at the times of need of one special ion (Nies and Silver 2007). The primary uptake systems are often inducible ABC-type or P-type APTases. Another transport systems operate at high metal concentrations, being constitutively expressed, with a lower specificity and is based on the chemiosmotic gradient across the plasma membrane. Exception to this rule is the NiCoT, which are inducible, membrane potential-dependent transport systems for Ni and Co uptake (Nies and Silver 2007). Both types of transport systems function in the uptake of essential metals. However, the high chemical similarity between divalent cations provokes the uptake of toxic metals via the systems evolved for the uptake of the essential metals. For instance, Cd^{2+}, Hg^{2+}, or Pb^{2+} can enter the cell via the transport systems used for essential metals like Mn^{2+}, Fe^{2+}, Co^{2+}, Ni^{2+}, Cu^{2+}, and Zn^{2+}. This also applies to oxianions; for instance, arsenate (AsO_4^{3-}), a chemical analogous of phosphate (PO_4^{3-}), can enter the cell using the phosphate transporters, whereas chromate (CrO_4^{2-}) enters the cell via the sulfate transporters (Silver and Phung 1996).

It is assumed that bacterial cells have lived in the presence of high concentrations of metals from the beginning of life, nearly four billion years ago. Microorganisms need resistance systems that allow them to maintain the homeostasis of essential metals and also to detoxify nonessential toxic metals. Microorganisms have evolved resistance mechanisms in order to cope with high concentrations of heavy metals and oxianions. The most common mechanism of heavy-metals resistance is the extrusion of heavy metals and oxianions from bacterial cell, avoiding accumulation to levels that possibly inhibit growth, or cause cell death (Silver 1996). This is achieved by the participation of different transport systems (Nies 2003). Some of the efflux resistance systems are ATPases and chemiosmotic ion/proton exchangers (Silver and Phung 2005). In addition, accumulation and complexation of the metal ions inside the cell, reduction of toxic metal to less toxic forms, methylations, precipitation, and chelation with S-rich ligands like metallotioneins, glutathione, etc. are other metal detoxification mechanisms adopted by microbes (Outten et al. 2000; Gusmão et al. 2006). Gram-negative bacteria can also store metals in the periplasm, associated to proteins or peptides, in order to keep metals out of the cytoplasm and membrane where the important reactions take place. For example, proteins such as CopC or SilE store heavy metals in the periplasm (Moore and Helmann 2005). Other passive resistance mechanisms include metal biosorption to cell surface (Volesky 2007; Malik 2004). These resistance mechanisms are not incompatible, and several of them can act simultaneously. In recent years, a great diversity of rhizobia resistant to heavy metals has been reported (Table 4.1). Most of them belonging to different species, such as *Mesorhizobium loti*, *Sinorhizobium meliloti*,

Table 4.1 Rhizobial strains resistant to heavy metals and metalloids isolated from contaminated soils

Rhizobium species	Metal (loid) resistance	Reference
Azorhizobium caulinodans	4–5 mM Cd	Zhengwei et al. (2005)
Bradyrhizobium sp. RM8	5.1 mM Ni	Wani et al. (2007a)
	21.4 mM Zn	
Bradyrhizobium sp. STM2464	15 mM Ni	Chaintreuil et al. (2007)
Mesorhizobium metallidurans	16–32 mM Zn	Vidal et al. (2009)
	0.3–0.5 mM Cd	
Mesorhizobium sp. RC1 and RC4	7.7 mM Cr	Wani et al. (2009)
Rhizobium leguminosarum bv. viciae E20-8	2 mM Cd	Figueira et al. (2005)
Rhizobium sp. RP5	6 mM Ni	Wani et al. (2008a)
	28.8 mM Zn	
Rhizobium sp. VMA301	2.8 mM AsO_4^{3-}	Mandal et al. (2008)
Sinorhizobium medicae MA11	10 mM AsO_2^-	Pajuelo et al. (2008)

S, fredii, Rhizobium leguminosarum, Bradyrhizobium sp. etc., have been isolated from polluted soils. For instance, *Mesorhizobium metallidurans* has recently been identified as a novel bacterium able to grow at high Zn (16–32 mM) and Cd (0.3–0.5 mM) concentrations (Vidal et al. 2009). Also, *Rhizobium selenireducens* sp. nov. isolated from laboratory reactor was found to reduce selenate to elemental red selenium (Hunter et al. 2007).

4.3.2   Examples of Heavy Metal Resistance in Rhizobia

4.3.2.1   Arsenic Resistance in *Rhizobium* Strains

The detoxifying mechanism for arsenate is widespread among microbial life (Oremland and Stolz 2003). Arsenate enters bacterial cells via the phosphate transporters. It is usually an inducible system that reduces arsenate to arsenite, which is then extruded from the bacterial cell. The transport system is able to transport As (III) but not As (V), so bacteria have evolved an arsenate reductase, which catalyzes the reduction of As (V) to As (III). Arsenic detoxification systems have been found in *S. meliloti*. This species shows a low level of resistance to arsenic compared to some other much more resistant genera of Gram-negative bacteria, like *Ochrobactrum, Pseudomonas, E. coli,* or Gram-positive bacteria, like *Staphylococcus* or *Bacillus* (Table 4.2). The arsenic-resistant operon of *S. meliloti* is located on one of the symbiotic plasmids (*psmed02*). The operon is composed of four genes: (1) *arsC:* codifies an arsenate reductase, which reduces arsenate to arsenite; (2) *aqpS:* codifies an aqua-glycerolporine; (3) *arsR*, whose gene product is a transcriptional regulator, a trans-acting repressor that senses As(III), and control the expression of both *arsB* and *arsC;* and (4) *arsH*, which codifies a NADPH-dependent FMN-reductase

Table 4.2 Comparison of the resistance to arsenic of *Sinorhizobium* sp. with other genera of Gram-negative bacteria

Bacteria	AsO_2^- (mM)	AsO_4^{3-} (mM)	*ars* genes	Reference
E. coli W3110 (plasmid R773)	4	5	*arsRDABC*	Carlin et al. (1995)
Mesorhizobium loti	n.d.	n.d.	*arsC, arsA*	Sá-Pereira et al. (2007)
Ochrobactrum tritici SCII24T	50	200	1. *arsRDAB* + CBS-domain protein 2.*arsR*(2)*arsC*(2)*Acr3arsH*	Branco et al. (2008)
Pseudomonas sp. As-1	n.d.	65	*arsRBC*	Patel et al. (2007)
Rhizobium sp. (vigna)	2.8 mM	n.d.	Only *arsR* identified	Mandal et al. (2008)
R. leguminosarum	n.d.	n.d.	*arsC*	Sá-Pereira et al. (2007)
Sinorhizobium medicae MA11	10	300	*arsHBCRTyrP*	Copeland et al. (2007), accession number
Sinorhizobium meliloti Rm1021	n.d.	40	*arsHCRAqpS*	Yang et al. (2005)

(Yang et al. 2005). The main difference between arsenic resistance operon and that of other more resistant bacteria is the absence of an energy-dependent efflux pump (ArsB), which extrudes arsenite out of the cell by functioning as an $As(OH)_3/H^+$ antiporter. Mutagenesis analysis has demonstrated the involvement of the aquaglicerolporine in the mechanism of arsenic resistance in *S. meliloti* (Yang et al. 2005). However, this porine is a transmembrane channel that probably does not extrude arsenite in as efficient way as the ArsB pump does. Recently, the gen *arsC* has been reported in strains of *R. leguminosarum*, *S. loti*, and *M. loti*, isolated from contaminated soils of Portugal (Sá-Pereira et al. 2007). The gen *arsC* seems to be more widely distributed than other arsenic resistance genes. In fact, several strains had one or several copies of the gene *arsC*, but not an *arsB* gene. An *arsR* gene has also been amplified in *S. fredii* VMA301, isolated from nodules of *Vigna mungo* grown in As-contaminated field (Mandal et al. 2008).

4.3.2.2 Cadmium Resistance in *Rhizobium* Strains

The resistance mechanism for cadmium in bacteria is based on the expulsion of Cd from the bacterial cell through more or less specific cation transporters. There are three different efflux mechanisms: a P-type ATPase (the *CadA* ATPase) found both in Gram-positive and Gram-negative bacteria; a chemiosmotic pump consisting of three polypeptides of the RDN (CBA) family also found in Gram-positive and Gram-negative bacteria; and a single polypeptide chemiosmotic efflux system of the cation diffusion facilitator (CDF) family (the *Czc* system involved in the efflux of both Cd and Zn), described for the first time in *Ralstonia metallidurans* (Anton et al. 1999). Cadmium resistance determinants have been also reported in *Rhizobium*; for instance, *loci* with similarity to both *cadA* and *cadC* determinants are found in the completely sequenced genome of *R. leguminosarum* vb. trifolii WSM2034 (Reeve et al. 2010) and *Mesorhizobium* sp. BNC1 (Copeland et al. 2006). However, although the function of *cadA* is known for bacteria like *Staphylococcus aureus* (Nucifora et al. 1989), its role in rhizobia is unknown. Together with *cadA*-like determinant, other cadmium resistance determinant like the protein NccN (presumably involved in Ni, Co, and Cd resistance) is present in rhizobia like *Mesorhizobium* sp. BCN1. Furthermore, the resistance to Cd in some species of *Rhizobium* has been associated to elevated levels of glutathione (GSH) (Gusmão et al. 2006). The GSH seems to play an important role in the detoxification of Cd by *R. leguminosarum*, suggesting the importance of glutathione in coping with metal stress. Moreover, glutathione was found as the main Cd chelator in *Rhizobium*, responsible for sequestering 75% of intracellular Cd in tolerant strain. In addition, metal could also bind to the cell surface, demonstrating an effective avoidance mechanism. However, the Cd biosorption capacity of both tolerant and sensitive *Rhizobium* strains did not differ significantly suggesting that the adsorption of Cd was not the basis of difference in metal tolerance among rhizobial strains (Gusmão et al. 2006).

4.3.2.3   Nickel Resistance Determinants in *Rhizobium* Strains

Nickel resistance determinants have also been identified in *Bradyrhizobium* strains, isolated from nodules of the endemic New Caledonia legume *Serianthes calycina*, growing in Ni-rich soils (Chaintreuil et al. 2007). The isolated *Bradyrhizobium* strains grew well in the presence of 15 mM NiCl$_2$. The Ni detoxification occurred through the extrusion mechanism. The genomes of these strains had two Ni resistance determinants, the *nre* and *cnr* operons. Of these, *nre* confers resistance toward moderate Ni concentrations, whereas *cnr* determines the extremely high Ni resistance in *Bradyrhizobium* strains.

4.3.2.4   Resistance Against Chromium in *Rhizobium*

Chromate enters the cell by the sulfate uptake system and is effluxed by the ChrAB proteins, which form a sulfate-chromate antiporter (Nies et al. 1998). Gene clusters have been recently identified in rhizobial strains that are regulated by heavy metals, particularly chromium. Recently, an ABC transporter involved in chromium efflux has been amplified in *R. leguminosarum, M. loti*, and *S. meliloti* strains (Sá-Pereira et al. 2009). The chromate efflux determinant can be also involved in the extrusion of some other metal cations, since it corresponds to a cation/multridug efflux pump, which belongs to the family of ABC transporter, confirming homology with an ATPase from PP super-family. The function of ABC transporters in the rhizobial strains possibly involves translocation of Cr through a pore formed by two integral membrane protein domains. Besides the efflux pump, a chromate sensor has been also identified. This gene shows homologous sequences to a hybrid sensor histidine kinase and a two-component sensor histidine kinase. This newly identified sensor may be a regulator of chromium uptake/efflux pump, serving, on the one side, in Cr sensing and at the same time, displaying a kinase activity, which may act in the activation of the efflux pump once the intracellular Cr concentration reaches a determined point. The identification of a two-component hybrid sensor kinase and a cation/multridug efflux pump in *S. meliloti* and *R. leguminosarum* suggests that it was identified as a newly different structure related to Cr resistance, and possibly, to other heavy metals (Sá-Pereira et al. 2009).

4.3.3   Nodulation Efficiency of Metal Resistant Rhizobia

It has long been known that the high concentration of heavy metals inhibits rhizobial growth (Broos et al. 2005). This is probably the main reason that explains why nodulation and the symbiotic nitrogen fixation (SNF) efficiency of rhizobia decrease in metal-contaminated soils (Broos et al. 2004; Giller et al. 1998). These and other reports demonstrated a diminution of two to four orders of magnitude in the most probable number (MPN) of *R. leguminosarum* bv. *trifolii* in polluted soils. Consequently, when the rhizobial population falls below a threshold, N_2-fixation is severely impaired. Furthermore, under metal stress, the genetic diversity of rhizobial populations is altered. Under the selective pressure of elevated metal concentrations, only the most resistant strains were able to survive, which were symbiotically ineffective and formed white nodules on roots (Broos et al. 2005; Lakzian et al. 2002; Castro et al. 1997; Chaudri et al. 1992). All these studies involved *R. leguminosarum* bv. *Trifolii*-inoculated clover plants where clover rhizobia was found more sensitive to heavy metals than host plants (Giller et al. 1998). Among rhizobia, *R. leguminosarum* bv. *trifolii* has been reported to be more sensitive to metals compared to *Sinorhizobium* (Giller et al. 1993). Despite these results, metal-resistant rhizobia are known that are fully effective in terms of nodulation and N_2-fixation. For example, symbiotically active rhizobia were isolated from several legume plants, such as *Medicago* sp., *Trifolium* sp., *Vicia sativa*, *Lupinus angustifolius*, etc., growing in polluted soils (Carrasco et al. 2005). *Sinorhizobium medicae* MA11 isolated from a *Medicago* sp. plant tolerated arsenite and arsenate up to a level of 10 and 300 mM, respectively, and were fully effective in nodulation and N_2-fixation of alfalfa. This strain was found more competitive than *S. meliloti* 1021 (Rm1021) in competition experiments, when inoculated together in the presence of As (Pajuelo et al. 2008). In other study, the strain (VM301) of *S. fredii* tolerated 2.8 mM arsenate and was symbiotically effective (Mandal et al. 2008). Yet in other similar experiment, Chaintreuil et al. (2007) isolated symbiotically effective *Bradyrhizobium* strains, which showed resistant to 15 mM Ni. Also, *Mesorhizobium metallidurans*, sp. *nova* isolated from nodules of a metallicolous ecotype of *Anthyllis vulneraria* growing on mine soils, have shown normal levels of N_2-fixation (Vidal et al. 2009). Interestingly, some of these strains were able to fix N to almost normal levels on polluted soils (Pajuelo et al. 2008). In this case, even though the number of nodules was reduced in the presence of heavy metals, the nodules formed on polluted soils were apparently normal and fixed about 75% N compared to those observed in non-polluted soils (Carrasco et al. 2005).

4.3.4 Non-rhizobial Bacteria Nodulates Legumes Under Heavy Metal Stress

Nodulation of legume plants by non-rhizobial strains has been reported both in polluted and nonpolluted soils (Zurdo-piñeiro et al. 2007; Sawada et al. 2003). For instance, several *Ochrobacterium* species, such as *O. lupini* (Trujillo et al. 2005) or *O. cytisi* (Zurdo-Piñeiro et al. 2007) capable of nodulating legumes, were isolated from nodules of the legume *Cytisus scoparius*, grown in heavy metal-polluted soils after a toxic mine spill. Nodules formed by *O. cytisi* were non-fixing white nodules. Nevertheless, the presence of a 1.6-Mb symbiotic plasmid in *O. cytisi* and amplification of *nodD* and *nifH* genes have been confirmed (Zurdo-Piñeiro et al. 2007). Interestingly, some genes involved in metal resistance were detected in the symbiotic plasmids. More recently, nodulation of the legume plant *Lespedeza cuneata* growing in mine soils of China by the copper-resistant *Agrobacterium tumefaciens* strain CCNWRS33-2 has been reported (Wei et al. 2009). This strain showed resistance to different metals like Co (2 mM), Cd (2 mM), Pb (2 mM), and Zn (3 mM). A copper resistance operon inducible by both copper and silver in *A. tumefaciens* was reported by Nawapan et al. (2009). This operon contained three genes, *copAZR*, where *copA* codes for a copper efflux pump of the ATPase transporters family, *copZ* encodes a copper chaperone, and *copR* codifies a regulator.

4.3.5 Legumes and Heavy Metals

Plants growing in polluted soils have developed mechanisms allowing them to tolerate, and even detoxify heavy metals. The plant responds to heavy metals in two ways: some plants accumulate heavy metals while others exclude heavy metals (Baker 1981). With regard to the heavy metal sensitivity, plants can be classified into (1) metal excluders, (2) metal indicators, and (3) metal hyperaccumulators (Ghosh and Singh 2005; Prasad and Freitas 2003; Bleeker et al. 2003). Of these, metal hyperaccumulators are characterized for tolerating extremely high concentrations of a particular metal. Furthermore, these plants accumulate metals in tissues without any toxicity symptoms. Hyperaccumulators also have the ability to translocate metals from roots to the aerial organs of the plants, so that the accumulation in shoots exceeds the concentration of metals in roots. Also, the accumulation in shoots exceeds the concentration of metal in the soil (bioconcentration factor »1). Indeed, the concentration of metal in shoots must exceed 1% of the dry weight for a plant to be considered as Ni, Mn, or Zn hyperaccumulator; 0.1% of the dry weight to be considered as a Cu, Co, or Cr hyperaccumulator; and 0.01% of the dry weight for a Cd or As hyperaccumulator (Callahan et al. 2006; Reeves and Baker 2000). Nickel hyperaccumulator, *Sebertia accuminata*, has, however, reported to accumulate up to 25% of Ni of the dry weight of the xylem sap. In general, hyperaccumulator plants can accumulate one or as maximum as two heavy metals. For example, *Thlaspi caerulescens*, has been found to hyperaccumulate Zn and Cd (Baker 1981; Pence et al. 2000). Besides their ability to grow in harsh environments, extreme metal accumulation ability of hyperaccumulators protects plants from fungal infection, and also from herbivores or insect attack, in contaminated soils (Callahan et al. 2006). Metal excluders on the contrary prevent the accumulation of metals in shoots by (1) preventing the uptake by plant roots, (2) keeping the metal in the root tissues, and (3) by preventing the translocation of metals to the aerial parts of the plant. As a result of these processes, the concentration of metals in the aerial parts remains at a very low level in plants grown in a wide range of metal-contaminated soil.

In phytoremediation practices, hyperaccumulator plants are preferred since the remediation strategy is based on phytoextraction. Prospecting and identification of new hyperaccumulators are emerging as a cutting-edge area of research and gaining commercial significance in the field of environmental biotechnology. For instance, a program for identifying metal hyperaccumulators has been launched in China, which led to the discovery of new hyperaccumulators, such as *Pteris vittata* (for As) (Ma et al. 2001), *Malva sinensis* (for Cd) (Zhang et al. 2010), etc. Ideal metal hyperaccumulators should be fast-growing, large biomass-producing, and deep-rooted plants in order to remediate deeper layers of polluted soil. However, several hyperaccumulators are slow-growing small herbs with poor root systems (Vangrosveld et al. 2009). Non-accumulator plants (excluders) are suitable for phytostabilization studies. These plants tolerate moderate concentrations of heavy metals and grow on polluted soils. However, no metals are accumulated in shoots. The main advantage of this technique is that the metals are immobilized in soil, which in turn avoids leaching, erosion, and metal transfer into the food chain. However, it requires continuous monitoring of the polluted areas, since metal is not completely removed but is immobilized (Kidd et al. 2009). In non-hyperaccumulators like most of the legumes, metals are taken up and stored in root cells. Metal ions are chelated with different molecules like glutathione, phytochelatins, metallothioneins, organic acids, histidine, nicotinamine, etc. (Callahan et al. 2006; Mejáre and Bülow 2001), and stored within the vacuoles or in the apoplast, far from the cytoplasm where most of the physiological reactions occur (Pilon-Smits 2005). In non-hyperaccumulators, therefore, only a small quantity of metal is loaded into the xylem and is translocated to the shoots. It seems clear that hyperaccumulators do have completely different mechanisms for metal accumulation. For example, in some hyperaccumulators, like the Zn- and Cd-hyperaccumulator *T. caerulescens*, metal accumulation did not correlate with increased levels of phytochelatins, suggesting a different mechanism for hyperaccumulation (Ebbs et al. 2002). High levels of the Zn-transporter *Znt1* both in roots and shoots are found in the hyperaccumulator *T. caerulencens* (Pence et al. 2000). By contrast, high concentrations of histidine residues have been described in the Ni-hyperaccumulator *Alyssum* (Ingle et al. 2005). In the As-hyperaccumulating fern *P. vittata*, arsenate is taken up through the phosphate transporters, reduced to arsenite, and sequestered in the fronds primarily as As(III) (Wang et al. 2002). A high level of arsenate reductase has been reported (Duan et al. 2005). Also, a phytochelatin synthase with differences at the 5' sequence has been described, containing a greater number of cysteine residues as compared to other known phytochelatin synthases from non-hyperaccumulators (Dong 2005).

4.3.5.1 Accumulation of Heavy Metals in Legume Plants

Some legume plants, including species of the genera *Vicia, Cytisus, Astragalus, Lupinus*, etc., are known to grow on soils polluted by relatively higher concentrations of heavy metals (Prasad and Freitas 2003). Legumes along with grasses have been reported as one of the first colonizers of degraded and polluted soils (Wei et al. 2009; Bleeker et al. 2003; del Rio et al. 2002). For instance, some *Medicago* species were one of the first colonizers of severely polluted soils after a mine spill (Carrasco et al. 2005). Also, leguminous trees such as *Acacia* spp. and *Prosopis* spp. have been reported as successfully colonizing mine tailings in the Western United States

(Day et al. 1980). In spite of some legumes being tolerant to heavy metals, most of these plants fall into the category of metal excluders and accumulate very low concentrations of heavy metals in shoots (Table 4.3), and almost undetectable in grains (Wani et al. 2008a). In most of the cases, the concentration of metals in shoots of legume plants grown on polluted soils remains normally below the limits established for animal grazing (Table 4.3), suggesting that using legume plants in metal remediation projects does not pose any risk to the food chain. The use of local metallicolous legume species, which can tolerate higher concentrations of heavy metals, also in association with grasses, may improve metal phytostabilization in particular areas (Frérot et al. 2006; Sharples et al. 2000). For instance, *Lupinus albus* has been proposed as a good candidate for phytostabilization of Cd- and As-contaminated soils (Vázquez et al. 2006). Also, *L. lutetus* has been proposed in phytostabilization of multi-metal polluted soils, since it behaves as an As and metal excluder (Dary et al. 2010). Only in the case of acidic soils polluted with Zn can the accumulation of this metal in *Lupinus* plants reach values above the threshold (up to 3,600 ppm) for herbivore consumption (Pastor et al. 2003).

Some legume species belonging to the genus *Astragalus*, like *A. sinicus* and *A. bisulcatus*, are Se-hyperaccumulators (Prasad and Freitas 2003). The latter one accumulates more than 6,000 ppm Se in leaves and up to 10,000 ppm Se in fruits and seeds (Freeman et al. 2006). Besides Se-phytoextraction by leaves, other phytoremediation processes also occur in *Astragalus* plants. Selenium is a chemical analogous from S and enters the plant root via the S assimilation pathway. Plants (and microbes) can take up inorganic and organic forms of Se like selenite and selenate. Once inside the plant, Se is enzymatically bound to cysteine or methionine, to form selenocysteine or selenomethionine, which are transported to the leaves and converted into the volatile forms, methylselcnide (MSe) and dimethylselenide (DMSe). Biological volatilization has the advantage of removing Se from a contaminated site in a relatively nontoxic form, since DMSe is 500–700 times less toxic than SeO_4^{-2} or SeO_3^{-2} (Le Duc and Terry 2005). The phytovolatilization of volatile organoselenide species, such as dimethylselenide and dimethyldiselenide, has been demonstrated in *Astragalus bisulcatus* (Freeman et al. 2006; Pickering et al. 2003).

4.4 Effect of Heavy Metals on the Legume–*Rhizobium* Symbiotic Interaction

In order to use legumes for soil bioremediation, it is important to know the effect of soil contaminants on the symbiotic interaction. The effect of heavy metals on the legume–*Rhizobium* symbiotic interaction has extensively been studied (Table 4.4). Of the various symbiotic stages, nodulation in general is more sensitive to heavy metals than the roots or shoot growth of legumes (Gupta et al. 2007). While, root growth and photosynthetic pigments (chlorophyll) are parameters commonly used to determine the metal toxicity to legumes. As an example, *S. medicae*–inoculated *Medicago sativa*, when grown in the presence or absence of 25 µM sodium arsenite

Table 4.3 Concentration of heavy metals and metalloids in green tissues of legume plants grown on heavy metal-polluted field soils

Common name	Botanical name	As	Cd	Cu	Zn	Pb	Reference
Common broom	Cytisus scoparius	n.d.	0.9	19	200	n.d.	Moreno-Jiménez et al. (2009)
Broom	C. oromediterraneus	n.d.	1.0	5	159	n.d.	Moreno-Jiménez et al. (2009)
Broom	Genista cinerascens	n.d.	1.1	6	108	n.d.	Moreno-Jiménez et al. (2009)
Bird's foot trefoil	Lotus tetraphylus	n.d.	0	14	99	20	Del Rio et al. (2002)
White lupin	Lupinus albus	4	n.d.	10	3,605	n.d.	Pastor et al. (2003)[a]
White lupin	L. albus	<1.5	<0.15	8	86	<1.5	Pajuelo (unpublished)
Narrowleaf lupin, blue lupin	L. angustifolius	n.d.	0	16	108	18	Del Rio et al. (2002)
Yellow lupin	L. luteus	<1.5	2.0	52	784	35	Dary et al. (2010)
Alfalfa, lucerne, medic	Medicago sativa	n.d.	0	18	82	21	Del Rio et al. (2002)
Burr medic	Medicago polymorpha	<1.5	<0.15	14	52	<1.5	Pajuelo (unpubl. shed)
Restharrows	Ononis viscosa	n.d.	0	13	81	16	Del Rio et al. (2002)
Common bean	Phaseolus vulgaris	n.d.	1	4	57	2	Gupta et al. (2007)
Scorpion tick trefoil, scorpion's tail	Scorpiurus sp.	5–19	0	14	114	19	Del Rio et al. (2002)
Subclover, subterranean clover	Trifolium subterraneum	<1.5	3	20	240	11	Pajuelo (unpublished)
Yellow vetch	Vicia lutea	n.d.	0	13	207	9	Del Rio et al. (2002)
Common vetch	Vicia sativa	<1.5	4	17	781	12	Pajuelo y col. (unpublished)
Max. metal concentration recommended for animal grazing		30	10	40	500	100	Méndez and Maier (2008)

nd not determined

[a]Acidic soils, pot experiments

Table 4.4 Effect of metals on the legume–*Rhizobium* symbiotic interaction

Legume–*Rhizobium* symbiosis	Metal/concentration	Effect	Reference
Medicago sativa	30 μM Cd	Oxidative burst, necrosis, root hair damage	Ortega-Villasante et al. (2005)
Medicago sativa	30 μM Hg	Oxidative burst, necrosis, root hair damage	Ortega-Villasante et al. (2005)
Glycine max-Bradyrhizobium japonicum 109	50–500 μM Al	Decreased number of nodules, activity of N-assimilatory enzymes, and N_2-fixation. Increased oxidative stress	Balestrasse et al. (2006)
Cicer arietinum	Pb (390 mg/kg soil)	Number of nodules reduced by 13.7%, reduced plant growth, chlorophyll content, and N content	Wani et al. (2007b)
Vigna unguiculata-Bradyrhizobium strain CB756	2 μM Cu	Complete inhibition of nodulation, shortening and swelling at the root tips, number of root hairs decreases	Kopittke et al. (2007)
Cicer arietinum	Cd (23 mg/kg soil)	Number of nodules reduced by 69%, reduced biomass, chlorophyll content, and N content	Wani et al. (2007b)
Pisum sativium- Rhizobium leguminosarum RP 5	Cu (1,338 mg/kg soil)	Number of nodules reduced by 25%, N_2-fixation was less affected, N content in shoots reduced by 20%	Wani et al. (2008a)
Medicago sativa-Sinorhizobium medicae	25 μM As	Number of nodules decreases to 25%, number of rhizobial infections decreased by 90%, root hair damage, N_2-fixation was not affected	Pajuelo et al. (2008)

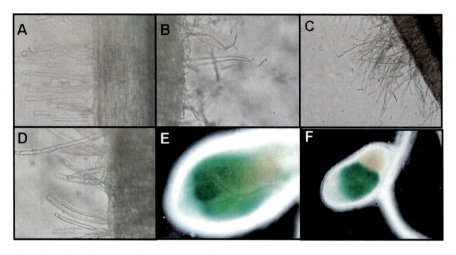

Fig. 4.2 Effect of arsenic (25 μM) on *Medicago* nodule development (**a**) root hairs of control plants three days post-inoculation, (**b, c**) root hairs of 5-days-old plants grown with arsenic where root hair tips are affected, (**d**) infection thread (*blue*) growing through a root hair in an As-grown plant, (**e**) mature nodule of control plant, and (**f**) mature N_2-fixing nodule of arsenic-treated plant (Reproduced from Pajuelo et al. 2008)

showed a 33% and 15% reduction in root length and chlorophyll content, respectively compared to plants grown without As (Pajuelo et al. 2008). A diminution of the number of nodules has been reported as a general effect of these contaminants although the sensitivity varies among legume species and with experimental designs. For instance, nodulation on *Vigna unguiculata* was completely abolished in the presence of 2 μM Cu in hydroponic cultures (Kopittke et al. 2007), whereas *M. truncatula* formed effective nodules even in the presence of 200 μM Cu when grown on sand:vermiculite (J. Delgadillo, personal communication). Different factors causing the decrease in nodule numbers includes atrophy of root hairs, decline in the total number of root hairs, shortening of the root zone susceptible to nodulation, and decrease in the number of infections events (Pajuelo et al. 2008). To validate the hypothesis that inhibition of root hair formation leads to reduction in nodulation, Brady et al. (1990) conducted an experiment to assess the impact of low activities of Al on soybean while Kopittke et al. (2007) analyzed the effect of varying concentrations of Cu on cowpea. Both of these studies concluded that reduction in nodulation with increasing metal activity was associated with an inhibition of root hair formation rather than to a reduction in the size of the *Rhizobium* population. Furthermore, in a microscopy study, Pajuelo et al. (2008) suggested that damages in root hairs, a shorter infective root zone, together with symptoms of necrosis, root tip swelling, and a 90% reduction in the number of rhizobial infections led to a decline in nodulation of alfalfa plants, grown in the presence of As (Fig. 4.2). However, once nodulation is established, N_2-fixing ability of legumes grown in the presence of heavy metals is affected, which depends on the type of metals and the legume species. An increase of oxidative stress and reactive oxygen species (ROS),

Table 4.5 Gene expression of nodulation marker genes in roots of alfalfa plants inoculated with *Sinorhizobium medicae* MA11 in the presence of arsenic

Gene	Marker for...	Days post inoculation	% expression (100% control without As)
Nork	Nod factor perception	1–5	20–40
NIN	Nodule initiation	1–5	60–80
Enod2	Infection thread progression and nodule organogenesis	2–10	10–25
N6	Infection thread formation and growing	2–10	20–60
Enod40	Nod factor responses and primordium initiation	1–10	96–102
ccs52	Cortical cell division	2–15	98–103
Legbrc	Nitrogen fixation	10–21	30–50

Adapted from Lafuente et al. (2010)

together with redox imbalance, has been reported in plants grown in the presence of metals like Al, Cd, or Hg (Ortega-Villasante et al. 2005). Furthermore, the levels of antioxidant enzymes like superoxide dismutase (SOD), catalase, and peroxidases have been found severely decreased in legume nodules, as in the case of soybean, grown in the presence of 500 μM Al, which resulted in greater oxidative stress, leghaemoglobin breakdown, and reduction in nitrogenase activity (Gupta et al. 2007). Nitrogen assimilatory enzymes including glutamate dehydrogenase (GDH), glutamine synthetase (GS), and glutamate synthase (GOGAT) of nodules are also reported to be adversely affected by metals, like Al (50–500 μM). On the contrary, the nitrogenase activity did not show an obvious decrease in moderately contaminated soils. Nodule development of alfalfa plants grown in the presence of low concentration (25 μM) of As was, however, normal, and mature nodules were functionally effective (Pajuelo et al. 2008). Nitrogen is another parameter commonly used to evaluate the N_2-fixing efficiency of legumes. Nitrogen content of chickpea plants grown in the presence of Cd or Pb was diminished by 30% and 10%, respectively (Wani et al. 2007b). Pea plants grown in Cu-treated soils showed a 9% decrease in shoot N (Wani et al. 2008a). However, a marginal increase in N content in legumes, grown in metal-contaminated soils has been reported (Wani et al. 2008a; Lasat 2000).

Only very recently, molecular approaches have been used to assess the effect of heavy metals on the establishment of the legume–*Rhizobium* interactions. For example, reduced nodulation in alfalfa induced by As has been correlated with altered expression of early nodulins (Lafuente et al. 2010). In this regard, seven well-known nodulin genes, markers for the different events leading to nodule formation, were analyzed by RT-PCR. A significant decrease in the expression of four early nodulins – the genes coding the Nod factor receptor (*nork*), the transcription factor *NIN* and the markers for infection progression (*N6*) and nodule organogenesis (*Enod2*) – especially 1–5 days after inoculation was observed. However, the expression of markers for primordium initiation (*Enod40*) and differentiation (*ccs52*) was not significantly altered. The expression of a marker for N_2-fixation (*Legbrc*, coding for leghemoglobin) was also reduced (Table 4.5). Results of the expression of *nork* and *Enod2* genes were confirmed by qPCR, which were influenced by As. Moreover, it was postulated from this study that As affects the expression of nodulation genes associated with processes that occur in the epidermis and the outer cortical cells, whereas the expression of genes associated with events that occur in the inner cortical cells is probably less affected.

4.5   Application of Legume–*Rhizobium* Symbioses in Metal-contaminated Soils

The simultaneous application of plants and rhizosphere microbes (rhizoremediation) improves the effectiveness of bioremediation. For this technique to be effective, a careful attention must be paid on to the selection of rhizosphere microorganisms displaying bioremediation potential and PGP properties, with high competitive ability. Accordingly, inoculants having multifarious activities can be developed for their ultimate use under conventional or derelict environment. In this context, many experiments have been accomplished using different legume–*Rhizobium* symbioses for remediation of metal-contaminated soils. Most of these researches have however, been done in pots/greenhouse environments. In some cases, results have not been found consistent and reproducible when tested under field trials (Vangrosveld et al. 2009). Therefore, it is generally suggested that when designing an in situ rhizoremediation strategy, the availability of metals in the soil, the viability of the inoculants, and the selection of the most adequate agronomical practices must be considered. Despite these problems, when symbiotically effective and metal-resistant rhizobia were used, they increased the fitness of plants and showed dual benefits: on one hand, they increased grain yields, plant biomass, and N content through N_2-fixation, and synthesizing growth-regulating substances (Dary et al. 2010; González and González-Chávez 2006). The positive effect of selected inoculants on plant yield and biomass improves the vegetal cover and helps the regeneration of polluted and degraded soils (Zhuang et al. 2007). On the other hand, inoculation with metal-resistant inoculants also affects metal solubility and bioavailability for plant uptake. In most cases, a decrease in metal accumulation in shoots of inoculated plants is reported (Frérot et al. 2006), although in some cases, an increase in metal accumulation occurs, when selected metal mobilizing inoculants are used in combination with hyperaccumulator plants (Ma et al. 2009; Zaidi et al. 2006; de Souza et al. 1999). Several mechanisms have been proposed for the observed decrease in metal content in inoculated legume plants. Such mechanisms includes biosorption of metal to bacterial cell surface (Rodríguez-Llorente et al. 2010), or plant roots, metal chelation, secretion of substances by plant roots or bacteria that immobilize metal in the soil, etc. All these processes can act together and therefore may enhance metal phytostabilization. For this strategy to be useful, metal-resistant bacteria (or bacterial consortia) as inoculants could be applied. For instance, the combination of *Rhizobium, Pseudomonas, Ochrobactrum, Bacillus*, mycorrhizal fungi, and mycorrhizal helper bacteria (MHB) have been used (Khan et al. 2009; Khan 2005). In a study, *Lupinus luteus* plants inoculated with a bacterial consortium

Fig. 4.3 Field trials of *Lupinus luteus* plants grown in metal-polluted soil after the toxic spill of the Aznalcóllar mine (Aznalcóllar, SW Spain) (**a**) non-inoculated control plants, (**b**) plants inoculated with *Bradyrhizobium*, (**c**) plants inoculated with a bacterial consortium resistant to heavy metals including *Bradyhizobium* and *Ochrobactrum*, and (**d**) plants inoculated with bacterial consortia (including *Bradyrhizobium, Ochrobactrum,* and *Pseudomonas*) resistant to heavy metals (Reproduced from Dary et al. 2010)

resistant to heavy metals have been used in metal phytostabilization of soils affected by a mine spill. This consortium included *Bradyrhizobium* and native *Ochrobactrum* and *Pseudomonas* resistant to heavy metals. The mixture of bacteria increased plant yields, biomass and N content together with a 2–4 fold reduction in metal concentration in shoots (Fig. 4.3). In other work, inoculation of greengram plants with the PGPR and metal-resistant *Bradyrhizobium* sp. RM8 protected the plants from the toxicity of Ni and Zn and consequently increased plant growth (Wani et al. 2007a). These and other associated studies thus suggest that metal-tolerant rhizobia could be used to offset the toxicity of heavy metal to legumes when grown in metal-poisoned soils.

4.6   Engineering Legume–*Rhizobium* Symbiosis for Improving Bioremediation

Recently, genetic engineering has been used to improve the legume–*Rhizobium* symbiosis for bioremediation purposes. The advent of high-throughput methods for DNA sequencing and analysis of gene expression (genomics) and function (proteomics), as well as advances in modeling microbial metabolism *in silico*, provides a global, rational approach to unravel the largely unexplored potentials of microorganisms in facilitating sustainable development (de Lorenzo 2008; Singh et al. 2008; Wood 2008). Since legume plants are used most frequently in metal phytostabilization, strategies aimed to increase metal accumulation in root nodules have been reported. For example, Sriprang et al. (2002) described the construction of a genetically modified strain of *Mesorhizobium huakii* subsp. *rengei* B3 by the expression of a gene encoding metal-binding protein, synthetic tetrameric metallothionein (MTL4) on the cell surface under the control of a bacteroid-specific promoter, *pnifH* or *pnolB*. Inoculation of *Astragalus sinicus* plants with this strain led to a 1.7-fold increase in Cd accumulation inside the nodule. The accumulation of Cd increased not only in nodules, but also in the roots of *A. sinicus* infected by the recombinant rhizobia (threefold increase). In a follow-up study, Sriprang et al. (2003) engineered a *M. huakuii* strain expressing a phytochelatin synthase from *A. thaliana* (cadmium accumulation in nodules increased up to 1.5-fold as compared to non-modified rhizobial partner). The combination of both strategies in the same rhizobial strain slightly increased the accumulation of Cd in *Astragalus* nodules (Ike et al. 2007). The release of genetically modified organisms in Europe is however tightly regulated. So, all these approaches have only been tested under laboratory conditions, since regulatory restrictions have prevented in situ application. Thus, more field trials are necessary to test the abilities of these modified organisms under "real conditions" (Vangrosveld et al. 2009). However, probably the utilization of genetically modified microorganisms (and transgenic plants) in bioremediation is likely to generate less public opposition than using in food production.

4.7 Conclusion

Legumes accumulate heavy metals mainly in roots, showing very low levels of metal translocation to the aerial parts of the plants. Thus, the preferable use of these plants is in the metal phytostabilization. However, it must be taken into account that zinc under acid soil conditions and selenium in legume species can be accumulated over the threshold recommended level for animal feeding. Several strains of rhizobia resistant to metalloids and heavy metals have been isolated from polluted soils. Some of these bacteria are fully effective in nodulation and nitrogen fixation, even in the presence of moderate metal concentrations in the soils. Furthermore, in some cases, a great diversity in rhizobial population is found in contaminated soils, although this fact depends on the *Rhizobium* species considered in the study. Moreover, some other genera of bacteria, distinct from rhizobia, inefficiently nodulate legumes under metal stress. Inoculation of legume plants (preferably native ecotypes from polluted soils) with bacterial consortia resistant to heavy metals (including *Rhizobium*, mycorrhiza, MHB, endophytes, and other PGPR) has proved to be a promising and cost-effective technology for metal phytostabilization, allowing the re-vegetation of metal-contaminated areas with moderate levels of pollution. For this purpose, inoculants with metal resistance ability, PGP-expressing capacity, and having greater competitive potential must be selected. However, research must also focus on how to increase the metal-detoxifying ability of both rhizobia and legumes so that the benefits of bioremediation technologies are realized under metal-stressed environments.

References

Abou-Shanab RA, Ghanem K, Ghanem N, Al-Kolaibe A (2008) The role of bacteria on heavy-metal extraction and uptake by plants growing on multi-metal-contaminated soils. World J Microbiol Biotechnol 24:253–262

Anderson TA, Guthrie EA, Walton BT (1993) Bioremediation. Environ Sci Technol 27:2630–2636

Anton A, Grosse C, Reissmann J, Privyl T, Nies DH (1999) CzcD is a heavy metal ion transporter involved in regulation of heavy metal resistance in *Ralstonia* sp. strain CH34. J Bacteriol 181:6876–6881

Baker AJM (1981) Accumulators and excluder-strategies in the response of plants to heavy metals. J Plant Nutr 3:643–654

Balestrasse KB, Gallego SM, Tomaro ML (2006) Aluminium stress affects nitrogen fixation and assimilation in soybean (*Glycine max* L.). Plant Growth Regul 48:271–281

Bleeker PM, Schat H, Vooijs R, Verkleij JAC, Ernst WHO (2003) Mechanisms of arsenate tolerance in *Cytisus striatus*. New Phytol 157:33–38

Brady DJ, Hecht-Buchholz CH, Asher CJ, Edwards DG (1990) Effects of low activities of aluminium on soybean (*Glycine max*). In: van Beusichem ML (ed.) Early growth and nodulation. Plant nutrition, physiology and applications. Kluwer Academic Publishers, Haren, pp 329–334

Branco R, Chung AP, Morais PV (2008) Sequencing and expression of two arsenic resistance operons with different functions in the highly arsenic-resistant strain *Ochrobactrum tritici* SCII24T. BMC Microbiol 8:95

Brewin NJ (2004) Plant cell wall remodelling in the *Rhizobium*-legume symbiosis. Crit Rev Plant Sci 23:293–316

Broos K, Uyttebroek M, Mertens J, Smolders E (2004) A survey of symbiotic nitrogen fixation by white clover grown on metal contaminated soils. Soil Biol Biochem 36:633–640

Broos K, Beyens H, Smolders E (2005) Survival of rhizobia in soil is sensitive to elevated zinc in the absence of the host plant. Soil Biol Biochem 37:573–579

Callahan DL, Baker AJM, Kolev SD, Wedd AG (2006) Metal ion ligands in hyperaccumulating plants. J Biol Inorg Chem 11:2–12

Carlin A, Shi W, Dey S, Rosen BP (1995) The ars operon of *Escherichia coli* confers arsenical and antimonial resistance. J Bacteriol 177:981–986

Carrasco JA, Armario P, Pajuelo E, Burgos A, Caviedes MA, López R, Chamber MA, Palomares AJ (2005) Isolation and characterisation of symbiotically effective *Rhizobium* resistant to arsenic and heavy metals after the toxic spill at the Aznalcóllar pyrite mine. Soil Biol Biochem 37:1131–1140

Castro IV, Ferreira E, McGrath SP (1997) Effectiveness and genetic diversity of *Rhizobium leguminosarum* biovar trifolii strains in Portuguese soils polluted by industrial effluents. Soil Biol Biochem 29:1209–1213

Cebolla A, Vinardell JM, Kiss E, Oláh B, Roudier F, Kondorosi A, Kondorosi E (1999) The mitotic inhibitor *ccs52* is required for endoreduplication and ploidy-dependent cell enlargement in plants. EMBO J 18:4476–84

Chaintreuil C, Rigault F, Moulin L, Jaffré T, Fardoux J, Giraud E, Dreyfus B, Bailly X (2007) Nickel resistance determinants in *Bradyrhizobium* strains from nodules of the endemic New Caledonia legume *Serianthes calycina*. Appl Environ Microbiol 73:8018–8022

Chaudri AM, McGrath SP, Giller KE (1992) Survival of the indigenous population of *Rhizobium leguminosarum* biovar *trifolii* in soil spiked with Cd, Zn, Cu and Ni salts. Soil Biol Biochem 24:625–632

Cooper JE (2007) Early interactions between legumes and rhizobia: disclosing complexity in a molecular dialogue. J Appl Microbiol 103:1355–1365

Copeland A, Lucas S, Lapidus A, Barry K, Detter JC, Glavina del Rio T, Hammon N, Israni S, Dalin E, Tice H, Pitluck S, Chertkov O, Brettin T, Bruce D, Han C, Tapia R, Gilna P, Schmutz J, Larimer F, Land M, Hauser L, Kyrpides N, Mikhailova NL, Richardson P (2006) Complete sequence of chromosome of *Mesorhizobium* sp. BNC1, NCBI genome project, accession number NC-008254

Copeland A, Lucas S, Lapidus A, Barry K, Glavina del Rio T, Dalin E, Tice H, Pitluck S, Chain P, Malfatti S, Shin M, Vergez L, Schmutz J, Larimer F, Land M, Hauser L, Kyrpides N, Mikhailova N, Reeve W, Richardson P (2007) Complete genome sequence of the *Medicago* microsymbiont *Ensifer* (*Sinorhizobium*) *medicae* strain WSM419, US DOE Joint Genome Institute, accession number CP000740

Dary M, Chamber-Pérez MA, Palomares AJ, Pajuelo E (2010) "In situ" phytostabilisation of heavy metal polluted soils using *Lupinus luteus* inoculated with metal resistant plant-growth promoting rhizobacteria. J Hazard Mater 177:323–330

Day AD, Ludeke KL, Tucker TC (1980) Plant response in vegetative reclamation of mine wastes. In: Vegetative reclamation of mine wastes and tailings in the Southwest. Arizona Mining and Mineral Resources Research Institute, Tucson, pp 1–3

De Hoff P, Hirsch AM (2003) Nitrogen comes down to earth: report from the 5th european nitrogen fixation conference. Mol Plant Microbe Interact 16:371–375

de Lorenzo V (2008) Systems biology approaches to bioremediation. Curr Opin Biotechnol 19:579–589

de Souza MP, Huang CP, Chee N, Terry N (1999) Rhizosphere bacteria enhance the accumulation of selenium and mercury in wetland plants. Planta 209:259–263

Del Rio M, Font F, Almela C, Vélez D, Montoro R, De Haro A (2002) Heavy metals and arsenic uptake by wild vegetation in the Guadiamar river area after the toxic spill of the aznalcóllar mine. J Biotechnol 98:125–137

Den Herder G, Schroeyers K, Holsters M, Goormachtig S (2006) Signaling and gene expression for water-tolerant legume nodulation. CRC Crit Rev Plant Sci 25:367–380

Dong R (2005) Molecular cloning and characterization of a phytochelatin synthase gene, PvPCS1, from *Pteris vittata* L. J Ind Microbiol Biotechnol 32:527–533

Doty SL (2008) Enhancing phytoremediation through the use of transgenics and endophytes. New Phytol 179:318–333

Doty SL, Shang QT, Wilson AM, Moore AL, Newman LA, Strand SE, Gordon MP (2003) Metabolism of the soil and groundwater contaminants, ethylene dibromide and trichloroethylene, by the tropical leguminous tree, *Leucaena leucocephala*. Water Res 37:441–449

Downie JA (2010) The roles of extracellular proteins, polysaccharides and signals in the interactions of rhizobia with legume roots. FEMS Microbiol Rev 34:150–170

Duan GL, Zhu YG, Tong YP, Cai C, Kneer R (2005) Characterization of arsenate reductase in the extract of roots and fronds of chinese brake fern, an arsenic hyperaccumulator. Plant Physiol 138:461–469

Ebbs S, Lau I, Ahner B, Kochian LV (2002) Phytochelatin synthesis is not responsible for Cd tolerance in the Zn/Cd hyperaccumulator *Thlaspi caerulescens* (J & C Presl). Planta 214:635–640

Figueira EM, Gusmão AI, Almeida SI (2005) Cadmium tolerance plasticity in *Rhizobium leguminosarum* bv. *viciae*: glutathione as a detoxifying agent. Can J Microbiol 51:7–14

Freeman JF, Zhang LH, Marcus MA, Fakra S, McGrath SP, Pilon-Smits EAH (2006) Spatial imaging, speciation and quantification of selenium in the hyperaccumulator plants *Astragalus bisulcatus* and *Stanleya pinnata*. Plant Physiol 142:124–134

Frérot H, Lefèbvre C, Gruber W, Collin C, Dos Santos A, Escarre J (2006) Specific interactions between local metallicolous plants improve the phytostabilization of mine soils. Plant Soil 282:53–65

Gage DJ (2004) Infection and invasion of roots by symbiotic nitrogen-fixing rhizobia during nodulation of temperate legumes. Microbiol Mol Biol Rev 68:280–300

Gerhardt KE, Huang XD, Glick BR, Greenberg BM (2009) Phytoremediation and rhizoremediation of organic soil contaminants: potential and challenges. Plant Sci 176:20–30

Geurts R, Fedorova E, Bisseling T (2005) Nod factor signalling genes and their function in the early stages of *Rhizobium* infection. Curr Opin Plant Biol 8:346–352

Ghosh M, Singh SP (2005) A review on phytoremediation of heavy metals and utilization of its byproducts. Appl Ecol Environ Res 3:1–18

Gibson KE, Kobayashi H, Walker GC (2008) Molecular determinants of a symbiotic chronic infection. Annu Rev Genet 42:413–441

Giller KE, Nussbaun R, Chaudri AM, McGrath SP (1993) *Rhizobium meliloti* is less sensitive to heavy-metal contamination in soil than *R. leguminosarum* bv. *trifolii* or *R. loti*. Soil Biol Biochem 25:273–278

Giller KE, Witter E, McGrath SP (1998) Toxicity of heavy metals to microorganisms and microbial processes in agricultural soils: a review. Soil Biol Biochem 30:1389–1414

Glick BR (2003) Phytoremediation: synergistic use of plants and bacteria to clean up the environment. Biotechnol Adv 21:383–393

González RC, González-Chávez MC (2006) Metal accumulation in wild plants surrounding mining wastes. Environ Pollut 144:84–92

Graham PH, Vance CP (2003) Legumes: importance and constraints to greater use. Plant Physiol 131:872–877

Gupta AK, Dwivedi S, Sinha S, Tripathi RD, Rai UN, Singh SN (2007) Metal accumulation and growth performance of *Phaseolus vulgaris* grown in ash amended soil. Bioresour Technol 98:3404–3407

Gusmão AI, Caçoilo S, Figueira EM (2006) Glutathione-mediated cadmium sequestration in *Rhizobium leguminosarum*. Enzyme Microb Technol 39:763–769

Hinsinger P, Gobran GR, Gregory PJ, Wenzel WW (2005) Rhizosphere geometry and heterogeneity arising from root-mediated physical and chemical processes. New Phytol 168:293–303

Hunter WJ, Kuykendall LD, Manter DK (2007) *Rhizobium selenireducens* sp. nov.: a selenite-reducing *a*-Proteobacteria isolated from a bioreactor. Curr Microbiol 55:455–460

Ike A, Sriprang R, Ono H, Murooka Y, Yamashita M (2007) Bioremediation of cadmium contaminated soil using symbiosis between leguminous plant and recombinant rhizobia with the MTL4 and the PCS genes. Chemosphere 66:1670–1676

Ingle RA, Mugford ST, Rees JB, Campbell MM, Smith JA (2005) Constitutively high expressed of the histidine biosynthetic pathway contributes to nickel tolerance in hyperaccumulator plants. Plant Cell 17:2089–2106

Jones KM, Kobayashi H, Davies BW, Taga ME, Walker GC (2007) How rhizobial symbionts invade plants: the *Sinorhizobium-Medicago* model. Nat Rev Microbiol 5:619–633

Jorgensen KS (2007) *In situ* bioremediation. Adv Appl Microbiol 61:285–305

Khan AG (2005) Role of soil microbes in the rhizospheres of plants growing on trace metal contaminated soils in phytoremediation. J Trace Elem Med Biol 18:355–364

Khan MS, Zaidi A, Wani PA, Oves M (2009) Role of plant growth promoting rhizobacteria in the remediation of heavy metal contaminated soils. Environ Chem Lett 7:1–19

Kidd P, Barceló J, Bernal MP, Navari-Izzo F, Poschenrieder C, Shilev S, Clemente R, Monterroso C (2009) Trace element behaviour at the root-soil interface: Implications in phytoremediation. Environ Exp Bot 67:243–259

Kopittke PM, Dart PJ, Menzies NW (2007) Toxic effects of low concentrations of Cu on nodulation of cowpea (*Vigna unguiculata*). Environ Pollut 145:309–315

Kuiper I, Lagendijk EL, Bloemberg GV, Lugtenberg BJJ (2004) Rhizoremediation: a beneficial plant–microbe interaction. Mol Plant Microbe Interact 17:6–15

Lafuente A, Pajuelo E, Caviedes MA, Rodríguez-Llorente ID (2010) Reduced nodulation in alfalfa induced by arsenic correlates with altered expression of early nodulins. J Plant Physiol 167:286–291

Lakzian A, Murphy P, Turner A, Beynon JL, Giller KE (2002) *Rhizobium leguminosarum* bv. *viciae* populations in soils with increasing heavy metal contamination: abundance, plasmid profiles, diversity and metal tolerance. Soil Biol Biochem 34:519–529

Lasat MM (2000) Phytoextraction of metals from contaminated soil: a review of plant/soil/metal interaction and assessment of pertinent agronomic issues. J Hazard Sub Res 2:1–21

Le Duc DL, Terry T (2005) Phytoremediation of toxic trace elements in soil and water. J Ind Microbiol Biotechnol 32:514–520

Ma LQ, Komar KMM, Tu C, Zhang W, Cai Y, Kennelley ED (2001) A fern that hyperaccumulates arsenic. Nature 409:579–581

Ma Y, Rajkumar M, Freitas H (2009) Isolation and characterization of Ni mobilizing PGPB from serpentine soils and their potential in promoting plant growth and Ni accumulation by *Brassica* spp. Chemosphere 75:719–725

Malik A (2004) Metal bioremediation through growing cells. Environ Int 30:261–278

Mandal SM, Pati BR, Das AK, Ghosh AK (2008) Characterization of a symbiotically effective *Rhizobium* resistant to arsenic: isolated from the root of *Vigna mungo* (L.) hepper grown in an arsenic-contaminated field. J Gen Appl Microbiol 54:93–99

Mejáre M, Bülow L (2001) Metal-binding proteins and peptides in bioremediation and phytoremediation of heavy metals. Trends Biotechnol 19:67–73

Méndez MO, Maier RM (2008) Phytostabilisation of mine tailings in arid and semiarid environments: an emerging remediation technology. Environ Health Perspect 116:278–283

Moore CM, Helmann JD (2005) Metal ion homeostasis in *Bacillus subtilis*. Curr Opin Microbiol 8:188–195

Moreno-Jiménez E, Peñalosa JM, Manzano R, Carpena-Ruíz RO, Gamarra R, Esteban E (2009) Heavy metals distribution in soils surrounding an abandoned mine in NW Madrid (Spain) and their transference to wild flora. J Hazard Mat 162:854–859

Nawapan S, Charoenlap N, Charoenwuttitam A, Saenkham P, Mongkolsuk S, Vattanaviboon P (2009) Functional and expression analyses of the *cop* operon, required for copper resistance in *Agrobacterium tumefaciens*. J Bacteriol 191:5159–5168

Nies DH (2003) Efflux-mediated heavy metal resistance in prokaryotes. FEMS Microbiol Rev 27:313–339

Nies DH, Silver S (2007) Molecular microbiology of heavy metals. Springer, Berlin/Heidelberg/New York

Nies DH, Koch S, Wachi S, Peitzsch N, Saier MH Jr (1998) CHR, a novel family of prokaryotic proton motive force-driven transporters probably containing chromate/sulfate antiporters. J Bacteriol 180:5799–5802

Nucifora G, Chu L, Misra TK, Silver S (1989) Cadmium resistance from *Staphylococcus aureus* plasmid pI258 cadA gene results from a cadmium-efflux ATPase. Proc Natl Acad Sci USA 86:3544–3548

Oldroyd ED, Downie JA (2006) Nuclear calcium changes at the core of symbiosis signalling. Curr Opin Plant Biol 9:351–357

Oldroyd ED, Downie JA (2008) Coordinating nodule morphogenesis with rhizobial infection in legumes. Annu Rev Plant Biol 59:519–546

Oremland SO, Stolz JF (2003) The ecology of arsenic. Science 300:939–943

Ortega-Villasante C, Rellán-Álvarez R, Del Campo FF, Carpena-Ruíz RO, Hernández LE (2005) Cellular damage induced by cadmium and mercury in *Medicago sativa*. J Exp Bot 56:2239–2251

Outten FW, Outten CE, Hale J, O'Halloran TV (2000) Transcriptional activation of an *Escherichia coli* copper efflux regulon by the chromosomal MerR homologue, cueR. J Biol Chem 275:31024–31029

Pajuelo E, Rodríguez-Llorente ID, Dary M, Palomares AJ (2008) Toxic effects of arsenic on *Sinorhizobium–Medicago sativa* symbiotic interaction. Environ Pollut 154:203–211

Pastor J, Hernandez AJ, Prieto N, Fernandez-Pascual M (2003) Accumulating behaviour of *Lupinus albus* L. growing in a normal and a decalcified calcic luvisol polluted with Zn. J Plant Physiol 160:1457–1465

Patel PC, Goulhen F, Boothman C, Gault AG, Charnock JM, Kalia K, Lloyd JR (2007) Arsenate detoxification in a Pseudomonad hypertolerant to arsenic. Arch Microbiol 187:171–183

Pence NS, Larsen PB, Ebbs SD, Letham DLD, Lasat MM, Garvin DF, Eide D, Kochian LV (2000) The molecular physiology of heavy metal transport in the Zn and Cd hyperaccumulator *Thlaspi caerulescens*. PNAS 97:4956–4960

Pickering IJ, Wright C, Bubner B, Ellis D, Persans NW, Yu EY, George GN, Prince RC, Salt DE (2003) Chemical form and distribution of selenium and sulphur in the selenium hyperaccumulator *Astragalus bisulcatus*. Plant Physiol 131:1460–1467

Piha MI, Vallack HW, Reeler BM, Michael N (1995) A low input approach to vegetation establishment on mine and coal ash wastes in semi-arid regions. 1. Tin mine tailings in Zimbabwe. J Appl Ecol 32:372–381

Pilon-Smits E (2005) Phytoremediation. Annu Rev Plant Biol 56:15–39

Prasad MNV, Freitas HM (2003) Metal hyperaccumulation in plants. Biodiversity prospecting for phytoremediation technology. Electr J Biotechnol 6:287–321

Radutoiu S, Madsen LH, Madsen EB, Felle HH, Umehara Y, Grønlund M, Sato S, Nakamura Y, Tabata S, Sandal N, Stougaard J (2003) Plant recognition of symbiotic bacteria requires two LysM receptor-like kinases. Nature 425:585–592

Reeve WG, O'Hara G, Chain P, Ardley J, Braeu L, Nandesena K, Tiwari R, Malfatti S, Kiss H, Lapidus A, Copeland A, Nolan M, Land M, Ivanova N, Mavromatis K, Markowitz V, Kyrpides NC, Melino V, Denton M, Yates R, Howieson J (2010) Complete genome sequence of *Rhizobium leguminosarum* bv. *trifolii* strain WSM2304, an effective microsymbiont of the South American clover *Trifolium polymorphum*. Stand Genomic Sci 2:66–76

Reeves RD, Baker AJM (2000) Metal accumulating plants. In: Raskin I, Ensley B (eds.) Phytoremediation of toxic metals: using plants to clean up the environment. Wiley, New York, pp 193–229

Requena N, Perez-Solis E, Azcón-Aguilar C, Jeffries P, Barea JM (2002) Management of indigenous plant-microbe symbioses aids restoration of desertified ecosystems. Appl Environ Microbiol 67:495–498

Rivera-Becerril F, Calantzis C, Turnau K, Caussanel JP, Belimov AA, Gianinazzi S, Strasser RJ, Gianinazzi-Pearson V (2002) Cadmium accumulation and buffering of cadmium-induced stress by arbuscular mycorrhiza in three *Pisum sativum* L. genotypes. J Exp Bot 53:1177–1185

Rodríguez-Llorente ID, Gamane D, Lafuente A, Dary M, El Hamdaoui A, Delgadillo J, Doukkali K, Caviedes MA, Pajuelo E (2010) Cadmium biosorption of the metal resistant *Ochrobactrum cytisi* Azn6.2. Eng Life Sci 10:49–56

Rodríguez-Navarro DN, Dardanelli MS, Ruíz-Sáinz JE (2007) Attachment of bacteria to the roots of higher plants. FEMS Microbiol Lett 272:127–136

Roth LE, Stacey G (1989) Bacterium release into host cells of nitrogen-fixing soybean nodules: the symbiosome membrane comes from three sources. Eur J Cell Biol 49:13–23

Sá-Pereira P, Rodrigues M, Videira e, Castro I, Simões F (2007) Identification of an arsenic resistance mechanism in rhizobial strains. World J Microbiol Biotechnol 23:1351–1356

Sá-Pereira P, Rodrigues M, Simões F, Domingues L, Videira e, Castro I (2009) Bacterial activity in heavy metals polluted soils: metal efflux systems in native rhizobial strains. Geomicrobiol J 26:281–288

Sawada H, Kuykendall LD, Young JM (2003) Changing concepts in the systematics of bacterial nitrogen-fixing legume symbionts. J Gen Appl Microbiol 49:155–179

Sharples JM, Meharg AA, Chambers SM, Cairney JWG (2000) Evolution: symbiotic solution to arsenic contamination. Nature 404:951–952

Sheng FX, Xia JJ (2006) Improvement of rape (*Brassica napus*) plant growth and cadmium uptake by cadmium-resistant bacteria. Chemosphere 64:1036–1042

Silver S (1996) Bacterial resistances to toxic metal ions: a review. Gene 179:9–19

Silver S, Phung LT (1996) Bacterial heavy metals resistance: new surprises. Annu Rev Microbiol 50:753–789

Silver S, Phung LT (2005) A microbial view of the periodic table: genes and proteins for toxic inorganic ions. J Ind Microbiol Biotechnol 32:587–605

Singh S, Kang SH, Mulchandani A, Chen W (2008) Bioremediation: environmental clean-up through pathway engineering. Curr Opin Biotechnol 19:437–444

Sprent JI (2007) Evolving ideas of legume evolution and diversity: a taxonomic perspective on the occurrence of nodulation. New Phytol 174:11–25

Sriprang R, Hayashi M, Yamashita M, Ono II, Saeki K, Murooka Y (2002) A novel bioremediation system for heavy metals using the symbiosis between leguminous plant and genetically engineered rhizobia. J Biotechnol 99:279–293

Sriprang R, Hayashi M, Ono H, Takagi M, Hirata K, Murooka Y (2003) Enhanced accumulation of Cd2+ by *Mesorhizobium* transformed with a gene for phytochelatin synthase from *Arabidopsis*. Appl Environ Microbiol 69:1791–1796

Stephanie M, Hélène F, Céline V, Antoine G, Karine H, Lucette M, Brigitte B, Claude L, José E, Jean-Claude CM (2011) *Anthyllis vulneraria/Mesorhizobium metallidurans*, an efficient symbiotic nitrogen fixing association able to grow in mine tailings highly contaminated by Zn. Pb Cd Plant Soil. doi:10.1007/s11104-010-0705-7

Trujillo ME, Willems A, Abril A, Planchuela AM, Rivas R, Luden D, Mateos PF, Martínez-Molina E, Velázquez E (2005) Nodulation of *Lupinus albus* by strains of *Ochrobactrum lupini* sp. nov. Appl Environ Microbiol 71:1318–1327

van Aken B (2008) Transgenic plants for phytoremediation: helping nature to clean up environmental pollution. Trends Biotechnol 26:225–227

van Batenburg FHD, Jonker R, Kijne JW (1986) *Rhizobium* induces marked root hair curling by redirection of tip growth: a computer simulation. Physiol Plant 66:476–480

van Brussel AAN, Bakhuizen R, Van Sprosen PC, Spaink HP, Tak T, Lutenberg BJJ, Kijne JW (1992) Induction of preinfection threads structures in the leguminous host plant by mitogenic lipo-oligosaccharides of *Rhizobium*. Science 257:70–72

Vangrosveld J, Herzig R, Weyens N, Boulet J, Adriansen K, Ruttens A, Thewys T, Vassiley A, Meers E, Nehnevajova E, van der Lelie D, Mench M (2009) Phytoremediation of contaminated soils and groundwater: lessons from the field. Environ Sci Pollut Res 16:765–794

Vázquez S, Agha R, Granado A, Sarro MJ, Esteban E, Peñalosa JM, Carpena RO (2006) Use of white lupin plant for phytostabilization of Cd and As pollute acid soil. Water Air Soil Pollut 177:349–365

Velázquez E, García-Fraile P, Ramírez-Bahena MH, Rivas R, Martínez-Molina E (2010) Bacteria involved in nitrogen-fixing legume symbiosis: current taxonomic perspective. In: Khan MS, Zaidi A, Musarrat J (eds.) Microbes for legume improvement. Springer, Wien

Vidal C, Chantreuil C, Berge O, Mauré L, Escarré J, Béna BB, Cleyet-Marel JC (2009) *Mesorhizobium metallidurans* sp. nov., a metal resistant symbiont of *Anthyllis vulneraria* growing on metallicolous soil in Languedoc, France. Int J Syst Evol Microbiol 59:850–855

Volesky B (2007) Biosorption and me. Water Res 41:4017–4029

Wang J, Zhao F-J, Meharg AA, Raab A, Feldmann J, McGrath SP (2002) Mechanisms of arsenic hyperaccumulation in *Pteris vittata*. Uptake kinetics, interactions with phosphate, and arsenic speciation. Plant Physiol 130:1552–1561

Wani PA, Khan MS, Zaidi A (2006) An evaluation of the effects of heavy metals on the growth, seed yield and grain protein of lentil in pots. Ann Appl Biol (Suppl TAC) 27:23–24

Wani PA, Khan MS, Zaidi A (2007a) Effect of metal tolerant plant growth promoting *Bradyrhizobium* sp. (vigna) on growth, symbiosis, seed yield and metal uptake by greengram plants. Chemosphere 70:36–45

Wani PA, Khan MS, Zaidi A (2007b) Impact of heavy metal toxicity on plant growth, symbiosis, seed yield and nitrogen and metal uptake in chickpea. Aust J Exp Agric 47:712–720

Wani PA, Khan MS, Zaidi A (2008a) Effect of heavy metal toxicity on growth, symbiosis, seed yield and metal uptake in pea grown in metal amended soil. Bull Environ Contam Toxicol 81:152–158

Wani PA, Khan MS, Zaidi A (2008b) Chromium reducing and plant growth promoting *Mesorhizobium* improves chickpea growth in chromium amended soil. Biotechnol Lett 30:159–163

Wani PA, Zaidi A, Khan MS (2009) Chromium reducing and plant growth promoting potential of *Mesorhizobium* species under chromium stress. Biorem J 13:121–129

Wei G, Fan L, Zhu W, Fu Y, Yu F, Tang M (2009) Isolation and characterization of the heavy metal resistant bacteria CCNWRS33-2 isolated from root nodule of *Lespedeza cuneata* in gold mine tailings in China. J Hazard Mat 162:50–56

Weir BS (2009) The current taxonomy of rhizobia. New Zealand rhizobia website. http://www.rhizobia.co.nz/taxonomy/rhizobia.html

Wood TK (2008) Molecular approaches in bioremediation. Curr Opin Biotechnol 19:572–578

Yang HC, Cheng J, Finan M, Rosen BP, Bhattacharjee H (2005) Novel pathway for arsenic detoxification in the legume symbiont *Sinorhizobium meliloti*. J Bacteriol 187:6991–6997

Zaidi S, Usmani S, Singh BR, Musarrat J (2006) Significance of *Bacillus subtilis* strain SJ 101 as a bioinoculant for concurrent plant growth promotion and nickel accumulation in *Brassica juncea*. Chemosphere 64:991–997

Zhang S, Chen M, Li T, Xu X, Deng L (2010) A newly found cadmium accumulator-*Malva sinensis* Cavan. J Hazard Mat 173:705–709

Zhengwei Z, Fang W, Lee HY, Yang Z (2005) Response of *Azhorhizobium caulinodans* to cadmium stress. FEMS Microbiol Ecol 54:455–461

Zhuang X, Chen J, Shim H, Bai Z (2007) New advances in plant growth-promoting rhizobacteria for bioremediation. Environ Int 33:406–413

Zurdo-Piñeiro JL, Rivas R, Trujillo ME, Vizcaíno N, Carrasco JA, Chamber MA, Palomares AJ, Mateos PF, Martínez-Molina E, Velázquez E (2007) *Ochrobactrum cytisi* sp. nov., isolated from nodules of *Cytisus scoparius* in Spain. Int J Syst Evol Microbiol 57:784–788

Chapter 5
Importance of Arbuscular Mycorrhizal Fungi in Phytoremediation of Heavy Metal Contaminated Soils

Arshad Javaid

Abstract Heavy metal contamination caused either by natural processes or by human activities is one of the most serious environmental problems. Physicochemical methods such as soil washing, excavation, and reburial for heavy metal removal from contaminated soils are expensive and disruptive. Phytoremediation in contrast is a low-cost environmentally friendly and potentially effective technology for the reclamation of polluted soils. Arbuscular mycorrhizal (AM) fungi provide an attractive system to advance plant-based environmental clean-up. They are critical in the establishment and fitness of plants in severely disturbed sites, including those contaminated by heavy metals. Mycorrhizal plants play an important role both in phytostabilization and phytoextraction. Strategies used by AM-fungi in phytostabilization includes immobilization of metals by precipitating polyphosphate granules in the soil, compounds secreted by the fungus, adsorption to fungal cell walls, and chelation of metals inside the fungus. By phytoextraction, AM-fungi make heavy metals more available for plant absorption, help plants to accumulate metals, facilitate plant growth and biomass production, and increase plant tolerance to metals. Since tolerance to heavy metals varies with the fungal genotype, efficacy of the hyperaccumulators in phytoremediation can be best exploited by selecting most suitable mycorrhizal culture. The importance of AM-fungi in enhancing phytoremediation of metal-contaminated soil is highlighted.

Keywords Arbuscular mycorrhizal fungi • Phytoremediation • Phytostabilization • Phytoextraction

A. Javaid (✉)
Institute of Agricultural Sciences, University of the Punjab, Quai-e-Azam Campus,
Lahore, Pakistan
e-mail: arshadjpk@yahoo.com

M.S. Khan et al. (eds.), *Biomanagement of Metal-Contaminated Soils*,
Environmental Pollution 20, DOI 10.1007/978-94-007-1914-9_5,
© Springer Science+Business Media B.V. 2011

5.1 Introduction

Heavy metals are a group of 53 elements with density higher than 5 g/cm^3 (Holleman and Wiberg 1985). Although some metals are essential for plant and animals, many are toxic at high concentrations. Trace elements like iron (Fe), copper (Cu), nickel (Ni), zinc (Zn), and manganese (Mn) are essential for normal growth and development of plants and are required in electron transfer, in numerous enzyme catalyzed or redox reactions, and have structural function in nucleic acids (Zenk 1996). Others like Cu and Zn are involved in plant growth, flowering, and seed production, especially when their availability is very low (Vamerali et al. 2010). In contrast, some heavy metals, such as, mercury (Hg), cadmium (Cd), arsenic (As), and lead (Pb) are not essential (Mertz 1981). Heavy metals occur mainly in terrestrial or aquatic ecosystems although they can be also emitted into the atmosphere (Gohre and Paszkowski 2006). The presence of heavy metals in soil may be natural or due to anthropogenic activities primarily associated with industrial processes such as mining, metallurgical and energy production, or agricultural practices (Cho et al. 2009; Gong et al. 2010). Among various organic and inorganic pollutants, greater concern worldwide about soil contamination is regarding heavy metals. For example, in the European Union, contamination by metals accounts for >37% of the cases, followed by mineral oil (33.7%), polycyclic aromatic hydrocarbons (13.3%), and others (Vamerali et al. 2010). In uncontaminated soils, heavy metal concentrations vary in magnitudes, but on average, the order of metal concentrations is: Cd 0.1– 0.5 ppm, Zn 80 ppm, and Pb 15 ppm. However, in polluted soil, the concentrations were: Cd >14,000 ppm, Zn >20,000 ppm, and Pb >7,000 ppm (Gohre and Paszkowski 2006). Contamination is often highly localized in industrialized countries, and hence, the pressure to use contaminated land and water for food production or for human consumption is minimal (Kramer 2005). However, such types of contaminations are widespread in Eastern Europe, and are increasingly recognized as a major threat in many parts of the developing world, especially in China and India (Cheng 2003; Meharg 2004).

Heavy metal pollution of the biosphere has received huge attention due to its toxicity, abundance, persistence, and subsequent accumulation in environment (Dong et al. 2010). Long-term use of industrial and municipal wastewater on agricultural lands contributes significantly to the buildup of these metals in soils and plants (Mapanda et al. 2005; Sharma et al. 2007), which is of course a serious concern. Excessive accumulation of heavy metals in agricultural soils following uptake by plants may result in poor food quality and safety (Muchuweti et al. 2006; Zhu et al. 2008; Gupta et al. 2010). According to their distinct chemical and physical properties, three different molecular mechanisms of heavy metal toxicity are recognized: (1) production of reactive oxygen species by autoxidation and Fenton reaction (Fe, Cu), (2) blocking of essential functional groups in biomolecules (Cd, Hg), and (3) displacement of essential metal ions from biomolecules (Schutzendubel and Polle 2002). Phytotoxicity is mainly associated with nonessential metals like Cd, As, Cr, and Pb, which generally have very low toxicity thresholds (Clemens 2006) and lower values for hyperaccumulation (especially for Cd) than the other metals

(Vamerali et al. 2010). At elevated concentrations, heavy metals interfere with essential enzymatic activities by modifying protein structure or by replacing a vital element resulting in deficiency symptoms (Gohre and Paszkowski 2006). The plasma membrane is particularly vulnerable to heavy metal toxicity since membrane permeability and thus functionality can be affected by alterations of important membrane intrinsic proteins such as H+-ATPases (Hall 2002).

5.2 Arbuscular Mycorrhizal Fungi: An Overview

Arbuscular mycorrhizal (AM) fungi are indigenous soil-borne microorganisms and are integral functioning parts of plant that live in mutualistic association with the roots of about 80% of all terrestrial land plants (Smith and Read 1997). AM-fungi are obligate biotrophs because they rely on their host plant to proliferate and survive. All AM-fungi belong to the Glomeromycota, an ancient group of fungi that was present about 450 million years ago, and were instrumental for plants to colonize land (Redecker et al. 2000; Schüssler et al. 2001).The phylum Glomeromycota comprises a single class Glomeromycetes having four orders and 13 families. Based on morphological and molecular characteristics, 19 genera like, *Acaulospora, Ambispora, Archaeospora, Cetraspora, Dentiscutata, Diversispora, Entrophospora, Fuscutata, Geosiphon, Gigaspora, Glomus, Intraspora, Kuklospora, Otospora, Pacispora, Paraglomus, Racocetra, Scutellospora,* and *Quatunica* comprising more than 200 species are recognized (Manoharachary et al. 2010). The AM-fungi consists of an internal phase inside the root and an external phase, or extraradical mycelium phase, which can form an extensive network within the soil. A key feature of these fungi is the formation of specialized haustoria-like structure within the root cortical cells called arbuscules where metabolite exchange takes place between the fungus and host cytoplasm (Parniske 2000). Characteristic vesicles usually also form later as terminal or intercalary swellings in the cortical cells and function as nutrient storage organs or as propagules in root fragments.

Arbuscular mycorrhizal fungi are important in natural and managed ecosystems due to their nutritional and nonnutritional benefits to their symbiotic partners. These symbionts can act as biofertilizers, bioprotectants, or biodegraders (Xavier and Boyetchko 2002).The fungi assist the host plant in the uptake of nutrients especially P in exchange for C substrates from host plant photosynthesis (Javaid 2009; Sharda and Koide 2010) through extensive and highly branched extra radical hyphae. The mycorrhizal colonization also improves plant N nutrition, which, however, has not been fully appreciated until recently (Read and Perez-Moreno 2003). Uptake of other nutrients such as Na, K, Mg, Ca, B, Fe, Mn, Cu, and Zn is influenced by mycorrhizal colonization (Cardoso and Kuyper 2006; Meding and Zasoski 2008). Enhanced mineral nutrition helps plants in increasing chlorophyll content and hence, higher photosynthetic rate (Feng et al. 2002). AM- fungi also improve soil structure by forming soil aggregates (Rillig and Mummey 2006) and thus result in enhanced plant growth and productivity. Root colonization by AM-fungi also induce important physiological and biochemical changes in the host plant, enabling it to better overcome biotic (Ozgonen and Erkilic 2007) and abiotic stresses such as metal toxicity (Leung et al. 2007), salinity, high soil temperature, drought (Khalvati et al. 2010), and allelopathy (Javaid and Bajwa 1999; Javaid 2007, 2008; Barto et al. 2010). AM-fungi provide other benefits as well to plants including enhanced enzyme production (Adriano-Anaya et al. 2006), synthesizing secondary metabolites (Schliemann et al. 2008), enhancing symbiotic N_2 fixation by symbiotic (Javaid et al. 1993, 1994; Kaschuk et al. 2010; Ray and Valsalakumar 2010) and associative N_2-fixing bacteria (Saini et al. 2004), and osmotic adjustment under drought stress (Ruiz-Lozano 2003). The AM symbiosis has a significant impact on plant interactions with other microorganisms and there are many documented reports on its role in controlling plant diseases, especially soil-borne plant pathogens including fungi, nematodes, and bacteria (Khaosaad et al. 2007; Oyekanmi et al. 2007).

5.3   AM-Fungi-Assisted Phytoremediation

Physicochemical methods for heavy metals removal from contaminated soils are expensive and disruptive (Gardea-Torresdey et al. 2005). The most highly developed remediation methods for metal-contaminated soils are physical or chemical, such as soil washing, excavation, and reburial. However, physical displacement, transport, and storage or alternatively soil washing are expensive procedures and leave a site behind, which may be devoid of any soil microflora. Recently, phytoremediation (the use of plants to remediate polluted soils) has emerged as a more reliable alternative (Cho et al. 2009; Franco-Hernandez et al. 2010). This biological approach is based on the capability of some plant species to take up and to concentrate pollutants in their roots and shoots, and is often simpler in design and is inexpensive (Petra et al. 2009; Rai 2010). Phytostabilization and phytoextraction are the most reliable categories of phytoremediation for heavy metals (Gohre and Paszkowski 2006). Phytostabilization is used to provide a cover of vegetation for a moderately to heavily contaminated site to prevent wind and water erosion. It is often performed using species from plant communities growing on local contaminated sites. These plants possess tolerance to the contaminant metals, develop an extensive root system, provide good soil cover, and ideally immobilize the contaminants in the rhizosphere (Kramer 2005; Dary et al. 2010). In phytostabilization, plants must be able to develop extended root systems and keep the translocation of metals from roots to shoots as low as possible (Mendez and Maier 2008). Phytoextraction involves the cultivation of tolerant plants that concentrate soil contaminants in their above-ground tissues. At the end of the growth period, plant biomass is harvested. Contaminated biomass may be used for energy production, whereas remaining ashes are dumped, included in construction materials, or subjected to metal extraction (Brooks et al. 1998). Plants most suitable for phytoextraction are able to hyperaccumulate contaminants, possess tolerance to these chemicals, have a high biomass, and have a short growing cycle (Cho et al. 2009). All these traits are difficult to combine, and there are basically two available phytoextraction strategies, which make use of hyperaccumulators or biomass plant species (Vamerali et al. 2010).

The majority of hyperaccumulators present a slow growth rate leading to a low annual biomass yield. Moreover, the use of hyperaccumulators species in continuous phytoextraction process is limited by the low bioavailability of these pollutants for uptake by roots (Peer et al. 2005). Many efforts have been made to improve phytoextraction of heavy metals. The use of both natural and synthetic chelating agents has been practiced in phytoextraction to increase bioavailability, uptake and translocation of metals (Quartacci et al. 2006). Among various synthetic chelators, ethylenediamine tetraacetate (EDTA) has been tested more intensively (Barrutia et al. 2010; Zaier et al. 2010). In this context, AM-fungi provide an attractive system to advance plant-based environmental clean-up. In some cases, mycorrhizal plants can show enhanced heavy metals uptake and root-to-shoot transport (phytoextraction) while in other cases AM-fungi contribute to heavy metal immobilization (phytostabilization) within the soil (Gohre and Paszkowski 2006).

5.3.1 Occurrence of AM-Fungi in Heavy Metal Contaminated Soil

Arbuscular mycorrhizal fungi are reported to be present on the roots of plants growing on heavy metal-contaminated soils and play an important role in metal accumulation and tolerance (Hildebrandt et al. 1999; Gaur and Adholeya 2004). For example, AM fungal colonization was detected in most of the plants growing on mining sites in Chenzhou City, Hunan Province, Southern China (Leung et al. 2007). Zak et al. (1982) reported 390–2070 spores 100g^{-1} substratum in mine spoils of Canada. Weissenhorn et al. (1995a, b) observed high levels of mycorrhizal colonization in agricultural soils contaminated with metals originating from smelter and sludge amendments. In ultramafic soils in South Africa, naturally occurring Ni-hyperaccumulating plants of the Asteraceae were heavily colonized by AM-fungi (Turnau and Mesjasz-Przybylowicz 2003). Heavy metals have shown positive, negative, or neutral effects on mycorrhizal colonization in soil or culture solution (Chen et al. 2005). For example, Diaz et al. (1996) showed that mycorrhizal infection of *Lygeum spartum* L. and *Anthyllis cytisoides* L. was not affected by Zn or Pb in soil. Similarly, Weissenhorn et al. (1995a) observed no correlation between AM abundance in maize (*Zea mays*) and the degree of metal (Cd, Ni, Zn, Cu, Pb, and Mn) pollution in a field trial. On the other hand, mycorrhizal colonization and growth of external hyphae were inhibited by sewage sludge-contaminated soil containing Pb, Zn, and Cd (Del Val et al. (1999). In a similar study, Chao and Wang (1990) found that mycorrhizal infection rate of maize was reduced by the addition of Pb, Cr, Ni, Zn, Cu, and Cd. Recently, Khade and Adholeya (2009) identified a total of six species of AM-fungi belonging to two genera, *Glomus* and *Scutellospora,* from soils adjoining Kanpur Tanneries, Uttar Pradesh, India. AM- fungi was maximum in the non-contaminated site (six species) compared to the metal-contaminated site (four species). They further reported that for a particular plant species, the root colonization levels and spore density were generally higher in chromium-contaminated soil compared to non-contaminated soils. The authors attributed higher AM colonization rates in metal-contaminated soil to favorable time for spore germination and rapid colonization of emerging roots of the plants. Bedini et al. (2010) found no spores in the rhizosphere soil of the dominant plant species of Sacca San Biagio, a polluted ash dump island, characterized with high levels of Cu, Pb, and Zn. In contrast, Tonin et al. (2001) found that Cd- and Zn-polluted soil enhanced mycorrhizal diversity index of clover (*Trifolium repens* L.) roots. Similarly, Turnau et al. (1996) reported that metal-tolerant *Oxalis acetosella* L. plants colonizing acid forest soils treated with Cd, Zn, and Pb containing industrial dust showed higher AM colonization than nontreated soils.

Spores and pre-symbiotic hyphae are generally sensitive to heavy metals in the absence of plants. Shalaby et al. (2003) isolated spores from heavy metal-polluted and unpolluted soils and assessed their germination and subsequent hyphal growth in the presence of Zn, Pb, and Cd. Germination and hyphal growth were inhibited by heavy metals in all cases. However, spores from polluted soils were more tolerant to elevated concentrations of the three metals than spores from uncontaminated soils. This naturally occurring resistance is likely due to phenotypic plasticity rather than genetic changes in the spores, because tolerance was lost after one generation in the absence of heavy metals. Studies examined spore counts and colonization efficiency of sewage sludge-treated sites and revealed that spores tolerant to increased heavy metal application readily colonized host roots despite low spore counts (Del Val et al. 1999; Jacquot-Plumey et al. 2001). In contrast, Hua et al. (2009) demonstrated that AM-fungi isolated from polluted soils were not effective than those from unpolluted soils when grown in symbiosis with tobacco (*Nicotiana tabacum*).

Tolerance to heavy metals varies with the fungal genotype (Biro et al. 2009). Generally, species of genus *Glomus* are predominant in the rhizosphere of plants growing in heavy metal-contaminated soils (Khade and Adholeya 2009; Bedini et al. 2010). Study on the effect of Pb, Zn, and Cd on pre-symbiotic (spore germination and hyphal extension), and symbiotic (extraradical mycelial growth and sporulation) life stages of two *Glomus* species demonstrated *Glomus intraradices* to be more tolerant to each of the metals than *Glomus etunicatum* (Pawlowska and Charvat 2004).

5.3.2   Does Mycorrhizal Plants Exhibit Enhanced Tolerance to Heavy Metals?

Mycorrhizal interactions with plants are widely recognized in enhancing plant growth in severely disturbed sites, including those contaminated with heavy metals (Leyval et al. 1997; Gaur and Adholeya 2004). For example, Hildebrandt et al. (1999) reported that mycorrhizae improved the plants of *Viola calaminaria* (DC.) Lejeune tolerance to Zn and Pb stress in polluted soils. However, it remains unclear whether the observed effect was a consequence of improved nutrition or the fungal impact on the plant's physiological stress reactions (Jentschke and Godbold 2000). Chen et al. (2005) showed that mycorrhizae enhanced significantly shoot P concentration and shoot biomass under elevated Pb concentration, suggesting that higher efficiency of P acquisition by mycorrhizae might be a mechanism of plant tolerance to Pb stress. Hildebrandt et al. (1999) reported that a *Glomus* isolate Br1 obtained from roots of *V. calaminaria* grown on heavy metal-contaminated soil colonized maize, alfalfa (*Medicago sativa*), barley (*Hordeum vulgare*), and *V. calaminaria* and allowed each plant species to complete their life cycle on highly polluted soil. An isolate of *G. intraradices,* isolated from a non-contaminated soil also increased growth, but to a lower extent, whereas non-colonized plants died on the same soil. It could be attributed to the fact that heavy metals are selectively retained in the inner parenchyma cells coinciding with fungal structures (Kaldort et al. 1999). Accumulation of heavy metals in colonized tissue may be the predominant detoxification mechanism in AM-fungi. To substantiate this, cadmium has been found stabilized in the root system of clover (*Trifolium pratense*) (Medina et al. 2005), pea (*Pisum sativum*) (Rivera-Becerril et al. 2002), and ribwort (*Plantago lanceolata*) (Hutchinson et al. 2004). Rivera-Becerril et al. (2002) suggested a "mycorrhiza-buffering" of Cd-stress, which they attributed to detoxification mechanisms. Later on, Paradi et al. (2003) hypothesized that alterations in polyamine content and ratio in AM plants lead to Cd tolerance. In a study, Chen et al. (2007) found that *G. mosseae* may protect alfalfa shoots from As toxicity by "dilution effects" resulting from growth stimulation of AM plants and reduced transport of As to shoots. Liang et al. (2009) on the other hand reported that *G. mosseae* inoculation enhanced crop growth and protected maize from the toxicity of Pb, Zn, and Cd by decreasing the uptake of these heavy metals at higher soil concentrations. Similar effect of *G. mosseae* inoculation has also been reported in rice (*Oryza sativa* L.) against Cu toxicity (Zhang et al. 2009).

If plants are indeed sensitive, heavy metals will interfere with vital metabolic activities of plants as well as antioxidant enzyme activities (Azcón et al. 2009). When present in excessive amounts, heavy metals actually cause uncontrolled redox reaction in cells, resulting in the formation of reactive oxygen species (ROS), as reported by Hall (2002). Under stressed conditions, AM-colonized root cells accumulate ROS (Hause and Fester 2005). Several genes in AM-fungi with putative roles in oxidative stress alleviation have been described (Lanfranco et al. 2001). Induction of oxidative stress-related genes in AM-fungi is observed in extraradical mycelium upon exposure to heavy metals (Ouziad et al. 2005). Thus, a major function of AM-fungi could be to protect plants against heavy metal-induced oxidative stress (Schutzendubel and Polle 2002).

5.3.3 Contribution of AM-Fungi in Phytostabilization

Arbuscular mycorrhizal fungi contribute to the immobilization of heavy metals in the soil beyond the plant rhizosphere and thereby improve phytostabilization.

These symbiotic fungi employ strategies similar to those adopted by their host for stabilizing heavy metals. The strategies adopted by AM-fungi for immobilizing metals include (1) precipitation in polyphosphate granules in the soil, (2) compounds secreted by the fungus, (3) adsorption to fungal cell walls, and (4) chelation of metals inside the fungus (Gaur and Adholeya 2004).

Glomalin, an insoluble glycoprotein, is produced abundantly and released by AM-fungi (Rillig 2004) and plays a critical role in soil stability (Rillig et al. 2002; Bedini et al. 2009). Though the structure of glomalin has not been completely defined, it appears to be a complex of repeated monomeric structures bound together by hydrophobic interactions (Nichols 2003) that attaches to soil to help stabilize aggregates. Glomalin also binds to heavy metals in the soil and can be extracted from soil together with a significant amount of bound heavy metals. For example, Gonzalez-Chavez et al. (2004) reported that up to 0.08 mg Cd, 4.3 mg Cu, and 1.12 mg Pb per gram glomalin could be extracted from polluted soils that were inoculated with laboratory cultures of AM-fungi. Moreover, glomalin from hyphae of an isolate of *Gigaspora rosea* sequestered up to 28 mg Cu g^{-1} in vitro. Similarly, Bedini et al. (2010) reported that the amount of Cu, Ni, Pb, and Co bound to glomalin was 2.3, 0.83, 0.24, 0.24%, respectively, of the total content of heavy metals. Since there is a correlation between the amount of glomalin in the soil and the amount of heavy metals bound, fungal strains with significant secretion of glomalin should be more suitable in biostabilization efforts (Gohre and Paszkowski 2006).

Fungal cell wall is made up of chitin that has an important metal-binding capacity (Zhou 1999). And therefore, this binding ability of chitin is likely to reduce the concentration of heavy metals in soil. Moreover, the AM mycelium has a high metal sorption capacity relative to other microorganisms, and a cation exchange capacity comparable to other fungi. Passive adsorption to the hyphae of a metal-tolerant *G. mosseae* isolate led to the binding of up to 0.5 mg Cd per mg dry biomass (Joner et al. 2000). Fungal cell-wall components, which contain free amino, hydroxyl, carboxyl, and other groups, can be excellent binding sites for Cu^{2+} ions in fungi and plants (Kapoor and Virarghavan 1995). Many studies exhibited repeatedly that the retention of heavy metals by fungal hyphae may involve adsorption to cell walls, thereby minimizing metal translocation to the shoots (Galli et al. 1994; Liang et al. 2009). This hypothesis was supported by Joner et al. (2000) who demonstrated that AM mycelia had a high metal sorption capacity. Later on, Chen et al. (2001) reported that Zn was accumulated to a concentration over 1,200 mg kg^{-1} (dry matter) in *G. mosseae* mycelium associated with maize plants. Similar to plant and fungal vacuoles, fungal vesicles may also be involved in storing toxic compounds and, thereby, could provide an additional detoxification mechanism (Gohre and Paszkowski 2006). In several studies, significantly greater amount of heavy metals such as Pb and Cd was recorded in roots than shoots of mycorrhizal plants indicating that metals were accumulated in the mycorrhizal fungal structures such as vesicles and hyphae (Joner and Leyval 1997). Whitfield et al. (2004) showed that the metal-contaminated soil with Cd, Pb, and Zn enhanced mycorrhizal vesicular numbers of *Thymus polytrichus*. Chen et al. (2005) also observed that the mycorrhizal vesicle was stimulated by lower Pb concentration (300 mg kg^{-1} sand) but was inhibited by higher Pb concentration (600 mg kg^{-1} sand). Moreover, higher vesicular numbers accorded with the higher root/shoot ratio of Pb concentration in 300 mg kg^{-1} sand, indicating the storage of the metal in vesicles. Kaldort et al. (1999) found that Fe and Ni accumulated in mycorrhizal vesicles of maize. In the study of Weiersbye et al. (1999), vesicles of *Cynodon dactylon* (L.) Pers. were found to accumulate Mn, Cu, Ni, and U (uranium). In general, AM-fungi immobilize heavy metals within the soil or within roots and reflect their suitability for phytostabilization applications. However, since the mycorrhiza–plant interaction is a complex system, the importance of AM-fungi in phytostabilization cannot be generalized. Each contaminated site may contain a specific pollutant, for which an appropriate combination of fungal and plant genotypes must be identified. Besides these, other factors or interactions that occur in soil may also influence positively or negatively the efficiency of heavy metals stabilization by AM-fungi.

5.3.4 Importance of AM-Fungi in Phytoextraction

Phytoextraction relies on plants with high root-to-shoot transfer, accumulating high amounts of heavy metals in their aerial parts. Alternatively, plants producing high biomass with normal concentrations of heavy metals can also be employed. However, bioavailability of heavy metals in the soil is one of the major constraints for rapid phytoremediation. Consequently, many years are required to decrease soil contamination by half (McGrath and Zhao 2003). However, the addition of chelating agents such as EDTA accelerates the clean-up process even in non-hyperaccumulators resulting in induced heavy metals accumulation by plants (Barrutia et al. 2010; Zaier et al. 2010). Mycorrhizal fungi improve phytoextraction by making metals more available for uptake by plants. Improved phytoextraction following mycorrhization may be achieved by several mechanisms like (1) better plant growth and biomass production, (2) increased plant tolerance to metals, and (3) greater metal concentrations in plant tissues (Vamerali et al. 2010). Additional mechanisms to account for improved uptake by mycorrhizal roots may include small fungal hyphae radii, different uptake kinetics, greater total absorptive surface area, faster extension rate, increased functional longevity, chemical alteration of the rhizosphere–hyphosphere, greater carbon-use efficiency, exploration of smaller pore spaces, and differences in associated rhizosphere populations (O'Keefe and Sylvia 1991).

Arbuscular mycorrhizal fungi are known to enhance phytoextraction both in hyperaccumulators and non-hyperaccumulators. In recent years, there has been increasing contamination of soil, water, and crops by As in many parts of the world (Tripathi et al. 2007), particularly in some countries of southern Asia (Meharg 2004). *Pteris vittata* L. (Chinese brake fern) was the first reported of the eight As hyperaccumulator plant species identified so far (Ma et al. 2001). It has been found to accumulate As in its fronds with extraordinary efficiency, primarily due to high translocation from roots to shoots and to effective detoxification mechanisms within the plant (Webb et al. 2003; Singh and Ma 2006). Low to moderate (4.2–12.8%)

levels of AM colonization have been observed in *P. vittata* growing at several As-contaminated sites (Wu et al. 2007). Due to the fact that arsenate acts as a phosphate analogue, AM-fungi are likely to have a strong influence on arsenate uptake due to their role in enhancing phosphate acquisition for the host plants (Smith and Read 1997). Studies have also shown that AM-fungi significantly increased aboveground biomass and As accumulation, translocation, and bioconcentration by Chinese brake fern (Wu et al. 2009). There is evidence that arsenic uptake by Chinese brake fern is via P transport systems (Al-Agely et al. 2005). Leung et al. (2006) reported that non-AM *P. vittata* plants accumulated 60.4 mg As per kg while plants colonized by AM-fungi isolated from an As mine accumulated 88.1 g As per kg accompanied by enhanced growth. Phosphate uptake was 36.3 mg per pot in non-colonized and 257 mg per pot in colonized plants. Recently, Liu et al. (2009) demonstrated that colonization with *G. mosseae* substantially increased frond and root dry weight, and P and As contents in *P. vittata*. Intra-specific differences have been reported in AM-fungi in their impacts on As accumulation by *P. vittata* (Wu et al. 2009). Non-hyperaccumulators such as tomato (*Lycopersicon esculentum*) when grown in soils treated with 75 mg As per kg soil, had at least 30% higher root and shoot biomass than non-colonized plants, which coincided with higher P uptake. A maximum of 39% (As in shoot/total As) was reached at 75 mg As per kg soil in colonized plants (Liu et al. 2005). Mycorrhizal hyphae and plants can modify plant uptake of As by means of changes in the biotransformation of As at the interface between roots and rhizosphere soil (Ultra et al. 2007), downregulation of arsenate/phosphate transporters in the epidermis and root hairs (Gonzalez-Chavez et al. 2002), retention of As in external mycelium and/or possibly increased efflux of As as arsenite from mycorrhizal roots (Wang et al. 2008), and alteration of the translocation of As from roots to shoots (Dong et al. 2008). Arbuscular mycorrhizal fungi can also induce the accumulation of other heavy in host roots. *Berkheya coddii* Roessler, a Ni-hyperaccumulator plant of family Asteraceae, for example is used for phytomining, that is, for the recovery of metals from plant tissues (Salt et al. 1998). Mycorrhizal inoculation enhanced the biomass of this plant twice as compared to non-mycorrhizal control. In addition, mycorrhizal plants accumulated 30% more Ni than non-mycorrhizal plants (Turnau and Mesjasz-Przybylowicz 2003). In contrast, Amir et al. (2007) reported a negative correlation between AM colonization and leaf Ni content of three Ni-hyperaccumulators, namely, *Sebertia acuminate* Pierre ex Baill, *Psychotria douarrei* (Beauv.) Däniker, and *Phyllanthus favieri* M. Schmid. Whitfield et al. (2004) demonstrated enhanced Zn concentration in shoots of *Thymus polytrichus* A. Kerner ex Borbás due to AM inoculation. However, the resulting tissue metal concentrations were not large enough to adversely affect plant growth. Addition of chelating agents enhanced the bioavailability of heavy metals and thus the efficiency of phytoextraction even in the absence of the AM colonization. Studies have shown that application of EDTA or EDDS (ethylene-diaminedisuccinate) had no negative effect on the infectivity of AM-fungi (Grcman et al. 2001, 2003). Addition of EDTA led to phytotoxic concentrations of Zn in maize plants resulting in reduced plant growth. Colonization by AM-fungi reduced the phytotoxic effect of higher Zn levels and thereby contributed significantly to increase mobilization of Zn from the soil (Chen et al. 2004).

5.4 Conclusion

Arbuscular mycorrhizal fungi confer tolerance to plants against heavy metal contamination. They improve stabilization of heavy metals in soil or enhance uptake and transfer of these metals to the host plants and increase biomass of plants in order to enhance phytoextraction. However, the effects of AM colonization on the heavy metals uptake by plants have been conflicting. The efficacy of AM inoculation in phytoremediation has been shown to vary among plants as well as AM species. Generally, indigenous fungi from contaminated soils are considered most suitable for phytoremediation. The genotypic variation makes it difficult to identify a suitable AM-fungi for the restoration of metal-contaminated soils. Therefore, it requires a sustained effort of the scientists around the world to screen and identify heavy metal-tolerant mycorrhizal strains for their ultimate application in the management of metal-contaminated soils. Further research work is also needed to develop methods to produce and deliver mycorrhizal inocula inexpensively and to fully understand the molecular basis of metal detoxification by AM-fungi, when applied under metal-stressed soils.

References

Adriano-Anaya ML, Salvador-Figueroa M, Ocampo JA, Garcia-Romera I (2006) Hydrolytic enzyme activities in maize (*Zea mays*) and sorghum (*Sorghum bicolor*) roots inoculated with *Gluconacetobacter diazotrophicus* and *Glomus intraradices*. Soil Biol Biochem 38:879–886

Al-Agely A, Sylvia DM, Ma LQ (2005) Mycorrhizae increase arsenic uptake by the hyperaccumulator Chinese Brake Fern (*Pteris vittata* L.). J Environ Qual 34:2181–2186

Amir H, Perrier N, Rigault F, Jaffre T (2007) Relationships between Ni- hyperaccumulation and mycorrhizal status of different endemic plant species from New Caledonian ultramafic soils. Plant Soil 293:23–35

Azcón R, Perálvarez MC, Biró B, Roldán A, Ruíz-Lozano JM (2009) Antioxidant activities and metal acquisition in mycorrhizal plants growing in a heavy-metal multicontaminated soil amended with treated lignocellulosic agrowaste. Appl Soil Ecol 41:168–177

Barrutia O, Garbisu C, Hernández-Allica J, García-Plazaola JI, Becerril JM (2010) Differences in EDTA-assisted metal phytoextraction between metallicolous and non-metallicolous accessions of *Rumex acetosa* L. Environ Pollut 158:1710–1715

Barto K, Friese C, Cipollini D (2010) Arbuscular mycorrhizal fungi protect a native plant from allelopathic effects of an invader. J Chem Ecol 36:351–360

Bedini S, Pellegrino E, Avio L, Pellegrini S, Bazzoffi P, Argese E, Giovannetti M (2009) Changes in soil aggregation and glomalin-related soil protein content as affected by the arbuscular mycorrhizal fungal species *Glomus mosseae* and *Glomus intraradices*. Soil Biol Biochem 41:1491–1496

Bedini S, Turrini A, Rigo C, Argese E, Giovannetti M (2010) Molecular characterization and glomalin production of arbuscular mycorrhizal fungi colonizing a heavy metal polluted ash disposal island, downtown Venice. Soil Biol Biochem 42:758–765

Biro I, Nemeth T, Takacs T (2009) Changes of parameters of infectivity and efficiency of different *Glomus mosseae* arbuscular mycorrhizal fungi strains in cadmium-loaded soils. Comm Soil Sci Plant Anal 40:227–239

Brooks RR, Chambers MF, Nicks LJ, Robinson BH (1998) Phytomining. Trends Plant Sci 3:359–362

Cardoso IM, Kuyper TW (2006) Mycorrhizas and tropical soil fertility. Agric Ecosyst Environ 116:72–84

Chao CC, Wang YP (1990) Effects of heavy-metals on the infection of vesicular–arbuscular mycorrhizae and the growth of maize. J Agric Assoc China 152:34–45

Chen BD, Christie P, Li XL (2001) A modified glass bead compartment cultivation system for studies on nutrient and trace metal uptake by arbuscular mycorrhiza. Chemosphere 42:185–192

Chen BD, Liu Y, Shen H, Li XL, Christie P (2004) Uptake of cadmium from an experimentally contaminated calcareous soil by arbuscular mycorrhizal maize (*Zea mays* L.). Mycorrhiza, 14: 347–354

Chen X, Wu C, Tang J, Hu S (2005) Arbuscular mycorrhizae enhance metal lead uptake and growth of host plants under a sand culture experiment. Chemosphere 60:665–671

Chen BD, Xiao XY, Zhu YG, Smith FA, Xie ZM, Smith SE (2007) The arbuscular mycorrhizal fungus *Glomus mosseae* gives contradictory effects on phosphorus and arsenic acquisition by *Medicago sativa* Linn. Sci Total Environ 379:226–234

Cheng S (2003) Heavy metal pollution in China: origin, pattern and control. Environ Sci Pollut Res Int 10:192–198

Cho Y, Bolick JA, Butcher DJ (2009) Phytoremediation of lead with green onions (*Allium fistulosum*) and uptake of arsenic compounds by moonlight ferns (*Pteris cretica* cv *Mayii*). Microchem J 91:6–8

Clemens S (2006) Toxic metal accumulation, responses to exposure and mechanisms of tolerance in plants. Biochimie 88:1707–1719

Dary M, Chamber-Pérez MA, Palomares AJ, Pajuelo E (2010) "*In situ*" phytostabilisation of heavy metal polluted soils using *Lupinus luteus* inoculated with metal resistant plant growth promoting rhizobacteria. J Hazard Mat 177:323–330

Del Val C, Barea JM, Azcon-Aguilar C (1999) Diversity of arbuscular mycorrhizal fungus populations in heavy-metal-contaminated soils. Appl Environ Microbiol 65:718–723

Diaz G, Azconaguilar C, Honrubia M (1996) Influence of arbuscular mycorrhizae on heavy metal (Zn and Pb) uptake and growth of *Lygeum spartum* and *Anthyllis cytisoides*. Plant Soil 180:241–249

Dong Y, Zhu YG, Smith FA, Wang YS, Chen BD (2008) Arbuscular mycorrhiza enhanced arsenic resistance of both white clover (*Trifolium repens* Linn.) and ryegrass (*Lolium perenne* L.) plants in an arsenic-contaminated soil. Environ Pollut 155:174–181

Dong X, Li C, Li J, Wang J, Liu S, Ye B (2010) A novel approach for soil contamination assessment from heavy metal pollution: a linkage between discharge and adsorption. J Hazard Mater 175:1022–1030

Feng G, Zhang FS, Li XL, Tian CY, Tang C, Rengel Z (2002) Improved tolerance of maize plants to salt stress by arbuscular mycorrhiza is related to higher accumulation of soluble sugars in roots. Mycorrhiza 12:185–190

Franco-Hernandez MO, Vasquez-Murrieta MS, Patino-Siciliano A, Dendooven L (2010) Heavy metals concentration in plants growing on mine tailings in central Mexico. Biores Technol 101:3864–3869

Galli U, Schepp H, Brunold C (1994) Heavy metal binding by mycorrhizal fungi. Physiol Plant 92:364–368

Gardea-Torresdey LJR, Peralta-Videab G, Rosaa DL, Parsons JG (2005) Phytoremediation of heavy metals and study of the metal coordination by X-ray absorption spectroscopy. Coord Chem 249:1797–1810

Gaur A, Adholeya A (2004) Prospects of arbuscular mycorrhizal fungi in phytoremediation of heavy metal contaminated soils. Curr Sci 86:528–534

Gohre V, Paszkowski U (2006) Contribution of the arbuscular mycorrhizal symbiosis to heavy metal phytoremediation. Planta 223:1115–1122

Gong X, Yao H, Zhang D, Qiao Y, Li L, Xu M (2010) Leaching characteristics of heavy metals in fly ash from a Chinese coal-fired power plant. Asia Pac J Chem Eng 5:330–336

Gonzalez-Chavez C, Harris PJ, Dodd J, Meharg AA (2002) Arbuscular mycorrhizal fungi confer enhanced arsenate resistance on *Holcus lanatus*. New Phytol 155:163–171

Gonzalez-Chavez MC, Carrillo-González R, Wright SF and Nichols KA (2004) The role of glomalin, a protein produced by arbuscular mycorrhizal fungi, in sequestering potentially toxic elements. Environ Poll 130:317–323

Grcman H, Velikonja-Bolta Š, Vodnic D, Leštan D (2001) EDTA enhanced heavy metal phytoextraction: metal accumulation, leaching and toxicity. Plant Soil 235:105–114

Grcman H, Vodnik D, Velikonja-Bolta S, Lestan D (2003) Ethylenediaminedissuccinate as a new chelate for environmentally safe enhanced lead phytoextraction. J Environ Qual 32:500–506

Gupta S, Satpati S, Nayek S, Garai D (2010) Effect of wastewater irrigation on vegetables in relation to bioaccumulation of heavy metals and biochemical changes. Environ Monit Assess 165:169–177

Hall JL (2002) Cellular mechanisms for heavy metal detoxification and tolerance. J Exp Bot 53:1–11

Hause B, Fester T (2005) Molecular and cell biology of arbuscular mycorrhizal symbiosis. Planta 221:184–196

Hildebrandt U, Kaldorf M, Bothe H (1999) The zinc violet and its colonization by arbuscular mycorrhizal fungi. J Plant Physiol 154:709–717

Holleman A, Wiberg E (1985) Lehrbuch der Anorganischen. Chemie, Berlin

Hua J, Lin X, Yin R, Jiang Q, Shao Y (2009) Effects of arbuscular mycorrhizal fungi inoculation on arsenic accumulation by tobacco (*Nicotiana tabacum* L.). J Environ Sci 21:1214–1220

Hutchinson JJ, Young SD, Black CR, West HM (2004) Determining uptake of radio-labile soil cadmium by arbuscular mycorrhizal hyphae using isotopic dilution in a compartmented- pot system. New Phytol 164:477–484

Jacquot-Plumey E, van Tuinen D, Chatagnier O, Gianinazzi S, Gianinazzi-Pearson V (2001) 25 S rDNA-based molecular monitoring of glomalean fungi in sewage sludge-treated field plots. Environ Microbiol 3:525–531

Javaid A (2007) Allelopathic interactions in mycorrhizal associations. Allelopathy J 20:29–42

Javaid A (2008) Allelopathy in mycorrhizal symbiosis in the Poaceae family. Allelopathy J 21:207–218

Javaid A (2009) Arbuscular mycorrhizal mediated nutrition in plants. J Plant Nutr 32:1595–1618

Javaid A, Bajwa R (1999) Allelopathy and VA mycorrhizaIV: tolerance to allelopathy by VA mycorrhiza in maize. Pak J Phytopathol 11:70–73

Javaid A, Hafeez FY, Iqbal SH (1993) Interaction between vesicular arbuscular (VA) mycorrhiza and *Rhizobium* and their effect on biomass, noduation and nitrogen fixation in *Vigna radiata* (L.) Wilczek. Sci Int Lahore 5:395–396

Javaid A, Iqbal SH, Hafeez FY (1994) Effect of different strains of *Bradyrhizobium* and two types of vesicular arbuscular mycorrhizae (VAM) on biomass and nitrogen fixation in *Vigna radiata* (L.) Wilczek var. NM 20–21. Sci Int Lahore 6:265–267

Jentschke G, Godbold DL (2000) Metal toxicity and ectomycorrhizas. Physiol Plant 109:107–116

Joner EJ, Leyval C (1997) Uptake of Cd by roots and hypae of a *Glomus mosseae/Trifolium subterraneum* mycorrhiza from soil amended with high and low concentration of cadmium. New Phytol 135:353–360

Joner EJ, Briones R, Leyval C (2000) Metal-binding capacity of arbuscular mycorrhizal mycelium. Plant Soil 226:227–234

Kaldort M, Kuhn AJ, Schroder WH, Hildebrandt U, Bothe H (1999) Selective element deposits in maize colonized by a heavy metal tolerance conferring arbuscular mycorrhizal fungus. J Plant Physiol 154:718–728

Kapoor A, Virarghavan T (1995) Fungal biosorption – an alternative treatment option for heavy metal bearing wastewater: a review. Biores Technol 53:195–206

Kaschuk G, Leffelaar PA, Giller KE, Alberton O, Hungria M, Kuyper TW (2010) Responses of legumes to rhizobia and arbuscular mycorrhizal fungi: a meta-analysis of potential photosynthate limitation of symbioses. Soil Biol Biochem 42:125–127

Khade HW, Adholeya A (2009) Arbuscular mycorrhizal association in plants growing on metal-contaminated and non-contaminated soils adjoining Kanpur tanneries, Uttar Pradesh, India. Water Air Soil Poll 202:45–56

Khalvati M, Bartha B, Dupigny A (2010) Arbuscular mycorrhizal association is beneficial for growth and detoxification of xenobiotics of barley under drought stress. J Soils Sediments 10.54–64

Khaosaad T, García-Garrido JM, Steinkellner S, Vierheilig H (2007) Take-all disease is systemically reduced in roots of mycorrhizal barley plants. Soil Biol Biochem 39:727–734

Kramer U (2005) Phytoremediation: novel approaches to cleaning up polluted soils. Curr Opin Biotechnol 16:133–141

Lanfranco L, Bianciotto V, Lumini E, Souza M, Morton JB, Bonfante P (2001) A combined morphological and molecular approach to characterize isolates of arbuscular mycorrhizal fungi in *Gigaspora* (Glomales). New Phytol 152:169–179

Leung HM, Ye ZH, Wong MH (2006) Interactions of mycorrhizal fungi with Pteris vittata (as hyperaccumulator) in as-contaminated soils. Environ Pollut 139:1–8

Leung HM, Ye ZH, Wong MH (2007) Survival strategies of plants associated with arbuscular mycorrhizal fungi on toxic mine tailings. Chemosphere 66:905–915

Leyval C, Turnau K, Haselwandter K (1997) Effect of heavy metal pollution on mycorrhizal colonization and function: physiological, ecological and applied aspects. Mycorrhiza 7:139–153

Liang CC, Li T, Xiao YP, Liu MJ, Zhang HB, Zhao ZW (2009) Effects of inoculation with arbuscular mycorrhizal fungi on maize grown in multi-metal contaminated soils. Int J Phytoremed 11:692–703

Liu Y, Zhu YG, Chen BD, Christie P, Li XL (2005) Yield and arsenate uptake of arbuscular mycorrhizal tomato colonized by *Glomus mosseae* BEG167 in as spiked soil under glasshouse conditions. Environ Int 31:867–873

Liu Y, Christie P, Zhang J, Li X (2009) Growth and arsenic uptake by Chinese brake fern inoculated with an arbuscular mycorrhizal fungus. Environ Exp Bot 66:435–441

Ma LQ, Komar KM, Tu C, Zhang WH, Cai Y, Kennelley ED (2001) A fern that hyperaccumulates arsenic. Nature 409:579–1579

Manoharachary C, Kunwar IK, Tilak KVBR, Adholeya A (2010) Arbuscular mycorrhizal fungi-taxonomy, diversity, conservation and multiplication. Proc Natl Acad Sci India B Biol Sci 80:1–13

Mapanda F, Mangwayana EN, Myamangara J, Giller KE (2005) The effect of long term irrigation using wastewater on heavy metal contents of soils under vegetables in Harare, Zimbabwe. Agric Ecosyst Environ 107:151–165

McGrath SP, Zhao FJ (2003) Phytoextraction of metals and metalloids from contaminated soils. Curr Opin Biotechnol 14:277–282

Medina A, Vassilev N, Barea JM, Azcon R (2005) Application of *Aspergillus niger*-treated agro-waste residue and *Glomus mosseae* for improving growth and nutrition of *Trifolium repens* in a Cd- contaminated soil. J Biotechnol 116:369–378

Meding SM, Zasoski RJ (2008) Hyphal-mediated transfer of nitrate, arsenic, cesium, rubidium, and strontium between arbuscular mycorrhizal forbs and grasses from a California oak woodland. Soil Biol Biochem 40:126–134

Meharg AA (2004) Arsenic in rice – understanding a new disaster for South-East Asia. Trends Plant Sci 9:415–417

Mendez MO, Maier RM (2008) Phytostabilization of mine tailings in arid and semiarid environments – an emerging remediation technology. Environ Health Perspect 116:278–283

Mertz W (1981) The essential trace elements. Science 213:1332–1338

Muchuweti M, Birkett JW, Chinyanga E, Zvauya R, Scrimshaw MD, Lester JN (2006) Heavy metal content of vegetables irrigated with mixtures of wastewater and sewage sludge in Zimbabwe: implications for human health. Agric Ecosyst Environ 112:41–48

Nichols K (2003) Characterization of glomalin – a glycoprotein produced by Arbuscular Mycorrhizal fungi. Ph.D. dissertation, University of Maryland, College Park

O'Keefe DM, Sylvia DM (1991) Mechanisms of the vesicular-arbuscular mycorrhizal plant-growth response. In: Arora DK, Rai B, Mukerji KG, Knudsen GR (eds) Handbook of applied mycology. Marcel Dekker, New York, pp 35–53

Ouziad F, Hildebrandt U, Schmelzer E, Bothe H (2005) Differential gene expressions in arbuscular mycorrhizal-colonized tomato grown under heavy metal stress. J Plant Physiol 162:634–649

Oyekanmi EO, Coyne DL, Fagade OE, Osonubi O (2007) Improving root-knot nematode management on two soybean genotypes through the application of *Bradyrhizobium japonicum*, *Trichoderma pseudokoningii* and *Glomus mosseae* in full factorial combinations. Crop Prot 26:1006–1012

Ozgonen H, Erkilic A (2007) Growth enhancement and phytophthora blight (*Phytophthora capsici* Leonian) control by arbuscular mycorrhizal fungal inoculation in pepper. Crop Prot 26:1682–1688

Paradi I, Berecz B, Hala´ SZ K, Bratek Z (2003) Influence of arbuscular mycorrhiza and cadmium on the polyamine contents of Ri T-DNA transformed *Daucus carota* L. root cultures. Acta Biol Szegediensis 47:31–36

Parniske M (2000) Intracellular accommodation of microbes by plants: a common developmental program for symbiosis and disease? Curr Opin Plant Biol 3:320–328

Pawlowska TE, Charvat I (2004) Heavy-metal stress and developmental patterns of arbuscular mycorrhizal fungi. Appl Environ Microbiol 70:6643–6649

Peer WA, Baxter IR, Richards EL, Freeman JL, Murphy AS (2005) Phytoremediation and hyperaccumulator plants. In: Tamas M, Martionoia E(eds) Moleculor biology of metal homeostosis detoxification. Topics in current genetics, vol 14. Springer, Berlin, pp 299–340

Petra K, Juan B, Bernal MP, Flavia N, Charlotte P, Stefan S, Rafael C, Carmela M (2009) Trace element behaviour at the root–soil interface. Implications in phytoremediation. Environ Exp Bot 67:243–259

Quartacci MF, Argilla A, Baker AJM, Navari-Izzo F (2006) Phytoextraction of metals from a multiply contaminated soil by Indian mustard. Chemosphere 63:918–925

Rai PK (2010) Phytoremediation of heavy metals in a tropical impoundment of industrial region. Environ Monit Assess 165:529–537

Ray JG, Valsalakumar N (2010) Arbuscular mycorrhizal fungi and *piriformospora indica* individually and in combination with *Rhizobium* on greengram. J Plant Nutr 33:285–298

Read DJ, Perez-Moreno J (2003) Mycorrhizas and nutrient cycling in ecosystems – a journey towards relevance. New Phytol 157:475–492

Redecker R, Kodner R, Graham LE (2000) Glomalean fungi from the Ordovician. Science 289:1920–1921

Rillig MC (2004) Arbuscular mycorrhizae, glomalin and soil quality. Can J Soil Sci 84:355–363

Rillig MC, Mummey DL (2006) Mycorrhizas and soil structure. New Phytol 171:41–53

Rillig MC, Wright SF, Eviner VT (2002) The role of arbuscular mycorrhizal fungi and glomalin in soil aggregation: comparing effects of five plant species. Plant Soil 238:325–333

Rivera-Becerril F, Calantzis C, Turnau K, Caussanel JP, Belimov AA, Gianinazzi S, Strasser RJ, Gianinazzi-Pearson V (2002) Cadmium accumulation and buffering of cadmium-induced stress by arbuscular mycorrhiza in three *Pisum sativum* L. genotypes. J Exp Bot 53:1177–1185

Ruiz-Lozano JM (2003) Arbuscular mycorrhizal symbiosis and alleviation of osmotic stress: new perspectives for molecular studies. Mycorrhiza 13:309–317

Saini VK, Bhandari SC, Tarafdar JC (2004) Comparison of crop yield, soil microbial C, N and P, N-fixation, nodulation and mycorrhizal infection in inoculated and non-inoculated sorghum and chickpea crops. Field Crops Res 89:39–47

Salt DE, Smith RD, Raskin I (1998) Phytoremediation. Ann Rev Plant Physiol Plant Mol Biol 49:643–668

Schliemann W, Ammer C, Strack D (2008) Metabolite profiling of mycorrhizal roots of *Medicago truncatula*. Phytochemistry 69:112–146

Schüssler A, Schwarzott D, Walker C (2001) A new fungal phylum, the *Glomeromycota*: phylogeny and evolution. Mycol Res 105:1413–1421

Schutzendubel A, Polle A (2002) Plant responses to abiotic stresses: heavy metal-induced oxidative stress and protection by mycorrhization. J Exp Bot 53:1351–1365

Shalaby M, Helmy HM, Kaindl R, Rahman HBA (2003) Genesis of the metamorphosed Um Zeriq Zn-Pb-As-Ag prospect, Sinai, Egypt. Proceedings of 7th Biennial SGA Meeting, Aug. 24–28, 2003 Athens Greece, "Mineral Exploration and Sustalnable Development" Volumes 1 & 2. Millprcss Science Publishers, Rotterdam, pp 15–18

Sharda JN, Koide RT (2010) Exploring the role of root anatomy in P-mediated control of colonization by arbuscular mycorrhizal fungi. Bot Botanique 88:165–173

Sharma RK, Agrawal M, Marshal F (2007) Heavy metal contamination of soil and vegetables in suburban areas of Varanasi, India. Ecotox Environ Safe 66:258–266

Singh N, Ma LQ (2006) Arsenic speciation, and arsenic and phosphate distribution in arsenic hyperaccumulator *Pteris vittata* L. and non-hyperaccumulator *Pteris ensiformis* L. Environ Pollut 141:238–246

Smith SE, Read DJ (1997) Mycorrhizal symbiosis. Academic, London

Tonin C, Vandenkoornhuyse P, Joner EJ (2001) Assessment of arbuscular mycorrhizal fungi diversity in the rhizosphere of *Viola calaminaria* and effect of these fungi on heavy metal uptake by clover. Mycorrhiza 10:161–168

Tripathi RD, Srivastava S, Mishra S, Singh N, Tuli R, Gupta DK, Maathuis FJM (2007) Arsenic hazards: strategies for tolerance and remediation by plants. Trends Biotechnol 25:158–165

Turnau K, Mesjasz-Przybylowicz J (2003) Arbuscular mycorrhiza of *Berkheya coddii* and other Ni-hyperaccumulating members of *Asteraceae* from ultramafic soils in South Africa. Mycorrhiza 13:185–190

Turnau K, Miszals Z, Trouvelot A, Bonfante P, Gianinazzi S (1996) *Oxalis acetosella* as monitoring plant on highly polluted soils. In: Azcon-Agiular C, Barea JM (eds) Mycorrhizas in integrated system: from genes to plant development. European commission, Luxembourg, pp 483–486

Ultra VU, Tanaka S, Sakurai K, Iwasaki K (2007) Effects of arbuscular mycorrhiza and phosphorus application on arsenic toxicity in sunflower (*Helianthus annuus* L.) and on the transformation of arsenic in the rhizosphere. Plant Soil 290:29–41

Vamerali T, Bandiera M, Mosca G (2010) Field crops for phytoremediation of metal-contaminated land. A review. Environ Chem Lett 8:1–17

Wang ZH, Zhang JL, Christie P, Li XL (2008) Influence of inoculation with *Glomus mosseae* or *Acaulospora morrowiae* on arsenic uptake and translocation by maize. Plant Soil 311:235–244

Webb SM, Gaillard JF, Ma LQ, Tu C (2003) XAS speciation of arsenic in a hyper-accumulating fern. Environ Sci Technol 37:754–760

Weiersbye IM, Straker CJ, Przybylowicz WJ (1999) Micro-PIXE Mapping of elemental distribution in arbuscular mycorrhizal roots of the grass, Cynodon dactylon, from gold and uranium mine tailings. Nucl Instrum Meth B 158:335–343

Weissenhorn I, Leyval C, Berthelin J (1995a) Bioavailability of heavy metals and arbuscular mycorrhiza in a soil polluted by atmospheric deposition from a smelter. Biol Fertil Soils 19:22–28

Weissenhorn I, Leyval C, Berthelin J (1995b) Bioavailability of heavy metals and abundance of arbuscular mycorrhiza in a sewage sludge amended sandy soil. Soil Biol Biochem 27:287–296

Whitfield L, Richards AJ, Rimmer DL (2004) Effects of mycorrhizal colonization on *Thymus polytrichus* from heavy-metal-contaminated sites in northern England. Mycorrhiza 14:47–54

Wu FY, Ye ZH, Wu SC, Wong MH (2007) Metal accumulation and arbuscular mycorrhizal status in metallicolous and nonmetallicolous populations of *Pteris vittata* L. and *Sedum alfredii* Hance. Planta 226:1363–1378

Wu FY, Ye ZH, Wong MH (2009) Intraspecific differences of arbuscular mycorrhizal fungi in their impacts on arsenic accumulation by *Pteris vittata* L. Chemosphere 76:1258–1264

Xavier IJ, Boyetchko SM (2002) Arbuscular mycorrhizal fungi as biostimulants and bioprotectants of crops. In: Khachatourians GG, Arora DK (eds) Applied mycology and biotechnology. vol 2: Agriculture and food production. Elsevier, Amsterdam

Zaier H, Ghnaya T, Rejeb KB, Lakhdar A, Rejeb S, Jemal F (2010) Effects of EDTA on phytoextraction of heavy metals (Zn, Mn and Pb) from sludge-amended soil with *Brassica napus*. Biores Technol 101:3978–3983

Zak JC, Daneilson RM, Parkinson D (1982) Mycorrhizal fungal spore numbers and species occurrence in two amended mine spoils in Alberta, Canada. Mycologia 74:785–792

Zenk MH (1996) Heavy metal detoxification in higher plants-A review. Gene 179:21–30

Zhang ZH, Lin AJ, Gao YL, Reid RJ, Wong MH, Zhu YG (2009) Arbuscular mycorrhizal colonisation increases copper binding capacity of root cell walls of *Oryza sativa* L. and reduces copper uptake. Soil Biol Biochem 41:930–935

Zhou JL (1999) Zn biosorption by *Rhizopus arrhizus* and other fungi. Appl Microbiol Biotechnol 51:686–693

Zhu YG, Williams PN, Meharg AA (2008) Exposure to inorganic arsenic from rice: a global health issue? Environ Pollut 154:169–171

Chapter 6
Research Advances in Bioremediation of Soils and Groundwater Using Plant-Based Systems: A Case for Enlarging and Updating Information and Knowledge in Environmental Pollution Management in Developing Countries

Andrew Agbontalor Erakhrumen

Abstract Soil and groundwater are important components of agricultural and renewable natural resource (RNR) production systems. These components and production systems are influenced directly and/or indirectly by anthropogenic activities. Many of these activities have series of impacts, the negative ones being through the generation and deposition of xenobiotics that are dangerous to life forms, onto and/or into the soil and groundwater. Although, it may be difficult and/or expensive to remove these toxic substances from the environment in most countries, most especially the developing ones and particularly those in sub-Saharan Africa (SSA) using the available remediation technologies, owing to different levels of economic constraints and/or quality of research. The documented researches have shown that the growth and physiological characteristics of certain species of plants can be applied in cheap, adoptable, and adaptable ways, for removing toxic substances from the environment through processes collectively known as bioremediation. Bioremediation has been identified as a feasible choice for removing the noxious substances. These production systems are central to livelihoods and survival in many developing countries, SSA in particular. The remediation technologies can be used for cleaning up the environment, soil, and groundwater, in ways that is expected to benefit the present and future environmental and socio-economic conditions of users. The present review is focused on the use of various methods of plant-assisted bioremediation processes for soil and groundwater remediation, in many parts of the world, for the benefit of and its adoption/adaptation in the developing countries.

Keywords Bioremediation • Developing country • Groundwater • Pollution • Research capacity

A.A. Erakhrumen (✉)
Department of Forest Resources Management, University of Ibadan,
Ibadan, Nigeria
e-mail: erakhrumen@yahoo.com

M.S. Khan et al. (eds.), *Biomanagement of Metal-Contaminated Soils*,
Environmental Pollution 20, DOI 10.1007/978-94-007-1914-9_6,
© Springer Science+Business Media B.V. 2011

6.1 Introduction

Soil can be described as the loose material that covers the land surfaces of the Earth and supports the growth of plants. In general, soil is an unconsolidated, or loose, combination of inorganic and organic materials. The inorganic components of soil are principally the products of rocks and minerals that have been gradually broken down by weather, chemical action, and other natural processes. The organic materials are composed of debris from plants and from the decomposition of many tiny life forms that inhabit the soil (King 2006). Soils vary widely from place to place and many factors determine the chemical composition and physical structure of the soil. The different kinds of rocks, minerals, and other geologic influences and materials from which the soil is originally formed play their roles. The kinds of plants or other vegetation that grow on the soil are also important. Topography that is, whether the terrain is steep, flat, or some combination, is another factor. In some cases, human activities such as farming or building have caused disruption. Soils also differ in color, texture, chemical makeup, and the kinds of plants they support (Microsoft Encarta 2006). Soil actually constitutes a living system, combining with air, water, and sunlight to sustain plant life. The essential process of photosynthesis, in which plants convert sunlight into energy, depends on exchanges that take place within the soil. Plants, in turn, serve as a vital part of the food chain for living organisms, including humans. Without soil there would be no vegetation, no crops for food, no forests, flowers, or grasslands. To a great extent, life on Earth depends on soil. Soil takes a great deal of time to develop, thousands or even millions of years. As such, it is effectively a non-renewable resource. Yet even now, in many areas of the world, soil is under siege. Deforestation, over-development, and pollution from human-made chemicals are just a few of the consequences of human activity and carelessness. It is on record that anthropogenic activities have been the major cause of environmental degradation particularly soil pollution (Rajakaruna et al. 2006; Erakhrumen 2007a). Thus, as the human population grows, its demand for food from crops increases, making soil conservation crucial (King 2006).

Groundwater can be defined as water found below the surface of the land. Such water exists in pores between sedimentary particles and in the fissures of more solid rocks. In arctic regions, groundwater may be frozen. In general, such water maintains a fairly even temperature very close to the mean annual temperature of the area. Very deep-lying groundwater can remain undisturbed for thousands or millions of years. Most groundwater lies at shallower depths, however, and plays a slow but steady part in the hydrologic cycle. Worldwide, groundwater accounts for about one third of 1% of the earth's water, or about 20 times more than the total of surface waters on continents and islands (Microsoft Encarta 2006). Groundwater is of major importance to civilization, because it is the largest reserve of drinkable water in regions where humans can live. Groundwater may appear at the surface in the form of springs, or it may be tapped by wells. During dry periods, it can also sustain the flow of surface water, and even where the latter is readily available; groundwater is often preferable because it tends to be less contaminated by wastes and organisms. Although, groundwater is less contaminated than surface waters, pollution of this major water supply has become an increasing concern in many countries particularly in industrialized countries. For instance, in the United States, thousands of wells have been closed in the late twentieth century because of contamination by various toxic substances (Microsoft Encarta 2006).

The present condition of soils and groundwater in many parts of the world might be worsened considering the increasing modern day needs, most especially in this part of the world, where many of the countries are formulating, adopting, and adapting growth and developmental processes aimed at catching up with the current and future advances in various human endeavors (Erakhrumen 2007b, 2008) with serious implications on the environment if these developmental processes are not properly conceived, executed, and managed. For instance, estimates have shown that widespread contamination of agricultural lands has significantly decreased the extent of arable land available for cultivation worldwide (Grêman et al. 2003). Therefore, there is the need for sustainable, relatively cheap, and easily adoptable means of utilizing and managing the environment for various purposes by the present and future generations in perpetuity most especially in this part of the world where most of the inhabitants are presently dependent on agricultural and RNR, a trend that is likely to continue in the foreseeable future. The present review highlights some salient documented information regarding the use of plants for the removal of pollutants from the environment, particularly the soil and groundwater, for the benefit of the inhabitants of this region and stakeholders in issues concerning the environment.

6.2 Types, Sources, and Effects of Soil and Groundwater Pollutants

Pollutants are either organic or inorganic. The types and sources of pollutants and the magnitude of environmental pollution, soil, surface water, and groundwater inclusive, vary from one clime to another owing to the differences in site-specific characteristics and/or anthropogenic activities in different places. Pollutants may be traced to a particular source (point source) or may result from a large area (non-point source). Nevertheless, anthropogenic activities have been identified as the main cause of environmental degradation, one of which is the environmental pollution, although, some inorganic pollutants have been identified to occur as natural elements in the Earth's crust. Inorganic pollutants can be plant macro-nutrients such as nitrates and phosphates, micro-nutrients such as Cr, Cu, Fe, Mn, Mo, Ni, and Zn, non-essential elements such as As, Cd, Co, F, Hg, Se, Pb, V, and W, and radionuclides such as ^{238}U, ^{137}Cs, and ^{90}Sr (Dushenkov 2003). Issues relating to environmental pollution, particularly soil, surface water and groundwater pollution are important because natural water systems comprise chemical and physical processes that affect both the distribution and circulation of chemicals on the Earth's surface. Thus, studies concerning aquatic systems, the atmosphere, sediments, soil, and biota are extremely helpful in understanding the relationships that exist in the interfacial chemistry of the environment (Yabe and de Oliveira 2003).

Anthropogenic activities in this regard are numberless, but most have been in the area of industrialization, manufacturing, construction, mining, domestic, and commercial burning of fossil fuels, control of pests and diseases in agriculture, and RNR production, among others (Arvin et al. 1988; Dey et al. 2004; Erakhrumen 2007a). For instance, rapid industrialization has led to increased disposal of heavy metals and other toxic substances into the environment. During industrial activities, diesel engines also tend to produce significant quantities of particulate matter (soot) and $NO_{(x)}$ (Heck and Farrauto 1995). The soot consists of both solid and liquid components and there is evidence that particulates from diesel engines are biologically more active than those from spark ignition engines and may be carcinogenic (Russell-Jones 1987).

Other toxic substances like cadmium and heavy metals are introduced into water from smelting, metal plating, cadmium nickel batteries, phosphate fertilizers, mining, crude oil exploration, exploitation, and associated activities, pigments, stabilizers, alloy industries, sewage sludge, among others (Banks et al. 1997; Petrisor et al. 2004; Rajakaruna et al. 2006). Toxic heavy metals like arsenic stem from various industrial wastes, including those from the manufacture of insecticides and pesticides, manufacture of fertilizers, mining and smelting, and tannery industries. Arsenic is another priority heavy metal pollutant found in soil and groundwater contaminated by arsenic pesticides and industrial wastes (Lin and Puls 2003). Creosote oil has also been used for wood preservation for over a century; spills and sludge deposits on creosote wood preservation sites have led to severe contamination of soil and groundwater. Coal tar is formed as a by-product in the production of gas from coal and creosote oil is formed when coal tar is distilled. It is a complex mixture of organic chemical compounds. For instance, groundwater leaching from creosote-contaminated sites contains hundreds of aromatic compounds, consisting of polycyclic aromatic hydrocarbons, phenols, and nitrogen/sulfur/oxygen containing heterocyclic aromatic compounds (Arvin et al. 1988) with potential toxicity, carcinogenicity, and mutagenicity (Richardson and Gangolli 1992).

Wastewater discharges from acid mine drainage, galvanizing plants, as a leachate from galvanized structures and natural ores, and from municipal wastewater treatment plant may contain heavy metals such as zinc. Also, owing to the varying degree of chemicals used, the dye wastewater contains appreciable concentrations of biochemical oxygen demand (BOD), chemical oxygen demand (COD), suspended solids, toxic compounds, and color (Dey et al. 2004). Sulfur oxide (SO_2) emissions related to industrial operations primarily occur from combustion sources and thermal processes, such as power plants (coal or oil fired), incinerators, steam generation equipment, process heaters, chemical reactors, and other similar equipment and processes (Wu et al. 2004). It is imperative to understand that new developments in the variety of fields to meet the ever-increasing requirements of human being have also led to the accumulation of compounds in the effluents of processing plants, which are not readily degraded by the conventional effluent treatment methods (Bauer and Fallmann 1997; Mantzavinos et al. 1997; Otal et al. 1997; Feigelson et al. 2000).

Pollutants exert variable effects on different organisms. For instance, a minor pH variation in natural waters due to anthropogenic interference may result in the liberation of metals adsorbed on colloidal particles, which after uptake may cause death of fish and other species of biota (Florence and Batley 1980). Many of these substances like the heavy metals are multivalent element, occurring in many valence states, are not biodegradable, and enter the food chain via bioaccumulation. Many of these compounds are not only toxic but can be carcinogenic and mutagenic (Richardson and Gangolli 1992). Some plants are also sensitive to these potentially toxic substances at different ages and growth stages. For example, the tolerance of alfalfa (*Medicago sativa*) plants to Cd, Cu, and Zn was positively correlated with the age of the plants, although, there exist a possibility of using the species, via transplant, to clean up soils having elevated concentration of Cd, Cu, or Zn (Peralta-Videa et al. 2004). In another study, Krupa and Moniak (1998) demonstrated that in rye seedlings, correlations occurred among the efficiency of the photosynthetic apparatus of leaves, the stage of the leaf maturity, and the sensitivity to Cd toxicity. It was also shown in some other studies (Skorzynska-Polit and Baszynski 1997; Tukendorf et al. 1997) that relationships exist among the Cu or Cd susceptibility and the growth stage of runner bean (*Phaseolus coccineus* L.) plants. The heavy metals such as Cd (II) reduce shoot growth by decreasing the chlorophyll content and the activity of photosystem I (Waldemar and Baszynski 1996) while Jiang et al. (2000) reported that Cd (II) substantially declined the root growth of *Allium sativum*. In some instances, the young plant is affected more than the older ones, as observed in a study by Skorzynska-Polit and Baszynski (1997) for *P. coccineus* plants. Many of these potentially toxic substances affect the macro-elemental uptake in several plants species and also have influence on plant metabolism in different manners. For example, Ouariti et al. (1997) observed that Cu affects the lipid metabolism in *Lycopersicon esculentum* more than cadmium.

6.3 Phytoremediation: A Type of Plant-Assisted Bioremediation

Bioremediation is simply defined as the elimination, attenuation, or transformation of polluting or contaminating substances by the use of biological processes. The plant-based biological processes collectively termed phytoremediation involve various plant processes that promote the removal of contaminants from contaminated media such as soil, water, and air. The term phytoremediation is a combination of the Greek prefix *phyto* (for plant) and the Latin root *remidium* (to correct or remove an evil). Broadly, phytoremediation can be defined as the utilization of vascular plants, algae, and fungi to control, breakdown, or remove wastes, or to encourage degradation of contaminants in the rhizosphere, or root region of the plant (McCutcheon and Schnoor 2003). Phytoremediation has been reported to be an environmentally friendly, potentially very effective, and less expensive method than the physical and chemical remediation techniques for the clean-up of a broad spectrum of hazardous organic and inorganic pollutants (Chaney et al. 2000; McGrath et al. 2002; Kamaludeen et al. 2003; Pilon-Smits 2005). The phytoremediation of soil and water can be direct or indirect. Direct processes include plant uptake into roots or shoots and transformation, storage, or transpiration of the contaminants while indirect plant processes involve the degradation of contaminants by microbial, soil, and root interactions within the rhizosphere (Hutchinson et al. 2003). Depending on the types of plant and the contaminant, direct uptake can be considered either a passive and/or an active process. The principal process is passive transport, with the primary transport medium, external water, and soil water, carrying the contaminant into the plant. Active transport requires the plant to expend energy and generally applies to nutrients and organic and inorganic ions required and extracted by the plant.

There are however, different mechanisms by which phytoremediation processes can be achieved (Table 6.1), although, some of these mechanisms may act simultaneously. For instance, the ability of plants to remove pollutants particularly metals and other compounds and translocate them into the above-ground biomass (leaves and other plant tissues) can be termed phytoextraction. This process by which some plants accumulate remarkable levels of heavy metals, for instance, in the range of 100–1,000-fold the levels normally found in most species is termed hyperaccumulation. The plants so applied for this type of phytoremediation process may later on be harvested and removed from site and in the case of valuable metals, the accumulated element can be recycled, a process termed phytomining.

In another process known as rhizofiltration, the contaminants are also removed from the polluted medium, but in this case, into the root system of the plant and pollutants are removed from the site when the plants are harvested. This process has been exploited for the remediation of polluted groundwater (either in situ or extracted), surface water, or wastewater for removal of metals or other inorganic compounds. Phytostabilization is another process that takes the advantage of the changes that the presence of the plant induces in soil chemistry and environment. The changes induced in soil chemistry by the presence of this plant may induce adsorption of contaminants onto such plant roots or surrounding soil or it may cause metals precipitation onto the plant root. Presence of this kind of plant may also lead to reduction in the mobility of contaminants of interest by reducing the potential for water and wind erosion. Another plant-assisted phytoremediation mechanism is known as rhizodegradation, a process that refers to the breakdown of contaminants within the plant root zone, or rhizosphere. This kind of phytoremediation process is believed to be carried out mainly by bacteria or other microorganisms (Fig. 6.1) whose numbers are believed to vary in the rhizosphere. The variation in microbial populations might be due to changes in sugars, amino acids, enzymes, and other compounds exuded by plants. It has also been observed that the roots provide additional surface area for microbes to grow on and a pathway for oxygen transfer from the environment.

The process of phytodegradation is achieved when contaminants are broken down after they have been taken up by the plant. Uptake of contaminants by plants occurs when the solubility and hydrophobicity of such contaminant fall into a certain acceptable range. Phytovolatilization, like phytoextraction and phytodegradation, also involves contaminants being taken up into the body of the plant but unlike the latter two processes, the contaminant, a volatile form thereof, or a volatile

Table 6.1 Brief overview of some phytoremediation processes

Mechanisms	Brief description of process	Examples of contaminants	Examples of plants used for demonstration/remediation	References
Phytoextraction (phytoaccumulation or phytomining)	Uptake and hyperaccumulation of pollutants into harvestable plant biomass. Mostly used for soil remediation	Phenol, creosote, bromacil, nitrobenzene, Zn, Cu, Ni, Cr, Cd, Pb, etc.	Wetland plants, barley, soya beans, *Cardaminopsis halleri, Dactylis glomerata, Empetrum nigrum, Echinochloa colona, Potamogeton crispus, Thlaspi caerulescens, Medicago sativa*, hybrid poplar trees, *Brassica pekinensis*, maize, wheat, duckweed, etc.	Briggs et al. (1982), McFarlane et al. (1987), Vojtechová and Leblová (1991), Brown et al. (1994), Polprasert et al. (1996), Burken and Schnoor (1998), Hafez et al. (1998), Xiong (1998), Zayed et al. (1998), Monni et al. (2000), Öncel et al. (2000), Rout et al. (2000), Neumann and zur Nieden (2001), Peralta et al. (2001), Schwartz et al. (2001), Rasmussen and Olsen (2004), Gosh and Singh (2005), and Claus et al. (2007)
Rhizofiltration (phytofiltration)	The use of hydroponically cultivated plant roots to remediate contaminated water/groundwater through absorption, concentration and precipitation of pollutants	Cd, Cr, Zn, Na, Cu, U, As, antimony, cesium, Pb, Mn, natural uranium, Ni, (226) Ra, etc.	*Phaseolus vulgaris, Brassica juncea, Helianthus annuus, Hydrocotyle umbellate, Phragmites australis, Elodea Canadensis*, etc.	Dushenkov et al. (1995, 1997a, 1997b), Pletsch et al. (1999), Dushenkov and Kapulnik (2002), Marash-Whitman (2006), Verma et al. (2006), Lee et al. (2007), Prasad (2007), Ghassemzadeh et al. (2008), Khilji and Bareen (2008), Tomé et al. (2008), and Yang et al. (2008)
Phytostabilization (immobilization)	The use of plants to immobilize contaminants in the soil and groundwater through absorption and accumulation by roots, adsorption onto roots, or precipitation within the root zone of plants	Cu, Cd, Cr, Ni, Pb, Zn, As, Ni, etc.	*Anthyllis vulneraria, Betula pendula, Festuca arvernensis, Koeleria vallesiana, Phaseolus vulgaris, Oryza sativa, Brassica oleracea, Zea mays, Armeria arenaria, Lupinus albus, Lupinus uncinatus*, hybrid poplar, grasses, etc.	Gussarsson (1994), Guo and Marschner (1995), Pierzynski et al. (2002), Ximenez-Embun et al. (2002), Zornoza et al. (2002), Vazquez et al. (2006), and Ehsan et al. (2009)

(continued)

Table 6.1 (continued)

Mechanisms	Brief description of process	Examples of contaminants	Examples of plants used for demonstration/remediation	References
Rhizodegradation (phytostimulation)	Plant-assisted bioremediation through the production of enzymes that degrade pollutants or stimulate microbial and fungal degradation by releasing exudates/enzymes into the rhizosphere	Phenols, phenanthrene, polyaromatic hydrocarbons (PAH), polychlorinated pesticides (PCP), polychlorinated biphenyls (PCBs), petroleum hydrocarbons, nitroaromatic compounds, explosives, surfactants, organophosphate insecticides, trichloroethylene, etc.	Alfalfa plants, hybrid poplars, etc.	Lee and Banks (1993), Dec and Bollag (1994), Donnelly et al. (1994), Schwab and Banks (1994), Schnoor et al. (1995), Aitken and Heck (1998), Carman et al. (1998), Pradhan et al. (1998), Mackova et al. (1998), Yateem et al. (1999), Duran and Esposito (2000), Husain and Jan (2000), Miya and Firestone (2000), Alkorta and Garbisu (2001), Dietz and Schnoor (2001), and Chaudhry et al. (2005)
Phytodegradation/phytotransformation	Uptake and degradation/destruction of pollutants in plant tissue through which plants and associated microorganisms degrade organic pollutants	Polychlorinated biphenyls, trichloroethylene, etc.	Alfalfa plants, Hybrid poplar trees, etc.	Shimp et al. (1993), Davis et al. (1994), Mackova et al. (1997), Newman et al. (1997), Salt et al. (1998), Burken et al. (2000), Garrison et al. (2000), Macek et al. (2000), Meagher (2000), Newman and Reynolds (2004), and Su-ramanian et al. (2006)
Phytovolatilization	Uptake of contaminants and volatilization by leaves (and other parts of plant) and release into the air	Selenium, CCl₄, naphthalene, trichloroethylene (TCE), petroleum hydrocarbon, etc.	*Stanleya pinnata*, *Zea mays*, *Brassica sp.*, etc.	Sandermann (1994), Schwab and Banks (1994), Watkins et al. (1994), Davis et al. (1996), Burken and Schnoor (1999), Macek et al. (2000), Rugh (2004), and Ayotamuno and Kogbara (2007)
Hydraulic Control (plume control)	The use of trees and plants that can transpire large volume of water in such a way as to reduce leaching of contaminants from the vadose zone	Carbon tetrachloride, trichloroethylene (TCE), methyl tert-butyl ether (MTBE), polycyclic aromatic hydrocarbon (PAH), etc.	Poplars, Cottonwood Trees, etc.	Licht and Schnoor (1993), Rock (1997), Cooke (1999), Jones et al. (1999), Wang et al. (1999), Hong et al. (2001), Robinson et al. (2003b) and Widdowson et al. (2005)

Fig. 6.1 Rhizodegradation of toxic contaminants by rhizospheric microorganisms (Adapted from Singh and Jain 2003)

degradation product is transpired with water vapor from leaves. Phytovolatilization may also entail the diffusion of contaminants from the stems or other plant parts that the contaminant travels through before reaching the leaves.

6.4   Necessity for Sustained Bioremediation Research and Development

There is the need for sustained bioremediation research and development worldwide particularly in the developing countries where the average standard of living is lower compared to the developed countries. Developing country is a term generally used to describe a nation with a low level of material well-being. This term is not to be confused with third world countries, a term that arose during the Cold War era to define countries that remained non-aligned or not moving at all with either capitalism and NATO (which along with its allies represented the First World) or communism and the Soviet Union (which along with its allies represented the Second World). This definition of third world countries provided a way of broadly categorizing the nations of the world into three groups based on social, political, and economic divisions. Although, the term continues to be used colloquially to describe the poorest countries in the world, this usage is widely disparaged since the term no longer holds any verifiable meaning after the fall of the Soviet Union deprecated the terms First World and Second World (Wikipedia 2010a, b). No single definition of the term developing country is recognized internationally, the levels of development may vary widely within the so-called developing countries, with some developing countries having high average standards of living (Sullivan and Sheffrin 2003; UN 2010a). According to UN (2010a), there is no established convention for the designation of "developed" and "developing" countries or areas, for example, in the United Nations system. The designations "developed" and "developing" are intended for statistical convenience and do not necessarily express a judgment about the stage reached by a particular country or area in the development process (UN 2010b). Countries with more advanced economies than other developing nations, but which have not yet fully demonstrated the signs of a developed country, are categorized under the term newly industrialized countries (Waugh 2000; Guillén 2003; Bożyk 2006; Mankiw 2007).

For the purpose of simplification, the International Monetary Fund (IMF) uses a flexible classification system that considers (1) per capita income level, (2) export diversification – so oil exporters that have high per capita GDP would not make the advanced classification because around 70% of its exports are oil, and (3) degree of integration into the global financial system (IMF 2010). Furthermore, the World Bank classifies countries into four income groups. Low-income countries have Gross National Income (GNI) per capita of US \$975 or less. Lower-middle-income countries have GNI per capita of US \$976–\$3,855. Upper-middle-income countries have GNI per capita between US \$3,856 and \$11,905. High-income countries have GNI above \$11,906 (World Bank 2010a). The World Bank classifies all low- and middle-income countries as developing but noted that "the use of the term is convenient; it is not intended to imply that all economies in the group are experiencing similar development or that other economies have reached a preferred or final stage of development. Classification by income does not necessarily reflect development status" (World Bank 2010a). The development of a country is measured with statistical indices such as income per capita (per person) (GDP), life expectancy, the rate of literacy, among others.

The UN has developed the Human Development Index (HDI), a compound indicator of the above statistics, to gauge the level of human development for countries where data is available. Developing countries are, in general, countries that have not achieved a significant degree of industrialization relative to their populations, and that have, in most cases a medium to low standard of living. There is a strong correlation between low income and high population growth (Wikipedia 2010a). The terms utilized when discussing developing countries refer to the intent and to the constructs of those who utilize these terms. Nevertheless, most of the countries termed "developing countries" are located in Africa, Asia, Latin America and the caribbean. These countries particularly those in SSA are still presently faced with socio-economic challenges. Many of their inhabitants are living below poverty line, although it is also necessary to note that poverty is also multi-dimensional in nature with different meaning to different people. For instance, poverty was defined by

World Bank (2001) as a pronounced deprivation of well-being that can be related to lack of material and income, low levels of education and health, vulnerability and exposure to risk, lack of opportunity to be heard, powerlessness, among others. Africa today is the poorest region in the world, where half of the population lives on less than one dollar a day (CFA 2005). Africa south of the Sahara has 21 of the 31 poorest countries in the world (Sayer and Palmer 1994).

In addition to these challenges, reports have also shown that the problem of environmental pollution is increasing in intensity and scale in many parts of the world including the developing countries, a trend that is likely to increase further in times to come. This problem of environmental pollution is likely to be worsened if the present and projected future demographic and socio-economic conditions are not addressed on a priority basis. According to the World Bank (2010b), the population of developing countries (low- and middle-income countries) will increase from 5,770,003,000 in 2010 to 7,879,731,000 in 2050 (Table 6.2). Similarly, population in Africa is also expected to increase from 922 million people in 2005 to 1,149.1 million people in 2015 (Table 6.3). The GDP in Africa as depicted in Table 6.4 is also low as compared to many developed countries. The issue of socio-economic and demographic characteristics is particularly important as experiences have shown that the costs associated with the clean-up of organic and inorganic pollutants can be staggering, even for developed countries. For instance, US $6–$8 billion is spent annually for environmental clean-up in the United States alone while US $25–$50 billion is spent per year worldwide (Tsao 2003). Estimates also suggest that to effectively clean up 1,200 of the United States' most contaminated and abandoned sites, the so-called Superfund sites, an estimated US $700 billion would be required (Glass 1999, 2000). The costs involved in remediation of more than 33,000 contaminated sites in Europe are equally overwhelming (Adriano 2001).

Consequently, proffering sustainable, cost-effective, easily adoptable, and adaptable solutions to this problem is imperative in many of these developing countries. The various remediation technologies currently commercially available and adopted to clean up contaminated medium and the environment ranged from in situ vitrification and soil incineration to excavation and land filling, soil washing, soil flushing, and solidification and stabilization by electrokinetic systems, chemical precipitation, electro-flotation, ion exchange, reverse osmosis and adsorption onto activated carbon, among others (Poon 1986; Glass 1999). Globally, the cost of non-biological technologies for cleaning up contaminated soils ranges from US $10 to $4,000 per cubic meter, or from US $100,000 to $3 million per hectare (Weiersbye 2007). Estimates from the United States of America indicated that excavation of metal contaminated soil from a one hectare area to a depth of 45 cm would result in approximately 5,000 ton of soil for treatment or land filling (Ensley 2000).

Apart from the challenges highlighted above, it has also been observed that these non-biological treatment methods and the recommendation that residues produced during industrial processes should have an adequate site for their final destination (Holmes et al. 1993) are mostly not applied in this and many parts of the world owing perhaps to inefficient enforcement of standard by regulatory bodies, high cost of procuring and maintaining some of the mitigation equipments, ignorance,

Table 6.2 Population projections for the developing countries between 2010 and 2050

	2010	2015	2020	2025	2030	2035	2040	2045	2050
Low-income countries	1,017,462,000	1,127,414,000	1,241,794,000	1,357,043,000	1,471,135,000	1,582,694,000	1,690,809,000	1,794,118,000	1,890,629,000
Middle-income countries	4,752,541,000	5,006,452,000	5,241,198,000	5,443,753,000	5,609,394,000	5,742,616,000	5,853,524,000	5,937,852,000	5,989,102,000
Total	5,770,003,000	6,133,866,000	6,482,992,000	6,800,796,000	7,080,529,000	7,325,310,000	7,544,333,000	7,731,970,000	7,879,731,000

Source: World Bank (2010b)

Table 6.3 Demographic projections in Africa by subregion between 2005 and 2015

	Year	Northern Africa	Western Africa	Central Africa	Eastern Africa	Southern Africa	Total Africa
Total Population (millions of people)	2005	189.6	272.5	112.5	292.5	54.9	922.0
	2010	206.3	307.4	129.6	332.1	56.6	1032.0
	2015	223.2	344.5	148.5	375.0	57.9	1149.1
Urban population (% of total)	2005	50.2	41.7	39.9	22.1	51.0	37.9
	2010	52.0	44.6	42.9	23.7	54.6	39.9
	2015	54.0	47.6	46.1	25.6	57.4	42.2
Total population growth (% increase per year)	2000–2005	1.7	2.6	2.8	2.6	1.1	2.3
	2005–2010	1.7	2.4	2.8	2.5	0.6	2.3
	2010–2015	1.6	2.3	2.7	2.4	0.5	2.2
Urban population growth (% increase per year)	2000–2005	2.4	4.0	4.2	3.9	2.0	3.4
	2005–2010	2.4	3.8	4.3	3.9	1.5	3.3
	2010–2015	2.3	3.6	4.1	4.0	1.3	3.2

Source: UN (2009)

Table 6.4 GDP for Africa by subregion as at 2006

Subregion	GDP, 2006 (US$ billion)	Share (%)	GDP growth 2006 (% per year)	GDP per capita, 2006 (US$)	GDP per capita growth, 2006 (% per year)
Northern Africa	409.1	36.9	7.21	2,098.6	5.34
Western Africa	215.4	19.4	4.99	779.8	2.17
Central Africa	64.0	5.78	3.16	547.9	0.69
Eastern Africa	76.5	6.90	5.74	350.4	2.83
Southern Africa	342.9	30.9	6.88	2,628.5	5.17
Africa	1,107.9	100	5.60	1,182.6	3.24

Source: Adapted from ITTO (2010) quoting World Bank (2007)
Note: Totals might not tally due to rounding

lack of vision, or carelessness, among others (Erakhrumen 2007a). Furthermore, it has also been observed that owing to the increasing presence of molecules, refractory to the microorganisms in the wastewater streams for instance, the conventional biological methods cannot be used for complete treatment of the effluent (Gogate and Pandit 2004a) thereby leading to the introduction of newer technologies. Few among these new technologies are the combination of oxidation processes operating at ambient conditions like cavitation, photocatalytic oxidation, Fenton's chemistry (belonging to the class of advanced oxidation processes) and ozonation, use of hydrogen peroxide (belonging to the class of chemical oxidation technologies) since it has been observed that none of the methods can be used individually in wastewater treatment applications with good economics and high degree of energy efficiency. Moreover, the knowledge required for the large-scale design and application is perhaps lacking (Gogate and Pandit 2004b) although, ambiguities still exist in terms of the selection of operating conditions for cost-effective operation. Likewise, the focus on waste minimization and water conservation in recent years has also resulted in the production of concentrated or toxic residues. It should be noted that some of the newly developed technologies, for example, cavitation may be more efficient on the laboratory scale and the knowledge required for the scale-up of the same and efficient large scale operation is presently lacking (Mason 2000; Adewuyi 2001; Gogate 2002). Thus, there is the need for cost-effective and reliable complementary and or alternative methods in this regard, one of which bioremediation has been suggested to be, in line with the outcomes of some studies.

The use of bioremediation technologies, like the use of plants to remediate contaminated soils and groundwater may be an old concept but is a fairly recent scientific development. It is an alternative technology capable of achieving permanent remediation, for instance, at waste sites without much associated problems (Sims et al. 1990); also, acceptance by the general public is another major advantage of this technology (Skladney and Metting 1993). Therefore, researches concerning plant-based environmental remediation have been widely pursued by academic and industrial scientists as a favorable low-impact clean-up technology applicable in both developed and developing countries (Raskin and Ensley 2000;

Robinson et al. 2003a, b). Companies specializing in phytoremediation have emerged in many developed and some developing countries to service a growing global market; the US market alone is estimated to be about US $150 million per year (Glass 1999, 2000). Given the low-cost and widely effective nature of phytore-mediation, it is likely that this green technology may be the only alternative for developing countries where clean-up is hindered by a lack of funding (Rajakaruna et al. 2006). It is estimated that phytotechnology costs range from US $0.02 to $10 per cubic meter, or US $200 to $100,000 per hectare (Weiersbye 2007). The tech-nology also has an advantage of being environmentally benign. For instance, the use of metal hyperaccumulator plants retains soil in situ, and results in a mere 25–35 ton of plant ash for disposal or metal recovery (Ensley 2000).

Phytoremediation can also be an income-generating technology, especially if metals removed from the contaminated medium can be used as bio-ore to extract useable metal (phytomining) (Brooks et al. 1998; Angle et al. 2001), and energy can be generated through biomass burning (Li et al. 2003). Phytomining is now a fast-developing field with the potential to generate income by exploiting low-grade ore bodies that are not economical to mine by conventional methods. The overall out-come of a carefully planned phytoremediation-phytomining operation would be a commercially viable metal product (metal-enriched bio-ore) and land better suited for agricultural operations or general habitation (Boominathan et al. 2004). However, the design of an efficient bioremediation system requires a set of careful studies of the local conditions (Gogoi et al. 2003). For example, the success of phytoremedia-tion depends on the availability of plant species, ideally those native to the region of interest and able to tolerate and accumulate high concentrations of pollutants (Baker and Whiting 2002). It is believed that such species are competitive under the local conditions and pose a lesser threat of becoming invasive, although it is noteworthy that phytoremediation techniques are best applied to areas that show low to moder-ate levels of contamination (Glass 2000).

The positive effects of these methods can be both direct and indirect (Rasmussen and Olsen 2004) as highlighted in Table 6.1, although presently, there appears to be a widening gap between the science and application of phytoremediation. The underlying biological mechanisms of phytoremediation and their interactions with associated biota also remain largely unknown. Some authors like Ernst (2000, 2005) even believe that phytoremediation is only "hype" and up to now phytoextraction of heavy metals, for instance, has been nothing more than transporting the harvest of metal-loaded plants from contaminated to clean sites. Furthermore, considerable expectations are now placed on genetic modification to generate model plants for commercial phytoremediation (Raskin 1996; Rugh 2004) while increasing public concern over the utilization of genetically modified organisms could force govern-ments to prohibit their use. Phytoremediation is also limited by the bioavailability of the pollutant. If only a fraction of the pollutant is bioavailable, but the regulatory clean-up standards require that all of the pollutant be removed, then "green clean" may not be sufficient. In such cases, bioavailability may be enhanced via soil amendments (Salt et al. 1998) or engineering-based technologies to enhance the efforts of the biological method. Such an integrated remediation effort requires a multidisciplinary team of scientists; a set of skills and expertise that may not always be locally available in some developing countries (Rajakaruna et al. 2006). Other gaps in research still exist particularly as it concerns the limitation to the use of this technology, some of which are highlighted below.

- The roots of many plants are short; thus, in order for remediation to be successful, contamination must be shallow enough and be within the rooting zone for the remediative plant roots to reach the contaminants or contamination must be brought to the plant. Trees have longer roots and can clean up slightly deeper contamination than smaller plants, but cannot remediate deep aquifers. Further design work may be necessary in order to remediate deep aquifers while ground surface at the site may have to be modified to prevent flooding or erosion.
- The time required to completely clean up contaminated sites is often long and may take many growing seasons, which may last several years, for instance, from the time a tree stand is established to when the process of phytoremediation is completed.
- Most phytoremediation processes are restricted to sites with low contaminants concentrations as extremely high contaminant concentrations may not allow plants to be used for remediation to grow or survive.
- Plants that absorb toxic materials may contaminate the food chain and may cause ecological exposure issues. Even if not eaten directly, contaminants may still enter the food chain through animals/insects that eat plant material containing contaminants. Likewise, harvested plant biomass from phytoextraction for instance, may be classified as a hazardous waste hence disposal should be proper.
- Climatic conditions are a limiting factor; climatic or hydrologic conditions may restrict the rate of growth of plants that can be utilized.
- Introduction of non-native species may affect biodiversity in the future since the consequences or otherwise of introducing them to the ecosystem may be presently unknown or unexpected.
- Phytovolatilization may remove contaminants from the subsurface, but might then cause increased airborne exposure, that is, transforming a soil or groundwater pollution problem to an air pollution problem.
- It is also believed that phytoremediation may be less efficient for hydrophobic contaminants, which bind tightly to soil.

As noted by Schwitzguébel (2000), phytoremediation was described as a nascent technology, the present status of phytoremediation research and technology application in many developing countries still appears to be a nascent technology. Very little information about awareness and practical applications of the technology is available in these countries and most of the studies concerning this technology and their practical demonstrations are currently carried out in the developed countries (Erakhrumen 2007a). Nevertheless, there is limited available information that an increased number of researchers in these countries are conducting studies/researches concerning bioremediation with some results already documented; although, most of these studies are still on laboratory scales, the outcomes of which are expected to contribute to documented information concerning applicability of this technology in developing countries in the near future.

6.5   Conclusion

Studies have shown that different types of plant-assisted bioremediation processes can be exploited for the removal of pollutants from the environment. Presently, these methods of in situ remediation technologies and techniques that utilizes the inherent abilities of living plants in an ecologically friendly and solar-driven manner based on the concept of using nature to cleanse nature has not become commercially available in many parts of the world particularly in the developing countries irrespective of myriad of documented research outcomes concerning these processes. The application of these processes in this regards, owing to the inherent multiple advantages obtainable from them, is expected to be beneficial to countries with developmental challenges in line with many of the documented research outcomes. Nevertheless, it is noteworthy that application of these processes should be considered in tandem with not only the plant species to be employed as it concerns its physiological characteristics but the type(s), chemical properties, and quantity of the contaminant, site-specific conditions and interactions of a pollutant with soil or other contaminated medium, water, and plants, among other factors. Therefore, increased research efforts in this area are necessary, in developing countries and other parts of the world. This is particularly important to the developing countries since they are expected to surmount many of the developmental challenges they presently encounter toward achieving sustainable development. In order to achieve this, many of the developing countries are formulating, adopting, and adapting growth and developmental processes and strategies aimed at catching up with the current and future advances in various human endeavors.

As a result of this quest, substances that have been identified as being largely responsible for environmental pollution are likely to be produced in large quantities and at increased intensities, thereby requiring proper management. Therefore, research efforts in the area of plant-assisted bioremediation concerning the scaling-up of already developed technologies for local adoption/adaptation including areas that are still considered as limitation or constraint to the application of this technology in cleaning up the environment are imperative.

References

Adewuyi YG (2001) Sonochemistry: environmental science and engineering applications. Ind Eng Chem Res 40:4681

Adriano DC (2001) Trace elements in the terrestrial environments: biogeochemistry, bioavailability, and risks of metals. Springer, New York, p 866

Aitken MD, Heck PE (1998) Turnover capacity of coprinus cinereus peroxidase for phenol and monosubstituted phenols. Biotechnol Prog 14:487–492

Alkorta I, Garbisu C (2001) Phytoremediation of organic contaminants in soils. Biores Technol 79:273–276

Angle JS, Chaney RL, Baker AJM, Li Y, Reeves R, Volk V, Roseberg R, Brewer E, Burke S, Nelkin J (2001) Developing commercial phytoextraction technologies: practical considerations. S Afr J Sci 97:619–623

Arvin E, Godsy EM, Grbic-Galic D, Jensen B (1988) Microbial degradation of oil and creosote related aromatic compounds under aerobic and anaerobic conditions. In: International conference on physicochemical and biological detoxification of hazardous waste, Technomic, Lancaster, PA, pp 828–847

Ayotamuno JM, Kogbara RB (2007) Determining the tolerance level of *Zea mays* (maize) to a crude oil polluted agricultural soil. Afr J Biotechnol 6:1332–1337

Baker AJM, Whiting SM (2002) In search for the holy grail- another step in understanding metal hyperaccumulation? New Phytol 155:1–7

Banks D, Younger PL, Arnesen RT, Iversen ER, Banks S (1997) Mine-water chemistry: the good, the bad, and the ugly. Environ Geol 32:157–174

Bauer R, Fallmann H (1997) The photo–Fenton oxidation – a cheap and efficient wastewater treatment method. Res Chem Intermed 23:341

Boominathan R, Saha-Chaudhury NM, Sahajwalla V, Doran PM (2004) Production of nickel bio-ore from hyperaccumulator plant biomass: applications in phytomining. Biotechnol Bioengg 86:243–250

Bożyk P (2006) Newly industrialized countries. In: Globalization and the transformation of foreign economic policy. Ashgate Publishing, Ltd, Aldershot, UK, ISBN 0-75-464638-6

Briggs GG, Bromilow RH, Evans AA (1982) Relationship between lipophilicity and root uptake and translocation of non-ionised chemicals by barley. Pestic Sci 13:495–504

Brooks RR, Chambers MF, Nicks LJ, Robinson BH (1998) Phytomining. Trends Plant Sci 3:359–362

Brown SL, Chaney RL, Angle JS, Baker JM (1994) Phytoremediation potential of *Thlaspi caerulescens* and bladder campion for zinc and cadmium-contaminated soil. J Environ Qual 23:1151–1157

Burken JG, Schnoor JL (1998) Predictive relationships for uptake of organic contaminants by hybrid poplar trees. Environ Sci Technol 32:3379–3385

Burken JG, Schnoor JL (1999) Distribution and volatilization of organic contaminants following uptake by hybrid poplar trees. Int J Phytoremed 1:139–152

Burken JG, Shanks JV, Thompson PL (2000) Phytoremediation and plant Metabolism of explosives and nitroaromatic compounds. In: Spain JC, Hughes JB, Knackmuss HJ (eds.) Biodegradation of nitroaromatic compounds and explosives. Lewis, Washington, DC, pp 239–275

Carman E, Crossman T, Gatliff E (1998) Phytoremediation of No. 2 fuel-oil contaminated soil. J Soil Contam 7:455–466

CFA (2005) Our common interest. A report of the Commission for Africa, Commission for Africa, London, p 461

Chaney R, Li Y, Angle S, Baker A, Reeves R, Brown S, Homer F, Malik M, Chin M (2000) Improving metal hyperaccumulator wild plants to develop phytoextraction systems: approaches and progress. In: Terry N, Banuelos G (eds.) Phytoremediation of contaminated soil and water. Lewis Publishers, Boca Raton, pp 129–158

Chaudhry Q, Blom-Zandstra M, Gupta S, Joner EJ (2005) Utilising the synergy between plants and rhizosphere microorganisms to enhance breakdown of organic pollutants in the environment. Environ Sci Pollut Res 12:34–48

Claus D, Dietze H, Gerth A, Grossser W, Hedner A (2007) Application of agronomic practice improves phytoextraction on a multipolluted site. J Environ Engg Landscape Manage 15:208–212

Cooke MT (1999) Phytoremediation: using plants to remediate groundwater contaminated with trichloroethylene (TCE). Pennsylvania State University, State College

Davis LC, Muralidharan N, Visser VP et al (1994) Alfalfa plants and associated microorganisms promote biodegradation rather than volatilization of organic-substances from ground-water. ACS Symp Ser 563:112–122

Davis LC, Banks ML, Schwab AP, Muralidharan N, Erickson LE, Tracy JC (1996) Plant based bioremediation. In: Sikdar I (ed.) Bioremediation. Technomics, Basel

Dec J, Bollag JM (1994) Use of plant-material for the decontamination of water polluted with phenols. Biotechnol Bioengg 44:1132–1139

Dey BK, Hashim MA, Hasan S, Gupta BS (2004) Microfiltration of water-based paint effluents. Adv Environ Res 8:455–466

Dietz AC, Schnoor JL (2001) Advances in phytoremediation. Environ Health Perspect 109:163–168

Donnelly P, Hegde R, Fletcher J (1994) Growth of PCB-degrading bacteria on compounds from photosynthetic plants. Chemosphere 28:981–988

Duran N, Esposito E (2000) Potential applications of oxidative enzymes and phenoloxidase-like compounds in wastewater and soil treatment: a review. Appl Catal B Environ 28:83–99

Dushenkov S (2003) Trends in phytoremediation of radionuclides. Plant Soil 249:167–175

Dushenkov S, Kapulnik Y (2002) Phytofilttration of metals. In: Raskin I, Ensley BD (eds.) Phytoremediation of toxic metals: using plants to clean up the environment. Wiley, New York, pp 89–106

Dushenkov V, Kumar PBAN, Motto H, Raskin I (1995) Rhizofiltration: the use of plants to remove heavy metals from aqueous streams. Environ Sci Technol 29:1239–1245

Dushenkov S, Kapulnik Y, Blaylock M, Sorochisky B, Raskin I, Ensley B (1997a) Phytoremediation: a novel approach to an old problem. In: Wise DL (ed.) Global environmental biotechnology. Elsevier Science B.V, Amsterdam, pp 563–572

Dushenkov S, Vasudev D, Kapulnik Y, Gleba D, Fleisher D, Ting KC, Ensley B (1997b) Removal of uranium from water using terrestrial plants. Environ Sci Technol 31:3468–3474

Ehsan M, Santamaría-Delgado K, Vázquez-Alarcón A, Alderete-Chavez A, De la Cruz-Landero N, Jaén-Contreras D, Molumeli PA (2009) Phytostabilization of cadmium contaminated soils by *Lupinus uncinatus* Schldl. Span J Agric Res 7:390–397

Ensley BD (2000) Rational use for phytoremediation. In: Raskin I, Ensley BD (eds.) Phytoremediation of toxic metals: using plants to clean-up the environment. Wiley., New York, pp 3–12

Erakhrumen AA (2007a) Phytoremediation: an environmentally sound technology for pollution prevention, control and remediation in developing countries. Educ Res Rev 2:151–156

Erakhrumen AA (2007b) State of forestry research and education in Nigeria and Sub-Saharan Africa: implications for sustained capacity building and renewable natural resources development. J Sustain Dev Afr 9:133–151

Erakhrumen AA (2008) Environment as a common property in the context of historical and contemporary anthropogenic activities: Influence on climate change, effects, and mitigation/adaptation strategies. In: Popoola L (ed.) Climate change and sustainable renewable natural resources management, Proceedings of the 32nd annual conference of the forestry association of Nigeria held in Umuahia, Abia State, Nigeria, 20–24 Oct 2008, pp 135–151

Ernst WHO (2000) Evolution of metal hyperaccumulation and the phytoremediation hype. New Phytol 146:357–58

Ernst WHO (2005) Phytoextraction of mine wastes – options and impossibilities. Chem Erde 65:29–42

Feigelson L, Muszkat L, Bir L, Muszkat KA (2000) Dye photo-enhancement of TiO_2-photocatalyzed degradation of organic pollutants: the organobromine herbicide bromacil. Water Sci Technol 42:275

Florence TM, Batley GE (1980) Chemical speciation in natural waters. CRC Anal Chem 9:219–296

Garrison AW, Nzengung VA, Avants JK, Ellington JJ, Jones EW, Rennels D, Wolfet NL (2000) Phytodegradation of p, p' – DDT and the enantiomers of o, p' – DDT. Environ Sci Technol 34:1663–1670

Ghassemzadeh F, Yousefzadeh H, Arbab-Zavar MH (2008) Removing arsenic and antimony by *Phragmites australis*: rhizofiltration technology. J Appl Sci 8:1668–1675

Glass DJ (1999) US and international markets for phytoremediation, 1999–2000. D. Glass Associates, Needham, 270 pp

Glass DJ (2000) Economic potential of phytoremediation. In: Raskin I, Ensley BD (eds.) Phytoremediation of toxic metals. Using plants to clean up the environment. Wiley, New York, pp 15–31

Gogate PR (2002) Cavitation: an auxiliary technique in wastewater treatment schemes. Adv Environ Res 6:335

Gogate PR, Pandit AB (2004a) A review of imperative technologies for wastewater treatment I: oxidation technologies at ambient conditions. Adv Environ Res 8:501–551

Gogate PR, Pandit AB (2004b) A review of imperative technologies for wastewater treatment II: hybrid methods. Adv Environ Res 8:553–597

Gogoi BK, Dutta NN, Goswami P, Krishna Mohan TR (2003) A case study of bioremediation of petroleum-hydrocarbon contaminated soil at a crude oil spill site. Adv Environ Res 7:767–782

Gosh M, Singh SP (2005) Comparative uptake and phytoextraction study of soil induced chromium by accumulator and high biomass weed species. Appl Ecol Environ Res 3:67–69

Grêman H, Vodnik D, Velikonja-Bolta Š, Leštan D (2003) Heavy metals in the environment. J Environ Qual 32:500–506

Guillén MF (2003) Multinationals, ideology, and organized labor. The limits of convergence. Princeton University Press, Princeton, New Jersey, US, ISBN 0-69-111633-4

Guo Y, Marschner H (1995) Uptake, distribution and binding of cadmium and nickel in different plant species. J Plant Nutr 18:2691–2706

Gussarsson M (1994) Cadmium-induced alterations in nutrient composition and growth of *Betula pendula* seedlings: the significance of fine roots as a primary target for cadmium toxicity. J Plant Nutr 17:2151–2156

Hafez N, Abdalla S, Ramadan YS (1998) Accumulation of phenol by *Potamogeton crispus* from aqueous industrial waste. Bull Environ Contam Toxicol 60:944–948

Heck RM, Farrauto RJ (1995) Catalytic air pollution control: commercial technology. Wiley, New York

Holmes G, Singh BR, Theodore L (1993) Handbook of environmental management and technology. Wiley, New York, pp 229–264

Hong MS, Farmayan WF et al (2001) Phytoremediation of MTBE from a groundwater plume. Environ Sci Technol 35:1231–1239

Husain Q, Jan U (2000) Detoxification of phenols and aromatic amines from polluted wastewater by using phenol oxidases. J Sci Ind Res 59:286–293

Hutchinson SL, Schwab AP, Banks MK (2003) Biodegradation of petroleum hydrocarbons in the rhizosphere. In: McCutcheon SC, Schnoor JL (eds.) Phytoremediation: transformation and control of contaminants. Wiley-Interscience, Hoboken, pp 355–386

IMF (2010) How does the world economic outlook categorize advanced versus emerging and developing economies? International Monetary Fund. http://www.imf.org/external/pubs/ft/weo/faq.htm#q4b

ITTO (2010) Good Neighbours. Promoting intra-African markets for timber products. International tropical timber organization technical series number 35, p 112

Jiang W, Liu D, Hou W (2000) Hyper-accumulation of cadmium by roots, bulbs and shoots of garlic (*Allium sativum* L.). Biores Technol 76:9–13

Jones SA, Lee RW, Kuniansky EL (1999) Phytoremediation of Trichloroethylene (TCE) using Cottonwood Trees. In: Leeson A, Alleman BC (eds.) Phytoremediation and Innovative Strategies for Specialized Remedial Applications. The Fifth International *In Situ* and On-Site Bioremediation Symposium, April 19–22, 1999, vol 6. Battelle Press, San Diego, pp 101–108

Kamaludeen SPBK, Arunkumar KR, Avudainayagam S, Ramasamy K (2003) bioremediation of chromium contaminated environments. Ind J Exp Biol 41:972–985

Khilji S, Bareen F (2008) Rhizofiltration of heavy metals from the tannery sludge by the anchored hydrophyte, *hydrocotyle umbellata* L. Afr J Biotechnol 7:3711–3717

King C (2006) Soil. In: Microsoft® student 2007 [DVD]. Microsoft Corporation, Redmond, WA

Krupa Z, Moniak M (1998) The stage of leaf maturity implicates the response of the photosynthetic apparatus to cadmium toxicity. Plant Sci 138:149–156

Lee E, Banks MK (1993) Bioremediation of petroleum contaminated soil using vegetation: a microbial study. J Environ Sci Health A28:2187–2198

Lee M, Yang M, Chang Y (2007) Rhizofiltration to remove uranium from groundwater by using *Phaseolus vulgaris var. Humilis, Brassica juncea (L.) Czern.*, and *Helianthus annuus* L. Geol Soc Am Abstr Programs 39(6):61

Li YM, Chaney R, Brewer E, Rosenberg R, Angle SJ, Baker AJM, Reeves RD, Nelkin J (2003) Development of technology for commercial phytoextraction of nickel: economic and technical considerations. Plant Soil 249:107–115

Licht LA, Schnoor JL (1993) Tree buffers protect shallow ground water at contaminated sites. EPA ground water currents, office of solid waste and emergency response. EPA/542/N-93/011

Lin Z, Puls RW (2003) Potential indicators for the assessment of arsenic natural attenuation in the subsurface. Adv Environ Res 7:825–834

Macek T, Mackova M, Kas J (2000) Exploitation of plants for the removal of organics in environmental remediation. Biotechnol Adv 18:23–34

Mackova M, Macek T, Ocenaskova J, Burkhard J, Demnerova K, Pazlarova J (1997) Biodegradation of polychlorinated biphenyls by plant cells. Int Biodeter Biodegrad 39:317–325

Mackova M, Macek T, Kucerova P, Burkhard J, Trisk J, Demnerova K (1998) Plant tissue cultures in model studies of transformation of polychlorinated biphenyls. Chem Pap 52:599–600

Mankiw NG (2007) Principles of economics, 4th edn. Thompson & South-Western, New York. ISBN 0-32-422472-9

Mantzavinos D, Hellenbrand R, Livingston AG, Metcalfe IS (1997) Reaction mechanisms and kinetics of chemical pre-treatment of bio-resistant organic molecules by wet air oxidation. Water Sci Technol 35:119

Marash-Whitman D (2006) Heavy metal trafficking: rhizofiltration efficacy of *Elodea canadensis* in copper contaminated effluents. http://www.usc.edu/CSSF/History/2006/Projects/S0810.pdf

Mason TJ (2000) Large scale sonochemical processing: aspiration and actuality. Ultrason Sonochem 7:145

McCutcheon SC, Schnoor JL (eds.) (2003) Phytoremediation: transformation and control of contaminants. Wiley-interscience, Hoboken

McFarlane JC, Pfleeger T, Fletcher J (1987) Transpiration effect on the uptake and distribution of bromacil, nitrobenzene, and phenol in soybean plants. J Environ Qual 16:372–376

McGrath S, Zhao F, Lombi E (2002) Phytoremediation of metals, metalloids and radionuclides. Adv Agron 75:1–56

Meagher RB (2000) Phytoremediation of toxic elements and organic pollutants. Curr Opin Plant Biol 3:153–162

Microsoft Encarta (2006) Groundwater. In: Microsoft® Student 2007 [DVD]. Microsoft Corporation, Redmond, WA

Miya RK, Firestone MK (2000) Phenanthrene-degrader community dynamics in rhizosphere soil from a common annual grass. J Environ Qual 29:584–592

Monni S, Salemaa M, Milar N (2000) The Tolerance of *Empetrum nigrum* to copper and nickel. Environ Poll 109:221–229

Neumann D, Zur Nieden U (2001) Silicon and heavy metal tolerance of higher plants. Phytochemistry 56:685–692

Newman LA, Reynolds CM (2004) Phytodegradation of organic compounds. Curr Opin Biotechnol 15:225–230

Newman LA, Strand SE, Choe N et al (1997) Uptake and biotransformation of trichloroethylene by hybrid poplars. Environ Sci Technol 31:1062–1067

Öncel I, Kele Y, Üstün AS (2000) Interactive effects of temperature and heavy metal stress on the growth and some biochemical compounds in wheat seedlings. Environ Poll 107:315–320

Otal E, Mantzavinos D, Delgado MV, Hellenbrand R, Lebrato J, Metcalfe IS et al (1997) Integrated wet air oxidation and biological treatment of polyethylene glycol-containing wastewaters. J Chem Tech Biotech 70:147

Ouariti O, Boussama N, Zarrouk M, Cherif A, Ghorbal MH (1997) Cadmium- and copper-induced changes in tomato membrane lipid. Phytochemistry 45:1343–1350

Peralta JR, Gardea-Torresdey JL, Tiemann KJ, Gomez E, Arteaga S, Rascon E, Carrillo G (2001) Uptake and effects of five heavy metals on seed germination and plant growth in alfalfa (*Medicago sativa*). Bull Environ Contam Toxicol 66:727–734

Peralta-Videa JR, de la Rosa G, Gonzalez JH, Gardea-Torresdey JL (2004) Effects of the growth stage on the heavy metal tolerance of alfalfa plants. Adv Environ Res 8:679–685

Petrisor IG, Dobrota S, Komitsas K, Lazar I, Kuperberg JM, Serban M (2004) Artificial inoculation-perspectives in tailings phytostabilization. Int J Phytorem 6:1–15

Pierzynski GM, Schnoor JL, Youngman A, Licht L, Erickson LE (2002) Poplar trees for phytostabilization of abandoned zinc-lead smelter. Pract Periodical Haz Tox Radioactive Waste Mgmt 6:177–183

Pilon-Smits EAH (2005) Phytoremediation. Annu Rev Plant Biol 56:15–39

Pletsch M, de Araujo BS, Charlwood BV (1999) Novel biotechnological approaches in environmental remediation research. Biotechnol Adv 17:679–687

Polprasert C, Dan NP, Thayalakumaran N (1996) Application of constructed wetlands to treat some toxic wastewaters under tropical conditions. Water Sci Technol 34:165–171

Poon CPC (1986) Removal of Cd (II) from wastewaters. In: Mislin H, Raverva O (eds.) Cadmium in the environment. Birkha User, Basel, pp 6–55

Pradhan SP, Conrad JR, Paterek JR, Srivastava VJ (1998) Potential of phytoremediation for treatment of PAHS in soil at MGP sites. J Soil Contam 7:467–480

Prasad MNV (2007) Sunflower (*Helianthus annuus* L.) – a potential crop for environmental industry. HELIA 30:167–174

Rajakaruna N, Tompkins KM, Pavicevic PG (2006) Phytoremediation: an affordable green technology for the clean-up of metal-contaminated sites in Sri Lanka. Cey J Sci Bio Sci 35:25–39

Raskin I (1996) Plant genetic engineering may help with environmental cleanup. Proc Natl Acad Sci 93:3164–3166

Raskin I, Ensley BD (2000) Phytoremediation of Toxic Metals. Using plants to clean up the environment. Wiley, New York, 304pp

Rasmussen G, Olsen RA (2004) Sorption and biological removal of creosote-contaminants from groundwater in soil/sand vegetated with orchard grass (*Dactylis glomerata*). Adv Environ Res 8:313–327

Richardson ML, Gangolli S (1992) The dictionary of substances and their effects. Royal Society of Chemistry, Cambridge

Robinson BH, Fernandez JE, Madejon P, Maranon P, Murillo JM, Green S, Clothier B (2003a) Phytoextraction: an assessment of biogeochemical and economic viability. Plant Soil 249:117–125

Robinson BH, Green S, Mills T, Clothier B, Velde M, Laplane R, Fung L, Deurer M, Hurst S, Thayalakumaran T, Dijssel C (2003b) Phytoremediation: using plants as biopumps to improve degraded environments. Aust J Soil Res 41:599–611

Rock SA (1997) The Standard handbook of hazardous waste treatment and disposal. In: Freeman H (ed.) Phytoremediation, 2nd edn. McGraw Hill, New York

Rout GR, Samantaray S, Das P (2000) Effects of chromium and nickel on germination and growth in tolerant and non-tolerant populations of *Echinochloa colona* (L.) Link. Chemosphere 40:855–859

Rugh CL (2004) Genetically engineered phytoremediation: one man's trash is another man's transgene. Trends Biotechnol 22:496–468

Russell-Jones R (1987) The Health effects of vehicle emissions. IMECHE conference on vehicle emissions and their impact on European air quality. Paper C355/87, pp 55–59

Salt DE, Smith RD, Raskin I (1998) Phytoremediation. Annu Rev Plant Physiol Plant Mol Biol 49:643–668

Sandermann H (1994) Higher plant metabolism of xenobiotics: the green liver concept. Pharmacogenetics 4:225–241

Sayer JA, Palmer JR (1994) Overview on forest research in Africa. CIFOR working paper No. 1, Sep, 1994. Center for International Forestry Research. Invited paper for the international symposium supporting capacity building in forestry research in Africa AAS/IFS in collaboration with FAO, Nairobi, Kenya, 28 June–1 July 1994, p 19

Schnoor JL, Licht LA, McCutcheon SC, Wolfe NL, Carreira LH (1995) Phytoremediation of organic and nutrient contaminants. Environ Sci Technol 29:A318–A323

Schwab AP, Banks MK (1994) Bioremediation through rhizosphere technology. American chemical society, (ACS) symposium series 563. In: Anderson TA, Coats JR (eds.) Biologically mediated dissipation of polyaromatic hydrocarbons in the root zone. American Chemical Society, Washington, DC, pp 132–141

Schwartz C, Gérard E, Perronnet JK, Morel JL (2001) Measurement of *in-situ* phytoextraction of zinc by spontaneous metallophytes growing on a former smelter site. Sci Total Environ 279:215–221

Schwitzguébel J-P (2000) Potential of phytoremediation, an emerging green technology. In: Ecosystem service and sustainable watershed management in North China, proceedings of the international conference, Beijing, P.R. China, 23–25 Aug 2000

Shimp JF, Tracy JC, Davis LC, Lee E, Huang W, Ericknon LE, Schnoor JL (1993) Beneficial effects of plants in the remediation of soil and groundwater contaminated with organic materials. Crit Rev Environ Sci Technol 23:41–77

Sims JL, Sims RC, Mathews JE (1990) Approach to bioremediation of contaminated soil. Hazard Waste Hazard Mater 7:117–145

Singh OV, Jain RK (2003) Phytoremediation of toxic aromatic pollutants from soil. Appl Microbiol Biotechnol 63:128–135

Skladney GJ, Metting FB (1993) Bioremediation of contaminated Soil. In: Metting FB Jr (ed.) Soil microbial ecology. Marcel-Dekker, New York, pp 483–510

Skorzynska-Polit E, Baszynski T (1997) Differences in sensitivity of the photosynthetic apparatus in cd-stressed runner bean plants in relation to their age. Plant Sci 128:11–21

Subramanian M, Oliver DO, Shanks JV (2006) TNT Phytotransformation pathway characteristics in arabidopsis: role of aromatic hydroxylamines. Biotechnol Program 22:208–216

Sullivan A, Sheffrin SM (2003) Economics: principles in action. Pearson Prentice Hall, Upper Saddle River, 471 pp. ISBN 0-13-063085-3

Tomé FV, Rodríguez PB, Lozano JC (2008) Elimination of natural uranium and (226)Ra from contaminated waters by rhizofiltration using *Helianthus annuus* L. Sci Total Environ 393:351–357

Tsao DT (2003) Phytoremediation. advances in biochemical engineering biotechnology 78. Springer, Berlin, 206pp

Tukendorf A, Skorzynska-Polit E, Baszynski T (1997) Homophytochelatin accumulation in cd-treated runner bean plants in related to their growth stage. Plant Sci 129:21–28

UN (2009) World Urbanization Prospects: the 2007 revision population database by United Nations. www.esa.un.org/unup/index.asp

UN (2010a) Composition of macro geographical (Continental) regions, geographical sub-regions, and selected economic and other groupings. United Nations Statistics Division. Revised 1 Apr 2010. http://unstats.un.org/unsd/methods/m49/m49regin.htm#ftnc

UN (2010b) Standard country or area codes for statistical use: standard country or area codes and geographical regions for statistical use. United Nations Statistics Division. http://unstats.un.org/unsd/methods/m49/m49.htm

Vazquez S, Agha A, Granado A, Sarro MJ, Esteban E, Peñalosa JM, Carpena RO (2006) Use of white lupin plant for phytostabilization of Cd and as polluted acid soil. Water Air Soil Pollut 177:349–365

Verma P, George KV, Singh HV, Singh SK, Juwarkar A, Singh RN (2006) Modeling rhizofiltration: heavy-metal uptake by plant roots. Environ Mod Assess 11:387–394

Vojtechová M, Leblová S (1991) Uptake of lead and cadmium by maize seedlings and the effect of heavy metals on the activity of phosphoenolpyruvate carboxylase isolated from maize. Biol Plant 33:386–394

Waldemar M, Baszynski T (1996) Different susceptibility of runner bean plants to excess of copper as a function of the growth stage of primary leaves. J Plant Physiol 149:217–221

Wang X, Newman L, Gordon M, Strand S (1999) Biodegradation of carbon tetrachloride by poplar trees: results from cell culture and field experiments. In: Leeson A, Alleman BC (eds.) Phytoremediation and innovative strategies for specialized remedial applications. Battelle Press, Columbus, pp 133–138

Watkins JW, Sorensen DL, Sims RC (1994) Volatilization and mineralization of naphthalene in soil–grass microcosms, Chapter 11. ACS symposium series 563. American Chemical Society, Washington, DC, pp 123–131

Waugh D (2000) Manufacturing industries (Chapter 19), World development (Chapter 22). Geography, an integrated approach, 3rd edn. Nelson Thornes Ltd., UK, pp 563, 576–579, 633, 640, ISBN 0-17-444706-X

Weiersbye IM (2007) Global review and cost comparison of conventional and phyto-technologies for mine closure. Plenary paper. In: Fourie AB, Tibbett M, Wiertz J (eds.) Mine closure 2007 – proceedings of the 2nd international seminar, Santiago, Chile. Australian Centre for Geomechanics and the University of Western Australia, Perth, ISBN 978-0-9804 185-0-7, pp 13–31

Widdowson MA, Shearer S et al (2005) Remediation of polycyclic aromatic hydrocarbon compounds in groundwater using poplar trees. Environ Sci Technol 39:1598–1605

Wikipedia (2010a) Developing country. Wikipedia: the free online encyclopaedia. http://en.wikipedia.org/wiki/Developing_country

Wikipedia (2010b) Third world. Wikipedia: the free online encyclopaedia. http://en.wikipedia.org/wiki/Third_World

World Bank (2001) World development report 200/2001. Attacking poverty. Oxford University Press, Oxford

World Bank (2010a) Country classifications. http://data.worldbank.org/about/country-classifications

World Bank (2010b) Population projection tables by country and group. Online database. www.worldbank.org

Wu C, Khang SJ, Keener TC, Lee SK (2004) A Model for dry sodium bicarbonate duct injection flue gas desulfurization. Adv Environ Res 8:655–666

Ximenez-Embun P, Rodriguez-Sanz B, Madridalbarran Y, Camara C (2002) Uptake of heavy metals by lupin plants in artificially contaminated sand: preliminary results. Int J Environ Anal Chem 82:805–813

Xiong ZT (1998) Lead uptake and effects on seed germination and plant growth in a Pb Hyperaccumulator *Brassica pekinensis* Rupr. Bull Environ Contam Toxicol 60:285–291

Yabe MJS, de Oliveira E (2003) Heavy metals removal in industrial effluents by sequential adsorbent treatment. Adv Environ Res 7:263–272

Yang M, Chang Y, Kim J, Shin J, Lee M (2008) Study of the rhizofiltration by using *Phaseolus vulgaris* var., *Brassica juncea* (L.) Czern., *Helianthus annuus* L. to remove cesium from groundwater. Geochim Cosmochim Acta 72:A1056

Yateem A, Balba MT, El-Nawawy AS, Al-Awadhi N (1999) Experiments in phytoremediation of gulf war contaminated soil. Soil Groundwater Cleanup 2:31–33

Zayed A, Gowthaman S, Terry N (1998) Phytoaccumulation of trace elements by wetland plants: I. Duckweed. J Environ Qual 27:715–721

Zornoza P, Vazquez S, Esteban E, Fernandezpascual M, Carpena R (2002) Cadmium stress in nodulated white lupin. Strategies to avoid toxicity. Plant Physiol Biochem 40:1003–1009

Chapter 7
The Bacterial Flora of the Nickel-Hyperaccumulator Plant *Alyssum bertolonii*

Alessio Mengoni, Francesco Pini, and Marco Bazzicalupo

Abstract Recent years have witnessed a considerable growth of microbiological researches in serpentine soils in relation to the presence of hyperaccumulating plants. Nickel-hyperaccumulating plants accumulate huge amounts of heavy metals in shoots, and therefore, provide a specific environment for bacterial populations and in particular for endophytic bacteria. Bacterial endophytes have been studied in many different plant species and in some cases they have been found to promote plant growth or to confer the plant higher tolerance to biotic and abiotic stress. Here, we report the data on presence, composition and possible roles of bacteria associated with *Alyssum bertolonii* Desv. (Brassicaceae), the first nickel-hyperaccumulator plant discovered endemic to serpentine outcrops of Central Italy. The analysis of both cultivable and total fraction of the soil bacterial community showed a very strong effect of the plant in shaping the community composition. Moreover, the plant harbors a complex and highly variable endophytic bacterial flora with many Ni-resistant strains. Endophytic bacteria were isolated from roots, stems, and leaves of several *A. bertolonii* plants and populations allowing providing a model of correlation between taxonomic compositions of bacterial communities from different organs, plants, populations, and surrounding soils. Some of the endophytic bacteria tested for plant tissue colonization ability, and for their influence on plant growth and nickel-hyperaccumulation, resulted in increased biomass production and metal accumulation. Ecological and evolutionary implications of such findings are also discussed.

Keywords *Alyssum bertolonii* • Heavy metal • Molecular microbial ecology • Plant-associated bacteria

A. Mengoni (✉) • F. Pini • M. Bazzicalupo
Department of Evolutionary Biology, University of Firenze,
via Romana 17, I-50125 Florence, Italy
e-mail: alessio.mengoni@unifi.it

M.S. Khan et al. (eds.), *Biomanagement of Metal-Contaminated Soils*,
Environmental Pollution 20, DOI 10.1007/978-94-007-1914-9_7,
© Springer Science+Business Media B.V. 2011

7.1 Botany and Life History of *Alyssum bertolonii*

Serpentine soils are one of the most famous examples of soils naturally enriched by heavy metals (Fig. 7.1). They are characterized by high levels of nickel, cobalt, and chromium, low levels of N, P, K, and Ca, and present a high Mg/Ca ratio, which, in addition, limits plant colonization of these sites (Brooks 1987). Since the sixteenth century (Vergnano Gambi 1992), several endemic taxa have been identified within the characteristic flora of serpentine soils throughout the world (Pichi Sermolli 1948; Kruckeberg 1954; Kruckeberg and Kruckeberg 1990). One of the most interesting features described in serpentine endemic taxa is metal hypertolerance or metal hyperaccumulation (Baker 1981), a puzzling phenotype consisting of extremely high foliar metal contents, probably as a defense against herbivory (Boyd 2007). In temperate latitudes, the hyperaccumulation trait is found mainly in members of the family *Brassicaceae* (especially in the genera *Alyssum* and *Thlaspi*). The first record of a metal hyperaccumulator was for *Alyssum bertolonii* in which up to 1.2% nickel was found in the leaves (Minguzzi and Vergnano 1948). Many taxa in genus *Alyssum* have subsequently been shown to accumulate nickel in their aerial parts (Brooks et al. 1979). *Alyssum* is a genus of about 175 species, mainly of Mediterranean Europe and Turkey, with a few species in North Africa, the Near East (Iran, Iraq, and Transcaucasia), and scattered across the Ukraine and Siberia into the northwest of the American continent (Alaska, Yukon). In Europe, it is confined to the southern half of the continent and it may well be a pre-glacial relic since its distribution is to the south of areas formerly covered by the ice sheet during the Ice Ages.

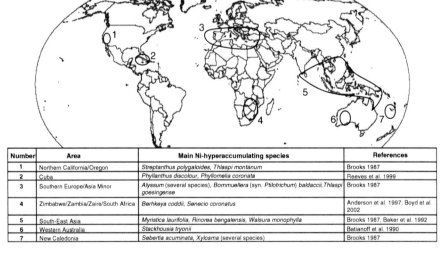

Number	Area	Main Ni-hyperaccumulating species	References
1	Northern California/Oregon	*Streptanthus polygaloides, Thlaspi montanum*	Brooks 1987
2	Cuba	*Phyllanthus discolour, Phyllomelia coronata*	Reeves et al. 1999
3	Southern Europe/Asia Minor	*Alyssum* (several species), *Bommuellera* (syn. *Ptilotrichum*) *baldaccii,Thlaspi goesingense*	Brooks 1987
4	Zimbabwe/Zambia/Zaire/South Africa	*Berhkeya coddii, Senecio coronatus*	Anderson et al. 1997; Boyd et al. 2002
5	South-East Asia	*Myristica laurifolia, Rinorea bengalensis, Walsura monophylla*	Brooks 1987, Baker et al. 1992
6	Western Australia	*Stackhousia tryonii*	Batianoff et al. 1990
7	New Caledonia	*Sebertia acuminata, Xylosma* (several species)	Brooks 1987

Fig. 7.1 Distribution of serpentine outcrops where Ni-hyperaccumulators have been found (Modified from Brooks 1987). Areas with serpentine outcrops are encircled with a *black line*. The table reports the name of the respective geographical areas and of main Ni-hyperaccumulating plant species (Modified from Mengoni et al. 2010)

One of the most investigated members of this genus is *Alyssum bertolonii* Desv. This is a diploid ($2n = 16$, Arrigoni et al. 1983) perennial plant, living exclusively on serpentine outcrops in Central Italy and particularly in Tuscany (Pignatti 1997). *A. bertolonii* is one of the 14 European species of *Alyssum* that hyperaccumulates nickel (Brooks and Radford 1978). The species has been suggested to be a useful indicator plant in prospecting for nickel (Brooks 1983). Moreover, cultivars of *Alyssum* have been proposed for phytoremediation (Salt et al. 1998) and patented for phytomining practices (Chaney et al. 1998). Phylogeny, population genetics, and physiological properties of this species have been deeply investigated (Mengoni et al. 2003a, b; Galardi et al. 2007a, b). In particular, it has been reported that, though nickel tolerance and hyperaccumulation are well-known constitutive species-level traits, the extent, or levels, of tolerance and Ni-accumulation are strongly variable among different populations. Variability of metal accumulation has been observed for other hyperaccumulating plants also (Assunção et al. 2003, 2008). The presence of populations or accessions of the same species having different tolerance and accumulation levels is an important feature for improvement of such traits through breeding and for identifying candidate genomic regions or genes responsible for the trait (Assunção et al. 2006). While in *A. bertolonii,* these studies are still in progress, for *Arabidopsis lyrata,* a species which present populations locally adapted to serpentine soils, a genome-wide map has recently been provided (Turner et al. 2010), which identify several candidate loci for serpentine adaptation. However, it is becoming more and more evident that, under field conditions, complex traits, which involves both specific genes and growth features as metal hyperaccumulation, strongly rely not only on the genetic background of the plant, but also on the interaction with soil mineral elemental composition and with the indigenous microbial flora. In particular, plant-associated bacteria have been claimed as important factors for the improvement of metal hyperaccumulation and consequently for improving phytoremediation of contaminated soils (Sessitsch and Puschenreiter 2008; Rajkumar et al. 2009b).

7.2   Plant-Associated Bacteria: Which, Why, for What?

A diverse range of bacteria, including pathogens, mutualists, and commensals are supported by plants. They grow in and around roots, in the vasculature, and in the aerial tissues and are known as rhizospheric, endophytic, and phyllospheric bacteria, respectively. Of these, rhizospheric and endophytic bacteria have been widely studied (Danhorn and Fuqua 2007). Most bacteria that are associated with plants are saprotrophic and do not harm the plant itself, and only a small number of them is able to cause disease (Jackson 2009). The rhizosphere is the important terrestrial habitat that contains living plant roots and closely associated soil where root exudates stimulate microbial metabolism and productivity. The activities of the rhizosphere microbial community significantly influence many aspects of plant physiology and growth, and therefore, play an important role in terrestrial ecosystems and sustainable agriculture. Plants provide rhizosphere microbes with a carbon source. In turn, microbes may provide nitrogen (N) and phosphorus (P) and also protect plants from parasites and pathogens. Root–microbe interactions thus play key roles in several other ecosystem functions, such as decomposition of organic matter, and the maintenance of soil structure and water relationships. The role of root-associated microbes in maintaining soil structure (i.e., aggregate stability) has also been documented (Singh et al. 2004). There is accumulating evidence that biotic interactions, occurring below ground, play an important role in determining plant diversity above ground by direct feedback on host growth and indirect effects on competing plants (Singh et al. 2004).

Endophytic bacteria can be defined as those bacteria that colonize the internal tissue of the plants showing no external sign of infection or negative effect on their host (Ryan et al. 2008). They can be classified as "obligate" or "facultative" endophytes in accordance with their life strategies. Obligate endophytes are strictly dependent on the host plant for their growth and survival; besides, transmission to other plants could occur only by seeds or via vectors, while facultative endophytes could grow outside host plants (Rajkumar et al. 2009a). In the last few years, there has been a considerable interest toward exploiting the potential of endophytic bacteria for plant growth promotion and for the improvement of phytoremediation.

Phyllospheric (epiphytic) bacteria inhabit the aerial parts of the plant (leaves, stems, buds, flowers, and fruits) possibly affecting plant fitness and productivity of agricultural crops (Whipps et al. 2008). Studies on the composition of bacterial communities of leaves have been numerous but rather limited in scope. It is generally believed that populations of culturable aerobic bacteria on leaves are dominated by a few genera, which are involved in processes as large in scale as the carbon cycle (intercepting carbon compounds released directly from plants or removed by sucking arthropods) and the nitrogen cycles (nitrification of ammonium pollutants intercepted by plants; nitrogen fixation) to processes affecting the health of individual plants (Lindow and Brandl 2003). To date, no studies have been conducted to specifically target the epiphytic bacterial flora of metallophytes.

Bacteria can have in fact a profound influence on plant health and productivity. Several studies have been conducted to explore the plant growth-promoting abilities of various rhizobacteria and endophytes that increase plant growth through the improved cycling of nutrients and minerals such as, N, P, and other nutrients (Ryan et al. 2008). Under N-stressed conditions, rhizobia, a paraphyletic group that falls into two classes of *Proteobacteria* (*alfa-* and *beta-Proteobacteria*), drive the formation of symbiotic nitrogen-fixing nodules on the roots or stems of leguminous hosts; the converted ammonia is then used by the plant as a N source (van Rhijn and Vanderleyden 1995). Moreover, plant growth, can be facilitated by endophytes altering the plant hormonal balance. Several bacteria such as strains of *Pseudomonas, Staphylococcus, Enterobacter, Azotobacter,* and *Azospirillum,* are able to produce phytohormones like auxins and cytokinins (Costacurta and Vanderleyden 1995; Lucy et al. 2004; Somers et al. 2004). Moreover, some bacterial strains, like, *Methylobacterium oryzae* CBMB20, *Pseudomonas fluorescens*, and strains of nitrogen-fixing symbiont *Sinorhizobium meliloti* and *Mesorhizobium loti*, can decrease the level of ethylene by cleaving its precursor through production of 1-aminocyclopropane-1-carboxylic acid (ACC) deaminase (Glick 2005; Rajkumar et al. 2009a). Endophytic bacteria also influence plant health by decreasing or preventing the pathogenic effects of certain parasitic microorganisms by producing antimicrobial compounds. For instance, in *Enterobacter* sp. 638, an endophyte of poplar, genes for the synthesis of the antimicrobial 4-hydroxybenzoate and 2-phenylethanol have been found (Taghavi et al. 2010). Many endophytes indeed are members of common soil bacterial genera, such as *Pseudomonas, Burkholderia,* and *Bacillus* (Rajkumar et al. 2009a). The bacterial strains associated with both metallophytes and non-metallophyte species are listed in Table 7.1.

Endophytes can also enhance plant growth and increase plant resistance to heavy-metal stress in several ways. Indirect mechanisms are similar to those described for

Table 7.1 Endophytic bacterial species recovered from different plants

Endophytes	Plant species
α-*Proteobacteria*	
Azorhizobium caulinodans	Rice
Azospirillum brasilense	Banana
Azospirillum amazonense	Banana, pineapple
Bradyrhizobium japonicum	Rice
Devosia sp.	*Thlaspi caerulescens*
Gluconacetobacter diazotrophicus	Sugarcane, coffee
Methylobacterium mesophilicum	Citrus plants; *Thlaspi goesingense*
Methylobacterium extorquens	Scots pine, citrus plants; *Thlaspi goesingense*
Methylobacterium populi BJ001	*Populus deltoides x nigra* DN34
Methylobacterium oryzae sp. *CBMB20*	*Oryza sativa*
Methylobacterium sp.	*Thlaspi caerulescens*
Phyllobacterium sp.	*Thlaspi caerulescens*
Rhizobium leguminosarum	Rice
Rhizobium (Agrobacterium) radiobacter	Carrot, rice
Sinorhizobium meliloti	Sweet potato
Sphingomonas paucimobilis	Rice
Sphingomonas sp.	*Thlaspi caerulescens; Thlaspi goesingense*
β-*Proteobacteria*	
Azoarcus sp.	Kallar grass, rice
Burkholderia pickettii	Maize
Burkholderia cepacia	Yellow lupine, citrus plants
Burkholderia sp.	Banana, pineapple, rice
Burkholderia sp. *Bu61 (pTOM-Bu-61)*	Poplar
Chromobacterium violaceum	Rice
Herbaspirillum seropedicae	Sugarcane, rice, maize, banana
Herbaspirillum rubrisulbalbicans	Sugarcane
Herbaspirillum sp.*K1*	Wheat
γ-*Proteobacteria*	
Citrobacter sp.	Banana
Enterobacter spp.	Maize; *Nicotiana tabacum*
Enterobacter sakazakii	Soybean
Enterobacter cloacae	Citrus plants, maize

(continued)

Table 7.1 (continued)

Endophytes	Plant species
Enterobacter agglomerans	Soybean
Enterobacter asburiae	Sweet potato
Erwinia sp.	Soybean
Escherichia coli	Lettuce
Klebsiella sp.	Wheat, sweet potato, rice
Klebsiella pneumoniae	Soybean
Klebsiella variicola	Banana, rice, maize, sugarcane
Klebsiella terrigena	Carrot
Klebsiella oxytoca	Soybean
Pantoea sp.	Rice, soybean
Pantoea agglomerans	Citrus plants, sweet potato
Pseudomonas chlororaphis	Marigold (*Tagetes* spp.), carrot
Pseudomonas putida	Carrot
Pseudomonas fluorescens	Carrot, *Brassica napus*
Pseudomonas citronellolis	Soybean
Pseudomonas synxantha	Scots pine
Pseudomonas viridiflava	Grass
Pseudomonas aeruginosa strain R75	Wild rye (*Elymus dauricus*)
Pseudomonas savastanoi strain CB35	Wild rye (*Elymus dauricus*)
P. putida VM1450	Poplar (*Populus*) and willow (*Salix*)
Pseudomonas fulva	*Nicotiana tabacum*
Pseudomonas sp.	Populus cv. Hazendans and cv. Hoogvorst; *Alyssum bertolonii, Nicotiana tabacum*
Salmonella enterica	Alfalfa, carrot, radish, tomato
Serratia sp.	Rice
Serratia marcescens	Rice, *Rhyncholacis penicillata*
Stenotrophomonas sp.	Dune grasses (*Ammophila arenaria* and *Elymus mollis*); *Nicotiana tabacum*
Firmicutes	
Bacillus spp.	*Citrus; Alyssum bertolonii; Thlaspi goesingense*
Bacillus megaterium	Maize, carrot, citrus plants
Clostridium	Grass, *Miscanthus sinensis*
Clostridium aminovalericum	*Nicotiana tabaccum*
Desulfitobacterium metallireductans	*Thlaspi goesingense*
Paenibacillus odorifer	Sweet potato
Paenibacillus polymyx	Wheat, Lodeg pine, green beans, *Arabidopsis thaliana, Canola*
Paenibacillus sp.	*Alyssum bertolonii*
Staphylococcus saprophyticus	Carrot
Staphylococcus sp.	*Alyssum bertolonii*
Bacteroidetes	
Flavobacterium sp.	*Thlaspi goesingense*
Sphingobacterium sp.	Rice
Sphingobacterium multivorum	*Thlaspi caerulescens*

(continued)

Table 7.1 (continued)

Endophytes	Plant species
Actinobacteria	
Arthrobacter globiformis	Maize
Arthrobacter sp.	*Alyssum bertolonii*
Blastococcus sp.	*Thlaspi goesingense*
Curtobacterium flaccumfaciens	Citrus plants
Curtobacterium sp.	*Alyssum bertolonii; Thlaspi goesingense*
Kocuria varians	Marigold
Leifsonia	*Alyssum bertolonii*
Microbacterium esteraromaticum	Marigold
Microbacterium testaceum	Maize
Microbacterium sp.	*Brassica napus, Alyssum bertolonii*
Mycobacterium sp.	Wheat, Scots pine
Nocardia sp.	Citrus plants
Plantibacter flavus	*Thlaspi goesingense*
Propionibacterium acnes	*Thlaspi goesingense*
Rhodococcus sp.	*Thlaspi caerulescens, Thlaspi goesingense*
Streptomyces	Wheat
Streptomyces griseus	*Kandelia candel*
Streptomyces NRRL 30562	*Kennedia nigriscans*
Streptomyces NRRL 30566	*Grevillea pteridifolia*
Streptomyces sp.	*Monstera* sp.
Sanguibacter sp.	*Nicotiana tabaccum*

Modified from Rosenblueth and Martinez-Romero (2006), Rajkumar et al. (2009b) and Ryan et al. (2008)

PGPR (Rajkumar et al. 2009a) such as nitrogen fixation, improving mineral nutrition (for instance the solubilization of P into plant-available forms), or increasing resistance or tolerance to biotic and abiotic stresses (Ryan et al. 2008). Directly, bacteria can increase heavy-metal mobilization or lessen heavy-metal toxicity by the production of bacterial siderophore that enhances the supply of iron to the plant (Sessitsch and Puschenreiter 2008). Siderophores are organic molecules that show high affinity for Fe(III) ions, and can also form complexes with other bivalent heavy metal ions that can be assimilated by the plant (Rajkumar et al. 2009a). Cadmium-resistant endophytes (e.g., *Sanguibacter* sp., *Pseudomonas* sp., and *Enterobacter* sp.) isolated from *Nicotiana tabacum* seeds decrease the Cd toxicity by increasing the uptake of trace elements (Zn and Fe) by plants (Mastretta et al. 2009). Other challenges, however have been faced to identify and engineer endophytic bacteria to enhance plant growth on polluted soil over phytotoxicity threshold. To solve this, the pTOM toluene degradation plasmid was inserted into the lupine endophyte *Burkholderia cepacia* G4, which upon inoculation improved the *in planta* degradation of toluene and concomitantly decreased its transpiration to the atmosphere (Barac et al. 2004). The use of these technologies is at the beginning. Therefore, the stability of the degradation capabilities within the endophytic community (Newman and Reynolds 2005) and the consistent production of secondary toxic metabolite involved in the degradation pathways are needed to be explored. However, the use of endophytic bacteria to improve phytoremediation shows great promises (Weyens et al. 2009) as presented in Fig. 7.2.

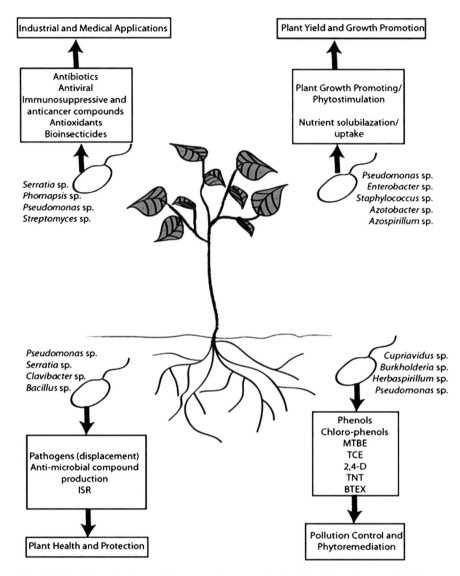

Fig. 7.2 Possible applications of plant-associated bacteria (Modified from Ryan et al. 2008)

7.3   Soil and Rhizosphere Bacteria Involved in Metal Detoxification

Serpentine soil bacteria were described by Lipman in 1926, who, in an attempt to identify the reasons for the low fertility of serpentine soils, wrote: "there is little diversity, as well as a general paucity, in the bacterial flora of the serpentine soils" (Lipman 1926). However, it is still not clear if certain bacterial taxonomic groups

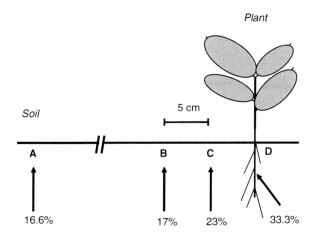

Fig. 7.3 Proportion of nickel-resistant bacteria at different distances from the Ni-hyperaccumulator *A. bertolonii. A* bulk soil, *B* 10 cm, *C* 5 cm, *D* rhizosphere soil. Values are percent of resistant bacteria over the total isolates (Adapted from Mengoni et al. 2001)

are inhibited or favored by the serpentine soil conditions (Mengoni et al. 2001; Lodewyckx et al. 2002; Oline 2006). Moreover, metal-hyperaccumulating plants have been proposed as a selective factor toward soil bacteria, increasing the level of metals in their close proximity. Actually, it has been found that the presence of some plants (i.e., the Ni-hyperaccumulating tree, *Sebertia acuminata*) positively correlated with the presence of Ni-resistant soil bacteria (Schlegel et al. 1991). A hypothetical "nickel cycle," driving the evolution of the bacterial community toward a higher percentage of nickel-resistant strains was suggested for such species. The "nickel cycle" leads to an increased nickel concentration in the upper soil layers in the proximity of the plant due to the "pumping" of nickel from deep soil performed by the roots, followed by the translocation of nickel to leaves and then, after the abscission of the leaves, the release of accumulated nickel from the litter. As a consequence of this cycle, top soil layers near the plant contain higher nickel concentrations than those far away from the plant, and consequently exert a stronger selective pressure for Ni-resistance toward soil bacteria. An increased number of Ni-resistant bacteria was also observed in the rhizosphere of the Ni-hyperaccumulators *A. bertolonii* (Mengoni et al. 2001), as outlined in Fig. 7.3. This finding was also confirmed in other species for example in *Thlaspi goesingense* and *A. serpyllifolium* susp. *lusitanicum* and *T. caerulescens* (Lodewyckx et al. 2002; Idris et al. 2004; Aboudrar et al. 2007; Becerra-Castro et al. 2009). However, due to the small size and shallow rooting of these plants (including *A. bertolonii*), it is probably not correct to invoke a real "metal cycle," that is, an increase of the top soil metal concentration due to the foliar hyperaccumulation of deep-soil metals and subsequent leaf fall. Recently, Mengoni et al. (2010) proposed that "root-foraging" could be the main cause of the increase in heavy-metal-tolerant bacteria. In other words, the presence of highly tolerant bacteria near the

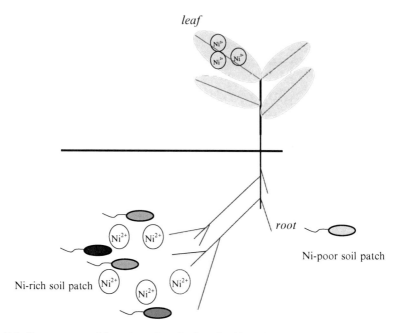

Fig. 7.4 Consequences of "metal root foraging" on the rhizosphere bacterial flora. Patches of soil rich in metals are already inhabited by a large fraction of Ni-resistant bacteria. Different gray tones suggest possibly different bacterial species (Modified from Mengoni et al. 2010)

roots of metal hyperaccumulators could be due to the effect of a concomitant specific tropism of roots of hyperaccumulating plants toward soil patches rich in metals (Whiting et al. 2000). Consequently, the presence of highly tolerant bacteria near *A. bertolonii* roots may not be due to plant activity but simply to the chemical properties of the soil patch that already selected a highly tolerant bacterial flora (Fig. 7.4). In agreement with such a model (Ni content of soil patches play the main role in the selection of Ni-resistant bacteria), in *A. bertolonii,* the proportion of resistant bacteria was variable in different outcrops and partially related to soil Ni content, that is, the higher the bioavailable Ni in soil, the higher the percentage of Ni-resistant bacteria in bulk soil (Mengoni et al. 2001). Despite the selective environment of serpentine soil and rhizosphere, a high genetic diversity was in general found, in contrast with the initial finding by Lipman (1926). However, probably due to the rich culture medium used (LB), mainly copiotrophic species particularly members of genera *Pseudomonas* and *Streptomyces* were recovered. Interestingly, *Pseudomonas* isolates were strongly present in the rhizosphere, while *Streptomyces* were predominant in the soil samples, in agreement with a "rhizosphere effect" which favors the presence of genera that include known plant growth-promoting rhizobacteria (PGPR). Rhizosphere effect was also shown in an analysis of total bacterial flora by cultivation-independent analysis (Mengoni et al. 2004) where the presence of other bacterial groups known to interact with plant roots was also detected (i.e., *alpha-Proteobacteria*). Another interesting finding of serpentine soil bacteria associated with *A. bertolonii* (Mengoni et al. 2001) was the prevalence of high phenotypic diversity for single or multiple metal tolerances. Interestingly, no correlation between genetic groupings and heavy-metal-tolerant phenotypes was found. Nevertheless, a higher proportion of *Pseudomonas* strains were resistant to high concentrations of nickel compared to *Streptomyces*, probably reflecting the highest bioavailable Ni present in rhizosphere soil.

7.3.1 Endophytic Bacteria

The increasing interest in the use of endophytic bacteria, that is bacteria intimately associated with plant tissues (Weyens et al. 2009), has opened up new perspectives on the study of metal-hyperaccumulating plants. Endophytes may colonize plant-internal environments that are less toxic than soil (that is with lower available metal content), or environments, such as xylem vessels, where toxic metals might be available at higher concentration than in soil (Smart et al. 2007). The Ni-hyperaccumulator *Thlaspi goesingense* was the first species to be investigated for its endophytic bacterial community composition (Idris et al. 2004). Results showed that majority of endophytic bacteria belonged to *Proteobacteria* division and had a high number of sequences related to the genus *Sphingomonas*. Moreover, members of the genus *Methylobacterium* were recovered and a new species, namely *Methylobacterium goesingense*, was found to be associated with *T. goesingense* (Idris et al. 2006). Bacteria associated with tissues of metal-hyperaccumulators (from genera *Thlaspi* (*Noccaea*) and *Alyssum*) are listed in Table 7.1. In *A. bertolonii,* most of the diversity was represented by Gram-positive bacteria (Barzanti et al. 2007). In particular, genera as *Bacillus*, *Paenibacillus*, *Leifsonia*, *Curtobacterium*, *Microbacterium*, *Micrococcus,* and *Staphylococcus* were found. While only few members of *Proteobacteria* (mainly belonging to the genus *Pseudomonas*) were reported. Similar to previous findings on soil bacteria (Mengoni et al. 2001), a high phenotypic diversity with regard to heavy-metal resistance was found, suggesting the occurrence of a high number of different "microenvironments" within plant tissues. Nevertheless, contrary to soil isolates, only few isolates showed co-resistance to Ni and Co. Furthermore, there was no relationship between taxonomic groups and resistance phenotype, which suggest the presence of highly transmissible genetic elements carrying the determinants for heavy-metal resistance (e.g., plasmids) as observed in the model metal-resistant strain *Cupriavidus metallidurans* CH34 (Janssen et al. 2009).

Recently, using cultivation-independent Terminal-Restriction Fragment Length Polymorphism (T-RFLP) Mengoni et al. (2009), characterized the leaf-associated bacterial flora of *A. bertolonii* plants, collected from three different populations. Interestingly, more than half of the taxonomical diversity (as Terminal-Restriction Fragments, TRFs) was assigned to *Alpha-* and *Gamma-Proteobacteria* and *Actino-bacteria.* Two TRFs were sequenced and matched with 16S rRNA gene sequences of methylobacteria. Methylobacteria were also found in the Ni-hyperaccumulator

T. goesingense (Idris et al. 2006) as well as in the Zn-hyperaccumulator *T. caerulescens* (Lodewyckx et al. 2002), associated with rhizosphere and plant tissues. However, it is not clear if they play any role in hyperaccumulation even though methylobacteria have been detected earlier in several other plant species (Lidstrom and Chistoserdova 2002). It is proposed that plant-by-plant variability of bacterial community composition is far higher than variability due to the sampling sites, suggesting that a large fraction of bacteria could be associated to the plant simply by chance and may not provide any positive (or negative) relevant effect toward plant phenotypes and fitness.

7.4   Conclusion and Perspectives

Plant-associated bacteria are promising partners which may increase plant fitness and performances and consequently yields. However, few studies have been carried out on the economical relevance of bacteria associated with plants living in metal-containing soil. Despite little studies on this aspect, some very promising strains have been isolated from hyperaccumulators and from bulk serpentine soil, as *Methylobacterium goesingense*, some strains from the rhizosphere of *Alyssum murale* (Abou-Shanab et al. 2003, 2007), *Serratia marcescens* C-1 (Marrero et al. 2007) or *Streptomyces yatensis* (Saintpierre et al. 2003). In *A. bertolonii,* two strains belonging to genera *Arthrobacter* and *Pseudomonas* were isolated as the endophytic community which showed plant growth-promoting activities and tolerance to nickel, and Ni-hyperaccumulation by *A. bertolonii* plantlets in hydroponic cultures (unpublished results). Genome sequencing of these strains are likely to provide better understanding of genetic basis of interactions of endophytes with hyperaccumulating plants. Positive effects of endophytes on plants, however, depend on several factors like metal uptake, phytotoxicity of metals to plants, rate of plant growth, etc. In particular, the first two are the major limiting factors in the application of phytoextraction (Weyens et al. 2009).The exploitation of plant-associated bacteria, however, could be a promising strategy to improve the efficiency of phytoextraction both via enrichment of the bacterial community present *in planta*, or through metabolic engineering and re-inoculation of suitable strains to improve metal availability, and hence, to reduce phytotoxicity. The reduction in the costs of complete genome sequencing and systems biology-based modeling of biotic and metabolic interactions will, therefore, greatly help in the development of effective bacterial inoculants, isolated from target plants, which may enhance metal phytoextraction by metal-hyperaccumulating plants and in particular by *A. bertolonii*.

Acknowledgments This work was partially supported by grants of University of Florence to AM and MB (Contributo di Ateneo anno 2009). FP performed part of his Ph.D. work on A. bertolonii endophytes sponsored by a fellowship of the Italian Ministry of Research and Education (MIUR).

References

Aboudrar W, Schwartz C, Benizri E, Morel JL, Boularbah A (2007) Soil microbial diversity as affected by the rhizosphere of the hyperaccumulator *Thlaspi caerulescens* under natural conditions. Intern J Phytoremed 9:41–52

Abou-Shanab RA, Angle JS, Delorme TA, Chaney RL, van Berkum P, Moawad H, Ghanem K, Ghozlan HA (2003) Rhizobacterial effects on nickel extraction from soil and uptake by *Alyssum murale*. New Phytologist 158:219–224

Abou-Shanab RA, van Berkum P, Angle JS (2007) Heavy metal resistance and genotypic analysis of metal resistance genes in Gram-positive and Gram-negative bacteria present in Ni-rich serpentine soil and in the rhizosphere of *Alyssum murale*. Chemosphere 68:360–367

Arrigoni PV, Ricceri C, Mazzanti A (1983) La vegetazione serpentinicola del Monte Ferrato di Prato in Toscana. Centro Scienze Naturali Prato, Pistoia

Assunção AGL, Bookum WM, Nelissen HJM, Vooijs R, Schat H, Ernst WHO (2003) Differential metal-specific tolerance and accumulation patterns among *Thlaspi caerulescens* populations originating from different soil types. New Phytologist 159:411–419

Assunção AGL, Pieper B, Vromans J, Lindhout P, Aarts MGM, Schat H (2006) Construction of a genetic linkage map of *Thlaspi caerulescens* and quantitative trait loci analysis of zinc accumulation. New Phytologist 170:21–32

Assunção AGL, Bleeker P, Ten Bookum WM, Vooijs R, Schat H (2008) Intraspecific variation of metal preference patterns for hyperaccumulation in *Thlaspi caerulescens*: evidence from binary metal exposures. Plant Soil 303:289–299

Baker AMJ (1981) Accumulators and excluders: strategies in the response of plants to heavy-metals. J Plant Nutr 3:643–654

Barac T, Taghavi S, Borremans B, Provoost A, Oeyen L, Colpaert JV, Vangronsveld J, van der Lelie D (2004) Engineered endophytic bacteria improve phytoremediation of water-soluble, volatile, organic pollutants. Nature Biotechnol 22:583–588

Barzanti R, Ozino F, Bazzicalupo M, Gabbrielli R, Galardi F, Gonnelli C, Mengoni A (2007) Isolation and characterization of endophytic bacteria from the nickel-hyperaccumulator plant *Alyssum bertolonii*. Microbial Ecol 53:306–316

Becerra-Castro C, Monterroso C, García-Lestón M, Prieto-Fernández A, Acea MJ, Kidd PS (2009) Rhizosphere microbial densities and trace metal tolerance of the nickel hyperaccumulator *Alyssum serpyllifolium* subsp. *lusitanicum*. Intern J Phytoremed 11:525–541

Boyd RS (2007) The defense hypothesis of elemental hyperaccumulation: status, challenges and new directions. Plant Soil 293:153–176

Brooks RR (1983) Biological methods of prospecting for minerals. Wiley, New York

Brooks RR (1987) Serpentine and its vegetation. A multidisciplinary approach. Dioscorides Press, Portland

Brooks RR, Radford CC (1978) Nickel accumulation by European species of the genus *Alyssum*. Proc R Soc Lond B 200:217–224

Brooks RR, Morrison RS, Reeves RD, Dudley TR, Akman Y (1979) Hyperaccumulation of nickel by *Alyssum* Linnaeus (Cruciferae). Proc R Soc Lond B 203:387–403

Chaney RL, Angle JS, Baker AJM, Li Y-M (1998) Method for phytomining of nickel, cobalt and other metals from soils. US Patent, No. 5: 711,784

Costacurta A, Vanderleyden J (1995) Synthesis of phytohormones by plant-associated bacteria. Crit Rev Microbiol 21:1–18

Danhorn T, Fuqua C (2007) Biofilm formation by plant-associated bacteria. Ann Rev Microbiol 61:401–422

Galardi F, Corrales I, Mengoni A, Pucci S, Barletti L, Arnetoli M, Gabbrielli R, Gonnelli C (2007a) Intra-specific differences in nickel tolerance and accumulation in the Ni-hyperaccumulator *Alyssum bertolonii*. Environ Exp Bot 60:377–384

Galardi F, Mengoni A, Pucci S, Barletti L, Massi L, Barzanti R, Gabbrielli R, Gonnelli C (2007b) Intra-specific differences in mineral element composition in the Ni-hyperaccumulator *Alyssum bertolonii*: a survey of populations in nature. Environ Exp Bot 60:50–56

Glick BR (2005) Modulation of plant ethylene levels by the bacterial enzyme ACC deaminase. FEMS Microbiol Lett 251:1–7

Idris R, Trifonova R, Puschenreiter M, Wenzel WW, Sessitsch A (2004) Bacterial communities associated with flowering plants of the Ni hyperaccumulator *Thlaspi goesingense*. Appl Environ Microbiol 70:2667–2677

Idris R, Kuffner M, Bodrossy L, Puschenreiter M, Monchy S, Wenzel WW, Sessitsch A (2006) Characterization of Ni-tolerant methylobacteria associated with the hyperaccumulating plant *Thlaspi goesingense* and description of *Methylobacterium goesingense* sp nov. Syst Appl Microbiol 29:634–644

Jackson RW (2009) Plant pathogenic bacteria: genomics and molecular biology. Caister Academic Press, Norwich

Janssen PJ, Van Houdt R, Moors H, Monsieurs P, Morin N, Michaux A, Benotmane MA, Leys N, Vallaeys T, Lapidus A, Monchy S, Médigue C, Taghavi S, McCorkle S, Dunn J, van der Lelie D, Mergeay M (2009) The complete genome sequence of *Cupriavidus metallidurans* strain CH34, a master survivalist in harsh and anthropogenic environments. PLoS One 5(5):e10433

Kruckeberg AR (1954) The ecology of serpentine soils. III. Plant species in relation to serpentine soils. Ecology 35:267–274

Kruckeberg AR, Kruckeberg AL (1990) Endemic metallophytes: their taxonomic, genetic and evolutionary attributes. In: Shaw AJ (ed.) Heavy metal tolerance in plants: evolutionary aspects. CRC Press Inc, Boca Raton, pp 301–312

Lidstrom ME, Chistoserdova L (2002) Plants in the pink: cytokinin production by *Methylobacterium*. J Bacteriol 184:1818

Lindow SE, Brandl MT (2003) Microbiology of the phyllosphere. Appl Environ Microbiol 69:1875–1883

Lipman CB (1926) The bacterial flora of serpentine soils. J Bacteriol 12:315–318

Lodewyckx C, Mergeay M, Vangronsveld J, Clijsters H, Van Der Lelie D (2002) Isolation, characterization, and identification of bacteria associated with the zinc hyperaccumulator *Thlaspi caerulescens* subsp *calaminaria*. Intern J Phytoremed 4:101–115

Lucy M, Reed E, Glick BR (2004) Applications of free living plant growth-promoting rhizobacteria. Antonie Van Leeuwenhoek 86:1–25

Marrero J, Auling G, Coto O, Nies DH (2007) High-level resistance to cobalt and nickel but probably no transenvelope efflux: metal resistance in the cuban *Serratia marcescens* strain C-1. Microbial Ecol 53:123–133

Mastretta C, Taghavi S, van der Lelie D, Mengoni A, Galardi F, Gonnelli C, Barac T, Boulet J, Weyens N, Vangronsveld J (2009) Endophytic bacteria from seeds of *Nicotiana tabacum* can reduce cadmium phytotoxicity. Intern J Phytoremed 11:251–267

Mengoni A, Barzanti R, Gonnelli C, Gabbrielli R, Bazzicalupo M (2001) Characterization of nickel-resistant bacteria isolated from serpentine soil. Environ Microbiol 3:691–698

Mengoni A, Baker AMJ, Bazzicalupo M, Reeves RD, Adigüzel N, Chianni E, Galardi F, Gabbrielli R, Gonnelli C (2003a) Evolutionary dynamics of nickel hyperaccumulation in *Alyssum* revealed by ITS nrDNA analysis. New Phytologist 159:691–699

Mengoni A, Gonnelli C, Brocchini E, Galardi F, Pucci S, Gabbrielli R, Bazzicalupo M (2003b) Chloroplast genetic diversity and biogeography in the serpentine endemic Ni-hyperaccumulator *Alyssum bertolonii*. New Phytologist 157:349–356

Mengoni A, Grassi E, Barzanti R, Biondi EG, Gonnelli C, Kim CK, Bazzicalupo M (2004) Genetic diversity of bacterial communities of serpentine soil and of rhizosphere of the nickel-hyperaccumulator plant *Alyssum bertolonii*. Microbial Ecol 48:209–217

Mengoni A, Pini F, Huang L-N, Shu W-S, Bazzicalupo M (2009) Plant-by-plant variations of bacterial communities associated with leaves of the nickel hyperaccumulator *Alyssum bertolonii* Desv. Microbial Ecol 58:660–667

Mengoni A, Schat H, Vangronsveld J (2010) Plants as extreme environments? Ni-resistant bacteria and Ni-hyperaccumulators of serpentine flora. Plant Soil 331:5–16

Minguzzi C, Vergnano Gambi O (1948) Il contenuto di nichel nelle ceneri di *Alyssum bertolonii* Desv. Memorie Società Toscana di Scienze Naturali 55:49–74

Newman LA, Reynolds CM (2005) Bacteria and phytoremediation: new uses for endophytic bacteria in plants. Trends Biotechnol 23:6–8

Oline DK (2006) Phylogenetic comparisons of bacterial communities from serpentine and nonserpentine soils. Appl Environ Microbiol 72:6965–6971

Pichi Sermolli R (1948) Flora e vegetazione delle serpentine e delle altre ofioliti dell'alta valle del Tevere (Toscana). Webbia 17:1–380

Pignatti S (1997) Flora d'Italia, vol 1. Agricole, Bologne

Rajkumar M, Ae N, Freitas H (2009a) Endophytic bacteria and their potential to enhance heavy metal phytoextraction. Chemosphere 77:153–160

Rajkumar M, Vara Prasad MN, Freitas H, Ae N (2009b) Biotechnological applications of serpentine soil bacteria for phytoremediation of trace metals. Crit Rev Biotechnol 29:120–130

Rosenblueth M, Martinez-Romero E (2006) Bacterial endophytes and their interactions with hosts. Mol Plant Microbe Interact 19:827–837

Ryan RP, Germaine K, Franks A, Ryan DJ, Dowling DN (2008) Bacterial endophytes: recent developments and applications. FEMS Microbiol Lett 278:1–9

Saintpierre D, Amir H, Pineau R, Sembiring L, Goodfellow M (2003) *Streptomyces yatensis* sp nov., a novel bioactive streptomycete isolated from a New-Caledonian ultramafic soil. Antonie Van Leeuwenhoek 83:21–26

Salt DE, Smith RD, Raskin I (1998) Phytoremediation. Ann Rev Plant Physiol Plant Mol Biol 49:643–668

Schlegel HG, Cosson JP, Baker AJM (1991) Nickel-hyperraccumulating plants provide a niche for nickel-resistant bacteria. Botanica Acta 104:18–25

Sessitsch A, Puschenreiter M (2008) Endophytes and rhizosphere bacteria of plants growing in heavy metal-containing soils. In: Dion P, Nautiyal CS (eds.) Microbiology of extreme soils, vol 1, Soil biology. Springer, Berlin/Heidelberg, pp 317–332

Singh BK, Millard P, Whiteley AS, Murrell JC (2004) Unravelling rhizosphere-microbial interactions: opportunities and limitations. Trends Microbiol 12:386–393

Smart KE, Kilburn MR, Salter CJ, Smith JAC, Grovenor CRM (2007) NanoSIMS and EPMA analysis of nickel localisation in leaves of the hyperaccumulator plant Alyssum lesbiacum. Int J Mass Spectrom 260:107–114

Somers E, Vanderleyden J, Srinivasan M (2004) Rhizosphere bacterial signalling: a love parade beneath our feet. Crit Rev Microbiol 30:205–240

Taghavi S, van der Lelie D, Hoffman A, Zhang YB, Walla MD, Vangronsveld J, Newman L, Monchy S (2010) Genome sequence of the plant growth promoting endophytic bacterium *Enterobacter* sp. 638. PLoS Genet 6(5):e1000943

Turner TL, Bourne EC, Von Wettberg EJ, Hu TT, Nuzhdin SV (2010) Population re-sequencing reveals local adaptation of *Arabidopsis lyrata* to serpentine soils. Nat Genet 42:260–263

Van Rhijn P, Vanderleyden J (1995) The *Rhizobium*-plant symbiosis. Microbiol Rev 59:124–142

Vergnano Gambi O (1992) The distribution and ecology of the vegetation of ultramafic soils in Italy. In: Roberts BA, Proctor J (eds.) The ecology of areas with serpentinized rocks – a world view. Kluwer, Dordrecht, pp 217–247

Weyens N, van der Lelie D, Taghavi S, Vangronsveld J (2009) Phytoremediation: plant-endophyte partnerships take the challenge. Curr Opin Biotechnol 20:248–254

Whipps JM, Hand P, Pink D, Bending GD (2008) Phyllosphere microbiology with special reference to diversity and plant genotype. J Appl Microbiol 105:1744–1755

Whiting SN, Leake JR, McGrath SP, Baker AJM (2000) Positive responses to Zn and Cd by roots of the Zn and Cd hyperaccumulator *Thlaspi caerulescens*. New Phytologist 145:199–210

Chapter 8
Use of Biosurfactants in the Removal of Heavy Metal Ions from Soils

Yeşim Sağ Açıkel

Abstract Heavy metal contamination of soils is one of the world's major environmental problems, posing significant risks to human health as well as to the ecosystems. Conventional treatment technologies for heavy metal polluted soils such as excavation and transport of contaminated soil to hazardous waste sites for landfilling have several disadvantages. They cannot completely remove metals, they can only immobilize them in the contaminated soil. Novel technologies involving microorganisms and their products to remove heavy metals have been successfully applied to waste streams such as sewage sludge, industrial effluents, and mine water. Biosorption of metal-contaminated soils presents a more complex separation problem. Use of biosurfactants to improve the removal of heavy metal contaminants from aqueous media and soils has received increasing attention in recent years. Surfactin produced by *Bacillus subtilis*, rhamnolipids from *Pseudomonas aeruginosa*, sophorolipids from *Torulopsis bombicola*, Aescin from *Aesculus hippocastanum*, and saponin from quillaja bark have been employed to remove metals from contaminated soils. The possible mechanisms for the removal of heavy metals by biosurfactants are ion exchange, precipitation–dissolution, and counter ion binding. Reports on the use of biosurfactants in metal removal are however scanty. Even though sorption isotherms have been widely used to measure the heavy metal accumulation in soils,the desorption of heavy metals and the possible hysteresis have been scarcely reported. This chapter highlights the use of biosurfactants of various origins in the removal of heavy metals from soils contaminated with metals.

Keywords Biosurfactant • Desorption • Heavy metal ions • Soil bioremediation • Sorption

Y.S. Açıkel (✉)
Department of Chemical Engineering, Hacettepe University, 06800 Beytepe, Ankara, Turkey
e-mail: yesims@hacettepe.edu.tr

M.S. Khan et al. (eds.), *Biomanagement of Metal-Contaminated Soils*,
Environmental Pollution 20, DOI 10.1007/978-94-007-1914-9_8,
© Springer Science+Business Media B.V. 2011

8.1 Introduction

Domestic and industrial wastes are increasingly disposed into the environment causing long-term effects on the ecosystem. Environmental poisoning by heavy metals (HMs) has increased in the last semicentennial due to extensive use of metals in agricultural and industrial processes, which in turn has become a serious threat to functional ecosystems. The main sources of HM pollution are mining, metallurgical, milling, electronic, electrolysis, electro-osmosis, photography, electroplating, metal finishing, tanneries industries, and the manufacture of paints, metal pipes, batteries, ammunition, porcelain enameling, energy and fuel production, fertilizer and pesticide industry and iron and steel industries and aerospace and atomic energy installation. Why is metal pollution so serious environmental pollution problem? This is due largely to the fact that metals cannot be degraded or destroyed. However, microorganisms used in bioremediation process can change only the speciation of metals and transform them into nontoxic form, but the same metal still persist in the environment. Heavy metals are dangerous because they tend to bioaccumulate. Heavy metal enters our bodies via food, drinking water, and air. Some heavy metals like Fe, Cu, Co, Ni, and Zn, are considered as "essential" elements for microbial growth while others like Cd, Hg, As, Ag, and Au, are "nonessential" elements. Essential heavy metals catalyze biochemical reactions, stabilize proteins, regulate gene expression, and control osmotic pressure across various microbial membranes. Some enzymes require metal such as Mg, Zn, Mn, or Fe as a cofactor. Essential transition metals like Fe, Cu, and Ni play a role in redox processes. Other essential metals like Mg and Zn stabilize various enzymes and DNA. However, whether they are essential or not, all metals at high concentration are toxic to living organisms. For example, metal toxicity to human can cause birth defects, obstruct lung disease, lung cancer, skin lesions, anemia, mental and physical retardation, learning disabilities, liver and kidney damage, stomach and intestinal irritation, circulatory and nerve tissue disease (Bruins et al. 2000).

The traditional remediation technologies for metal-contaminated soils include excavation, landfilling, isolation, immobilization, toxicity reduction, physical separation, and extraction (Mulligan et al. 2001a). Of these, excavation and landfilling have been the most extensively used conventional methods. High cost of excavation, final disposal of landfills, and lack of available landfill sites are the disadvantages of these techniques. Moreover, there is always a risk of HM release into the environment. Another remediation method is solidification/stabilization (Shawabkeh 2005). Solidification is physical encapsulation of the HM pollutants in a solid matrix while stabilization involves chemical reactions to decrease metal mobility. The size selection processes remove the larger and cleaner particles from the smaller, more polluted ones. Mechanical separation processes include hydrocyclones, fluidized bed separation, and flotation (Peng et al. 2009). Electrokinetic process is based on passing a low-density electric current between a cathode and an anode imbedded in the HM-polluted soil. Ions and small charged particles together with water move between the electrodes. Soil washing, in situ soil flushing, bioleaching, phytoremediation, and bioremediation are other promising metal-removing techniques. Soil washing and in situ flushing comprise the addition of water with or without admixture. Soil washing processes are classified in three main groups: (1) physical separation, which includes hydrodynamic classification, gravity concentration, froth flotation, magnetic and electrostatic separations, attrition scrubbing; (2) chemical extraction, which includes acid extraction, salt solutions and high-concentration chloride solutions, chelant extraction, surfactant-enhanced solubilization, reducing and oxidizing agents; and (3) combination of both (Dermont et al. 2008). To recover HMs from soils, apart from biosurfactants, inorganic acids such as H_2SO_4 and HCl with pH less than 2, organic acids including acetic and citric acids (pH not less than 5), chelating agents such as EDTA, nitrilotriacetate (NTA), NaOH, and various combinations of the chemical agents, are used (Chaturvedi et al. 2006). Soil washing may be an effective alternative to solidification/stabilization and landfilling. However, these processes require a detailed soil characterization, a deep understanding of metal speciation and fractionation and interactive relation between the soil matrix and metals. Moreover, additive agents such as EDTA, although effective, are not only nonbiodegradable but are also highly toxic (Chen et al. 2004).

Removal of HMs from soils by microorganisms or plants has not been extensively studied. The techniques available so far include bioleaching, biosorption, and phytoremediation. A variety of microorganisms, for example, autotrophic *Thiobacillus* species, heterotrophic *Aspergillus* and *Penicillium* species, catalyze leaching of metals from ore deposits and mine tailings. The leaching of metals from soils includes (1) redox reactions, (2) the formation of organic or inorganic acids, and (3) the excretion of complexing agents (Krebs et al. 1997). The mediation by redox reactions is based either on electron transfer from minerals to microorganisms or on bacterial oxidation of metals, for example, Fe^{2+} to Fe^{3+} where ferric iron subsequently catalyzes metal solubilization as an oxidizing agent. *Thiobacillus* sp. reduce sulfur compounds under aerobic and acidic conditions. Use of plants such as *Thlaspi, Urtica, Chenopodium, Polygonum sachalase,* and *Alyssim,* trees, herbs, grasses, and other crops to remove metals from soils and ground waters is commonly known as phytoremediation (Lasat 2000; Römkens et al. 2002). This method is, however, restricted to shallow depths of soils with low levels (2.5–100 mg/kg) of metals and requires longer treatment times compared to other methods. Microorganisms and microbial products have attracted attention as alternative technologies for metal removal from soils.

Biosorption is defined as the microbial uptake of organic and inorganic metal species by physicochemical mechanisms, such as adsorption, ion-exchange, complexation, chelating, and surface precipitation (Sağ 2001). This "passive uptake" method is independent of the vital activity of microorganism. The HMs can also be transported into the cell across the cell membrane through the cell metabolic cycle. This type of uptake performed via growing cells is known as "active uptake." The metal uptake involving both active and passive modes is defined as "bioaccumulation" (Malik 2004). Microorganisms produce a range of specific and non specific metal-binding compounds. Bacteria, algae, and fungi produce extracellular polymeric substances (EPS), a mixture of polysaccharides, mucopolysaccharides, and proteins.

Metal ions and/or particulate matters such as precipitated metal sulfides and oxides are adsorbed or entrapped by EPS (Gadd 2004). Although the metal removal by microorganisms and their products has been extensively applied in the treatment of industrial and domestic wastewaters, the biosorption of metal ions from soils and sediments presents a more complex separation problem (Lebeau et al. 2002; Vig et al. 2003; Zoubolis et al. 2004; Lin and Lin 2005). Movement of metals in soils, on the other hand, is restricted by soil texture, structure, and organic matter content. The size of a bacterial cell is nearly as large as 0.2 mm in diameter, whereas soil pores change greatly in size ranging from less than 2 μm. Metal–cell complexes could be filtered out by the smallest pores and hinder transfer through the soil. Although bacterial and algal EPS bind to a variety of metals, they show strong affinities for oil–water interfaces, differ from biosurfactants as they are large, have molecular weight around 10^6, and have minimal surface activity. Biosurfactants offer a distinct advantage over EPS in the remediation of soils because of their relatively small size, (generally <1,500 Da) (Tan et al. 1994; Herman et al. 1995; Miller 1995). Microbial compounds that are produced by microorganisms and plants and show high surface and emulsifying activities are defined as biosurfactant. Biosurfactants have been used in the bioremediation of numerous types of hydrophobic hydrocarbon-organic contaminants. Only recently, it has been proved that biosurfactants can be used to enhance metal removal. This chapter discusses all aspects of the use of biosurfactants for the removal and recovery of heavy metals from soils.

8.2 Biosurfactants

Surfactants are substances that adsorb to and alter conditions prevailing at interfaces. They lower surface and interfacial tensions. Emulsifiers are a subclass of surfactants that stabilize dispersions of one liquid in another, for example, oil-in-water emulsions. Certain bioemulsifiers increase the growth of bacteria on hydrophobic water-insoluble substrates, by increasing their surface area, desorbing them from surfaces and enhancing their apparent solubility. Bioemulsifiers also regulate the attachment–detachment of microorganisms to and from surfaces (Ron and Rosenberg 2001). A surfactant's effectiveness is strongly related to its ability to lower surface tension, to increase solubility, its good detergency properties, wetting and complex foaming capacity. Biosurfactants are biological surfactants that are produced extracellularly or as part of the cell membrane by yeast, bacteria, fungi, or marine microorganisms inhabiting various substrates including sugars, oils, alkanes, and wastes. Molecular mass of biosurfactants ranges from 500 to 1,500 Da. Biosurfactants can be divided into (1) low-molecular-weight (LMW) molecules that lower surface and interfacial tensions efficiently and (2) high-molecular-weight (HMW) polymers that bind tightly to surfaces. The LMW types are generally glycolipids or peptidyl lipids (lipopeptides). The best known glycolipid bioemulsifiers are trehalose tetraesters and dicarynomycolates, fructose lipids, sophorolipids, and rhamnolipids.

Glycolipids are produced by microorganisms such as *Arthrobacter paraffineus* (trehalose lipids), *Rhodococcus erythropolis* (trehalose dimycolates), *Candida bombicola* (formerly *Torulopsis)* (sophorolipids), *Pseudomonas* (rhamnolipids), and *Alcanivorax borkumensis* (glucose lipids) (Ron and Rosenberg 2001; Christofi and Ivshina 2002). Sophorolipids consist of two glucose units linked ß-1,2 and a lipid portion connected to the reducing end through a glycosidic linkage. A group of biosurfactants that has been studied extensively is the rhamnolipids. Two types of rhamnolipids contain either two rhamnoses attached to ß-hydroxydecanoic acid or one rhamnose connected to the identical fatty acid (Mulligan 2005). Peptidyl lipids include surfactin from *Bacillus subtilis*, streptofactin from *Streptomyces tendae*, gramicidin S from *Bacillus brevis*, polymyxins from *Bacillus polymyxa* and related bacilli, viscosin from *Pseudomonas* strains that are effective antibiotics as well as potent surface active materials (Ron and Rosenberg 2001). The potential advantages of using surfactin include its biodegradability, effectiveness as a surfactant, and extensive biological properties including affecting the growth of tumors, bacteria, fungi, viruses, and mycoplasmas (Mulligan et al. 2001b). High-molecular-weight biosurfactants are amphiphilic (lipo)polysaccharides, (lipo)proteins, or complex mixtures of these biopolymers produced by numerous bacterial species belonging to different genera, as exocellular polymeric surfactants. These biosurfactants produce stable emulsions but do not lower the surface tension. Different isolates of *Acinetobacter* were found to produce HMW emulsifiers (Christofi and Ivshina 2002). Polysaccharide HMW emulsifiers can also bind with metals. For example, emulsan produced by *A. calcoaceticus* was demonstrated to bind uranium (Ron and Rosenberg 2001).

Biosurfactants have the following advantages (1) lower toxicity and higher biodegradability, (2) better environmental compatibility, (3) higher foaming, (4) higher selectivity for metal ions and organic compounds, (5) effectiveness at enhancing biodegradation and solubilization of low-solubility compounds, and (6) less expensive. In addition, they are less sensitive to pH, salt, and temperature variations. However, biosurfactants have two major disadvantages. If a commercial biosurfactant is added to a soil, groundwater system or wastewater system externally, it may be difficult to distribute the biosurfactant uniformly. In this case, the desired bioremediation process cannot be performed. Biosurfactants can sometimes be more biodegradable than the substance, which is planned to be treated. In this case, the biosurfactant itself may become a more favorable biodegradable material than the pollutant to be removed (UTTU 2005). However, to date, the persistency of rhamnolipid in soil has been rarely investigated. The rate of rhamnolipid degradation in soils with single and co-contaminated with Cd(II) and Zn(II) ions was investigated (Wen et al. 2009). Rhamnolipid, a metal sequestering agent produced by *Pseudomonas* sp., was found as more biodegradable than EDTA but more stable in the soil than citric acid. The degradation of rhamnolipid, citric acid, and EDTA was inhibited by Cd or/and Zn contamination in two uncontaminated soils and a rice (*Oryza sativa*) soil with previous contamination from mining. Single Cd-contamination had a less inhibitory effect whereas the biodegradation was retarded by co-contamination of Cd and Zn. Due to the co-existence of Cd and Zn in soils, there may have been an increase in metal toxicity caused by Cd and Zn interaction effects. Although the biodegradation of rhamnolipids strongly depends on soil properties such as organic matter content, cation exchange capacity (CEC), fertility, extent of contamination, and metal toxicity, rhamnolipid may persist in soil and consequently increase metal bioremoval but not remain long enough to raise concerns regarding metal transport in the long term.

8.2.1 Critical Micelle Concentration

Surfactants reduce the surface tension of a liquid medium. Surface tension is a measure of the surface free energy per unit area required to bring a molecule from the bulk phase to the surface. Surface tension of distilled water is 73 dyn/cm. An effective biosurfactant can reduce this value to <30 dyn/cm. The amount of surfactant needed to obtain the lowest possible surface tension is defined as the critical micelle concentration (CMC). After CMC is reached, surface tension remains constant, and surfactants begin to form micelles (Zhang and Miller 1992). The CMCs of biosurfactants typically range from 1 to 200 mg/L. Salinity, hydrocarbon chain length, and surfactant types affect the CMC.

8.2.2 Rhamnolipids

Of the various biosurfactants, rhamnolipids produced by *Pseudomonas aeruginosa* have been extensively studied. Six rhamnolipid homologues produced by a single strain of *Pseudomonas* sp. growing on soapstock, a waste product of vegetable oil manufacturing, were described (Van Hamme et al. 2006). Up to 11 rhamnolipid homologues in *P. aeruginosa* 47 T2 growing on waste frying oil was identified (Haba et al. 2003). Surface tensions of 29 mN/m and interfacial tensions of 0.25 mN/m are characteristic of these compounds (Christofi and Ivshina 2002). The CMCs of rhamnolipids varied between 50 and 200 mg/L. Four types of rhamnolipids were identified. Two major types of rhamnolipids, RLL (R1) and RRLL (R2), have a molecular mass of 504 g mol^{-1} and 650 g mol^{-1}, respectively. RLL ($C_{26}H_{48}O_9$) is L-rhamnosyl-ß-hydroxydecanoyl-ß-hydroxydecanoate. RRLL ($C_{32}H_{58}O_{13}$) is L-rhamnosyl-ß-L-rhamnosyl-ß-hydroxydecanoyl-ß-hydroxydecanoate (Fig. 8.1). The other two types of rhamnolipids contain either two rhamnoses connected to ß-hydroxydecanoic acid or one rhamnose connected to the identical fatty acid. Rhamnolipids RLL, RRLL, and a mixture of the mono- and di-rhamnolipid forms are especially used for soil washing to remove hydrocarbons and HMs, as wastewater treatment to remove hydrocarbons and HMs, and as chelating agent and oil slick dispersant in environmental bioremediation. Although the RLL form was reported to be far superior at metal complexation (Ochoa-Loza et al. 2001), a mixture of RLL and RRLL (RL = RLL/RRLL = 1.1, JBR 425, lot.no = 030126), forms was generally

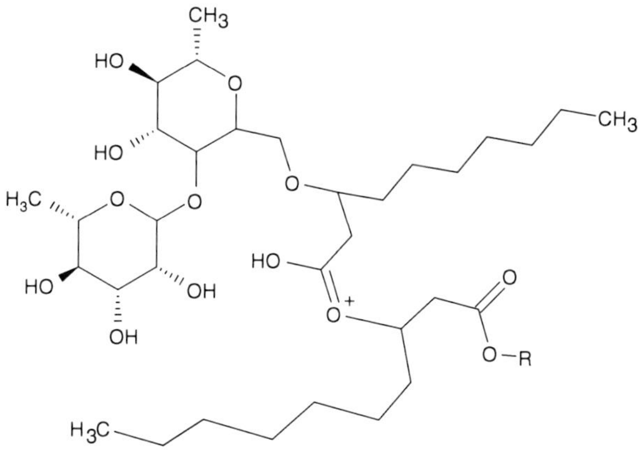

Fig. 8.1 Structure of di-rhamnolipid

used in environmental applications. The CMC of monorhamnolipid and a mixture of RLL and RRLL produced by *P. aeruginosa* is 0.1 mM (Ochoa-Loza et al. 2007). This low CMC points to the strong surface activity shown at low concentrations. It is characterized by low surface tension for water and electrolyte solutions and very low interfacial tensions for water/hydrocarbon systems. The pK_a value of RL is determined as 5.6 (Ishigami et al. 1987).

8.2.3   *Surfactin*

Surfactin, a cyclic peptide antibiotic that contains seven amino acids bond to the carboxyl and hydroxyl groups of a 14-carbon acid (Fig. 8.2) is produced by *Bacillus subtilis* and *B. pumilus* and *B. licheniformis*. The primary structure of surfactin is a heptapeptide with a ß-hydroxy fatty acid within a lactone ring structure (Kakinuma et al. 1969). The three-dimensional structure of surfactin has a ß-sheet structure. It looks like a horse saddle at the air/water interface and in aqueous solutions (Bonmatin et al. 1995; Mulligan 2005). Surfactin has an amphiphilic structure and has extensive antibiotic properties that may affect the growth of tumors, bacteria, fungi, viruses, and mycoplasmas (Christofi and Ivshina 2002). For swarming motility in *B. subtilis*, both flagella biosynthesis and surfactin production are important. *B. subtilis* mutants, unable to produce surfactin and deficient in extracellular proteolytic activity, could neither swarm nor form biofilms (Van Hamme et al. 2006). Individual surfactin molecules have a molecular mass of approximately 1,050 Da. As surfactin reduces the surface tension of water from 72 to 27 mN/m at a concentration as low as 0.005%, it is considered as one of the most effective biosurfactants. Interfacial tension and CMC of surfactin are 1 mN/m and 23 mg/L, respectively. Surfactin involves the glutamic and aspartic amino acids, where glutamate residues are

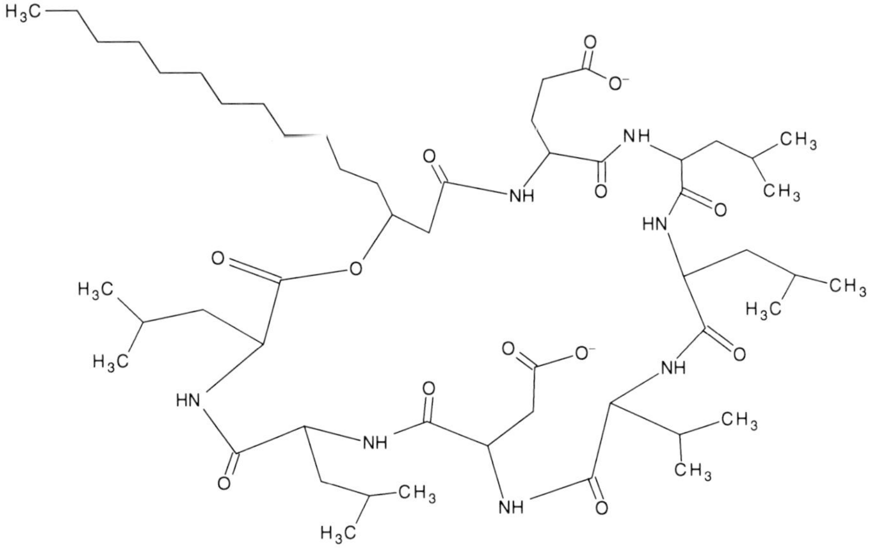

Fig. 8.2 Structure of surfactin

reported to bind metals such as Mg, Mn, Ca, Ba, Li, and rubidium (Thimon et al. 1992; Singh et al. 2007). The theoretical ratio of metals to the surfactin is 1 mol metal: 1 mol surfactin due to the two charges on the surfactin molecule; however, this ratio was found to be 1.2:1 in experimental studies (Mulligan et al. 1999b). Heavy metals are generally found associated with carbonate, oxides, and organic fractions in the contaminated soil. These are removed using a combination of 0.25% surfactin and 1% NaOH. Proposed metal recovery mechanism is the attachment of surfactin at the soil interface and metal removal through lowering the interfacial tension and micellar complexation (Christofi and Ivshina 2002).

8.2.4 Saponin

Saponin, a nonionic biosurfactant, is a triterpene glycoside obtained from quillaja bark and includes ß-D-glucuronic acid with carboxyl group of sugar moiety in hydrophilic fraction (Fig. 8.3). The triterpene portion of saponin backbone chain, the sapogenin ($C_{30}H_{46}O_5$), is 13.9% (wt) of the total hydrolyzed saponin. The chemical structure of saponin comprises one hydrophobic fused-ring of triterpenes, which does not resemble the hydrophobic tail of common surfactants having a long, straight hydrocarbon chain. Two hydrophilic sugar chains are connected to the two ends, C-3 and C-28, of the hydrophobic triterpene backbone, in which one end carries ß-D-glucuronic acid with anionic carboxyl group and the other end carries nonionic glycoside groups. The CMC of saponin at pH 6.5 is 100–200 mg

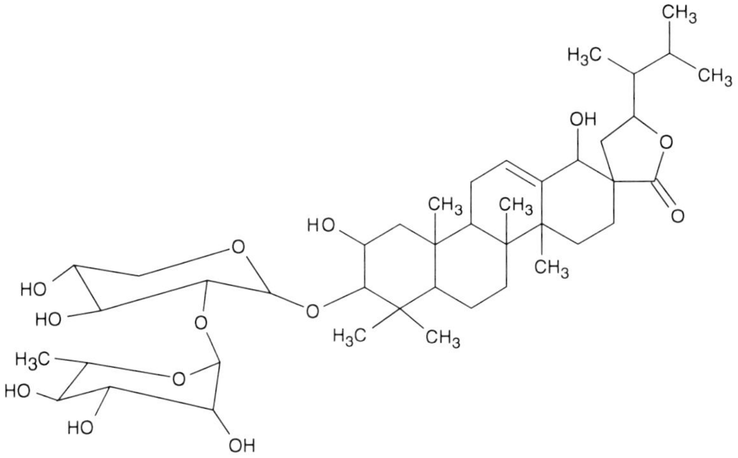

Fig. 8.3 Structure of saponin

L^{-1} (0.1 mass-%), surface tension is 36–39 mMm^{-1}, and interfacial tension is 6.0 mMm^{-1}. Saponin is weakly acidic (pH 4.6) due to the hydrolysis of glycosides. Elemental analysis of saponin exhibits organic elements, 42–44% C, 6–6.2% H, 51% O_2, and inorganic elements, 13.9% sulfated ash (Hong et al. 2002; Urum and Pekdemir 2004; Chen et al. 2008).

8.2.5 Sophorolipids

Sophorolipids are obtained from the yeast *Torulopsis bombicola*. They are produced in the fermentation medium containing soybean oil and glucose (0.35 g/g substrate), and obtained from the medium directly as no foam is produced (Fig. 8.4). Ethyl acetate is used for the extraction of sophorolipids from the fermentation medium. Sophorolipids reduce the surface tension to 34 mN m^{-1}. The CMC of sophorolipids is 0.80 g L^{-1}. They are generally used for the release of bitumen from tar sands.

8.2.6 Aescin

Aescin ($C_{54}H_{84}O_{23}$) with molecular weight 1,101 g mol^{-1} (Fig. 8.5) is commercially provided from the seeds of the horse chestnut tree: *Aesculus hippocastanum L.* (*Hippocatanacea*), by percolation with 60–80% ethanol. Aescin consists of aglycones protoaescigenin or barringtogenol C, 3-*O*-[ß-D-glucopyranosyl-(1,2)-ß-D-glucopyranosyl(1,4)-ß-D glucopyranosyl]-21ß-tigloyl-22α-acetyl-protoaescigenin,

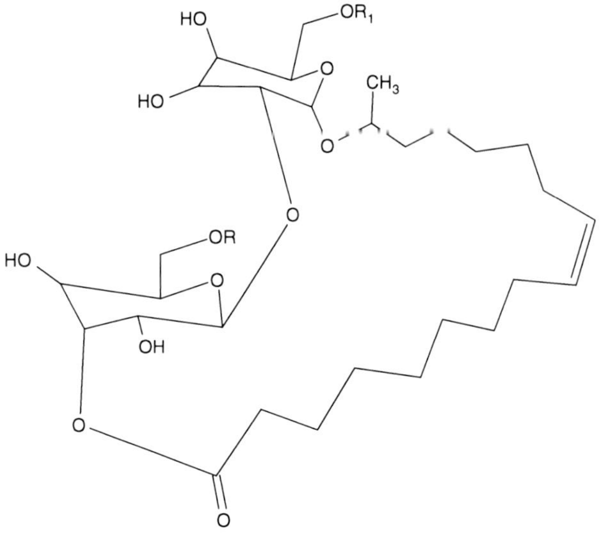

Fig. 8.4 Structure of sophorolipis

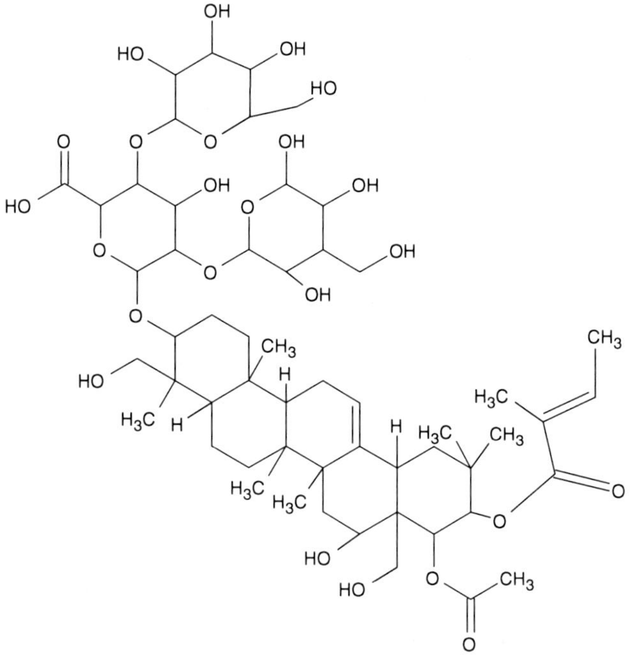

Fig. 8.5 Structure of aescin

and 21ß-angeloyl analog (Hong et al. 1998). The CMC of aescin is 0.1 mass-%, surface tension is 444 mMm^{-1}, and interfacial tension is 7.0 mMm^{-1} (Urum and Pekdemir 2004).

8.3   Heavy Metal Sorption on Soils

The retention mechanism of metal ions at soil surfaces includes adsorption, surface precipitation, and fixation (Bradl 2004). Adsorption can be defined as a two-dimensional accumulation of metal at the solid/water interface. Intermolecular interactions consist of surface complexation reactions, electrostatic interactions, hydrophobic expulsion of metal complexes, and surfactant adsorption metal–polyelectrolyte complexes due to reduced surface tension (Sposito 1984). Heavy-metal adsorption may be specific and nonspecific (or ion exchange). Specific adsorption is selective, strong, and less reversible reactions involving chemisorbed inner-sphere complexes (McBride 1994). Ion exchange is an electrostatic phenomenon and is less selective and more reversible involving outer-sphere complexation with only weak covalent bonding between metals and charged surfaces (Reed and Cline 1994). Specific adsorption is explained by a surface complexation model. This model describes surface complex formation as a reaction between functional surface groups such as silanol, inorganic hydroxyl groups, or organic functional groups and an ion in a solution, which form a stable unit. This type of adsorption occurs by adsorption reactions at OH$^-$ groups at the soil surfaces and edges, which are negatively charged. The sorbed metal ions are connected by an inner sphere mechanism to atoms at the surface. These reactions are represented as follows for a metal ion Me and a surface S:

$$S-OH+Me^{2+}+H_2O \leftrightarrow S-O-MeOH_2^+ +H^+ \tag{8.1}$$

In surface precipitation, a new solid phase grows and repeats itself in three dimensions and forms a 3-D network. Metals precipitate as oxides, hydroxides, carbonates, sulfides, or phosphates onto soils. Surface precipitation depends on pH and the amounts of metals and anions present. The surface complexation model cannot define the adsorption curves at high cation concentrations. In the first case, a saturation of the adsorption capacity is reached, which is represented better by a Langmuir isotherm. In the second case, a continuous increase of the adsorption capacity without saturation at the soil surface is observed, which is modeled by a Freundlich isotherm. The surface precipitation model postulating a multilayer sorption process consider precipitation reactions in addition to adsorption reactions at the surface, and is defined by two reactions: (1) surface complex formation of metal ion (Me) and surface (S) as given by Eq. 8.1 (2) the precipitation of metal (Me) at the surface (S) (Farley et al. 1985; Robertson and Leckie 1997; Bradl 2004):

$$S-O-MeOH_2^+ +Me^{2+}+H_2O \leftrightarrow S-O-MeOH_2^+ +Me(OH)_{2(s)}+2H^+ \tag{8.2}$$

This model obeys Langmuir model at low metal concentrations and Freundlich model for increasing metal concentrations. If the metal concentration continues to increase, solid solution precipitation controls. The third mechanism of sorption is known as fixation or absorption. Heavy metals adsorbed onto clay minerals and metal oxides diffuse into the lattice structures of these minerals. The metals are then fixed into the pore spaces of the minerals by a process called solid-state diffusion. Surface functional groups of soils include a variety of hydrous oxide minerals, organic matter (carboxyl (–COOH), carbonyl, and phenolic groups), alumosilicates (clay minerals, micas, zeolites, and most Mn oxides). Alumina surfaces have terminal –OH groups that resist dissociation to the anionic $\equiv$Al–H$^-$ form. For that reason, it will form a positively charged $\equiv$Al–OH$_2^+$ site. Once deprotonated, the terminal –OH group binds more strongly to metals than the bridging –OH group (McBride 1994). Alumosilicates exhibit both aluminol ($\equiv$A–OH) and silanol ($\equiv$Si–OH) edge-surface groups. The deprotonated aluminol group (i.e., ($\equiv$Al–O$^-$)) binds metals more strongly (Bradl 2004).

Characterization of the metal-contaminated soil matrix used in biosurfactant washing tests should follow the guidelines of EPA or ASTM, which include (1) soil pH and moisture content, (2) particle size distribution, (3) oil and grease content, (4) organic matter content, (5) chemical oxygen demand (COD), and (6) cation exchange capacity (Mulligan et al. 1999b). The bioavailability and mobility of metals in soil strongly depend upon sorption and desorption of the HM with different soils and/or soil constituents. Dispersion and partitioning of HMs between solid and aqueous phases are subject to soil properties such as surface area and charge, pH, ionic strength, and concentration of complexing ligands. To date, most studies on HM sorption and related binding mechanisms in soils have been focused on individual synthetic sorbents or combination of sorbents or soil components or real soils. Use of non-humus and humus soil formed as a result of leaf litter decay in a certain ratio of 1:3 gave effective results in situ soil bioremediation (Misra and Pandey 2004). Humic acids among the humic substances included in humus soil are natural organic macromolecules with multiple properties and high structural complexity. Phosphate, apatite mineral such as hydroxyapatite, and phosphatic clay, a by-product of the phosphate mining industry, were used as promising immobilizing agents to remediate HM-contaminated soils, sediments, and wastewaters (Arey et al. 1999; Hettiarchchi et al. 2000; Singh et al. 2001). The retention mechanism of Pb(II) on the non-humus-humus soil and hydroxyapatite was due to sorption, and immobilization of Zn(II) and Cd(II) was co-precipitation and ion-exchange. Metal sorption rate and equilibrium capacity order depend on type of soil and/or soil component, as well as physico-chemical properties of metal ions such as, atomic weight, electronic configuration, electronegativities, ionic radius, reduction potential, hydrated ion radius, crystal radius, equilibrium constant, p_K, and covalent binding. The smaller the ionic radius and the greater the valance, the more tightly and intensively is the ion adsorbed onto the clay. For the cationic metal ions, it was also reported to be a direct relationship between the valance/ionic radius ratio and the adsorption rate constant. Soils with CEC of 50–100 meq/kg and particle sizes of 0.25–2 mm, with contaminant solubility in water of greater than 1,000 mg/L and with low contents of cyanide, fluoride, and sulfide, with less than 10–20% clay and organic content (i.e., sandy soils), can be most effectively treated by soil washing (Mulligan et al. 2001a).

8.3.1 Comparison of Metal Sorption on Various Soils and/or Components

The sorption is usually measured by the parameter q (mmol or mg of metal accumulated per g or kg of soil or soil component). The two most common types of adsorption models for assessing this system are the Langmuir (L) and Freundlich (F) models. The Langmuir parameter, Q^o, represents the highest experimentally observed value of the specific sorption (sorbed metal ion quantity per unit weight of dry soil or soil component at equilibrium). A large value of Langmuir constant, K, implies strong bonding. The Freundlich constants K_F and $1/N_{sorp}$ are an indicator of the sorption capacity of the sorbent and sorption intensity, respectively. It has been observed that values of Q^o and K_F are not comparable with other values reported for the same metal. It may depend not only on different sorption abilities of soils and/or soil components, but also on not exactly equal operating conditions. In fact, works of different authors cannot be compared directly: operating conditions are often different even if they are nominally equal. Comparing Langmuir parameter, Q^o, obtained by various soils and/or soil components listed in Table 8.1, sorption preference for Cd(II) decreases in the following order: Soil A[1] (Smectite-moderate soil) > Soil C[2] (Smectite-moderate-dominant soil) > Soil B[3] (Smectite-dominant soil) > Sepiolite > Kaolin > K-feldspar (Aşçı et al. 2007, 2008a, b). In terms of Freundlich constant K_F, preference order changes slightly. Soil C (Smectite-moderate-dominant soil) > Soil B (Smectite-dominant soil) > Sepiolite > Soil A (Smectite-moderate soil) > Kaolin > K-feldspar > Quartz (Aşçı et al. 2007, 2010). The presence of smectite as the dominant clay in soil ensures high metal sorption capacity. The structure and chemical composition, exchangeable ion type, and small crystal size of smectite provide a large chemically active surface area, a high CEC, and inter-lamellar surface having unusual hydration characteristics (Miranda-Trevino and Coles 2003; Singh et al. 2006). Smectite-moderate-dominant soil with higher clay content (70%) had the greatest sorption efficiency and sorption capacity as estimated by the maximum sorption capacity (K_F) and intensity (N_{Sorp}) of the Freundlich equation. The clay fraction of smectite-moderate-dominant soil was dominated by well-crystallized smectite and an ample proportion of feldspar and illite that provides the surface charge to soil. Both smectite-dominant and smectite-moderate-dominant soils had similar proportions of smectite and feldspar in the clay fraction and so had similar

---

[1] Soil A Smectite, serpentine, amphibole, feldspar-moderate.
[2] Soil C Smectite-moderate-dominant, feldspar-moderate, illite-moderate.
[3] Soil B Smectite-dominant, feldspar-moderate.

Table 8.1 Comparison of heavy metal sorption capacities of various soils and/or components in terms of the Langmuir and Freundlich model constants

Soil type	Metal	Langmuir $Q°$	Model K	Freundlich K_F	Model n	Reference
Sepiolite (Orera)	Cd(II)	8.26 mg g^{-1}	1.67 L mg^{-1}	4.44 mg^n g^{-1} L^n	3.83	Garcia-Sanchez et al. (1999)
Smectite, Serpentine, Amphibole, Feldspar-moderate soil	Cd(II)	51.03 mmol kg^{-1}	0.411 L mmol^{-1}	14.90 mmol^n kg^{-1} L^n	1.786	Aşçı et al. (2008b)
Urban soil	Cd(II)	2.321 mg g^{-1a} 2.294 mg g^{-1b}	3.646 L mg^{-1a} 1.449 L mg^{-1b}	–	–	Markiewicz-Patkowska et al. (2005)
Illite (Ballclay)	Cr(III)	69.0 mmol kg^{-1}	12.90 L mmol^{-1}	23.1 mmol^n kg^{-1} L^n	2.44	Chantawong et al. (2003)
Kaolin (Thai)	Cr(III)	34.9 mmol kg^{-1}	0.72 L mmol^{-1}	1.13 mmol^n kg^{-1} L^n	1.567	Chantawong et al. (2003)
Kaolin	Cu(II)	0.72 mg g^{-1}	1.82 L mg^{-1}	0.438 mg^n g^{-1} L^n	2.281	Chen et al. (2008)
Kaolinite	Cu(II)	10.787 mg g^{-1}	0.155 L mg^{-1}	–	–	Yavuz et al. (2003)
Na-montmorillonite modified with rhamnolipid (low conc.)	Cu(II)	48.3 mg g^{-1}	0.020 L mg^{-1}	–	–	Özdemir and Yapar (2009)
Palygorskite clay	Pb(II)	62.11 mg g^{-1}	0.75 L mg^{-1}	23.89 mg^n g^{-1} L^n	1.80	Potgieter et al. (2006)
Phosphatic clay	Pb(II)	37.2 mg g^{-1}	0.125 L mg^{-1}	–	–	Singh et al. (2001)
Smectite (well crystallized) and a sizable proportion of illite	Pb(II)	2.52×10^4 μmol kg^{-1}	166.7 L μmol^{-1}	–	–	Serrano et al. (2005)
Smectite (well crystallized) and a sizable proportion of illite	Pb(II)(+Cd(II))	1.50×10^4 μmol kg^{-1}	76.9 L μmol^{-1}	–	–	Serrano et al. (2005)
Soil-amended humus soil and hydroxyapatite	Pb(II)	149.5 mg g^{-1}	4.54 L mg^{-1}	–	–	Chaturvedi et al. (2006)
Urban soil	Pb(II)	0.615 mg g^{-1a} 0.579 mg g^{-1b}	3.249 L mg^{-1a} 0.645 L mg^{-1b}	–	–	Markiewicz-Patkowska et al. (2005)
Quartz	Zn(II)	4.209 mmol kg^{-1}	0.245 L mmol^{-1}	0.834 mmol^n kg^{-1} L^n	1.368	Aşçı et al. (2010)

[a] Single metal solutions
[b] Multi-metal solutions containing Cd(II), Cr(VI), Cu(II), Pb(II), Zn(II)

sorption properties. On the other hand, smectite-moderate soil contained less smectite and illite, and had the lower clay content (30%) than smectite-moderate-dominant soil. As a result, it had the lowest sorption and ion-exchange capacity.

The most important clay mineral groups which are used for environmental purposes are kaolins, smectites, illites, and chlorites. The sorption capacities of clay minerals decrease in the order: smectites > chlorites > illites > kaolins. The kaolin, low-permeability clayey soil, belongs to the two-layer minerals. Kaolinite, the most known kaolin mineral comprises a single-silica tetrahedral sheet and a single-alumina octahedral sheet which form the kaolin unit layer (Serrano et al. 2005). Smectite is a member of the three-layer minerals and consists of two silica tetrahedral sheets with a central alumina octahedral sheet. The lattice has an unbalanced charge due to isomorphic substitution of alumina for silica in the tetrahedral sheet and of Fe and Mg for alumina in the octahedral sheet. For this reason, the attractive force between the unit layers in the stacks is weak. The cations and polar molecules can enter between the layers and hence, the layers expand (Ayari et al. 2005). For mica-like clay, illite term is generally used. Basic structural unit of illite is similar to that of montmorillonite, generally known as smectite. As there is a large replacement of silica for alumina in the tetrahedral sheet, illites are typically characterized by a charge deficiency that is balanced by K ions that bridge the unit layers. As a result, illites are nonexpandable clay minerals (Gu and Evans 2007). Other clay mineral groups like chlorites and the mixed-layer clays comprise of mixtures of the unit layers, for example, illite-smectite, smectite-chlorite, illite-chlorite, etc. Two-layer minerals like kaolins have no additional ions between their silicate layers. On the contrary, kaolins, the silicate layers of three-layer minerals, carry an electric charge due to isomorphic substitution (Krawczyk-Barsch et al. 2004).

Sepiolite [$Mg_4Si_6O_{15}(OH) \cdot 6H_2O$], a zeolite-like clay mineral, is a hydrous magnesium silicate. It has fibrous morphology and intracrystalline channels. Sepiolite comprises a continuous two-dimensional tetrahedral sheet of T_2O_5 (T = Si, Al, Be) and uncontinuous octahedral sheets. Molecular size of channels of sepiolite is $3.6 \times 106 \,\text{Å}$ and specific surface area is more than 200 $m^2 \, g^{-1}$ (Garcia-Sanchez et al. 1999; Vico 2003). Because of the fibrous structure, organic and inorganic ions can penetrate into sepiolite, which makes it an exquisite metal accumulator. Because of the crystal-chemical features, HM removal by the sepiolite occurs by adsorption and/or cation exchange mechanisms. Adsorption occurs on the oxygen ions of the tetrahedral sheets, on the water molecules at the edges of the octahedral sheet and on Si–OH groups along the direction of fibers. Ion exchange arises by substituting cations inside the channels and/or inside the octahedra at the edges of the channels. In the ion exchange, bivalent metal cations replace Na(I) and/or Mg(II) at the edges of octahedral sheet (Brigatti et al. 2000).

Feldspars, usually found in rocks, sediments, and soils, are the common name of an important group of rock-forming minerals, which constitute perhaps as much as 60% of the Earth's crust. K-feldspar ($KAlSi_3O_3$; microcline or orthoclase) and Na-feldspar (albite; $NaAlSi_3O_8$) in a significant proportion of feldspar ores exist in the same matrix usually in quantities of about 3–5% Na_2O and K_2O. On the other hand, studies on HM sorption mechanisms with feldspars are scarce. Cadmium(II), for example, was reported to be physically sorbed by perthitic feldspar where outer-sphere complexation played an important role in Cd(II) removal (Farquhar et al. 1997). Quartz is a ubiquitous mineral of relatively simple structure and is the predominant (up to 70%) constituent of the sand and silt fractions in many soils; however, it shows weak HM sorption characteristics. As surface charge of specific crystals varies with pH, the medium pH is the dominant parameter controlling the sorption of metal ions (Taqvi et al. 2007). The point of zero charge is basically important to many processes occurring at the mineral–water interface. These processes include dissolution rates and sorption processes. Above the pH_{pzc}, minerals exhibit negative surface charge, whereas below the pH_{pzc}, a positive charge takes place. For example, the point of zero charge (pH_{pzc}) of quartz is 3. The quartz has a tetrahedral structure with oxygen atoms occupying the four corners of a tetrahedron. The presence of negative charge in the quartz in the form of oxides provides affinity for the positively charged Cd(II) ions (Ledin et al. 1999; Aşçı et al. 2010).

$$Cd^{2+} + O_2^- \rightarrow O^- ... Cd^{2+} ... O^- \tag{8.3}$$

The electrostatic attractive forces between Cd(II) ions and the negatively charged surface of the quartz are likely to control the retention of Cd(II) ions onto sorbent surface.

To improve adsorption and desorption capacity of soil components, rhamnolipids were also used as surface modification agents. The effect of an anionic biosurfactant rhamnolipid on the adsorption of Cu(II) ions by a Na-montmorillonite was investigated (Özdemir and Yapar 2009). Carboxylate groups of rhamnolipids are involved in an interaction electrostatically with the positively charged edges and layer sites of the Na-montmorillonite. Rhamnolipid moieties having a polyalcohol structure through –OH groups build hydrogen bridges with the faces of the Na-montmorillonite platelets. Clay expands through the insertion of rhamnolipid molecules. Clay platelets distribute water by attaching the rhamnolipid molecules on the edge groups. This causes a relative decrease in the mass transfer resistance in adsorption and desorption via the dispersion of the Na-montmorillonite platelets in water. In a study, Serrano et al. (2005) determined the competitive sorption of Pb and Cd, kinetics, and equilibrium sorption in surface soils from central Spain using single and binary metal solutions. Soils S2 containing less kaolinite and more smectite and illite and S4 containing well crystallized smectite and a sizable proportion of illite, with higher pH and clay content, showed the greatest metal sorption capacity. The sorption capacity of the soils for Pb, as estimated by $Q°$ parameter from Langmuir equation, was always greater than for Cd(II). The co-existence of both metals reduced greatly the sorption capacity of Cd(II) than Pb(II). The binding strength K from Langmuir equation was always greater for Pb(II) than for Cd(II). As competition for sorption sites could promote the sorption of both metals on more specific sorption positions, the simultaneous presence of both metals increased their corresponding K values. The Langmuir parameter, $Q°$, for the non-humus soil reclaimed with (1:3) humus soil and 1% hydroxyapatite decreased in the order Pb(II) > Zn(II) > Cd(II) (Chaturvedi et al. 2006). The maximum adsorption capacity of the Orera sepiolite was reported for Cd(II) (8.3 mg g⁻¹), followed by Cu(II)

(6.9 mg g^{-1}), and by Zn(II) (5.7 mg g^{-1}) (Garcia-Sanchez et al. 1999). The sorption of HM ions on kaolinite followed the Langmuir adsorption model and the resulting adsorption affinity order was: Cu(II) > Ni(II) > Co(II) > Mn(II) (Yavuz et al. 2003). For HM sorption by palygorskite clay, the maximum monolayer adsorption capacity (Q^{o}) diminished in the order: Pb(II) > Cr(VI) > Ni(II) > Cu(II) (Potgieter et al. 2006). Amounts of metal ions sorbed onto phosphatic clay in terms of Langmuir constants (Q^{o} and K) followed the order: Pb(II) > Cd(II) > Zn(II) (Singh et al. 2001). The order of metal adsorption by kaolin was found as Cr(III) > Zn(II) > Cu(II) ≈ Cd (II) ≈ Ni(II) > Pb(II) and by illite (ballclay) Cr(III) > Zn(II) > Cu(II) ≈ Cd(II) ≈ Pb(II) > Ni(II) (Chantawong et al. 2003). According to the Lewis hard–soft acid base principle, hard Lewis acids prefer to react with hard Lewis bases, and soft acids with soft bases (Puls and Bohn 1988). Kaolin is a 1:1 clay type and illite 2:1 clay type. Main surface adsorption sites on kaolin and illite show soft and hard, respectively, Lewis base characteristics. Illite has excess negative charges due to the spread of isomorphous substitution in tetrahedra and octahedra sheets. Both physical structure and hard Lewis base property of illite tend to result in the formation of outer-sphere complexes. It should not be forgotten that metal sorption is also affected by the speciation in solution and the organic matter (OM) content. Comparing the two clays with respect to the Q^{o} values from Langmuir model, it is seen that illite had about 1 order higher sorption capacity than kaolin. Illite is a 2:1 clay type, which has a higher CEC than kaolin, 1:1 type, and has a higher OM content. The average equilibrium adsorption of stronger acidic Cu(II) ions on kaolin was reported to be four-fold higher than weaker acidic Ni(II) ions in binary metal system (Chen et al. 2008). This competitive behavior of Cu(II) and Ni(II) ions can be explained by the Hard and Soft Acid/Base (HSAB) theory. The hard Ni(II) ions having low polarizability cannot compete with soft Cu(II) ions having high polarizability for the soft surface sites of kaolin, and Cu(II) ions are selectively adsorbed from the Cu(II)–Ni(II) binary metal system on kaolin. The study on sorption capacity of selected HMs from single and multiple metal solutions on urban soil containing a mix of mineral soil and residue materials (e.g., brick, concrete, wood) has received little attention (Markiewicz-Patkowska et al. 2005). The sorption capacity from single-metal solutions followed the order: Cd(II) at pH 7 > Cr(VI) at pH 2 > Cu(II) at pH 2 > Zn(II) at pH 7 > Pb(II) at pH 7. In multi-metal solutions, the values of Langmuir adsorption constants of all metals decreased, and varied in the following descending order: Cd (II) > Cr(VI) > Zn(II) > Pb(II) > Cu(II). Even though individual sorbed metal ion quantity per unit mass of soil with respect to single-metal solutions decreased, the total sorbed metal ion quantity per unit mass of soil from multi-metal solutions increased.

8.4 Heavy Metal Binding Mechanisms of Biosurfactants from Soil

The possible mechanisms for the extraction of HMs by biosurfactants are electrostatic interactions, ion exchange, precipitation–dissolution, and counter ion binding. For example, nonionic metals form complexes with biosurfactants, which in turn decreases the solution phase activity of the metal and, therefore, promotes desorption. Under conditions of reduced interfacial tension, biosurfactants can bind to sorbed metals directly, and can accumulate metals at solid solution interface (Singh and Cameotra 2004). Anionic surfactants cause an increase in association of metal with surfaces by sorption of the metal-surfactant combination formed or precipitation of the complexes. On the other hand, cationic surfactants decrease the association of metals by competition for some but not all negatively charged surfaces (Christofi and Ivshina 2002). Heavy metal removal is also influenced by concentrations and types of biosurfactants. As an example, at concentrations above the CMC, the rhamnolipid forms a variety of micellar (≈5 nm in diameter), and vesicular structures, generally <50 nm in diameter, which depend on solution pH. Above pH 6.8, the surfactant molecules themselves spontaneously aggregate into complex structures such as micelles (Zhang and Miller 1992). The anionic biosurfactant such as rhamnolipid carries a negative charge, so when the molecule encounters a cationic metal such as Cd(II), Zn(II) that carries a positive charge, an ionic bond is formed. This bond is stronger than the metal's bond with the soil. The polar head groups of micelles can bind metals and make the metals more soluble in water. Surfactant monomers also solubilize adsorbed metals through formation of dissolved complexes (Miller 1995). In addition, binding of some metal may occur onto the anionic exterior of rhamnolipid micelles. Metal ions are bound to opposite charged ions or can be replaced with same charged ions or complex with agents forming chelates on micelle surface. The micelles help recover the metals from soil surfaces and transport them into solution, making it easier to recover metals by flushing (Frazer 2000; Aşçı et al. 2007). It is also postulated that the metals bound onto the soil surface can be detached into the soil solution by the lowering of the interfacial tension. The surface tension of rhamnolipid solutions for instance is also quite sensitive to pH (Tan et al. 1994; Herman et al. 1995).

To explain the nature of the rhamnolipid–metal complexes, stability constants were determined by an ion-exchange resin technique (Ochoa-Loza et al. 2001). Cations of highest to lowest affinity for rhamnolipid were: Al(III) > Cu(II) > Pb(II) > Cd(II) > Zn(II) > Fe(III) > Hg(II) > Ca(II) > Co(II) > Ni(II) > Mn(II) > Mg(II) > K(I). The affinities were approximately the same or higher than those that acetic, citric, fulvic, and oxalic acids have for metals. Molar ratios of the rhamnolipid to HMs were 2.31 for Cu(II), 2.37 for Pb(II), 1.91 for Cd(II), 1.58 for Zn(II), and 0.93 for Ni(II) while for common soil cations, the ratios were 0.84 for Mg(II) and 0.57 for K(I) (Ochoa-Loza et al. 2001). Rhamnolipids form complex selectively with HM such as Cd and Pb while they have a much lower affinity for natural soil metal cations like Ca and Mg. Rhamnolipids however, do not work well in contaminated soils with a high clay or iron oxide content.

Metal removal by surfactin in general includes three stages: (1) accumulation of surfactant as hemimicelles (interfacial surface monolayers) or admicelles (interfacial surface bilayers) at soil interface; (2) removal of metal by lowering of soil–water interfacial tension, electrostatic attraction, and fluid forces; and (3) complexation of the metal with the micelles. For example, the removal of Cd(II) and Pb(II) was reported due to the complexation of aescin, rarely used biosurfactant, with metal ions adsorbed on soil surfaces (Hong et al. 1998). The anionic polar head group of (COO⁻) of aescin complex with cations adsorbed on the soil, while hydrophobic interactions occur between the nonpolar tails of aescin and organic matter in the soil. Carboxylate peaks in the infrared spectra of aescin indicate ionic and covalent bonding character of aescin with Cd(II) and Pb(II). Molar ratio of the aescin to Cd(II) was 2:1 while it was 3:1 for Pb(II) suggesting that the carboxylic and saccharide moieties of aescin may have higher binding capacities for Cd(II).

Although biosurfactants obtained from microorganisms have been used in the remediation of HMs from contaminated soils, plant-derived biosurfactants have been rarely used. However, complexation of HMs with saponin was demonstrated by Fourier Transform Infrared Spectroscopy (FTIR) analysis (Hong et al. 1998). The metal desorption by saponins as reported by Chen et al. (2008) involves three steps: (1) biosurfactant molecules at the surfactant concentrations above the CMC value from a dissociating micelle adsorb at a receptive interface. Because of reversible dynamic equilibrium, they desorb and re-orient back into a micelle. Lewis acid–base interactions and electrostatic charge attractions occurring either between the biosurfactant hydrophilic anions and acidic cationic metal-spiked surface sites or between nonionic hydrophilic polar groups of saponin and the nonmetal-spiked surfaces cause the first step. (2) The perpetual competitive sorption between the adsorbing surfactant and the presorbed metal ions occurs at the soil surface and/or soil constituent. The adsorbing biosurfactant films in a tail-to-tail and head-to-head shape consistently generate ion pairs with the presorbed metal ions toward the primary surface sites. (3) Orientational rearrangement of the saponin films at solid–liquid interface results in float-out of metal ions, self-assembly of metal–biosurfactant complexes by aggregation of lattice-like hemimicelle on the top of monolayer coating, and the release of the micellar metal–biosurfactant complexes. Saponin when used in bioremediation has the advantages like higher biodegradability and foaming, low toxicity, possibility of reuse, and easy isolation from plants. In addition, saponin has the ability to increase the aqueous dispersion of organic contaminants that is often found in HM-polluted soils by solubilization and mobilization. Another interesting result obtained is that microorganisms can respond to metal toxicity by producing biosurfactant. In a study, exogenously added rhamnolipid was reported to reduce Cd(II) toxicity for *Burkholderia* sp. growing on either naphthalene or glucose as sole C source. The reduction in toxicity was suggested that rhamnolipid after complexation with Cd(II) induced lipopolysaccharide removal from the cell surface, followed by its interaction with the cell surface to alter Cd(II) uptake (Sandrin et al. 2000).

8.4.1 Effect of pH

The type and size of aggregates formed depend on the structure of surfactant and the solution pH. At low pH, rhamnolipids form liposome-like vesicles, which are similar in structure to biological membranes. Size of vesicles ranges from 10 to more than 500 nm in diameter. The addition of Cd(II) to rhamnolipid solutions at pH 6.8 stabilized the formation of small (20–30 nm) vesicles (Tan et al. 1994) while between pH 6 and 6.6, rhamnolipids form either lamella-like structures or lipid aggregates. When the rhamnosyl moiety is negatively charged above pH 6.8, micelles, the most effective structure for metal immobilization, are formed. The surface activity of the rhamnolipid is highest between pH 7 and 7.5. As the pH increases above 7.5, the surface activity decreases slightly leading to an increase in surface tension from 30 to 32 mN/m. After increasing to 32 mN/m at pH 8, the surface tension of rhamnolipid solutions remains comparatively stable, even at pH 11. As the pH is decreased from 7 to 5, surface activity decreases significantly, resulting in a considerable increase in surface tension from 30 to >40 mN/m (Zhang and Miller 1992). At pH 5, rhamnolipid begins to visibly precipitate out of solution. To separate metals from rhamnolipid, metal sorbed rhamnolipid samples are acidified to a pH < 2.0 using 0.1 mL of concentrated HNO_3 and are centrifuged to pellet the rhamnolipid. To control that, metals are recovered from the rhamnolipid pellet by washing twice with 1% HNO_3 (Aşçı et al. 2007; Aşçı et al. 2008b). The removal of HMs from soils also increases with decreasing saponin pH. In this context, the pH 5–5.5 was found to be the most suitable pH for soil remediation with saponin. Because of the increased electrostatic attraction between saponin and soil, the amount of saponin sorbed onto soils increased with decreasing pH. For minimizing saponin sorption to soils, pH 3 was preferred as the final pH of saponin solution. Following NaOH precipitation method, HMs were efficiently recovered from the soil leachates after saponin treatment. The precipitation efficiency of HMs was 86, 80, 90, and 91% of sorbed Cd, Cu, Pb, and Zn, respectively at pH 10.7 (Hong et al. 2002). In other study, the CMC values of about 100–200 mg/L for the anionic saponin extracted from the tree *Quillaja saponaria* were detected at pH 6.5 (Chen et al. 2008). At the 100–200 mg/L CMC, saponin forms micelles, which is generally less than 5 nm in diameter. At pH 10, saponin however, forms micellar aggregates at the CMC of 2,000 mg/L, about tenfold greater than the CMC range at pH 6.5. Increase in micelle formation with using less *quillaja* saponin, as marked by its CMC variation, can be obtained at lower temperature, lower pH, and higher salt concentrations (Mitra and Dungan 1997). The desorption efficienceis of Ni(II) and Cu(II) ions from binary metal-spiked kaolin using 2,000 mg/L of saponin at room temperature were ~85% of the sorbed Ni(II) and ~83% of the sorbed Cu(II) at pH 5–8 (pH 6.5 optimum) (Chen et al. 2008). Decrease in metal desorption efficiency by saponin at pH 9–10 has been found more obvious compared to decrease at pH 5–8.

8.5   Biosurfactant Sorption onto Soils

Biosurfactants used for soil treatment should exhibit minimal sorptive interactions when applied to the soil system or soil-component matrix. Thus, most of the biosurfactant should remain in the liquid phase. Biosurfactant sorption in general is likely the reason that high rhamnolipid concentrations are needed for effective metal removal. However, there are many reasons to avoid injection of excess biosurfactant into soils. Firstly, use of excess biosurfactant is expensive, even if the degree of biodegradability and toxicity of the biosurfactant fulfills the EPA requirements; secondly, use of excess biosurfactant may lead to other environmental problems like higher concentrations of biosurfactants can plug the soil pores by the dispersion of fine materials, or by the formation of viscous emulsions (Wang and Mulligan 2004a). The major factors influencing the movement of particles of less than 50 nm in diameter through soil are advection, dispersion, and adsorption by soil surfaces. Little is presently known about the sorption of microbial surfactant monomers such as rhamnolipids or aggregate structures by soil or soil constituents. However, analogous to bacteriophage, viral particles, or microspheres behavior, sorption of biosurfactant depends on its molecular characteristics, for example, charge and hydrophobicity, as well as soil characteristics.

The rhamnolipid sorption mechanism involves the cation bridging between the anionic polar head group and sorbed cations on soil component. Hydrophobic interactions between the nonpolar tails and hydrophobic regions in the soil component also play important role in rhamnolipid sorption (Torrens et al. 1998; Ochoa-Loza et al. 2007). However, the mechanism of rhamnolipid sorption and the role of cation bridging in rhamnolipid sorption have not been investigated in detail. A characteristic S-shaped isotherm for the sorption of anionic surfactants is observed because of the combination of electrostatic and hydrophobic sorption forces (Torrens et al. 1998). Electrostatic attraction forces between individual anionic surfactant ions and positively charged sites on soil surfaces control the first stage of the S-shaped sorption isotherm. As the surfactant concentration is increased, surfactant ions show an increased tendency for self-aggregation, an analogous process to micelle formation. In the second stage of sorption, hemimicelle formation together with a rapid increase in surfactant sorption to soil surfaces is observed. This hemimicelle formation neutralizes solid surface charge. After neutralizing effects, surfactant sorption begins to slow down. When surfactant concentration is increased further, actual micelle formation appears, and the surface charge of soil surface changes from positive to negative. Repulsive forces become effective on soil surface and further sorption of surfactant is inhibited. A plateau is reached in the third stage of sorption and a considerable amount of surfactant unadsorbed remains in the liquid phase.

The most important clay minerals used for environmental purposes are kaolin, smectite, sepiolite, K-feldspar, Na-feldspar, and quartz. The rhamnolipid sorption on soil or soil-component matrix decreases with increasing rhamnolipid concentrations. The rhamnolipid sorption capacities of clay minerals and some soils in the absence of metal ions are reported to decrease in the order of soil A (smectite-moderate, sorption efficiency 100%)[4] > sepiolite (100–68.2%) > soil C[5] (smectite-moderate-dominant, 74.8–33.2%) > soil B[6] (smectite-dominant, 100–23.2%)

---

[4]Soil A smectite, serpentine, amphibole, feldspar-moderate.

[5]Soil C smectite-moderate-dominant, feldspar-moderate, illite-moderate.

[6]Soil B smectite-dominant, feldspar-moderate.

(Aşçı et al. 2008a, b) > kaolin (33.2–13.2%) (Aşçı et al. 2007) > K-feldspar (18.2–4.9%) (Aşçı et al. 2008a) > Na-feldspar (13%) (Unpublished data) > quartz (0%) (Aşçı et al. 2010). In the presence of 1 mM Cd(II), rhamnolipid sorption capacity of the soils was of the order: sepiolite (91.5–49.9%) > soil A (smectite-moderate, 75–35%) (Aşçı et al. 2008a, b) > soil B (smectite-dominant, 43.2%) > soil C (smectite-moderate-dominant, 31.5%) (Aşçı et al. 2008b) > K-feldspar (14.9%) (Aşçı et al. 2008a) > kaolin (0%) (Aşçı et al. 2007) > Na-feldspar (0%) (Unpublished data) ≈ quartz (0%) (Aşçı et al. 2010). Because of poor sorption properties of rhamnolipids, quartz-dominated soils gave better results than the other soils during bioremoval/recovery of metals. Sorption of rhamnolipids by soils also depends on the iron-oxide (Fe_2O_3) content, the clay content, and clay type. Soils with low content of aluminosilicate minerals and iron oxides reveal relatively low sorption of rhamnolipids. The contribution of soil constituents like OM, metal oxides, and clays to sorption of the rhamnolipids (monorhamnolipid, R1, and a mixture of R1 and R2, di-rhamnolipid) was investigated (Ochoa-Loza et al. 2007). Monorhamnolipid sorption at low R1 concentrations decreased in the order of hematite > kaolinite > MnO_2 ≈ illite ≈ Ca-montmorillonite > gibbsite($Al(OH)_3$) > humic acid-coated silica. Rhamnolipid sorption capacity of clays, metal oxides, and OM at high R1 concentrations followed the order: illite >> humic acid-coated silica > Ca-montmorillonite > hematite > MnO_2 > gibbsite ≈ kaolinite. Although the R1 form in certain studies was found as more effective than the R2 form in metal removal, the application of rhamnolipids to soil has been found most effective in a mixed R1/R2 system. Addition of R1 alone or increasing the amount of R1 in a R1/R2 mixture is likely to increase the aqueous phase concentration of the R1 and therefore, increases the efficiency of rhamnolipid in bioremediation. In other study, the sorption of certain surfactants to soil was compared, and was observed that the sorption of surfactant solutions to soil decreased in the order: aescin (80%) > rhamnolipid (75%) > saponin (67%) > tannin (60%) > lecithin (56%) > sodium dodecyl sulfate (SDS) (33%) (Urum and Pekdemir 2004).

8.6 Examples of Heavy Metal Removal from Soils Using Biosurfactants

The results obtained so far on HM recovery from soil and/or soil components using biosurfactants are presented in Table 8.2. In a study, in order to enhance HM binding capacity of rhamnolipid particularly at low concentrations, rhamnolipid matrix was loaded to KNO_3 (Herman et al. 1995). For this, the metal-containing soil was suspended in 0.1 M KNO_3 containing rhamnolipid at different concentrations or in a control solution to determine the potential for metal removal. In the presence of K^+ in the rhamnolipid matrix, the removal of metals ranged between 16% and 48% of the sorbed Cd(II) and Zn(II), at 12.5 and 25 mM rhamnolipid concentration. In the absence of K^+, less than 11% of sorbed Cd(II) and Zn(II) was desorbed. Desorption efficiency of metals by the control solution that included the same molar concentration

Table 8.2 Comparison of amount of metal desorbed per unit weight of sorbent (q_{desorp}) and desorption efficiencies as a function of heavy metal accumulated in soils (q_{sorp}) and/or components, surfactant concentration, operating conditions, and reactor type

Surfactant	Soil type	Metal	pH	Reactor type	q_{sorp} (mg kg⁻¹)	Surfactant conc.	q_{desorp} (mg kg⁻¹)	Recovery efficiency (%)	Reference
Rhamnolipid[a]	Oxidized Pb–Zn mine tailings (Sandy soil)	As(V)[b]	11	BSR[c]	2,180	25–1,200 mg L⁻¹ (10 mg rhamnolipid/g mine tailings)	119	5.5	Wang and Mulligan (2009)
Rhamnolipid[a]	Mine tailings	As(V)[b]	11	COCM[d]	2,180	0.1%	148	6.8	Wang and Mulligan (2009)
Aescin	Loam soil	Cd(II)	7.8	BSR	3,420	30 mM		41	Hong et al. (1998)
Di-rhamnolipid	Artificially contaminated garden soil	Cd(II)	6.3–6.8	COCM	435.4	0.1%	394.9	92	Juwarkar et al. (2007)
Monorhamnolipid	Hayhook sandy loam	Cd(II)	6.8	BSR	164.1	80 mM		≈60	Herman et al. (1995)
Rhamnolipid[a]	Smectite-moderate soil	Cd(II)	6.8	BSR	1,379	80 mM	729.5	52.9	Aşçı et al. (2008b)
Rhamnolipid	Vinton soil Sandy loam soil	Cd(II)	7.7	Column	592	5 mM		54	Torrens et al. (1998)
Rhamnolipid liquid solution	Sandy soil	Cd(II)	10.0	Column	1,706	0.5%		61.7	Mulligan and Wang (2006)
Rhamnolipid foam	Sandy soil	Cd(II)	10.0	Column	1,706	0.5%		73	Wang and Mulligan (2004a)
Rhamnolipid[a]	K-feldspar	Cd(II)	6.8	BSR	210.2	50–80 mM	201.8	96.0	Aşçı et al. (2008a)
Saponin	Sandy clay loam	Cd(II)	5.0–5.5	BSR	701	3%		90–100	Hong et al. (2002)
Rhamnolipid[a]	Mine tailings	Cu(II)[b]	11	COCM	1,100	0.1%	74	6.7	Wang and Mulligan (2009)
Saponin	Kaolin	Cu(II)	5.0–8.0 (6.5 optimum)	BSR	450	2,000 mg L⁻¹	373.5	83	Chen et al. (2008)
Saponin	Sandy clay loam	Cu(II)	5.0–5.5	BSR	1,521	3%		62	Hong et al. (2002)

(continued)

Table 8.2 (continued)

Surfactant	Soil type	Metal	pH	Reactor type	q_{sorp} (mg kg⁻¹)	Surfactant conc.	q_{desorp} (mg kg⁻¹)	Recovery efficiency (%)	Reference
Sophorolipid	Sandy soil with 12.6% oil and grease content	Cu(II)	5.5	BSR	420	4% with 0.7% HCl		37	Mulligan et al. (1999a)
Rhamnolipid Liquid solution	Sandy soil	Ni(II)	10.0	Column	2,010	0.5%		51.0	Mulligan and Wang (2006)
Rhamnolipid Foam	Sandy soil	Ni(II)	10.0	Column	2,010	0.5%		68.1	Mulligan and Wang (2006)
Saponin	Kaolin	Ni(II)	5.0–8.0 (6.5 optimum)	BSR	140	2,000 mg L⁻¹	119	85	Chen et al. (2008)
Aescin	Loam soil	Pb(II)	2.8	BSR	7,190	30 mM		25	Hong et al. (1998)
Monorhamnolipid	Hayhook sandy loam	Pb(II)	6.8	BSR	406.1	80 mM		41.6	Herman et al. (1995)
Di-rhamnolipid	Artificially contaminated garden soil	Pb(II)	6.3–6.8	COCM	905.4	0.1%	784.9	88	Juwarkar et al. (2007)
Rhamnolipid[a]	Mine tailings	Pb(II)[b]	11	COCM	12,860	0.1%	2,379	18.5	Wang and Mulligan (2009)
Saponin	Sandy clay loam	Pb (II)	5.0–5.5	BSR	5,253	3%		58	Hong et al. (2002)
Rhamnolipid[a]	Mine tailings	Zn(II)[b]	11	COCM	5,075	0.1%	259	5.1	Wang and Mulligan (2009)
Rhamnolipid	Sandy soil with 12.6% oil and grease content	Zn(II)	7.0	BSR	890	12%		20	Mulligan et al. (1999a)
Rhamnolipid[a]	Na-feldspar	Zn(II)	6.8	BSR	143.2	25 mM	141.2	98.8	Aşçı et al. (2008c)
Rhamnolipid[a]	Quartz	Zn(II)	6.8–7.2	BSR	43.9	25 mM	38.6	87.2	Aşçı et al. (2010)
Saponin	Sandy clay loam	Zn(II)	5.0–5.5	BSR	472	3%		98	Hong et al. (2002)

[a] A mixture of R1 and R2 forms
[b] Multi-metal solutions containing As, Cu, Pb, and Zn
[c] Batch-stirred reactor
[d] Column-operated continuous mode

of K^+ as the rhamnolipid solution was between 15.6% and 18.8% of sorbed Cd(II). At 50 and 80 mM rhamnolipid concentrations, Cd(II) desorption was about three-fold greater than the removal by ion exchange as rhamnolipid sorption by soil decreased with increasing rhamnolipid concentrations. In contrast to Cd(II) and Zn(II), less than 2% of sorbed Pb(II) was desorbed by ion exchange. Desorption efficiencies of 27.5% and 41.6% of sorbed Pb(II) were obtained by 50 and 80 mM rhamnolipid in the absence of K^+.

The biosurfactants have largely been used to remediate soils contaminated with single metal. The use of binary metal solutions for preparation of artificially contaminated soil is a newer and realistic approach. The use of binary metal solutions constitutes a model of competitive sorption in soil. Few column studies have also been conducted for the HM removal using biosurfactant. For example, column studies were performed to remove Cd(II) and Pb(II) together from artificially contaminated soil using di-rhamnolipid biosurfactant produced by *Pseudomonas aeruginosa* strain BS2 (Juwarkar et al. 2007). The sorption capacity of the soils for Pb(II) was higher (91%) than that of Cd(II) (87%). The results revealed that 92% of sorbed Cd(II) was removed by di-rhamnolipid as compared to only 88% of sorbed Pb(II). On the other hand, washing of artificially contaminated soil with tap water removed only ≈2.7% of sorbed Cd(II) and 9.8% of sorbed Pb(II). Di-rhamnolipid removed only 18.8% of sorbed Cd(II) and 8.4% of sorbed Pb(II) from the natural soil. Treatment of the soils with 0.1% di-rhamnolipid, however, did not show any toxic effect against bacteria, fungi, actinomycetes, and nitrogen fixers before and after rhamnolipid treatment.

The binding capacities of biosurfactants to remove Zn(II) and Cu(II) from 12.6% oil, grease, Zn(II) (890 mg kg^{-1}), Cu(II) (420 mg kg^{-1}), Pb(II) (102 mg kg^{-1}), Cd(II) (below the detection limit) -contaminated soil were compared by a batch wash and a series of five washings with surfactin, rhamnolipid, and sophorolipid (Mulligan et al. 1999a). The cumulative metal removal efficiencies after the five washes increased significantly. After a series of five batch washes, the Cu(II) removal efficiencies of certain biosurfactants and chemical agents decreased by 70% with 0.1% surfactin/1% NaOH > 50% with 4% sophorolipid/0.7%HCl ≈ 4% sophorolipid/2% Triton X-100 > 40% with 0.7% HCl > 38% with 0.1% rhamnolipid/1% NaOH > 20% with 1% NaOH. As the nonionic surfactant, Triton X-100 helps to solubilize the anionic surfactant, sophorolipid, which no longer forms a layer on top of the solution, using sophorolipid and Triton X-100 together gives better results. However, Cu(II) uptake by the Triton X-100 singly is not detectable. As the pH values of 0.7% HCl singly and the mixture of sophorolipid/HCl were approximately same (pH 5.5), the most considerable increase in both Cu(II) and Zn(II) removal was obtained by addition of HCl. On the contrary, the addition of HCl to rhamnolipid and surfactin caused precipitation of biosurfactants, thus making them unavailable for the metal removal. Zn(II) removal efficiencies followed the order: 100% with 4% sophorolipid/0.7% HCl > 80% with 0.7% HCl > 50% with 4% sophorolipid/2% Triton X-100 > 25% with 0.1% surfactin/1% NaOH > 17% with 0.1% rhamnolipid/1% NaOH > 10% with 1% NaOH. Surfactin or rhamnolipid with 1% NaOH removes the organically bound copper and the sophorolipid with HCl removes the carbonate and oxide-bound zinc (Mulligan et al. 1999a).

The success of the metal removal process for soil depends on the capacity of biosurfactant and the components constituting soil. Little is currently, however, known about the desorption of HM ions by biosurfactants from various clay minerals and other soil components. In a study, desorption of Cd(II) from various soils and/or soil constituents using rhamnolipid biosurfactant at approximately similar operating conditions was compared. It was observed that Cd(II) recovery using rhamnolipid decreased in the order of K-feldspar > Quartz > Kaolin > Soil A[7] (Smectite-moderate soil) > Soil B[8] (Smectite-dominant soil) > Soil C[9] (Smectite-moderate-dominant soil) > Sepiolite (Aşçı et al. 2007, 2008a, b, 2010). This order was almost the reverse of the Cd(II) sorption efficiency order on the soils and/or components. The more the metal sorption efficiency of soil and/or component increases, the more the metal desorption efficiency decreases.

Another example where biosurfactant has been used is the removal of arsenic. Among the various forms of arsenic, arsenite [As(III)] is more toxic and mobile than arsenate [As(V)]. The rhamnolipid biosurfactant with a mass ratio of 10 mg rhamnolipid/g mine tailings at pH 11 in batch system mobilized 119 mg As/kg from the mine tailings (Wang and Mulligan 2009). The rhamnolipid biosurfactant increases As mobilization through anion exchange. The addition of biosurfactant increases the negative zeta potential substantially, which produces a greater repulsive interaction. As mobilization is enhanced, the re-adsorption of As to the mine tailings is hindered. The mobilization of As by the rhamnolipid increased significantly with pH increase from 7 to 11. Increasing pH enhances the ionization of the carboxyl group of the rhamnolipids and metal solubility. High pH results in a more repulsive hydrophilic head group and increases the effective size of the head group and generates high-curvature micelles. In basic condition, addition of 1% NaOH can result in the formation of large aggregates (>2,000 Å) plus micelles in the range of 15–17 Å develop. In acidic condition, by addition of 1%NaCl, large polydisperse vesicles with a radius about 550–600 Å occur. In both conditions, the size of the aggregates permits the flow of the rhamnolipid solution through the porous media with the pore sizes of 200 nm. The accumulative removal of As, Cu, Pb, and Zn simultaneously using 0.1% rhamnolipid solution at pH 11 in column experiments was found to be 148, 74, 2,379, and 259 mg/kg after a 70-pore-volume flushing, respectively (Wang and Mulligan 2009). The presence of rhamnolipids hindered the formation of Fe hydroxide precipitate. While co-mobilization of the metals in the presence of rhamnolipids promoted transfer of As into aqueous organic complexes or micelles through metal-bridging mechanisms. Furthermore, the effect of rhamnolipids on the adsorption and desorption of Cu(II) ions on Na-montmorillonite (Na-rich smectite) was investigated (Özdemir and Yapar 2009). For this, copper(II) ions were mixed with pure-and/or rhamnolipid-modified Na-montmorillonite. Presence of rhamnolipids in the medium distributes clay platelets in water through

---

[7]Soil A Smectite, serpentine, amphibole, feldspar-moderate.

[8]Soil B Smectite-dominant, feldspar-moderate.

[9]Soil C Smectite-moderate-dominant, feldspar-moderate, illite-moderate.

the interactions of surfactant molecules with the positively and negatively charged surfaces. Distribution of the platelets in water causes to decrease in the external and internal mass transfer resistance in the diffusion of Cu(II) ions and increase the adsorption rate. The adsorption process was suggested to proceed principally via diffusion between the interlayers of rhamnolipid-modified clay by the ion exchange and specific adsorption mechanisms. Because of the reasons mentioned above, a considerable increase in the pseudo-second order rate constant of the clay modified with rhamnolipids was observed. The maximum monolayer adsorption capacity, Q°, was obtained for the clay modified with low concentration of rhamnolipids as 48.3 mg g^{-1}; this value was comparable with that of the activated carbon. Increasing the rhamnolipid concentration to 0.017 M proceeded their adsorption as Cu–rhamnolipid complexes onto clay until almost no Cu(II) ions remained in the solution. Then, after the addition of the excess rhamnolipid concentrations (from 0.0017 to 0.0021 M), the Cu(II) desorption began and reached 9.5 mg in solution.

Researchers have also focused their attention on the economics of biosurfactant use in metal removal from contaminated soils. In this context, ultrafiltration was used to concentrate the biosurfactants recovery and its subsequent reuse, and consequently to decrease the amount of biosurfactant needed. As an example, the removal of metals from water with surfactin by a 50,000 Da molecular weight cut-off ultrafiltration membrane was searched using a technique called micellar enhanced ultrafiltration. Cadmium(II) and Zn(II) rejection ratios were found to be nearly 100% at pH 8.3 and 11 while Cu(II) retention decreased with increasing pH and was determined as 85% at pH 6.7. More metals were associated with the surfactin micelles, remained in the retentate phase above the ultrafiltration membrane, and less metals passed through into the permeate phase.

Use of biosurfactants in the phytoremediation process is a new approach. In order to increase phytoremediation efficiency of high biomass plants, chelators and surfactants have been used to enhance the solubility of soil-bound synthetic organic compounds and heavy metals. For example, addition of tea (*Camellia sinensis*) saponin to the soil remarkably enhanced polychlorinated biphenyls accumulation (nearly 2.4 times higher than that of without adding biosurfactant) in root of corn (*Zea mays*) seedling and in shoots and roots by sugarcane (*Saccharum officinarum*) (Xia et al. 2009). Cadmium(II) concentration using 0.3% tea saponin was increased by 97% in roots, 157% in stems, and 30% in leaves compared with those observed in the absence of biosurfactant.

8.7   Biosurfactant Foam Technologies

Injection of aqueous solutions including biosurfactants and complexing or chelating agents into the soils or groundwaters results in some risks. The ability to control the migration of the fluids containing HM and toxic surfactant residuals can be improved by using foam technology. Foam is an emulsion-like two-phase system where the mass of gas or air cells is dispersed in a liquid and separated by thin liquid films.

To compose aqueous biosurfactant foam, non-wetting gas is dispersed within a continuous biosurfactant-laden liquid phase (Wang and Mulligan 2004b). The foaming ability of surfactant solutions depends on solution concentration (mass %), and time. The foaming ability of five biosurfactant solutions and a well-known chemical surfactant, SDS, was compared (Urum and Pekdemir 2004). The foaming ability of SDS increased sharply with increasing solution concentration; on the other hand, saponin generally exhibited a greater foaming ability initially than other surfactants at a wide range of solution concentration (0.004–0.5 mass-%). However, its foaming ability decreased rapidly after 5 min and was less than that of SDS. At a solution concentration of 0.1 mass-%, the initial foam heights of surfactant solutions decreased in the order of Saponin > SDS > Aescin > Rhamnolipid > Lecithin. At the same solution concentration, the foam heights of surfactant solutions after 5 min were of the order SDS > Aescin > Saponin. At this time, rhamnolipid and lecithin demonstrate no foaming. Tannin has not foaming ability. Anionic surfactants were generally recorded to have more stability than the nonionic surfactants.

Foams increase the flooding efficiency of surfactant flushing even in a heterogeneous porous medium; it results in higher metal removal efficiencies. Increasing the rhamnolipid concentration from 0.5% to 1.5% increases the foam stability. Foam quality of the rhamnolipid was indicated to vary between 90% and 99% with stabilities from 17 to 41 min at pH 8.0 (Mulligan and Wang 2006). The Cd(II) and Ni(II) removal efficiency by rhamnolipid foam increased 11% for Cd(II) and 15% for Ni(II) in comparison with that by rhamnolipid solution at the same concentration.

8.8   Role of Biosurfactants in Biofilm Formation and Use of Biofilms in Heavy Metal Bioremediation

An interface is any boundary between air and liquid, liquid and liquid, and solid and liquid phases. Microbial life arising at interfaces is demonstrated by microbial films, surface films, and aggregates. In this regard, biosurfactants do play a role whenever microorganisms come into contact with an interface. For instance, biosurfactants play an important role in gliding and swarming motility, de-adhesion from surfaces, cell–cell interactions such as biofilm formation, maintenance and maturation, quorum sensing, amensalism (microbial competition mediated by inhibitors), pathogenicity, cellular differentiation, substrate access, and avoidance of toxic chemicals (Van Hamme et al. 2006). Biosurfactants may further be used as C and energy storage molecules as a protective mechanism against high ionic strength. They are also by-products secreted in response to environmental changes. Quorum sensing, a process induced by genetic factor, depends on a critical cell density and plays a significant role in swarming motility and biofilm formation. Biosurfactants induced by quorum sensing signal molecules impact biofilm structure. As an example, rhamnolipids are reported to play a vital role in the maintenance of biofilm structure overtime by generating *rhlA* mutants lacking the rhamnosyltransferase enzyme, mediating rhamnolipid production (Davey et al. 2003). Mutant biofilms without rhamnolipid did not maintain open channels over time and formed thick cell mats. Rhamnolipid has also been found crucial for cell detachment, for example, in *P. aeruginosa* biofilm centers, returning cells to the planktonic phenotype, which involved sensitivity to antibiotics (Boles et al. 2005).

Vital role of biosurfactants in regulating the attachment–detachment of microorganisms to and from surfaces in quorum sensing and in biofilm formation has provided an insight with respect to metal bioremediation. Considering the importance of biofilms in metal removal, both pure and mixed culture of sulfate-reducing bacteria (SRB) biofilms, grown in continuous culture were treated with 20–200 μM Cd. It was found that eventhough both SRB cultures accumulated Cd, the mixed culture accumulated more and continued to accumulate Cd over a period of 14 days, while accumulation by the pure cultures stopped after 4–6 days. As accumulation of Cd within the mixed biofilm occurred by simultaneous accumulation of both protein and EPS, the pure culture biofilm accumulated only 25–30% of the amount of Cd accumulated by the mixed culture (White and Gadd 1998). In other study, immobilized biofilm of *Citrobacter* sp. was used in the removal of uranium and lead from aqueous flow and the treated metals were bioaccumulated in the form of insoluble metal phosphate (Macaskie and Dean 1987). Lead accumulation by *Burkholderia cepacia* biofilms was observed as nanoscale crystals of pyromorphite [$Pb_5(PO_4)_3(OH)$] adjacent to the outer membrane of a fraction of the total population of *B. cepacia* cells (Templeton et al. 2003). *P. aeruginosa* rhamnolipid was shown to reduce Cd toxicity while it led to an increased naphthalena biodegradation by a *Burkholderia* species. The reduction mechanism of metal toxicity of rhamnolipid might include a combination of rhamnolipid complexation of Cd–rhamnolipid interaction with the cell surface to change Cd uptake resulting in increased rates of bioremediation (Todd et al. 2000). An increased accumulation of cytoplasmic crystals of Au(III) and cell wall associated La(III) in biofilm relative to crystal formation during planktonic growth was observed. It was proposed that physiology and physicochemical conditions around cells in biofilms facilitate the removal of HM ions (Langley and Beveridge 1999). Interestingly, biofilms withstand the toxicity of HMs when compared with logarithmically grown or stationary phase cells. For example, biofilm of *P. aeruginosa* was more resistant to the toxicity of Zn, Cu, and Pb compared with equal number of free cells (Teitzel and Parsek 2003). However, the degree of HM resistance varies with the type and concentration of metals. Species diversity of biofilms also affects the HM resistance and removal efficiency. Mixed culture biofilms exhibit high metal uptake efficiency and are not affected by rapid increases or continuously high metal concentrations (Singh and Cameotra 2004).

8.9 Recent Use of Biosurfactants in Nanotechnology

The most exciting developments in the biosurfactant technology have recently been recorded in the area of nanotechnology. Rhamnolipids as a biosurfactant have been used in an eco-friendly manner for the production of nanomaterials. Nickel oxide nanorods have been produced using a solution based water-in-oil microemulsion technique (Palanisamy 2008). In this technique, rhamnolipid biosurfactant was rendered disperse in n-heptane hydrocarbon phase. Using this technique, the nanorods with nearly 22 nm in diameter and 150–250 nm in length at pH 9.6 were obtained. The morphology of the nanoparticle was adjusted by changing the pH of the solution without harmful effect on the environment. At lower pH, Ni(OH)$_2$ had flaky morphology. Mixed flaky and spherical particles were produced by increasing the pH of the solution from 8 to 10. In a follow-up study, spherical nanoparticles of NiO at pH greater than 10 have been synthesized (Palanisamy and Raichur 2009). Increase in the pH of the solution from 11.6 to 12.5 decreased the size of nanoparticles from 86 ± 8 nm to 47 ± 5 nm. Nanoparticles were characterized by SEM, XRD, TEM, and TG-DTA. In the synthesis of silver nanoparticles having unprecedented physical, chemical, magnetic, and structural properties, surfactin has been shown to act as a renewable, low-toxicity, and biodegradable stabilizing agent (Reddy et al. 2009). Surfactin extracted from the cell-free culture of *Bacillus natto* TK-1 has currently been used to stabilize superparamagnetic iron oxide nanoparticles (SPION) as contrast agents for magnetic resonance imaging (MRI) (Liao et al. 2010). The organic magnetic nanoparticles with an average diameter of 8 nm were transferred into water by the surfactin. Particles aggregation and size change were not observed. Despite the fact that the biosurfactants can be used in nanotechnology, synthesis of biosurfactant-added nanoparticles in the bioremediation of water and soil pollution has not yet been performed. Rhamnolipid has just been used to improve the electrokinetic and rheological behavior of nanozirconia particles (Biswas and Raichur 2008). The rhamnolipid adsorbs onto the zirconia with the increasing concentration. Zeta potential measurements indicated that the iso-electric point of zirconia with increasing rhamnolipid concentration shifted and the surface of zirconia became more electronegative. Maximum surface charge was reached at 230 mg L^{-1} rhamnolipid concentration. The zirconia suspension is viscous at high solids loading (>50 wt%). Addition of rhamnolipid decreased the viscosity markedly and increased the dispersion of zirconia particles at pH 7 and above. Zeta potential measurements, sedimentation, and viscosity tests proved that rhamnolipid acts as a good dispersant for flocculation and dispersion of high solid amounts of zirconia microparticles. Future research should focus on the use of biosurfactant-added nanoparticles in the treatment and remediation of environmental pollution.

8.10   Kinetic Modeling of Desorption

The time-dependent desorption data are fit to some frequently used kinetic models (Table 8.3). These models are used for both the adsorption and desorption kinetic data. A rate equation for the sorption of solutes from a liquid solution was developed by Lagergren (1898). This pseudo-first order rate equation is

$$\frac{dq}{dt} = k_1 (q_{eq} - q_t) \qquad (8.4)$$

Table 8.3 Kinetic models of sorption/desorption

Kinetic model	Equation	Model parameters	Reference
Zero order	$q = q_{eq} - k_0 t$	q is solid phase conc. of solute at time t, q_{eq} saturation capacity of sorbent for solute, k_0 intrinsic desorption constant	Saha et al. (2005)
First order	$\ln q = \ln q_{eq} - k_1 t$	k_1 intrinsic desorption constant	Chang et al. (2006)
Pseudo-first order (Lagergren equation)	$q_t = q_{eq}\left(1 - e^{-k_1 t}\right)$	q_t is the amount of metal desorbed at time t	Azizian (2004)
Pseudo-first order desorption model	$D = 100 \times \left(\dfrac{k^* - 1}{k}\right)\left[1 - e^{-k\alpha t}\right]$	$D = \left(1 - \dfrac{q}{q_O}\right) \times 100$ q_o and q are the amount of sorbed metal per unit weight of sorbent at time $t=0$ and at any time $t=t$, respectively	Purkait et al. (2005)
Second order	$\dfrac{1}{q} = \dfrac{1}{q_{eq}} + k_2 t$	k_2 intrinsic desorption constant	Saha et al. (2005)
Pseudo-second order	$q_t = \dfrac{q_{eq}^2 k_2 t}{1 + q_{eq} k_2 t}$	q_{eq} is the amount of metal desorbed at equilibrium	Ho and McKay (1999)
Elovich	$q_t = (1/\beta^*)\ln(\alpha^*\beta) + (1/\beta)\ln t$	β is a constant related to the initial rate of desorption	Chien and Clayton (1980)
Power function	$q_t = a^* t^{b^*}$		Shirvani et al. (2007)
Parabolic diffusion	$q_t = D t^{1/2} + c^*$	D is apparent diffusion rate coefficient	Shirvani et al. (2007)
Modified Freundlich equation	$q_t = k_d C_i t^{1/m^*}$	C_i is initial metal conc., k_d desorption rate coefficient	Kuo and Lotse (1973)
Prediction of the maximum amount of eluted solute	$\omega = \dfrac{w_{max} t}{k_s + t}$	w is the amount of eluted solute at time t, w_{max} the maximum amount of eluted solute, k_s the time required for the eluted solute to become half its maximum value	Chang et al. (2006)

*a, b, c, k, k_d, m, α, and β are constants

Integrating Eq. 8.6 for the boundary conditions $t=0$ to $t=t$ and $q=0$ to $q=q_t$ gives

$$\ln\frac{(q_{eq}-q_t)}{q_{eq}} = -k_1t \tag{8.5}$$

where q_t and q_{eq} are the amount of solute sorbed/desorbed per unit weight of sorbent at any time and at equilibrium, respectively, and k_1 is the rate constant of first-order sorption/desorption.

Another model used extensively to describe the sorption/desorption kinetics is pseudo-second order. The rate law (Ho and McKay 1999; Azizian 2004) for this system is explained as:

$$\frac{dq}{dt} = k_2(q_{eq}-q_t)^2 \tag{8.6}$$

The pseudo-second order and pseudo-first order sorption kinetics are frequently adopted to define HM sorption kinetics on soils and/or soil components (Table 8.4). The values of first-order adsorption rate constants of Cu(II), Ni(II), Co(II), and Mn(II) ions by raw kaolinite decreased in the order: Cu(II) > Ni(II) > Co(II) > Mn(II) (Yavuz et al. 2003). The magnitude of pseudo-first-order adsorption rate constants for palygorskite clay decreased in the order: Cr(VI) > Pb(II) > Ni(II) > Cu(II) (Potgieter et al. 2006). These kinetic models have never been employed on the desorption of HMs from soils and/or components using biosurfactants. Some models are, however, developed only for desorption process. The desorption rate at any instant is proportional to the difference between the initial (at $t=0$) amount of the sorbed metal on sorbent and the metal concentration in the solution at any time t. This is given by the following equation (Purkait et al. 2005)

$$\frac{dq}{dt} = \alpha(q_o - kq) \tag{8.7}$$

where α and k are the constants ($k \neq 1$), q_o and q are the amount of sorbed metal per unit weight of sorbent at time $t=0$ and at any time $t=t$, respectively. Integration of Eq. 8.9 between $t=0$ and any time t gives the following relation for percentage desorption.

$$D = 100 \times \left(\frac{k-1}{k}\right)\left[1 - e^{-k\alpha t}\right] \tag{8.8}$$

where D is described as $D = \left(1 - \dfrac{q}{q_o}\right) \times 100$.

Three different types of soils, andosol (clayloam), cambisol (loam), and regosol (sandy clay loam) were treated by saponin in batch system and desorption of HMs was shown to follow a first-order reaction kinetics within 30 min. The magnitude of

Table 8.4 For various soils and/or components–metal system, the best fit kinetic models of sorption and the values of sorption rate constants

Soil	Metal	Concentration (mg l^{-1})	Model	First-order rate constant (min^{-1})	Second-order rate constant (g mg^{-1}-min^{-1})	Reference
Soil amended humus soil and hydroxyapatite	Cd(II)	20–200	Pseudo-second order	0.0048–0.0057	0.0011–0.0051	Chaturvedi et al. (2006)
Kaolinite	Cu(II)	–	First order	5.10×10^3	–	Yavuz et al. (2003)
Palygorskite clay	Cr(VI)	20–100	Pseudo-first order (Lagergren)	0.40 ± 0.06	–	Potgieter et al. (2006)
Na-montmorillonite modified with Rhamnolipid	Cu(II)	600	Pseudo-second order		0.438	Özdemir and Yapar (2009)
Soil-amended humus soil and hydroxyapatite	Pb(II)	20–200	Pseudo-second order	0.0182–0.0105	0.0023–0.0017	Chaturvedi et al. (2006)
Soil-amended humus soil and hydroxyapatite	Zn(II)	20–200	Pseudo-second order	0.0038–0.0026	0.0195–0.0007	Chaturvedi et al. (2006)

desorption rate constant was Pb < Cu < Cd < Zn. In case of regosol, maximum rate constants for these metals were obtained (Hong et al. 2002). To evaluate the biosurfactant complexation affinity for soil and water cations and for HM ions, the conditional stability constants (log K) for metals were determined using an ion-exchange resin technique. This technique is based on the equilibrium complexation of a metal with an organic ligand and of a metal with a cation-exchange resin. The equilibrium reactions of a metal with an organic ligand and an ion-exchange resin are given as follows (Schubert 1948; Schubert and Richter 1948; Cheng et al. 1975):

$$M_L + \chi L \Leftrightarrow ML_\chi$$
$$M_R + R \Leftrightarrow MR \tag{8.9}$$

For a conditional stability constant, log K, a linear relationship was obtained (Ochoa-Loza et al. 2001).

$$\log\left(\frac{\lambda_o}{\lambda} - 1\right) = \log K + \chi \log L \tag{8.10}$$

A plot of $\log[(\lambda_o/\lambda)-1]$ versus log L, where L is the biosurfactant concentration, gives the values of χ and log K individually for each metal from the slope and intercept, respectively. This relationship is valid only if the organic ligand is not bound by the ion exchanger and the metal concentration is lower than that of the complexing agent.

8.10.1 Sorption–Desorption Equilibrium

Although sorption characteristics of various soils and/or components and metals have been often studied through sorption isotherms, those of desorption isotherms are quite limited. The desorption isotherm is prepared by plotting the amount of metal remained in the solid phase after desorption versus the corresponding equilibrium metal concentration in solution (Table 8.5). The sorption and desorption reactions may not provide the same isotherm equation, marking that hysteresis occurred in metal sorption–desorption processes. The desorption isotherms of Cd(II) and Zn(II) from quartz using rhamnolipid have been shown to well fit to Freundlich-desorption model (Aşçı et al. 2010). A desorption hysteresis index based on Freundlich exponent and an irreversibility index based on metal distribution coefficient have been calculated to quantify hysteretic behavior observed in the systems. The ratios of Freundlich exponents were 4.34 and 1.67 for Zn(II) and Cd(II) ions, respectively[10].

---

[10] Where M_L is the free metal concentration in solution at equilibrium in an organic ligand-containing system (mol l^{-1}), M_R is the free metal concentration in solution at equilibrium in an organic ligand-free system (mol l^{-1}), is the moles number of organic ligand that connect with one mole of metal ion (mol mol^{-1}), L is the soluble organic ligand concentration (mol l^{-1}), ML is the complexed metal-organic ligand concentration in solution at equilibrium (mol l^{-1}), R is the concentration of ion-exchange resin (kg l^{-1}), and MR is the amount of metal bound to the ion-exchange resin at equilibrium per a unit weight (mol kg^{-1}).

Table 8.5 The commonly used sorption/desorption equilibrium models in the heavy metal sorption/desorption on soils and hysteresis indices for quantification of the non singularity of metal sorption/desorption isotherms

Adsorption equilibrium model	Equation	Model parameters	Reference
Langmuir Model (Monolayer sorption model for homogeneous systems)	$q_{eq} = \dfrac{Q^o K C_{eq}}{1 + K C_{eq}}$	Q^o is the maximum amount of sorbed material required to give a complete monolayer on the surface; K a constant related to the energy of sorption	Langmuir (1916)
Freundlich model (Monolayer sorption model for heterogeneous systems)	$q_{eq} = K_F C_{eq}^{1/N} = K_F C_{eq}^{N_{sorp}}$	K_F is an indication of the sorption capacity of the sorbent; N_{sorp} indicates the sorption intensity	Freundlich (1907)
Redlich–Peterson	$q_{eq} = \dfrac{K_R C_{eq}}{1 + a_R C_{eq}^{\beta}}$		Redlich and Peterson (1959)
Koble–Corrigan sorption model	$q_{eq} = \dfrac{A C_{eq}^b}{1 + B C_{eq}^b}$		Koble and Corrigan (1952)
Desorption equilibrium model			
Freundlich type-desorption isotherm	$q_{eq} = K_{Fdesorp} C_{eq}^{Ndesorp}$	N_{sorp}, N_{desorp}, the sorption and desorption Freundlich isotherm slopes, respectively	Freundlich (1907)
A desorption hysteresis (irreversibility) index based on Freundlich exponent	$HI = \left(\dfrac{N_{desorp}}{N_{sorp}} \right)$		Cox et al. (1997)
Hysteresis index based on the quantity of metal sorbed	$w = \dfrac{\max \left(q_{desorb} - q_{sorb}\right)}{q_{sorb}} \times 100$	$\left(q_{desorb} - q_{sorb}\right)$ is the maximum difference between the sorption and desorption	Shirvani et al. (2006)
Hysteresis or irreversibility index based on metal distribution coefficient	$HC = \dfrac{k_{d(desorp)} - k_{d(sorp)}}{k_{d(desorp)}} \times 100$	$k_{d(sorp)}$, $k_{d(desorp)}$ are metal distribution coefficients $k_d = \dfrac{q_{eq}}{C_{eq}}$ based on sorption and desorption, respectively	Sander et al. (2005)

†For sorption, q_{eq} is the amount of solutes sorbed per unit weight of sorbent at equilibrium concentration; For desorption, q_{eq} is the quantity of retained metal per unit weight of sorbent after desorption (mg g^{-1}, mmol g^{-1}). C_{eq} is the corresponding metal concentration in solution at equilibrium (mg l^{-1}, mmol l^{-1})

8.11 Conclusion and Future Prospect

Biosurfactant technology as soil-washing process can be used successfully in the bioremediation of heavy metal-contaminated soils. However, more information is required to understand the structure of biosurfactants, to discover novel biosurfactants, secretion of biosurfactants, metabolic route, primary cell metabolism, scale up, and cost for biosurfactant production. If the HM remediation is performed in situ, production of the biosurfactants can also be in situ. Little work has been performed on the larger scale or field remediation of HM-contaminated soils due to high production costs of biosurfactants. Fermentation processes using cheap or waste substrates can be developed to obtain higher yields, rates, and recovery. Further fermentation process optimization at the biotechnological and engineering level are needed. Moreover, cost of downstream processing for the recovery of biosurfactant need to be minimized. To decrease the amount of biosurfactant utilized, for example, ultrafiltration can be used to concentrate the biosurfactants for recovery and subsequent reuse. As biosurfactant foam technology decreases the treatment costs due to the low usage of biosurfactants and other chemicals, it might be used as an additional process for the in situ pump-and-treatment methods. When foam technology is used, continuous kinetic studies in column systems can be performed to increase heavy metal removal and recovery efficiencies. Development of kinetic and equilibrium models could also be helpful to predict high metal removal and recovery efficiencies. Removal of single metal, multiple-metal, or hydrocarbon contaminants singly from soils using biosurfactants have been investigated extensively. Future studies might focus on more realistic systems such as mixed organic and HM contamination. As biosurfactants are involved in the processes of biofilm formation, hybrid systems including metal-remediating viable microorganisms forming biofilm and biosurfactants can be cultivated. Biofilms can also be processed to collect the sorbed metal and, in this way, downstream process cost can be reduced. Biosurfactants are being used in the synthesis of nanomaterials as nontoxic, renewable, biodegradable, "green" dispersant and stabilizer. As a matter of fact, there is however, no environmental application yet. And hence, innovative research techniques should focus on the use of novel-designed nanoparticles stabilized by biosurfactants in the remediation of HM-polluted soils.

References

Arey SJ, Seaman JC, Bertsen PM (1999) Immobilization of uranium in contaminated sediments by hydroxyapatite. Environ Sci Technol 33:337–342

Aşçı Y, Nurbaş M, Sağ Açıkel Y (2007) Sorption of Cd(II) onto kaolin as a soil component and desorption of Cd(II) from kaolin using rhamnolipid biosurfactant. J Hazard Mater 139:50–56

Aşçı Y, Nurbaş M, Sağ Açıkel Y (2008a) A comparative study for the sorption of Cd(II) by K-feldspar and sepiolite as soil components, and the recovery of Cd(II) using rhamnolipid biosurfactant. J Environ Manage 88:383–392

Aşçı Y, Nurbaş M, Sağ Açıkel Y (2008b) A comparative study for the sorption of Cd(II) by soils with different clay contents and minerology and the recovery of Cd(II) using rhamnolipid biosurfactant. J Hazard Mater 154:663–673

Aşçı Y, Nurbaş M, Sağ Açıkel Y (2008c) Removal of zinc ions from a soil component Na-feldspar by a rhamnolipid biosurfactant. Desalination 223:361–365

Aşçı Y, Nurbaş M, Sağ Açıkel Y (2010) Investigation of sorption/desorption equilibria of heavy metal ions on/from quartz using rhamnolipid biosurfactant. J Environ Manage 91:724–731

Ayari F, Srasra E, Trabelsi-Ayadi M (2005) Characterization of bentonitic clays and their use as adsorbent. Desalination 185:391–397

Azizian S (2004) Kinetic models of sorption: a theoretical analysis. J Colloid Interface Sci 276:47–52

Biswas M, Raichur AM (2008) Electrokinetic and rheological properties of nano zirconia in the presence of rhamnolipid biosurfactant. J Am Ceram Soc 91:3197–3201

Boles BR, Thoendel M, Singh PK (2005) Rhamnolipids mediate detachment of *Pseudomonas aeruginosa* from biofilms. Mol Microbiol 57:1210–1223

Bonmatin JM, Genest M, Labbe H, Grangemard I, Peypoux F, Maget-Dana R, Ptak M, Michel G (1995) Production, isolation and characterization of [Leu4] and [Ile4] surfactins from *Bacillus subtilis*. Lett Pept Sci 2:41–47

Bradl HB (2004) Adsorption of heavy metal ions on soils and soils constituents. J Colloid Interface Sci 277:1–18

Brigatti MF, Lugli C, Poppi L (2000) Kinetics of heavy-metal removal and recovery in sepiolite. Appl Clay Sci 16:45–57

Bruins MR, Kapil S, Oehme W (2000) Microbial resistance to metals in the environment. Ecotoxicol Environ Saf 45:198–207

Chang Y-K, Chu L, Tsai J-C, Chiu S-J (2006) Kinetic study of immobilized lysozyme on the extrudate-shaped NaY zeolite. Process Biochem 41:1864–1874

Chantawong V, Harvey NW, Bashkin VN (2003) Comparison of heavy metal adsorptions by Thai kaolin and ballclay. Water Air Soil Pollut 148:111–125

Chaturvedi PK, Seth CS, Misra V (2006) Sorption kinetics and leachability of heavy metal from the contaminated soil amended with immobilizing agent (humus soil and hydroxyapatite). Chemosphere 64:1109–1114

Chen Y, Shen Z, Li X (2004) The use of vetiver grass (*Vetiveria zizaniodies*) in the phytoremediation of soils contaminated with heavy metals. Appl Geochem 19:1553–1565

Chen W-J, Hsiao L-C, Chen KK-Y (2008) Metal desorption from copper(II)/nickel(II)-spiked kaolin as a soil component using plant-derived saponin biosurfactant. Process Biochem 43:488–498

Cheng MH, Patterson JW, Minear RA (1975) Heavy metal uptake by activated sludge. J Water Pollut Control Fed 47:362–376

Chien SH, Clayton WR (1980) Application of Elovich equation to the kinetics of phosphate release and sorption in soils. Soil Sci Soc Am J 44:265–268

Christofi N, Ivshina IB (2002) Microbial surfactants and their use in field studies of soil remediation. J Appl Microbiol 93:915–929

Cox L, Koskinen WC, Yen PY (1997) Sorption-desorption of imidacloprid and its metabolites in soils. J Agric Food Chem 45:1468–1472

Davey ME, Caiazza NC, O' Toole GA (2003) Rhamnolipid surfactant production affects biofilm architecture in *Pseudomonas aeruginosa* PAO1. J Bacteriol 185:1027–1036

Dermont G, Bergeron M, Mercier G, Richer-Lafleche M (2008) Soil washing for metal removal: a review of physical/chemical technologies and field applications. J Hazard Mater 152:1–31

Farley KJ, Dzombak DA, Morel FMM (1985) A surface precipitation model for the sorption of cations on metal oxides. J Colloid Interface Sci 106:226–242

Farquhar ML, Vaughan DJ, Hughes CR, Charnock JM, England KER (1997) Experimental studies of the interaction of aqueous metal cations with mineral substrates: lead, cadmium, and copper with perthitic feldspar, muscovite, and biotite. Geochim Cosmochim Acta 61:3051–3064

Frazer L (2000) Innovations: lipid lather removes metals. Environ Health Perspect 108:A320–A323

Freundlich H (1907) Ueber die adsorption in Loesungen. Z Phys Chem 57A;385–470

Gadd GM (2004) Microbial influence on metal mobility and application for bioremediation. Geoderma 122:109–119

Garcia-Sanchez A, Alastuey A, Querol X (1999) Heavy metal adsorption by different minerals: application to the remediation of polluted soils. Sci Total Environ 242:179–188

Gu X, Evans LJ (2007) Modelling the adsorption of Cd(II), Cu(II), Ni(II), Pb(II), and Zn(II) onto Fithian illite. J Colloid Interface Sci 307:317–325

Haba E, Pinazo A, Jauregui O, Espuny MJ, Infante MR, Manresa A (2003) Physicochemical characterization and antimicrobial properties of rhamnolipids produced by *Pseudomonas aeruginosa* 47 T2 NCBIM 40044. Biotechnol Bioeng 81:316–322

Herman DC, Artiola JF, Miller RM (1995) Removal of cadmium, lead, and zinc from soil by a rhamnolipid biosurfactant. Environ Sci Technol 29:2280–2285

Hettiarchchi GM, Pierzynski GM, Ransom MD (2000) In situ stabilization of soil lead using phosphorous and manganese oxide. Environ Sci Technol 34:4614–4619

Ho YS, McKay G (1999) Pseudo-second order model for sorption processes. Process Biochem 34:451–465

Hong K-J, Choi Y-K, Tokunaga S, Ishigami Y, Kajiuchi T (1998) Removal of cadmium and lead from soil using Aescin as a biosurfactant. J Surfactants Deterg 1:247–250

Hong K-J, Tokunaga S, Kajiuchi T (2002) Evaluation of remediation process with plant-derived biosurfactant for recovery of heavy metals from contaminated soils. Chemosphere 49:379–438

Ishigami Y, Bama Y, Nagahora H, Yamaguchi M, Makahara H, Kamata T (1987) The pH-sensitive conversion of molecular aggregates of rhamnolipid biosurfactants. Chem Lett 5:763–766

Juwarkar AA, Nair A, Dubey KV, Singh SK, Devotta S (2007) Biosurfactant technology for remediation of cadmium and lead contaminated soils. Chemosphere 68:1996–2002

Kakinuma A, Oachida A, Shima T, Sugino H, Isano M, Tamura G, Arima K (1969) Confirmation of the structure of surfactin by mass spectrometry. Agric Biol Chem 33:1669–1672

Koble RA, Corrigan TE (1952) Adsorption isotherms for pure hydrocarbons. Ind Eng Chem 44:383–387

Krawczyk-Barsch E, Arnold T, Reuther H, Brandt F, Bosbach D, Bernhard G (2004) Formation of secondary Fe-oxyhydroxide phases during the dissolution of chlorite-effects on uranium sorption. Appl Geochem 19:1403–1412

Krebs W, Brombacher C, Bosshard PP, Bachofen R, Brandl H (1997) Microbial recovery of metals from solids. FEMS Microbiol Rev 20:605–617

Kuo S, Lotse EG (1973) Kinetics of phosphate adsorption and desorption by hematic and gibbsite. Soil Sci 116:400–406

Lagergren S (1898) Zur theorie der sogenannten adsorption gelöster stoffe. Kungliga, Svenska Vetenskapsakademiens. Handlingar 24:1–39

Langley S, Beveridge TJ (1999) Metal binding by *Pseudomonas aeruginosa* PAO1 is influenced by growth of the cells as a biofilm. Can J Microbiol 45:616–622

Langmuir L (1916) The constitution and fundamental properties of solids and liquids. J Am Chem Soc 38:2221–2295

Lasat MM (2000) Phytoextraction of metals from contaminated soil: a review of plant/soil/metal interaction and assessment of pertinent agronomic issues. J Hazard Subst Res 2:5.1–5.25

Lebeau T, Bagot D, Jezequel K, Fabre B (2002) Cadmium biosorption by free and immobilised microorganisms cultivated in a liquid soil extract medium: effects of Cd, pH and techniques of culture. Sci Total Environ 291:73–83

Ledin M, Krantz-Rülcker C, Allard B (1999) Microorganisms as metal sorbents: comparison with other soil constituents in multi-component systems. Soil Biol Biochem 31:1639–1648

Liao Z, Wang H, Wang X, Wang C, Hu X, CaO X, Chang J (2010) Biocompatible surfactin-stabilized superparamagnetic iron oxide nanoparticles as contrast agents for magnetic resonance imaging. Colloid Surf A Physicochem Eng Aspects 370:1–5

Lin C-C, Lin H-L (2005) Remediation of soil-contaminated with the heavy metal. J Hazard Mater A122:7–15

Macaskie LE, Dean ACR (1987) Use of immobilized biofilm of *Citrobacter sp.* for the removal of uranium and lead from aqueous flows. Enzyme Microb Technol 9:2–4

Malik A (2004) Metal bioremediation through growing cells. Environ Int 30:261–278

Markiewicz-Patkowska J, Hursthouse A, Przybyla-Kij H (2005) The interaction of heavy metals with urban soils: sorption behaviour of Cd, Cu, Cr, Pb and Zn with a typical mixed brownfield deposit. Environ Int 31:513–521

McBride MB (1994) Environmental chemistry of soils. Oxford University Press, New York

Miller RM (1995) Biosurfactant-facilitated remediation of metal-contaminated soils. Environ Health Perspect 103:59–62

Miranda-Trevino JC, Coles CA (2003) Kaolinite properties, structure and influence of metal retention on pH. Appl Clay Sci 23:133–139

Misra V, Pandey SD (2004) Remediation of contaminated soil by amendment of non humus soil with humus rich soil for better metal immobilization. Bull Environ Contam Toxicol 73:561–567

Mitra S, Dungan SR (1997) Micellar properties of quillaja saponin. 1. Effects of temperature, salt, and pH on solution properties. J Agric Food Chem 45:1587–1595

Mulligan CN (2005) Environmental applications for biosurfactants. Environ Pollut 133:183–198

Mulligan CN, Wang S (2006) Remediation of a heavy metal-contaminated soil by a rhamnolipid foam. Eng Geol 85:75–81

Mulligan CN, Yong RN, Gibbs BF (1999a) On the use of biosurfactants for the removal of heavy metals from oil-contaminated soil. Process Saf Prog 18:50–54

Mulligan CN, Yong RN, Gibbs BF, James S, Bennett HPJ (1999b) Metal removal from contaminated soil and sediments by the biosurfactant surfactin. Environ Sci Technol 33:3812–3820

Mulligan CN, Yong RN, Gibbs BF (2001a) Remediation technologies for metal-contaminated soils and groundwater: an evaluation. Eng Geol 60:193–207

Mulligan CN, Yong RN, Gibbs BF (2001b) Surfactant-enhanced remediation of contaminated soil: a review. Eng Geol 60:371–380

Ochoa-Loza FJ, Artiola JF, Maier RM (2001) Stability constants for the complexation of various metals with a rhamnolipid biosurfactant. J Environ Qual 30:479–485

Ochoa-Loza FJ, Noordman WH, Jannsen DB, Brusseau ML, Maier RM (2007) Effect of clays, metal oxides, and organic matter on rhamnolipid biosurfactant sorption by soil. Chemosphere 66:1634–1642

Özdemir G, Yapar S (2009) Adsorption and desorption behavior of copper ions on Na-montmorillonite: effect of rhamnolipids and pH. J Hazard Mater 166:1307–1313

Palanisamy P (2008) Biosurfactant mediated synthesis of NiO nanorods. Mater Lett 62:743–746

Palanisamy P, Raichur AM (2009) Synthesis of spherical NiO nanoparticles through a novel biosurfactant mediated emulsion technique. Mater Sci Eng C 29:199–204

Peng J-f, Song Y-h, Yuan P, Cui X-Y, G-l Q (2009) The remediation of heavy metals contaminated sediment. J Hazard Mater 161:633–640

Potgieter JH, Potgieter-Vermaak SS, Kalibantonga PD (2006) Heavy metals removal from solution by palygorskite clay. Miner Eng 19:463–470

Puls RW, Bohn HL (1988) Sorption of cadmium, nickel and zinc by kaolinite and montmorillonite suspensions. Soil Sci Soc Am J 52:1289–1292

Purkait MK, DasGupta S, De S (2005) Adsorption of eosin dye on activated carbon and its surfactant based desorption. J Environ Manage 76:135–142

Reddy AS, Chen C-Y, Baker SC, Chen C-C, Jean J-S, Fan C-W, Chen H-R, Wang J-C (2009) Synthesis of silver nanoparticles using surfactin: a biosurfactant stabilizing agent. Mater Lett 63:1227–1230

Redlich O, Peterson DL (1959) A useful adsorption isotherm. J Phys Chem 63:1024–1029

Reed BE, Cline SR (1994) Retention and release of lead by a very fine sandy loam. I. Isotherm modeling. Sep Sci Technol 29:1529–1551

Robertson AP, Leckie JO (1997) Cation binding predictions of surface complexation models: effects of pH, ionic strength, cation loading, surface complex, and model fit. J Colloid Interface Sci 188:444–472

Römkens P, Bowwman L, Japenga J, Draaisma C (2002) Potentials and drawbacks of chelate-enhanced phytoremediation of soils. Environ Pollut 116:109–121

Ron EZ, Rosenberg E (2001) Natural roles of biosurfactants. Environ Microbiol 3:229–236

Sağ Y (2001) Biosorption of heavy metals by fungal biomass and modeling of fungal biosorption: a review. Sep Purif Methods 30:1–48

Saha UK, Liu C, Kozak LM, Huang PM (2005) Kinetics of selenite desorption by phosphate from hydroxyaluminum-and hydroxyaluminosilicate-montmorillonite complexes. Geoderma 124:105–119

Sander M, Lu Y, Pignatello JJ (2005) A thermodynamically based method to quantify true sorption hysteresis. J Environ Qual 34:1063–1072

Sandrin TR, Chech AM, Maier RM (2000) A rhamnolipid biosurfactant reduces cadmium toxicity during naphthalene biodegradation. Appl Environ Microbiol 66:4585–4588

Schubert J (1948) The use of ion exchangers for the determination of physical-chemical properties of substances, particularly radio-tracers, in solution; theoretical. J Phys Colloid Chem 52:340–350

Schubert J, Richter JW (1948) The use of ion exchangers for the determination of physical-chemical properties of substances, particularly radio-tracers, in solution; the dissociation constants of strontium citrate and strontium tartrate. J Phys Colloid Chem 52:350–357

Serrano S, Garrido F, Campbell CG, Garcia-Gonzalez MT (2005) Competitive sorption of cadmium and lead in acid soils of Central Spain. Geoderma 124:91–104

Shawabkeh RA (2005) Soludification and stabilization of cadmium ions in sand-cement-clay mixture. J Hazard Mater B125:237–243

Shirvani M, Kalbasi M, Shariatmadari H, Nourbakhsh F, Najafi B (2006) Sorption-desorption of cadmium in aqueous palygorskite,sepiolite, and calcite suspensions. Isotherm hysteresis. Chemosphere 65:2178–2184

Shirvani M, Shariatmadari H, Kalbasi M (2007) Kinetics of cadmium desorption from fibrous silicate clay minerals: influence of organic ligands and aging. Appl Clay Sci 37:175–184

Singh P, Cameotra SS (2004) Enhancement of metal bioremediation by use of microbial surfactants. Biochem Biophys Res Commun 319:291–297

Singh SP, Ma LQ, Harris WG (2001) Heavy metal interactions with phosphatic clay: sorption and desorption behaviour. J Environ Qual 30:1961–1968

Singh SP, Ma LQ, Hendry MJ (2006) Characterization of aqueous lead removal by phosphatic clay: equilibrium and kinetic studies. J Hazard Mater 136:654–662

Singh A, Van Hamme JD, Ward OP (2007) Surfactants in microbiology and biotechnology: Part 2. Application aspects. Biotechnol Adv 25:99–121

Sposito G (1984) The surface chemistry of soils. Oxford University Press, New York

Tan H, Champion JT, Artiola JF, Brusseau ML, Miller RM (1994) Complexation of cadmium by a rhamnolipid biosurfactant. Environ Sci Technol 28:2402–2406

Taqvi SIH, Hasany SM, Bhanger MI (2007) Sorption profile of Cd(II) ions onto beach sand from aqueous solutions. J Hazard Mater 141:37–44

Teitzel GM, Parsek MR (2003) Heavy metal resistance of biofilm and planktonic P. aeruginosa. Appl Environ Microbiol 69:2313–2320

Templeton AS, Trainor TP, Spormann AM, Newville M, Sutton SR, Dohnalkova A (2003) Sorption versus biomineralization of Pd(II) within Burkholderia cepacia biofilms. Environ Sci Technol 37:300–307

Thimon L, Peypoux F, Michel G (1992) Interactions of surfactin, a biosurfactant from Bacillus subtilis with inorganic cations. Biotechnol Lett 14:713–718

Todd RS, Andrea MC, Maier RM (2000) A rhamnolipid biosurfactant reduces cadmium toxicity during naphthalene biodegredation. Appl Environ Microbiol 66:4585–4588

Torrens JL, Herman DC, Miller-Maier RM (1998) Biosurfactant (Rhamnolipid) sorption and the impact on rhamnolipid-facilitated removal of cadmium from various soils under saturated flow conditions. Environ Sci Technol 32:776–781

Urum K, Pekdemir T (2004) Evaluation of biosurfactants for crude oil contaminated soil washing. Chemosphere 57:1139–1150

UTTU (2005, July/August) Hydrocarbon biosurfactants. Underground Tank Technology Update, pp. 7–10

Van Hamme JD, Singh A, Ward OP (2006) Physiological aspects Part 1 in a series of papers devoted to surfactants in microbiology and biotechnology. Biotechnol Adv 24:604–620

Vico LI (2003) Acid-base behaviour and Cu^{2+} and Zn^{2+} complexation properties of the sepiolite/water interface. Chem Geol 198:213–222

Vig K, Megharaj M, Sethunathan N, Naidu R (2003) Bioavailability and toxicity of cadmium to microorganisms and their activities in soil: a review. Adv Environ Res 8:121–135

Wang S, Mulligan CN (2004a) Rhamnolipid foam enhanced remediation of cadmium and nickel contaminated soil. Water Air Soil Pollut 157:315–330

Wang S, Mulligan CN (2004b) An evaluation of surfactant foam technology in remediation of contaminated soil. Chemosphere 57:1079–1089

Wang S, Mulligan CN (2009) Arsenic mobilization from mine tailings in the presence of a biosurfactant. Appl Geochem 24:928–935

Wen J, Stacey SP, McLaughlin MJ, Kirby JK (2009) Biodegradation of rhamnolipid, EDTA and citric acid in cadmium and zinc contaminated soils. Soil Biol Biochem 41:2214–2221

White C, Gadd GM (1998) Accumulation and effects of cadmium on sulfate-reducing bacterial biofilms. Microbiology 144:1407–1415

Xia H, Chi X, Yan Z, Cheng W (2009) Enhancing plant uptake of polychlorinated biphenyls and cadmium using tea saponin. Bioresour Technol 100:4649–4653

Yavuz Ö, Altunkaynak Y, Güzel F (2003) Removal of copper, nickel, cobalt and manganese from aqueous solution by kaolinite. Water Res 37:948–952

Zhang Y, Miller RM (1992) Enhanced octadecane dispersion and biodegradation by a *Pseudomonas* rhamnolipid surfactant (biosurfactant). Appl Environ Microbiol 58:3276–3282

Zoubolis AI, Loukidou MX, Matis KA (2004) Biosorption of toxic metals from aqueous solutions by bacteria strains isolated from metal-polluted soils. Process Biochem 39:909–916

Chapter 9
Mechanism of Metal Tolerance and Detoxification in Mycorrhizal Fungi

Shweta Saraswat and Jai Prakash Narain Rai

Abstract Mycorrhizal fungi, obligate biotrophs, form mutualistic associations with plants and provide mainly phosphorus to plants. Mycorrhizal fungi colonize the roots of many plants growing on metal-contaminated soils and play an important role in metal tolerance and accumulation. Even though mycorrhizae are known to inhabit metal contaminated sites; the exact mechanism of colonization is unclear. For example, how mycorrhizal fungi tolerate and maintain homeostasis to toxic metals? Could metal tolerance be transferred to host plants? If so, how do mycorrhizal associations enhance metal accumulation in plants? Mycorrhiza possesses the same constitutive mechanisms as do the higher plants to circumvent metal toxicity. The adaptive tolerance is acquired by expressing genes that confer enhanced metal tolerance under stressed conditions. Various mechanisms adopted by mycorrhizal symbionts to overcome metal toxicity are highlighted. The metal detoxification mechanisms discussed here are likely to serve as a base for developing transgenic plants with abilities of increased metal tolerance and uptake, for decontamination and restoration of the metal polluted sites.

Keywords Mycorrhiza • Heavy metals • Glutathiones • Metallothioneins • Transporter proteins

S. Saraswat • J.P.N. Rai (✉)
Ecotechnology Laboratory, Department of Environmental Sciences,
Govind Ballabh Pant University of Agriculture & Technology,
Pantnagar 263145, Uttarakhand, India
e-mail: jamesraionline@gmail.com

M.S. Khan et al. (eds.), *Biomanagement of Metal-Contaminated Soils*,
Environmental Pollution 20, DOI 10.1007/978-94-007-1914-9_9,
© Springer Science+Business Media B.V. 2011

9.1 Introduction

The development of symbioses between mycorrhizae and most terrestrial plants has been found beneficial to both interacting partners and hence to the agro-ecosystem. Among such beneficial activities, the ability of mycorrhizal fungi to overcome the undesirable effects of heavy metals onto plants and microbes is of special importance. Taking these into consideration, different mycorrhizal fungi like ericoid (ERM) (Martino et al. 2003), ectomycorrhiza (ECM) (Ramesh et al. 2009), and arbuscular mycorrhizal (AM) fungi (Redon et al. 2008) have been reported to significantly accumulate heavy metals (Joner et al. 2000) and to protect host plants from metal toxicity (Arriagada et al. 2007). Metal remediation by mycorrhizal fungi is influenced by its species, host genotypes, and metal species. Metal uptake by plants is, however, contradictory. For example, AMF inoculation is reported to enhance metal accumulation in *Helianthus annuus* (Awotoye et al. 2009) and *Cannabis sativa* L. (Citterio et al. 2005). Others have found that reduced metal concentrations in plants protect them from phytotoxic effects (Joner and Leyval 1997; Chen et al. 2007). Likewise, ECM fungi reduced metal uptake by the plant through immobilization in the fungal biomass (Li and Christie 2000; Zhu et al. 2001). Thus, mycorrhizal plants in some cases can exhibit enhanced metal uptake and root-to-shoot transport (phytoextraction), while in other cases, MF can immobilize metal within the soil (phytostabilization).

The ability of MF in improving plant biomass (Gamalero et al. 2004; Berta et al. 2005) by alleviating metal toxicity to the host plants (Rivera-Becerril et al. 2002) plays an important role in decontamination of polluted sites. Several reports have suggested that the mycorrhizal fungi isolated from metal contaminated sites have shown metal tolerance greater than fungi recovered from non-contaminated sites (Gaur and Adholeya 2004; Sudova et al. 2008). This was presumably due to the homeostasis and constitutive mechanisms, as also adopted by plants. Like plants, MF also possesses a range of detoxification and/or tolerance mechanism at the cellular and subcellular level. In addition, mycorrhiza also prevents the transport of metals to plants by binding the metals at the electronegative sites of the mycelial cell wall (Frey et al. 2000).

Experiments on the differential gene expression in extra-radical mycelium of an AMF, *Glomus intraradices* Sy167, spiked with cadmium, copper, or zinc under in vitro conditions indicated the synthesis of proteins, involved in metal tolerance, for example, a Zn transporter, metallothionein (MT), 90-kDa heat shock protein, and glutathione S-transferase (GST) (Hildebrandt et al. 2007). The gene expression, however, varies in response to different metals. For example, ECM fungi when grown in the presence of cadmium and copper increased the production of glutathione (GSH), γ-glutamylcysteine, and a Cd-binding MT (Ramesh et al. 2009). Efforts have also been directed to further untangle the regulatory genes coding for MTs and phytochelatins (PCs), specific metal transporter proteins, and synthesis of chelators in different mycorrhizal ecotypes. These proteins are known to increase tolerance and accumulation of metals. The mycorrhizal fungi also alleviate metal-induced oxidative stress caused to plant by antioxidant enzymes. As an example, Azcon et al. (2009) observed the enhanced catalase (CAT), ascorbate peroxidase (APX), and glutathione reductase (GR) activities in AMF-inoculated plants grown in metal-stressed soil. Thus, mycorrhizal fungi capable of adopting different strategies to overcome metal stress raise several questions regarding their role in metal remediation. For example, what mechanisms they follow at extra- and intracellular level to protect themselves from metal toxicity and the plants they colonize? Are these mechanisms metal, host, and/or species specific? The efforts have been made here to address these problems by synthesizing recent findings on the subject.

9.2 Metal Tolerance/Detoxification in Mycorrhizal Fungi

In any organism, the nonessential and essential metal ions are transported through nonspecific ion uptake systems (Nies and Silver 1995). However, when metal ions are in excess, AM fungi have evolved several mechanisms for maintaining homeostasis to virtually all toxic metals (Rouch et al. 1995; Turnau et al. 2001). Essential metal tolerance mechanisms are usually chromosome mediated, whereas nonessential are plasmid encoded, which have been found specifically under stress conditions (Silver and Walderhaug 1992). These mechanisms operate both at extracellular and intracellular levels. Extracellular mechanism involves the avoidance of metal entry into the cell (Fig. 9.1) and, hence, results in the reduction of ion influx and increases the efflux (Hall 2002; Drager et al. 2004). The intracellular mechanism, on the other hand, involves metal detoxification and sequestration and is mediated by complexation of metals with cytosolic peptides, like, GSHs, PCs, and MTs (Cobbett and Goldsbrough 2002), and polyphosphate granules or compartmentalization of metals into the vacuoles (Tomsett 1993). The synthesis of antioxidant enzymes namely superoxide dismutase (SOD) and glutathione reductase (GR), in contrast, are the other defense mechanisms that reduce the oxidative stress generated by metal-induced reactive oxygen species (ROS) (Azcon et al. 2009). Nevertheless, the information regarding whether extra- and intracellular mechanisms occur simultaneously or in isolation is scarce, and to resolve this further, both have been discussed in the following section.

9.2.1 Extracellular Mechanisms

The basic principles underlying metal tolerance include the extracellular chelation of HMs by root exudates and/or binding of HMs to the rhizodermal cell walls. As a result of complex formation, the uptake of HM is restricted. However, mycorrhizal fungi also adopt other physiological strategies to reduce the entry of toxic metals inside the cells (Meharg 2003).

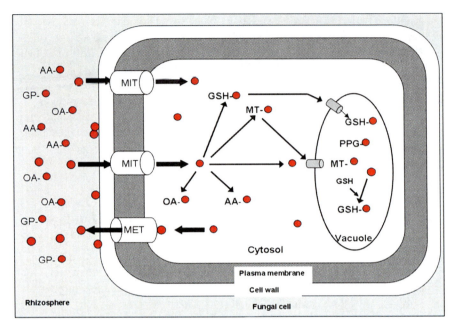

Fig. 9.1 Schematic representation of extra- and intra-cellular metal complexation through extruded organic ligands in the rhizosphere and metal-chelating agents in cytosol of MF in response to metal exposure. Red dots indicate metal ion, *OA* organic acid, *AA* amino acid, *GP* glomalin protein, *MIT* metal influx transporters, *MET* metal efflux transporters, *GSH* glutathiones, *MT* metallothioneins, *PPG* polyphosphate granules

9.2.1.1 Heavy Metal Chelation by Organic and/or Inorganic Ligands

Previous studies have indicated that the fungal cell wall binds to approximately 50% of the metal ions (Joner et al. 2000); most of which bind to negatively charged components of the cell wall, such as chitin, melanin, and in case of AM fungi, especially the glomalin (Ferrol et al. 2009). An important process in the maintenance of metal homeostasis is chelation by extracellular and intracellular organic compounds (Clemens et al. 2001). Mycorrhizal fungi, like certain plant hosts, exude organic acids such as citric, malic, and oxalic acid and amino acids into the rhizosphere (Jones 1998) to resist metal toxicity (Fig. 9.1). The deprotonation of organic acids acidifies the rhizosphere (Landeweert et al. 2001; Fomina et al. 2005) and increases the mobility of metal ions in soil, or immobilizes and detoxifies them through precipitation and complexation. In a follow-up study, Gonzalez-Chavez et al. (2004) reported an insoluble glycoprotein (glomalin) excreted by AM fungi to sequester metal ions especially Cu, Pb, and Cd from highly polluted soils and the amount of glomalin excretion and metal sequestration was significantly correlated. The reduction in the bioavailability of the contaminants occurs through the formation of glomalin–metal complexes in soil, which could assist in detoxifying the metal and concomitantly protect the plant and colonizing fungus from adverse effects of metals.

In a study, Driver et al. (2005) and Purin and Rillig (2008) observed that glomalin protein is found in about 80% fungal mycelium and has a primary function in the living hyphae of MF. Therefore, it is suggested that fungal strains with the ability to secrete glomalin be identified so that metal sequestration from polluted sites could be enhanced.

In addition to organic ligands, metals can also bind to inorganic binding sites present on AMF hyphae. The fungal cell wall possesses negatively charged free carboxyl, amino, hydroxyl, phosphate, and mercapto groups that intercept metal cations. Taking these facts into consideration, a range of metals has been shown to accumulate in the fungal mantle and rhizomorphs in *Suillus luteus–Pinus sylvestris* associations (Turnau et al. 2001), cell walls of *Paxillus involutus* (Blaudez et al. 2000), and in cortical cells of both mycorrhizal and non-mycorrhizal *Picea abies* (Jentschke et al. 1991). The pattern of metal chelation is, however, variable among mycorrhizae.

9.2.2 Intracellular Mechanisms

Despite extracellular chelation and cell-wall binding capacities of mycorrhizal fungi, large amounts of metal may enter into the cells through nonspecific ion uptake systems. To acquire micronutrients that are toxic at higher concentrations, fungi have evolved mechanisms to maintain cellular homeostasis for such elements, mechanisms to detoxify excess metals and repair mechanisms to counteract damage caused by metals (Meharg and Macnair 1994; Sanita di Toppi et al. 2002). Adaptation to a toxicant could involve alteration of one or more of these pathways, such as transformation, complexation or reduced transport into the cell, or localization within the cell. In addition, the metal that enters into the cell may also be altered to another species by pH or redox potential, or through complexation with plant biomolecules for either transport within the plant, or storage purposes.

9.2.2.1 Metal Complexation by Polyphosphate Granules

Electron micrographic studies on HM-treated ECM fungi have shown the presence of metal-phosphate deposits in vacuoles. These polyphosphate granules act as metal-chelating agents and detoxify excess metal ions (Vare 1990; Leyval et al. 1997). As an example, Hartley et al. (1997) reported that polyphosphates produced excessively by ECM fungi form an important intracellular storage material in the form of phosphorus in the vacuole. In this context, there are several reports which suggest that vacuolar polyphosphates exist in the form of insoluble granules complexed with a variety of cations (Ashford et al. 1986; Martin et al. 1994). The presence of functional groups within polyphosphate granules indicates the possibility for binding of metals with them, and thus forms the basis of intracellular metal detoxification in ECM fungi.

9.2.2.2 Vacuolar Compartmentalization by PCs and MTs

After cytosol complexation, the vacuolar compartmentalization is the major mechanism adopted by MF to remove metals intracellularly through chelation of metal ions with thiol-containing compounds, such as, GSH, PCs, and MTs, for maintaining homeostasis of toxic metals within the cytoplasm (Fig. 9.1). Reduced GSH (γ-glu-cys-gly), the most abundant non-protein thiol, acts as a metal chelator (Pocsi et al. 2004), scavenges free radicals, and repairs damage caused by the oxidative stress (Ferrol et al. 2009). It also protects the cell and its subcellular components from metal-induced damage by chelating and sequestering the metal ions. An increased production of GSH and its precursor γ-glutamylcysteine in *Paxillus involutus* under Cd exposure has been observed by Ott et al. (2002) and Courbot et al. (2004). Putative gene sequences coding for enzymes involved in glutathione and γ-glutamylcysteine synthesis has also been identified in expression sequence tag (EST) databases obtained from the ECM fungi *Hebeloma cylindrosporum* and *Paxillus involutus*.

Phytochelatins are intracellular metal-chelating agents comprised of a family of small cysteine-rich peptides having general structure (γ-glutamyl-cysteinyl)n-glycine ($n = 2$-ll) and the variants with the repeated γ-glutamylcysteinyl units are formed in some plants and yeast. They are capable of binding to various metals including Cd, Cu, Zn, or As via the sulfhydryl and carboxyl residues, but their biosyntheses are controlled preferentially by the metal. Phytochelatins are synthesized from reduced GSH by the transpeptidation of γ-glutamyl-cysteinyl dipeptides mediated by a constitutively synthesized enzyme, phytochelatin synthase (Schmoger et al. 2000; Vatamaniuk et al. 2001). However, the PC production in MF in response to metal stress is not reported. The MTs on the contrary are low molecular weight peptides that chelate metal ions by thiolate coordination and play a crucial role in cellular HM detoxification and homeostasis (Gadd 1993; Cobbett and Goldsbrough 2002). They are characterized by their small size (<7 kDa), a high content of amino acid cysteine (Cys, up to 33%), and a high degeneracy in the remaining residues. They are encoded by a multigene family and contain metal binding Cys-rich domains. Transcription of MTs is typically induced by the same metal ion(s) that bind to the protein, thus providing a direct activation of their protective function (Waalkes and Goering 1990). Metallothioneins are classified into two classes, based on the arrangement of cysteine residues (Fowler et al. 1987; Kojima 1991). Class I MTs are widespread in vertebrates, whereas class II MTs are found in plants and fungi. Until now, three glomeromycotan MTs have been identified: (1) *GrosMT1*, found in *Gigaspora rosea* (Stommel et al. 2001), (2) *GmarMT1*, found in *G. margarita* (Lanfranco et al. 2002), and (3) *GintMT1* in *G. intraradices* (González-Guerrero et al. 2007). To date, only two fungi have been reported to be able to synthesize both MTs and PCs: *Candida glabrata* produces MTs when exposed to toxic concentrations of Cu, but under Cd stress, it produces PCs only (Mehra et al. 1988, 1989). *Schizosaccharomyces pombe* produces HM-chelating PC peptides through a plant-like PC-synthase enzyme (SpPCS), and has a putative MT (Ha et al. 1999). In a study, Lanfranco et al. (2002) identified a gene (designated as *GmarMT1*) encoding a MT-like protein in *G. margarita*. Cloning of this gene into hypersensitive *S. pombe* strain enhanced Cd and Cu resistance compared to the nontransformed strain. *GmarMT1* gene was differentially expressed in pre-symbiotic spores compared to the symbiotic one, with down-regulation in the latter stage (Courbot et al. 2004). This was due to the fact that the organism may be more stressed in the pre-symbiotic stage. However, only Cu exposure has been found to up-regulate *GmarMT1* in the symbiotic stage and not in the pre-symbiotic stage. This study further unravels the difficulty in understanding the mechanisms involved in metal resistances in plants and fungi. Previous researchers reported that complexation of cadmium by MTs is a key mechanism for Cd tolerance in the ECM fungus, *Paxillus involutus* (Jacob et al. 2004; Courbot et al. 2004). Recently, Ramesh et al. (2009) characterized two MT genes, *HcMT1* and *HcMT2,* from the ectomycorrhizal fungus *Hebeloma cylindrosporum* and determined their expression in *H. cylindrosporum* under metal-stressed conditions by competitive RT-PCR analysis. The full length cDNAs were used to perform functional complementation in yeast mutant strains. The findings of this study assessed by heterologous complementation assays in yeast demonstrated that *HcMT1* and *HcMT2* encode a functional polypeptide capable of conferring increased tolerance against Cd and Cu, respectively. Based on this study, it was concluded that ECM fungi codes for different MTs; each of which has a particular pattern of expression, suggesting that they could play an important and specific role in improving the survival and growth of ectomycorrhizal trees growing in varied ecosystems contaminated by heavy metals. Besides facilitating metal binding, MTs are also known to protect cell from oxidative damage (Tamai et al. 1993; Achard-Joris et al. 2007) by reoxidizing the thiolate groups in the presence of metal-induced ROS and SOD (Maret 2003; González-Guerrero et al. 2007). In addition to protection against metal (loid)s, PCs and MTs also have other roles in cell function such as in sulfur storage and metabolism (Cobbett and Goldsbrough 2002). Cytosolic mechanism for metal-induced GSH production and formation of low molecular weight and less toxic form PC–metal complex (proposed) is presented in Fig. 9.2.

9.2.2.3 Transporter Proteins Involved in Metal Tolerance

Transporter proteins involved in metal tolerance facilitate the efflux of toxic metal ions from the cytosol or they allow metal sequestration into intracellular compartments, for example, vacuoles (Williams et al. 2000; Hall 2002). Transporter protein–metal complex and PC–metal complex, usually present in the apoplast and tonoplast, have been found to maintain cytoplasmic toxic metal ion concentrations by pushing metal ions outside the cell and into the vacuole, respectively (Ortiz et al. 1995). For example, permease (glutathione S-conjugate transporter)-mediated accumulation of cadmium in *Paxillus involutus* vacuoles is reported (Blaudez et al. 2000). This specific permease has been encoded from yeast cadmium factor (*Ycf1*) gene for the vacuolar sequestration of bis(glutathionato)-Cd (GS_2-Cd) (Li et al. 1997) as well as bis(glutathionato)-Hg (GS_2-Hg) (Gueldry et al. 2003). The hypothesis supported

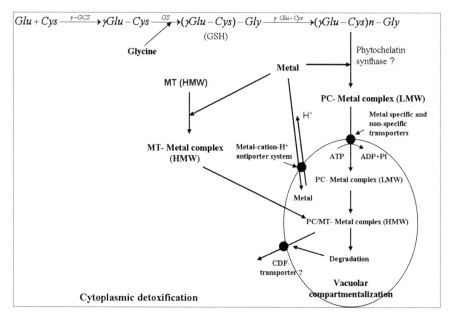

Fig. 9.2 Possible cytosolic mechanism for metal-induced glutathione production and formation of low molecular weight and less toxic form PC–metal complex (proposed) and/or MT–metal complex intracellularly and their compartmentalization into vacuole as high molecular weight complex and their further transformation to least toxic form. *γ-GCS* γ-glutamylcysteine synthetase, *GS* glutathione synthetase, *PC* phytochelatin, *MT* metallothionein, *LMW* low molecular weight, *HMW* high molecular weight, *CDF* cation diffusion facilitator

further by X-ray microanalysis has confirmed that Cd accumulated along with the accumulation of sulfur in electron-dense bodies in the vacuolar compartment (Ott et al. 2002). For other metals, the Zn-transporter gene *MtZIP2* from *Medicago truncatula* was up-regulated in the presence of Zn, while it was down-regulated by mycorrhizal colonization, leading to a lower content of Zn within the host plant tissues (Burleigh et al. 2003). Zinc transporters of the ZIP family are known to facilitate metal uptake from extracellular media or they mobilize metals from intracellular stores (Gaither and Eide 2001). In a similar study, Gonzalez-Guerrero et al. (2005) also observed increased transcript levels of a putative Zn transporter gene (*GintZnT1*) of the CDF family in the mycelium of *Glomus intraradices* when it was exposed to Zn, indicating a possible role of this gene product in Zn homeostasis and protection against Zn stress. In yet another study, Gonzalez-Guerrero et al. (2006) reported a Cd- and Cu-dependent up-regulation of a putative ABC transporter gene (*GintABC1*) in the extra-radical mycelium of *G. intraradices*. The gene encodes a polypeptide with homology to the N-terminal region of the Multidrug-Resistance-Protein (MRP) subfamily of ABC transporters and suggested to be involved in Cd and Cu detoxification in the extra-radical mycelium of *G. intraradices*. However, enhanced Zn efflux may also act as a potential tolerance mechanism, as observed in the ECM fungus *Suillus bovinus* (Adriaensen 2005). Alternatively, down-regulation of transporter genes involved in the uptake of metal at the plasma membrane may also be part of tolerance mechanisms, as observed in other fungi (Eide 2003). The precise roles of metal transporter genes during influx or efflux in different mycorrhizal fungi are, however, not clear. Moreover, whether these mechanisms are metal specific or concentration/toxicity dependent is not well explained.

9.2.2.4 Antioxidative Mechanisms to Combat Metal-Induced Oxidative Stress

Besides adopting avoidance or compartmentalization strategies, the organisms may also have mechanisms to combat metal-induced oxidative stress, and to repair damaged proteins, which may be caused by redox active elements (Amor et al. 1998). Redox active elements can also cause severe damage to other cellular components. Smirnoff (1993) has divided these systems into two categories: one that interacts with active forms of O_2 and maintains them at low levels. Some of the enzymes involved here are superoxide dismutases (SODs), catalases (CATs), and ascorbate peroxidases (APX). In other system, oxidized antioxidants like glutathiones (GSHs), glutathione reductases (GRs), ascorbate, and mono- and dihydroascorbate reductases are regenerated. The first group of enzymes is involved in the detoxification of $O_2^{\cdot-}$ radicals and H_2O_2 and consequently prevents the formation of $OH^{\cdot-}$ radicals. For instance, GR and GSH are important components of the ascorbate–glutathione pathway and cause the removal of H_2O_2 in different cellular compartments (Dalton 1995). Superoxide dismutases on the contrary are metalloproteins which convert superoxide to H_2O_2 and molecular oxygen (O_2) and act as a primary defense during oxidative stress by protecting cell membranes from ROS damage. Ott et al. (2002), while studying the antioxidative systems of ECM fungus *P. involutus* generated in response to Cd, observed that the induction of SOD and higher accumulation of GSH, GSH-dependent peroxidase, and glutathione reductases prevented the accumulation of H_2O_2 in the fungus. Lanfranco et al. (2005) identified a gene encoding a functional Cu/Zn SOD (*GmarCuZnSOD*), which deactivated the ROS induced by Cu and Zn to avoid oxidative stress. Azcon et al. (2009) in a study observed the enhancement of CAT, APX, and GR activities in AMF-inoculated plants, which in turn, protected the plants from oxidative damage. González-Guerrero et al. (2007) reported a gene *GintMT1* in *G. intraradices*, encoding a functional MT that responds to oxidative stress caused by Cu. Further, a suppression subtractive hybridization (SSH) library (Diatchenko et al. 1996) prepared from hyphae of *G. intraradices,* grown on varying Zn concentrations (Ouziad et al. 2005), was found to have several EST-sequences, which putatively coded for enzymes like GST, SOD, cytochrome P450, and thioredoxin, involved in the detoxification of ROS. Their differential expression later confirmed by reverse Northern analysis suggested that the primary function of the fungal cells was to cope with the heavy metal-induced oxidative stress. Similarly, glutathione S-transferases catalyze the conjugation of glutathione with a variety of reactive electrophilic compounds and may provide protection against oxidative stress (Moons 2003; Smith et al. 2004). In a SSH library obtained from *G. intraradices* grown under heavy metal stress, several ESTs had significant sequence homologies to GST-encoding genes from other organisms (Rhody 2002). This finding on the transcriptional up-regulation of the GST gene (4b07) by Cd, Cu, or Zn could well indicate that GSTs of symbiotic mycelium participated in the removal of heavy metal toxicity (Hildebrandt et al. 2007). Recently, Benabdellah et al. (2009) also observed the production of reactive oxygen radicals by *G. intraradices* upon high levels of Cu exposure. To date, only a few genes encoding proteins putatively involved in ROS homeostasis have been identified and characterized in AM fungi. For example, three SODs (González-Guerrero et al. 2005; Lanfranco et al. 2005), ten genes putatively encoding GSTs (Waschke et al. 2006), a glutare-doxin (Benabdellah et al. 2009), and an MT (González-Guerrero et al. 2007) have been reported for combating oxidative stress to MF.

9.3   Prospects of Genetic Engineering in Metal Remediation

The indigenous mycorrhizal fungi recovered from metal contaminated sites have attracted attention of the people engaged in remediation of metal-polluted soils. This is largely due to their extraordinary physiological and variable genetic abilities to survive in metal-rich environments. And therefore, in recent times, focus has been directed toward exploiting the potential of metal-tolerant mycorrhizal species in heavy metal removal from contaminated sites. For this, there is urgent need to identify candidate genes for high metal tolerance and accumulation. In this context, genes encoding metallothioneins, metal transporters, and other antioxidant enzymes putatively involved in metal tolerance and uptake have been identified in different mycorrhizal fungi (Table 9.1). Mycorrhizal symbiont facilitates HM uptake by the plant and also improves inorganic P nutrition to the host plants. Mycorrhiza-induced high-affinity plant Pi transporter genes have been identified in plants (Maldonado-Mendoza et al. 2001; Benedetto et al. 2005). The over-expression or induction of Pi transporters genes in mycorrhizal symbiont and their expression into corresponding host could be advantageous for plants. Apart from this, plants take up arsenic as arsenate (AsO_3^-) via Pi transporter systems (Meharg and Macnair 1994), and it is likely that such Pi transporters could contribute to arsenic removal from the polluted soil. The exploitation of transgenic approaches to improve the ability of shoots to take up more and more arsenate is likely to help to generate plant lines endowed with enhanced phytoextraction properties. This, in turn, may increase arsenic mobilization, acquisition, and "deposition" in above-ground tissues. Likewise, the genes responsible for SO_4^{-2} absorption and assimilation could be introduced in mycorrhizal fungi for accelerating the production of sulfur-rich compounds, like cysteine. Such compounds are known to influence GSHs and MTs metabolism and concomitantly reduce metal toxicity (Cobbett and Goldsbrough 2002; Ferrol et al. 2009). Since the metal-induced PCs are reported to decrease cellular levels of GSH, there exists a

Table 9.1 Genes encoding metallothioneins, metal transporters, and other antioxidant enzymes putatively involved in metal tolerance and uptake identified in different mycorrhizal fungi

Genes	Product	Isolated from	Function	References
GrosMT1	Metallothionein	Gigaspora rosea	Metal binding and chelation	Stommel et al. (2001)
GintZnT1	Zn transporter (CDF family)	Glomus intraradices	vacuolar Zn compartmentalization	Gonzalez-Guerrero et al. (2005)
GmarCuZn-SOD1	Cu.Zn-SOD	Gigaspora margarita	Reduce oxidative stress	Lanfranco et al. (2005)
GintSOD1	Cu.Zn-SOD	G. intraradices	Reduce oxidative stress	Gonzalez-Guerrero et al. 2005
GintABC1	ABC transporter (MRP Subfamily)	G. intraradices	Cu and Cd detoxification and reduce oxidative stress	Gonzalez-Guerrero et al. (2006, 2010)
GintMT1	Functional Metallothionein	G. intraradices	Increase tolerance against Cu, Cd, and reduce oxidative stress	Gonzàlez-Guerrero et al. (2007)
HcMT1	Functional polypeptides	Hebeloma cylindrosporum (ECM)	Increase tolerance against Cd	Ramesh et al. (2009)
HcMT2	Functional polypeptides	H. cylindrosporum (ECM)	Increase tolerance against Cu	

possibility of increasing the level of metal-binding peptides in mycorrhizal fungi. This can be achieved by increasing the level of GSH by up-regulating the expression of the enzymes responsible for GSH synthesis. Therefore, in order to enhance the metal remediation abilities of mycorrhizae, genetic engineering can play an important role in (1) over-expressing extra- and intracellular enzymes and (2) cell-wall biosynthesis and modification, so that more metal binding groups could be introduced. Furthermore, the modification in the genetic makeup of MF for enhanced production of metal transporter proteins, which pumps out the ions from the cytoplasm to the apoplastic region or to the vacuole, could be targeted for successful phytoremediation of metal-polluted sites.

9.4 Conclusion

Considering the immense metal tolerance and accumulation potential of MF, the plant-mycorrhizal symbiosis could profitably be exploited for decontaminating and restoring metal-polluted sites. Due to mycorrhizal efficiency in maintaining metal homeostasis and buffering metal stress in both partners, mycorrhizal fungi should be mass produced and recommended as an alternative biotechnological tool for reclamation of metal-contaminated soils. It is, therefore, of great practical importance to inoculate the plants by efficient and effective mycorrhizal fungal strains that can adapt better to a particular set of conditions and/or host plant to expedite the process of metal remediation and successful restoration of degraded ecosystems. In this context, application of genetic engineering could prove an asset for enhancing the efficiency and improving the adaptability of fungal symbionts to variously polluted sites through alterations in metal uptake pathways leading to high metal tolerance and/or detoxification. Also, by employing genetic engineering, the promising genes causing over production of metal-chelating agents for enhanced metal binding at target sites could be incorporated into other plants. In this endeavor, issues regarding how metal tolerance and accumulation at the whole plant level are affected and governed by the establishment of plant-mycorrhizal symbiosis need to be addressed. Additionally, the better understanding of mechanistic basis of metal restriction, translocation, and distribution among different plant organs mediated through mycorrhizae could fine-tune the process of phytoremediation for its wider and sustainable application.

Acknowledgment The authors gratefully acknowledge the concept and literature provided on the theme by various workers, which have been cited in the manuscript.

References

Achard-Joris M, Moreau JL, Lucas M, Baudrimont M, Mesmer-Dudons N, Gonzalez P, Boudou A, Bourdineaud JP (2007) Role of metallothioneins in superoxide radical generation during copper redox cycling: defining the fundamental function of metallothioneins. Biochimie 89:1474–1488

Adriaensen K (2005) Adaptive heavy metal tolerance in the ectomycorrhizal fungi *Suillus bovinus* and *Suillus luteus*. Ph.D. thesis, Limburgs Universitair Centrum, Diepenbeek

Amor Y, Babiychuk E, Inze D, Levine A (1998) The involvement of poly(ADP-ribose) polymerase in the oxidative stress responses in plants. FEBS Lett 440:1–7

Arriagada CA, Herrera MA, Ocampo JA (2007) Beneficial effect of saprobe and arbuscular mycorrhizal fungi on growth of *Eucalyptus globulus* co-cultured with Glycine max in soil contaminated with heavy metals. J Environ Manage 84:93–99

Ashford AE, Peterson RL, Dwarte D, Chilvers GA (1986) Polyphosphate granules in eucalypt mycorrhizas: Determination by energy dispersive x-ray microanalysis. Can J Bot 64:677–687

Awotoye OO, Adewole MB, Salami AO, Ohiembor MO (2009) Arbuscular mycorrhiza contribution to the growth performance and heavy metal uptake of *Helianthus annuus* Linn in pot culture. Afr J Environ Sci Technol 3:157–163

Azcon R, Peralvarez MDC, Biro B, Roldan A, Lozano JMR (2009) Antioxidant activities and metal acquisition in mycorrhizal plants growing in a heavy-metal multi-contaminated soil amended with treated lignocellulosic agrowaste. Appl Soil Ecol 41:168–177

Benabdellah K, Merlos MA, Azcón-Aguilar C, Ferrol N (2009) GintGRX1, the first characterized glomeromycotan glutaredoxin, is a multifunctional enzyme that responds to oxidative stress. Fungal Genet Biol 46:94–103

Benedetto A, Magurno F, Bonfante P, Lanfranco L (2005) Expression profiles of a phosphate transporter gene (*GmosPT*) from the endomycorrhizal fungus *Glomus mosseae*. Mycorrhiza 15:1–8

Berta G, Sampo S, Gamalero E, Massa N, Lemanceau P (2005) Suppression of *Rhizoctonia* root-rot of tomato by *Glomus mossae* BEG12 and *Pseudomonas fluorescens* A6RI is associated with their effect on the pathogen growth and on the root morphogenesis. Eur J Plant Pathol 111:279–288

Blaudez D, Botton B, Chalot M (2000) Cadmium uptake and subcellular compartmentation in the ectomycorrhizal fungus *Paxillus involutus*. Microbiology 146:1109–1117

Burleigh SH, Kristensen BK, Bechmann IE (2003) A plasma membrane zinc transporter from *Medicago truncatula* is up-regulated in roots by Zn fertilization, yet down-regulated by arbuscular mycorrhizal colonization. Plant Mol Biol 52:1077–1088

Chen BD, Thu YG, Duan J, Xiao XY, Smith SE (2007) Effects of the arbuscular mycorrhizal fungus *Glomus mosseae* on growth and metal uptake by four plant species in copper mine tailings. Environ Pollut 147:374–380

Citterio S, Prato N, Fumagalli P, Aina R, Massa N, Santagostino A, Sgorbati S, Berta G (2005) The arbuscular mycorrhizal fungus *Glomus mosseae* induces growth and metal accumulation changes in *Cannabis sativa* L. Chemosphere 59:21–29

Clemens S, Schroeder JI, Degenkolb T (2001) *Caenorhabditis elegans* express a functional phytochelatin synthase. Eur J Biochem 268:3640–3643

Cobbett C, Goldsbrough P (2002) Phytochelatins and metallothioneins: roles in heavy metal detoxification and homeostasis. Annu Rev Plant Biol 53:159–182

Courbot M, Diez L, Ruotolo R, Chalot M, Leroy P (2004) Cadmium-responsive thiols in the ectomycorrhizal fungus *Paxillus involutus*. Appl Environ Microbiol 70:7413–7417

Dalton DA (1995) Antioxidant defenses of plants and fungi. In: Ahmad S (ed) Oxidative stress and antioxidant defenses in biology. Chapman & Hall, New York, pp 298–355

Diatchenko L, Lau YF, Campbell AP, Chenchik A, Moqadam F, Huang B, Lukyanov S, Lukyanov K, Gurskaya N, Sverdlov ED, Siebert PD (1996) Suppression subtractive hybridization: a method for generating differentially regulated or tissue-specific cDNA probes and libraries. Proc Natl Acad Sci USA 93:6025–6030

Drager DB, Desbrosses-Fonrouge AG, Krach C, Chardonnens AN, Meyer RC, Saumitou-Laprade P, Krämer U (2004) Two genes encoding *Arabidopsis halleri* MTP1 metal transport proteins co-segregate with zinc tolerance and account for high MTP1 transcript levels. Plant J 39: 425–439

Driver JD, Holben WE, Rilling MC (2005) Characterization of glomalin as a hyphal wall component of arbuscular mycorrhizal fungi. Soil Biol Biochem 37:101–106

Eide DJ (2003) Multiple regulatory mechanisms maintain zinc homeostasis in *Saccharomyces cerevisiae*. J Nutr 133:1532–1535

Ferrol N, González-Guerrero M, Valderas A, Benabdellah K, Azcón-Aguilar C (2009) Survival strategies of arbuscular mycorrhizal fungi in Cu-polluted environments. Phytochem Rev 8: 551–559

Fomina M, Alexander IJ, Colpaert JV, Gadd GM (2005) Solubilization of toxic metal minerals and metal tolerance of mycorrhizal fungi. Soil Biol Biochem 37:851–866

Fowler BA, Hildebrand CE, Kojima Y, Webb M (1987) Nomenclature of metallothionein. Experientia Suppl 52:19–22

Frey B, Zierold K, Brunner I (2000) Extracellular complexation of Cd in the Hartig net and cytosolic Zn sequestration in the fungal mantle of *Picea abies-Hebeloma crustuliniforme* ectomycorrhizas. Plant Cell Environ 23:1257–1265

Gadd GM (1993) Interactions of fungi with toxic metals. New Phytol 124:25–60

Gaither IA, Eide DJ (2001) Eukaryotic zinc transporters and their regulation. Biometals 14: 251–270

Gamalero E, Trotta A, Massa N, Copetta A, Martinotti MG, Berta G (2004) Impact of two fluorescent pseudomonads and an arbuscular mycorrhizal fungus on tomato plant growth, root architecture, and P acquisition. Mycorrhiza 14:185–192

Gaur A, Adholeya A (2004) Prospects of arbuscular mycorrhizal fungi in phytoremediation of heavy metal contaminated soils. Curr Sci 86:528–534

Gonzalez-Chavez MC, Carrillo-Gonzalez R, Wright SF, Nichols KA (2004) The role of glomalin, a protein produced by arbuscular mycorrhizal fungi in sequestering potentially toxic elements. Environ Pollut 130:317–323

Gonzalez-Guerrero M, Azcon-Aguilar C, Mooney M, Valderas A, MacDiarmid CW, Eide DJ, Ferrol N (2005) Characterization of a *Glomus intraradices* gene encoding a putative Zn transporter of the cation diffusion facilitator family. Fungal Genet Biol 42:130–140

Gonzalez-Guerrero M, Azcon-Aguilar C, Ferrol N (2006) *GintABC1* and *GintMT1* are involved in Cu and Cd homeostasis in *Glomus intraradices*. In: Abstracts of the 5th international conference on Mycorrhiza, Granada

González-Guerrero M, Cano C, Azcón-Aguilar C, Ferrrol N (2007) *GintMT1* encodes a functional metallothionein in *Glomus intraradices* that responds to oxidative stress. Mycorrhiza 17:327–335

González-Guerrero M, Benabdellah K, Valderas A, Azcón-Aguilar C, Ferrol N (2010) *GintABC1* encodes a putative ABC transporter of the MRP subfamily induced by Cu, Cd, and oxidative stress in *Glomus intraradices*. Mycorrhiza 20:137–146

Gueldry O, Lazard M, Delort F, Dauplais M, Grigoras I, Blanquet S, Plateau P (2003) Ycf1p-dependent Hg(II) detoxification in *Saccharomyces cerevisiae*. Eur J Biochem 270:2486–2496

Ha SB, Smith AP, Howden R, Dietrich WM, Bugg S, O'Connell MJ, Goldsbrough PB, Cobbett CS (1999) Phytochelatin synthase genes from *Arabidopsis* and the yeast *Schizosaccharomyces pombe*. Plant Cell 11:1153–1164

Hall JL (2002) Cellular mechanisms for heavy metal detoxification and tolerance. J Exp Bot 53:1–11

Hartley J, Cairney JWG, Meharg AA (1997) Do ectomycorrhizal fungi exhibit adaptive tolerance to potentially toxic metals in the environment? Plant Soil 189:303–319

Hildebrandt U, Regvar M, Bothe H (2007) Arbuscular mycorrhiza and heavy metal tolerance. Phytochemistry 68:139–146

Jacob C, Courbot M, Martin F, Brun A, Chalot M (2004) Transcriptomic response to cadmium in the ectomycorrhizal fungus *Paxillus involutus*. FEBS Lett 576:423–427

Jentschke G, Fritz E, Godbold DL (1991) Distribution of lead in mycorrhizal and non-mycorrhizal Norway spruce seedlings. Physiol Plant 81:417–422

Joner EJ, Leyval C (1997) Uptake of [109]Cd by roots and hyphae of a *Glomus mosseae/Trifolium subterraneum* mycorrhiza from soil amended with high and low concentrations of cadmium. New Phytol 135:353–360

Joner EJ, Leyval C, Briones R (2000) Metal binding capacity of arbuscular mycorrhizal mycelium. Biol Fertil Soil 226:227–234

Jones DL (1998) Organic acids in the rhizosphere – a critical review. Plant Soil 205:25–44

Kojima Y (1991) Definitions and nomenclature of metallothioneins. Meth Enzymol 205:8–10

Landeweert R, Hoffland E, Finlay RD, Kuyper TW, Van Breemen N (2001) Linking plants to rocks: ectomycorrhizal fungi mobilize nutrients from minerals. Trends Ecol Evol 16:248–254

Lanfranco L, Bolchi A, Ros EC, Ottonello S, Bonfante P (2002) Differential expression of a metallothionein gene during the presymbiotic versus the symbiotic phase of an arbuscular mycorrhizal fungus. Plant Physiol 130:58–67

Lanfranco L, Novero M, Bonfante P (2005) The mycorrhizal fungus *Gigaspora margarita* possesses a Cu/Zn superoxide dismutase that is up-regulated during symbiosis with legume hosts. Plant Physiol 137:1319–1330

Leyval C, Turnau K, Haselwandter K (1997) Effect of heavy metal pollution on mycorrhizal colonization and function: physiological, ecological and applied aspects. Mycorrhiza 7:139–153

Li XL, Christie P (2000) Changes in soil solution Zn and pH and uptake of Zn by arbuscular mycorrhizal red clover in Zn-contaminated soil. Chemosphere 42:201–207

Li ZS, Lu YP, Zhen RG, Szczypka M, Thiele DJ, Rea PA (1997) A new pathway for vacuolar cadmium sequestration in *Saccharomyces cerevisiae*: YCF1-catalyzed transport of bis(glutathionato) cadmium. Proc Natl Acad Sci USA 94:42–47

Maldonado-Mendoza IE, Dewbre GR, Harrison MJ (2001) A phosphate transporter gene from the extra-radical mycelium of an arbuscular mycorrhizal fungus *Glomus intraradices* is regulated in response to phosphate in the environment. Mol Plant Microbe Interact 14:1140–1148

Maret W (2003) Cellular zinc and redox states converge in the metallothionein/thionein pair. J Nutr 133:1460–1462

Martin F, Rubini P, Côté R, Kottke I (1994) Aluminium polyphosphate complexes in the mycorrhizal basidiomycete Laccaria bicolor: A $_{27}$Al-nuclear magnetic resonance study. Planta 194: 241–246

Martino E, Perottoa S, Parsons R, Gadd GM (2003) Solubilization of insoluble inorganic zinc compounds by ericoid mycorrhizal fungi derived from heavy metal polluted sites. Soil Biol Biochem 35:133–141

Meharg AA (2003) The mechanistic basis of interactions between mycorrhizal associations and toxic metal cations. Mycol Res 107:1253–1265

Meharg A, Macnair M (1994) Relationship between plant phosphorus status and the kinetics of arsenate influx in clones of *Deschamsia caespitosa (L.) Beauv.* that differ in their tolerance of arsenate. Plant Soil 162:99–106

Mehra RK, Tarbet EB, Gray WR, Winge DR (1988) Metal-specific synthesis of two metallothioneins and γ-glutamil peptides in *Candida glabrata*. Proc Natl Acad Sci USA 85:8815–8819

Mehra RK, Garey JR, Butt TR, Gray WR, Winge DR (1989) *Candida glabrata* metallothioneins: cloning and sequence of the genes and characterization of proteins. J Biol Chem 264: 19747–19753

Moons A (2003) Osgstu3 and osgstu4, encoding tau class glutathione S-transferases, are heavy metal- and hypoxic stress-induced and differentially salt stress-responsive in rice roots. FEBS Lett 553:427–432

Nies DH, Silver S (1995) Ion efflux systems involved in bacterial metal resistances. J Indust Microbiol 14:186–199

Ortiz DF, Ruscitti T, McCue KF, Ow DW (1995) Transport of metal binding peptides by HMT1, a fission yeast ABC-type vacuolar membrane protein. J Biol Chem 270:4721–4728

Ott T, Fritz E, Polle A, Schutzendubel A (2002) Characterisation of antioxidative stress systems in the ectomycorhiza-building basidiomycete *Paxillus involutus* (Bartsch) Fr. and its reaction to cadmium. FEMS Microbiol Ecol 42:359–366

Ouziad F, Hildebrandt U, Schmelzer E, Bothe H (2005) Differential gene expressions in arbuscular mycorrhizal-colonized tomato grown under heavy metal stress. J Plant Physiol 162:634–649

Pocsi I, Prade RA, Penninckx MJ (2004) Glutathione, altruistic metabolite in fungi. Adv Microbiol Physiol 49:1–76

Purin S, Rillig MC (2008) Immuno-cytolocalization of glomalin in the mycelium of the arbuscular mycorrhizal fungus *Glomus intraradices*. Soil Biol Biochem 40:1000–1003

Ramesh G, Podila GK, Gay G, Marmeisse R, Reddy MS (2009) Copper and cadmium metallothioneins of the ectomycorrhizal fungus *Hebeloma cylindrosporum* have different patterns of regulation. Appl Environ Microbiol 75:2266–2274

Redon PO, Beguiristain T, Leyval C (2008) Influence of *Glomus intraradices* on Cd partitioning in a pot experiment with *Medicago truncatula* in four contaminated soils. Soil Biol Biochem 40:2710–2712

Rhody D (2002) Erste Schritte zur Etablierung und Verbesserung von Transformations systemen fu r wurzelbesiedelnde Pflanzen. Ph.D. thesis, The University of Marburg, Marburg

Rivera-Becerril F, Calantzis C, Turnau K, Caussanel JP, Belimov AA, Gianinazzi S, Strasser RJ, Gianinazzi-Pearson V (2002) Cadmium accumulation and buffering of cadmium-induced stress by arbuscular mycorrhiza in three *Pisum sativum L. genotypes*. J Exp Bot 53:1177–1185

Rouch DA, Lee BTD, Morby AP (1995) Understanding cellular responses to toxic agents: a model for mechanism choice in bacterial metal resistance. J Ind Microbiol 14:132–141

Sanita` L, Prasad MNV, Ottonello S (2002) Metal chelating peptides and proteins in plants. In: Prasad MNV, Strzalka K (eds) Physiology and biochemistry of metal toxicity and tolerance in plants. Kluwer, Dordrecht, pp 59–94

Schmoger MEV, Oven M, Grill E (2000) Detoxification of arsenic by phytochelatins in plants. Plant Physiol 122:793–802

Silver S, Walderhaug M (1992) Gene regulation of plasmid and chromosome determined inorganic ion transport in bacteria. Microbiol Rev 56:195–228

Smirnoff N (1993) The role of active oxygen in the response of plants to water deficit and desiccation (Tansley Review No. 52). New Phytol 125:27–58

Smith AP, DeRidder BP, Guo WJ, Seeley EH, Regnier FE, Goldsbrough PB (2004) Proteomic analysis of *Arabidopsis* glutathione S-transferases from benoxacor- and copper-treated seedlings. J Biol Chem 279:26098–26104

Stommel M, Mann P, Franken P (2001) EST-library construction using spore RNA of the arbuscular mycorrhizal fungus *Gigaspora rosea*. Mycorrhiza 10:281–285

Sudová R, Doubková P, Vosátka M (2008) Mycorrhizal association of *Agrostis capillaris* and *Glomus intraradices* under heavy metal stress: combination of plant clones and fungal isolates from contaminated and uncontaminated substrates. Appl Soil Ecol 40:19–29

Tamai KT, Gralla EB, Ellerby LM, Valentine JS, Thiele DJ (1993) Yeast and mammalian metallothioneins functionally substitute for yeast copper-zinc superoxide dismutase. Proc Natl Acad Sci USA 90:8013–8017

Tomsett AB (1993) Genetics and molecular biology of metal tolerance in fungi. In: Jennings DH (ed.) Stress tolerance of fungi., pp 69–95

Turnau K, Ryszka P, Gianinazzi-Pearson V, van Tuinen D (2001) Identification of arbuscular mycorrhizal fungi in soils and roots of plants colonizing zinc wastes in southern Poland. Mycorrhiza 10:169–174

Vare H (1990) Aluminium polyphosphate in the ectomycorrhizal fungus *Suillus variegatus* (Fr.) O. Kunze as revealed by energy dispersive spectrometry. New Phytol 116:663–668

Vatamaniuk OK, Bucher EA, Ward JT, Rea PA (2001) A new pathway for heavy metal detoxification in animals: phytochelatins synthase is required for cadmium tolerance in *Caenorhabditis elegans*. J Boil Chem 276:20817–20820

Waalkes MP, Goering PL (1990) Metallothionein and other cadmium-binding proteins: recent developments. Chem Res Toxicol 3:281–288

Waschke A, Sieh D, Tamasloukht M, Fischer K, Mann P, Franken P (2006) Identification of heavy metal-induced genes encoding glutathione S-transferases in the arbuscular mycorrhizal fungus *Glomus intraradices*. Mycorrhiza 17:1–10

Williams LE, Pittman JK, Hall JL (2000) Emerging mechanisms for heavy metal transport in plants. BBA Biomembr 1465:104–126

Zhu YG, Christie P, Laidlaw AS (2001) Uptake of Zn by arbuscular mycorrhizal white clover from Zn-contaminated soil. Chemosphere 42:193–199

Chapter 10
Metal Signaling in Plants: New Possibilities for Crop Management Under Cadmium-Contaminated Soils

Ana Lima and Etelvina Figueira

Abstract Metal stress restricts plant growth and distribution and has become a widespread problem. Plants can respond to toxic metals in a variety of ways, but the most important of them is the production of phytochelatins (PC). The knowledge of how plants perceive metal presence and switch on or off the PC synthesis pathway could help understanding the metal tolerance mechanisms in plants. This knowledge can be used for enhancing crop tolerance in metal-polluted soils and for metal phytoremediation techniques. However, the signaling mediators that trigger metal tolerance mechanisms such as synthesis of phytochelatins are still largely unknown. Here, we discuss the importance of signal transduction in phytochelatin synthesis and cadmium tolerance, identifying specific signal transducers that may be involved in increasing PC production or reducing metal uptake in plants by analyzing the role of calcium signals, protein phosphatases, and reactive oxygen species induction during metal detection and response in plants. The understanding of signaling networks can open new possibilities to design crops with abilities to better adapt to excess metal conditions. Therefore, the process of PC synthesis and Cd absorption was analyzed in *Arabidopsis thaliana* cells, using different pharmacological modulators of the cytoplasmatic calcium levels and PP1 activity, as well as the addition of ROS. With these procedures, we expect to show a possible pathway for Cd signaling and PC induction in plants that can be used for regulating Cd uptake and tolerance in plants and thus could be used as a tool in the development of rational breeding programs and transgenic approaches.

Keywords Phytochelatins • Metal signaling • Protein phosphatase • Cadmium stress

A. Lima • E. Figueira (✉)
Centre for Cell Biology, Biology Department,
University of Aveiro, Universidade de Aveiro, Aveiro, Portugal
e-mail: efigueira@ua.pt

M.S. Khan et al. (eds.), *Biomanagement of Metal-Contaminated Soils*,
Environmental Pollution 20, DOI 10.1007/978-94-007-1914-9_10,
© Springer Science+Business Media B.V. 2011

10.1 Introduction

10.1.1 Stress Signal Transduction Mechanisms in Plants

Unlike animals, plants are sessile organisms that cannot move away from adverse environmental conditions and, therefore, require high sensitivity detection and adaptation mechanisms to withstand environmental perturbations. In nature, plants are exposed to various environmental stimuli that affect their physiology, morphology, and development. These factors are biotic such as fungal or pathogenic attack or abiotic such as climatic alterations or metal and pesticide contamination (Clark et al. 2001). A rapid and precise perception of many of these alterations by plants is important in order to adapt to changing environments, as it allows them to rapidly perceive environmental alterations and to trigger mechanisms that avoid the deleterious effects of any specific stress. Plants perceive the environmental alterations in different ways, such as by plasma membrane located receptors and intracellular and/or cytoskeleton-associated proteins. Subsequently, the imposed signal is recognized and a complex cascade of events involving several interacting components that recognize such signal is triggered, leading to altered gene expression and metabolic activities (Kaur and Gupta 2005). This cascade of events, called signal transduction, normally acts through second messengers and triggers the molecular events leading to the physiological response. Various signal pathways can operate independently from each other or can modulate other pathways positively or negatively. Different signaling pathways may also share components and second messengers to achieve their objectives. As a result, many signals interact in a cooperative manner with each other (Knight 2000). Differences in stress tolerance between genotypes may arise from variations in signal perception and transduction mechanisms (Hare et al. 1997). Thus, while most of the biochemical factors necessary for stress tolerance are present in all species, subtle differences in signal transducers hold the key to improve plant tolerance to distinct abiotic and biotic stresses.

10.1.2 Most Stress Signaling Mechanisms Share the Same Components

Studies on molecular responses of plants to various types of stresses indicate that different types of constraints provide different information to the cells. The multiplicity of this information makes the response of plants and hence the stress signaling pathway more complex (Knight 2000). It is now widely accepted that plants in general use similar transduction mechanisms to cope with different stresses (Mithöfer et al. 2004). Such examples were found during mechanical wounding and pathogen attack (Schaller and Weiler 2002), as well as salt, cold, and drought stress (Xiong et al. 2002).

Most signal transduction pathways share a generic signal perception, such as the modulation of intracellular Ca^{2+} levels, which initiates a protein phosphorylation cascade that finally targets proteins directly involved in cellular protection or transcription factors controlling specific sets of stress-regulated genes (Kaur and Gupta 2005). Recent advances have identified some novel specific signal transducers that are exclusive to the plant kingdom. Most of them function as Ca^{2+} sensors, namely the recently discovered family of novel calcium sensors CBLs from *Arabidopsis* and their target proteins, the calcium-induced protein kinases, (CIPKs), involved in various abiotic stresses, such as salt, drought, cold, and heat (Kudla et al. 1999; Luan 2004). Nevertheless, reactive oxygen species (ROS) have also emerged as important signaling molecules that control various processes including pathogen defense, programmed cell death, and stomatal behavior (Mithöfer et al. 2004; Maksymiec 2007). Overall, these components can act together in a multiplicity of ways, according to the final metabolic adjustment. Since stress tolerance mechanisms can involve several physiological responses, a complex cross talk can occur between different pathways, which act together to attain the same tolerance.

10.1.3   Metal Stress in Plants

Toxic metals have become one of the main abiotic stresses for living organisms because of their increased use in industry and agro-practices. Over the last decades, there has been an increasing awareness of how metals act as environmental pollutants (Baker and Walker 1989) and their effects on plants (Rauser 2000; Cobbett and Goldsbrough 2002; Hall 2002; Clemens 2006). Cadmium is one of the most important environmental pollutants, particularly in areas of high anthropogenic pressure. Its presence in the atmosphere, soil, and water can cause serious toxicity to organisms, and its bioaccumulation in the food chain can be highly dangerous (Wagner 1993; Sanitá di Toppi and Gabrielli 1999). In plants, Cd is known to inhibit seed germination and root growth, induce chromosomal aberrations, and disrupt micronucleus formation (Fojtová and Kovařík 2000). Cadmium can also cause membrane depolarization and cytoplasmic acidification leading to the disruption of cellular homeostasis (Pinto et al. 2003).

Plants can respond to metal toxicity in different ways. Such responses include immobilization, exclusion, chelation and compartmentalization of the metal ions, as well as the expression of more general stress response mechanisms such as synthesis of ethylene and stress proteins (Cobbett 2000; Clemens 2006). One recurrent general mechanism for toxic metal detoxification in plants and other organisms is the chelation of the metal ions by a specific ligand (Rauser 2000). A number of metal-binding ligands have now been recognized in plants, the most important of which is phytochelatins. Phytochelatins (PCs) are a family of Cys-rich, small non-protein thiol peptides with the general structure $(\gamma\text{-Glu-Cys})_n\text{-Gly}$ and are synthesized in a wide variety of plant species, algae, yeast, and nematodes (Rauser 2000; Cobbett and Goldsbrough 2002). These peptides are exclusively formed in the presence of metals, by the transpeptidation of the tripeptide glutathione (GSH), through the action of a constitutive enzyme, known as PC synthase (PCS, EC 2.3.2.15) (Zenk 1996; Rauser 1999). In the presence of toxic metal concentrations, particularly cadmium, PCs form complexes with metal ions and prevent toxic metals from interfering with the cellular metabolism (Vögeli-Lange and Wagner 1990;

Ortiz et al. 1995). These complexes are then stored in the vacuole, where they are rendered harmless to the cell. Therefore, the knowledge of how plants perceive the metal presence and switch on or off the PC synthesis pathway can be of crucial importance to better understand the metal tolerance mechanisms in plants, to improve crop tolerance in metal-polluted soils, and also to enhance metal phytore-mediation techniques. Nonetheless, the signal transduction pathways that involve Cd signaling and the subsequent PC production pathway are still at the infancy.

10.1.4   Metal Signal Transduction in Plants – Possible Pathways

The analysis of the mechanisms behind metal tolerance has become an important aspect of research, especially in the case of cadmium. Several groups suggest that Cd tolerance can be achieved through the same signal transduction pathways that plants use for other abiotic stresses (Mithöfer et al. 2005). Previous studies showed that a pre-exposure to metals also induces enhanced tolerance to biotic factors, suggesting a sort of chemical memory, attained by enhanced signaling pathways that can be triggered by stress (Trewavas 1999). For example, Ghoshroy et al. (1998) and Mittra et al. (2004) showed that a mild dose of Cd pre-exposure increased plant resistance to viral and fungal infections. Also, other works observed that pre-exposure to a metal can enhance tolerance to other metals. For instance, a pre-exposure to Hg enhanced Cd accumulation in *Euglena gracilis* (Avilés et al. 2003). Metal stress can also induce alterations in ROS accumulation and glutathione pools, two important signaling mediators in many abiotic stresses, such as salt, osmotic, and temperature (Clark et al. 2001; Maksymiec 2007). Collectively, these data indicate the existence of a signal transduction pathway underlying metal tolerance that shares at least some signaling components with other biotic and abiotic stresses. But how they can be related to phytochelatin induction remains to be elucidated.

10.1.5   Possible Points in PC Signaling Regulation

The signaling pathways involved in phytochelatins synthesis are mediated by enzymes. Some reports showed that increasing the activity of the enzymes involved in the GSH pathway (γ-glutamylcysteine synthetase and glutathione synthetase) have increased PC synthesis and enhanced Cd tolerance. In fact, increasing both γ-glu-tamylcysteine synthetase (γ-ECS) and glutathione synthetase (GSHS) activity enhances Cd tolerance and PC synthesis (Noctor and Foyer 1998; Schafer et al. 1998; Xiang and Oliver 1998). On the other hand, specific signaling components have already been identified as regulators of Cd stress. He et al. (2005) observed that calcium can play an important role in Cd tolerance, reducing the toxic effects of this metal, by directly affecting PC synthase. Protein phosphorylation, carried out mostly through MAPK kinases, is also emerging as an important mediator in Cd signaling

(Nakagami et al. 2004; Rios-Barrera et al. 2009). There are also reports suggesting that ROS production is involved in GSH synthesis and calcium signaling (Xing et al. 1997; Grant et al. 2000; Yang and Poovaiah 2002), and hence, ROS may be related with PC synthesis. These findings have provided new possibilities to understand the signaling pathway involved in PC synthesis and Cd tolerance, which, however, requires further studies to identify possible Cd-induced signaling pathways.

10.2   Calcium Signaling in Cadmium Stress

10.2.1   The Role of Calcium

Stress-induced changes in the cytosolic concentration of calcium (Ca) occur as a result of influx of Ca^{2+} from outside the cell, or release of Ca from intracellular stores (Hong-Bo et al. 2008). The calcium alterations then target specific proteins that act as calcium sensors and carry on the signal to other molecules, such as enzymes or transcription factors, which are the tolerance response *per se* (Luan 2004; Hedrich and Kudla 2006). In this way, calcium serves as an important second messenger during abiotic stresses and is a major point of signaling and cross talk, because it can be elicited by numerous stress cues, being particularly important in osmotic, heat, salt, and water stress (Cheong et al. 2003; Rentel and Knight 2004; Hong-Bo et al. 2008). In the last few years, calcium signaling has been better explained by the observation of calcium oscillations across the cell, referred to as the calcium signatures (Luan et al. 2002). According to some authors, each signature represents a cellular expression that confers specificity to calcium signals. In order to fully understand the Ca signaling pathways, one must understand the "combination code" that consists of calcium oscillations, calcium sensors, and downstream target proteins (Clark et al. 2001). Until recently, little was known about the *in vivo* targets and the downstream outputs of stress signaling pathways, but some calcium sensor proteins have now been identified and are well characterized in plants. Few of them are well-conserved proteins, known in animal tissues; others are novel calcium sensors that exist only in the plant kingdom (Kim et al. 2000). Calcium signaling serves as an important second messenger during abiotic stresses and provides a major point of signaling cross talk. However, very few reports highlight the relation between metal stress and Ca signals. Although there is lack of information on Cd stress signaling and calcium, few reports have shown that calcium is important for Cd tolerance.

10.2.2   Calcium Influences Cadmium Uptake
but Also PC Synthesis

In plants, two kinds of calcium stores contribute to calcium modulation: extracellular (apoplastic) stores in the cell wall and intracellular stores in the vacuole and

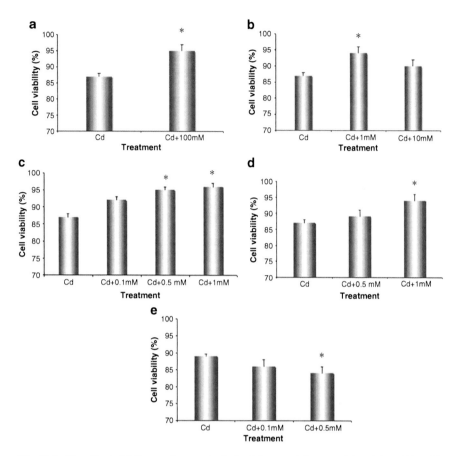

Fig. 10.1 The effect of different pharmacological calcium modulators: calcium (**a**), caffeine (**b**), EGTA (**c**), lanthanum (**d**), and ruthenium red (**e**) on *A. thaliana* Cd tolerance, during a 2-h exposure period. Cell viability is expressed as a percentage of controls. Values are the mean of three replicates ± SE. Values significantly ($P < 0.05$) different from controls are marked with asterisk

endoplasmic reticulum (Bush 1995). In our study, Cd-induced stress was combined with chemical agents known to modulate cellular Ca concentration, such as caffeine, which causes the release and depletion of Ca from internal stores, lanthanum (La), a Ca channel blocker, EGTA (ethylene glycol–bis(2-qminoethylether)-*N*,*N*,*N'*,*N'*–tetra acetic acid) an extracellular Ca chelator, and ruthenium red, an inhibitor of Ca release from internal stores (Shimazaki et al. 1999; Cessna and Low 2001; Navazio et al. 2001; Taylor et al. 2001). The effect of calcium modulation on Cd tolerance was analyzed, as presented in Fig. 10.1. In the presence of Cd, the addition of Ca (Fig. 10.1a) or caffeine (Fig. 10.1b) significantly increased Cd tolerance, when compared to Cd application alone. The calcium blockers EGTA (Fig. 10.1c) and La (Fig. 10.1d) induced similar responses to calcium induction and increased cell viability, when compared to Cd alone. This increment in Cd tolerance

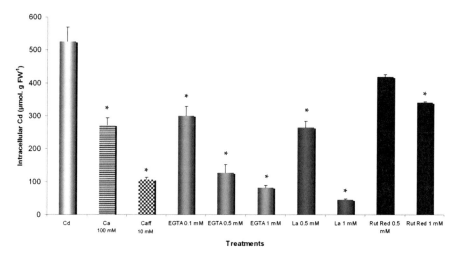

Fig. 10.2 The effect of Ca and Caffeine, and the Ca blockers EGTA, La, and Rut Red on *A. thaliana* Cd absorption after 1 h of 100-µM Cd exposure. The cells were incubated with each treatment 1 h before Cd. Values are the mean of three replicates ± SE. Values significantly ($P < 0.05$) different from controls are marked with asterisk

was higher with increasing EGTA and La concentrations. On the contrary, the addition of ruthenium red under Cd exposure significantly reduced the cell viability when compared to Cd control. Calcium influence on Cd absorption was also investigated (Fig. 10.2). The different calcium modulators altered very significantly the amount of Cd absorbed by cells. After application of Ca and caffeine, Cd uptake was markedly reduced, particularly with caffeine. With EGTA and La, Cd absorption was also reduced relative to control, and a dose-dependent response was observed. Although ruthenium red at the highest concentration significantly reduced Cd absorption, the effect on Cd uptake was less pronounced.

These pharmacological tests suggest that Ca levels, either by intracellular release or by extracellular sources, induce Cd tolerance, possibly by reducing Cd uptake. Previous works have already demonstrated that Cd can compete with Ca for membrane transporters and also for Ca-binding proteins (Rivetta et al. 1997), suggesting that calcium can reduce Cd absorption. Perfus-Barbeoch et al. (2002), in patch-clamp studies with *V. faba* guard cell protoplasts, showed that Ca channels were permeable to Cd. The observed Cd uptake inhibition by Ca channel blockers can be associated to the alleviation of Cd toxicity, observed in radish (Rivetta et al. 1997), rice roots (Kim et al. 2002), and *Arabidopsis* seedlings (Suzuki 2005). Nonetheless, the inhibition of Ca release from intracellular stores by ruthenium red application showed that Ca is indeed important for Cd tolerance. Because ruthenium red blocks Ca release from intracellular stores, it did not interfere as much with extracellular Cd absorption as did with the other Ca blockers. The combined application of ruthenium red and cadmium effectively reduced the cell viability, compared to the sole application of Cd, corroborating that Ca is indeed important for Cd tolerance, but not only by reducing Cd uptake.

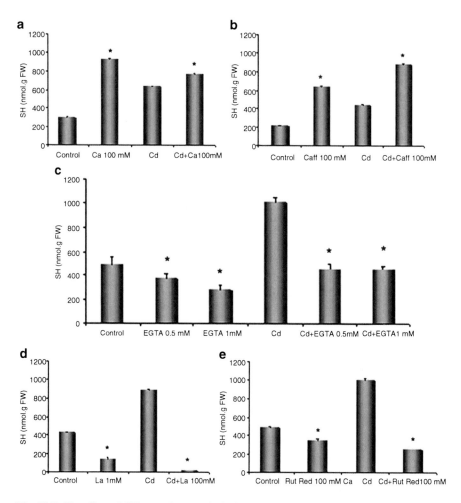

Fig. 10.3 The effect of different pharmacological calcium modulators: calcium (**a**), caffeine (**b**), EGTA (**c**), lanthanum (**d**), and ruthenium red (**e**) on *A. thaliana* GSH synthesis after 1 h of 100-μM Cd exposure. The cells were incubated with the modulators 1 h before Cd exposure. Values are the mean of three replicates ± SE. Values significantly (*P* < 0.05) different from controls are marked with asterisk

Ca modulation had no effect on PC synthesis (data not shown) but had a strong influence on GSH production under the different Ca modulations and Cd exposures (Fig. 10.3). In the absence of Cd, Ca addition to the growth media increased GSH production by threefolds (Fig. 10.3a). In the presence of Cd, Ca addition also significantly increased GSH levels (Fig. 10.3a). The effect of Ca release from intracellular stores stimulated by caffeine (Fig. 10.3b) induced similar effects to those observed with the addition of Ca. In the absence of Cd, caffeine promoted a significant increase in GSH concentrations, nearly twofold increase over control levels. Under Cd exposure, caffeine also increased GSH, which was even more prominent than with Cd alone. The calcium inhibitors in general reduced the GSH production. Increasing EGTA concentrations reduced the amount of GSH in the cells (Fig. 10.3c). This trend was consistent with a scenario where Ca is important for GSH synthesis. When Ca channels were blocked with lanthanum (Fig. 10.3d), both GSH levels and Cd absorption were also reduced. The La-induced reduction in Cd absorption was more increased with higher La concentrations, corroborating that Ca channels are important for Cd absorption in cells. Similar results were obtained with ruthenium red (Fig. 10.3e), with a reduction in GSH levels after Ca blocking, particularly when ruthenium red and cadmium were simultaneously provided. Overall, results presented here introduced novel findings on the role of Ca in Cd stress. Firstly, they corroborate the notion that Cd uptake in cells can occur through Ca channels. This knowledge can be used for modulating the levels of Cd uptake in plants. Increasing Ca levels in the soils can ultimately be used to allow plants to grow under contaminated environments with lower Cd accumulation and thus less physiological disturbances, metal accumulation in the edible tissues, and metal transfer to the food chain. Secondly, Ca can have an important modulator effect on GSH synthesis. Since GSH is an important metabolite, especially in the case of metal exposures because it is the precursor of PCs, GSH regulation through Ca exposures can be used to enhance plant tolerance to Cd, which can be important for phytoremediation studies.

10.3 Protein Phosphorylation Signaling in Metal Stress

10.3.1 The Role of Protein Phosphatases

The reversible phosphorylation of proteins regulates many aspects of viable cell and is also an important part of signal transduction. It is well recognized that protein phosphorylation/dephosphorylation is a key step in most stress signal transduction pathways, which is mediated both by kinases and phosphatases. Protein phosphatases, found in all eukaryotes, play an important role in signaling processes. Although little is known about protein phosphatases in higher plants, evidence on their functional involvement is overwhelming. Using substrates for mammalian phosphatases and pharmacological agents (e.g., okadaic acid), protein phosphatases such as PP1 and PP2A have been detected in several plant species (Mackintosh et al. 1991; Luan et al. 2002). In addition to substrate specificity and pharmacological properties, the primary structure of plant phosphatases is also highly similar to that of the mammalian enzymes. Nevertheless, different structural domains and unique functions have been identified through studies on plant enzymes. In recent years, a great deal of research has been concentrated on identifying genes and elucidating signal transduction pathways involved in the plant response to abiotic stresses, mostly salt, osmotic, temperature, wounding, and pathogen attack. Although one family of plant PPs, the PP2C, has been extensively studied in plant responses to stress (Luan 2004), the roles of PP1 and PP2A in abiotic stress signaling have been largely neglected. These PPPs are the most ubiquitous protein phosphatases in eukaryotes and have been identified as important signal transducers of several metabolic pathways. Our knowledge of PP1 and PP2A functions in higher plants comes mostly from studies using specific inhibitors such as okadaic acid and others (Mackintosh and Cohen 1989 Mackintosh et al. 1991). Because it is difficult to distinguish PP1 from PP2A by using pharmacological inhibitors, most studies defined the inhibitor-sensitive process as involving PP1/PP2A. Generally, inhibitor analyses have indicated that PP1/2A is involved in ion channel regulation, gene expression, and developmental processes. Also, expression of cold responsive genes is enhanced at normal temperature by the protein phosphatase 1 and 2A inhibitor (Monroy et al. 1997, 1998). These inhibitors also caused an increase in freezing tolerance at normal growth temperature (Sangwan et al. 2001). Moreover, low temperature caused a rapid and dramatic decrease in protein phosphatase 2A (PP2A) activity, which is dependent on Ca^{2+} influx. Studies with an unspecific inhibitor of PP1 and PP2A also suggest a role for PP1 and PP2A in regulation of ABA signaling and expression of ABA and cold-responsive genes in *Arabidopsis* (Wu et al. 1997).

Since PP1 and PP2A have been related to abiotic stress signaling, it is likely that they could also be associated with metal signaling. Nonetheless, the only reports focusing on protein phosphorylation and Cd have focused only on protein kinases and not phosphatases. Few recent reports on the importance of protein phosphorylation in metal stress have increased the interest in this area. The work of Jonak et al. (2004) showed that several MAP kinases (mitogen activated kinases) are induced upon metal signal transduction, and that they are differently involved with different metals (Nakagami 2004). Rios-Barrera et al. (2009) also observed that in *Euglena gracilis,* the inhibition of MAPKs induced a reduction in PC synthesis after Cd exposure. However, the pathway suggested by these authors was activated by $CuCl_2$ but not by $CdCl_2$. This fact allows inferring that the diversity of signal transducers might be of greater importance for signal specificity in exposures to different metals. Since plant PPs are much fewer in number, their importance shall be much easier to investigate than kinases. Therefore, if protein phosphorylation can regulate PC synthase, then a valid approach would be to test which plant PP is involved.

10.3.2 Protein Phosphatase Inhibition Increases Cadmium Tolerance by Enhancing PC Synthesis

The availability of permeant cell membrane inhibitors like okadaic acid (Tachibana et al. 1981) and cantharidin (Honkanen 1993) has facilitated the study on the functional role of PP1 and PP2A. Using these inhibitors, PPs have gained interest as potentially important regulators of cellular function. In this work, we used cantharidin, a well-known PP1 and PP2A inhibitor in order to assess the importance of

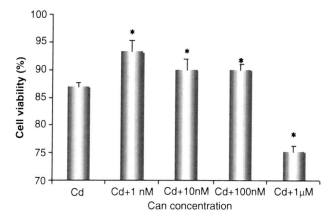

Fig. 10.4 The effect of PP1/PP2A inhibitor cantharidin on Cd tolerance of *Arabidopsis* cells exposed for 1 h to 1 nM, 10 nM, 100 nM, and 1 μM of cantharidin (Can). Can exposures were performed 1 h prior to exposure to 100-μM Cd. Values are the mean of five replicates ± SE. Values significantly (P < 0.05) different from controls are marked with asterisk

protein phosphorylation in PC synthesis and Cd tolerance. It was observed that the inhibition of PP1/PP2A enhanced Cd tolerance, while inducing an increase in Cd uptake. The influence of cantharidin (Can) on Cd tolerance was observed, after 1-h exposure (Fig. 10.4). The lower concentrations of cantharidin significantly increased Cd tolerance while the highest Can concentration (1 μM) reduced the number of viable cells, which was lower than cells obtained even with Cd alone. The highest reduction in the percentage of viable cells, even without the addition of Cd (data not shown), made clear that the highest Can concentration used is very toxic to cells, and for this reason it was excluded from the rest of the experiments. Cantharidin exposure increased Cd absorption in a dose–response manner. In the presence of Cd, exposure to cantharidin significantly increased PCs levels, also in a dose-dependent manner. This increase was accompanied by a high depletion of the GSH pools (data not shown). Furthermore, data show that the observed increase in PC production is, at least partially, responsible for the increase in Cd tolerance despite an increase in intracellular Cd levels. These results are significant, since they suggest that PC synthesis can be regulated by protein phosphorylation and that this regulation can be used to modulate Cd tolerance, even in the presence of elevated intracellular Cd concentrations. More importantly, PP1, PP2A, or both can block PC synthesis. It has been widely accepted that phytochelatin synthase (PCs) is a constitutive enzyme that may be controlled by post-translational modifications (Vatamaniuk et al. 2004). Thus, there should be an efficient mechanism that blocks PCS activity during normal conditions and that can be switched off in the presence of metals such as Cd. Results presented suggest that PP1/PP2A, or both, would be responsible for maintaining PCS inactivity under low intracellular concentration of metal(s). A model for Cd-induced signaling pathways involved in the regulation of PC is presented in Fig. 10.5.

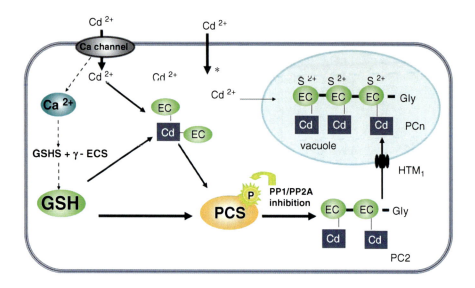

Fig. 10.5 A model for Cd-induced signaling pathways. The build-up of GS–Cd conjugates activates PCS that suffers a phosphorylation during activation, after PP1/PP2A inactivation. The increase in PCs yields a high Cd binding and GSH depletion. During PC synthesis, PC–Cd complexes are formed in the cytosol and are transported to the vacuole, where more Cd ions and sulfide are added, forming more stable complexes. Solid arrows represent events previously describe in other works, whereas dashed arrows represent proposed events in the signaling of PC formation in plant cells under Cd stress

If PPs are blocking the PC synthesis, then in order to trigger their production under Cd stress, a specific pathway with a specific kinase is expected. So far, the very few reports on protein phosphorylation involvement in metal stress point toward the involvement of MAP kinases. Further, exposure of *Medicago sativa* seedlings to excess copper or cadmium ions was shown to activate four distinct mitogen-activated protein kinases (MAPKs.). Rios-Barrera et al. (2009) also found that p38 MAPK-like activity was stimulated by acute or chronic metal exposure, and its inhibition by a p38 MAPK inhibitor slightly diminished the accumulation of PCs. However, further work is required in order to understand how protein phosphorylation affects Cd tolerance, since it is not clear which step of the PC synthesis requires phosphorylation to induce PC synthesis. Nevertheless, Wang et al. (2009) reported the importance of serine threonine phosphorylation in PC synthesis. These authors showed that PCS activity was increased after phosphorylation by casein kinase 2 (CK2) and decreased in the presence of alkaline phosphatase. Taken together, our results and those reported by others (Wang et al. 2009) contribute to elucidate the route leading to the activation of PC synthesis. These findings are of extreme importance for the understanding of the PCS activation because they demonstrate that PCS is not only triggered by GSH conjugates and by Cd ions, but its activity is also regulated by protein phosphorylation.

10.4   Conclusion

Metal tolerance in plants is achieved by phytochelatins. The ability to induce PCs and effectively chelate metal ions is a recognized key factor for plant survival in metal-contaminated soils and is pointed out as the basis for the inter and intraspecific differences in plant tolerance to metals. Taken together, most important alterations and signaling events found evoke that (1) Cd can be taken up by calcium channels and be regulated by calcium levels, (2) calcium can be an important positive regulator of GSH synthesis, and (3) protein phosphatases PP1 or PP2A can regulate PCS activity. Understanding the signaling mechanisms underlying metal stress, therefore, can be very important not only to explain the tolerance mechanisms, but to find out new possibilities to regulate metal uptake, translocation, and tolerance in plants, without genetic modification of plants. The genetic engineering of plants for specific traits is, however, controversial due to its negative impact on surrounding environment and poor acceptance by the public opinion. The modulation of Cd uptake can result in an efficient way to reduce Cd absorption in plants and may lead to higher crop productivity, which can be achieved by maintaining the soil Ca levels higher, for example, through Ca amendments, already widely used in agronomical practices. This could decrease the accumulation of metals in plants, and, thus, is likely to reduce the health risk to humans, consuming crops grown under Cd-affected soils.

References

Avilés C, Loza-Tavera H, Terry R, Moreno-Sánchez N (2003) Mercury pretreatment selects an enhanced cadmium-accumulating phenotype in *Euglena gracilis*. Arch Microbiol 180:1–10

Baker AJM, Walker PL (1989) Review: physiological response of plants to heavy metal and the quantification of tolerance and toxicity. Sci Technol Lett 1:7–17

Bush DS (1995) Calcium regulation in plant cells and its role in signaling. Annu Rev Plant Physiol Plant Mol Biol 46:95–122

Cessna SG, Low PS (2001) An apoplastic Ca^{2+} sensor regulates internal Ca^{2+} release in aequorin-transformed tobacco cells. J Biol Chem 276:10655–10662

Cheong YH, Kim K-N, Pandey GK, Gupta R, Grant JJ, Luan S (2003) CBL1, a calcium sensor that differentially regulates salt, drought, and cold responses in *Arabidopsis*. Plant Cell 5:1833–1845

Clark GB, JrG T, Roux SJ (2001) Signal transduction mechanisms in plants: an overview. Curr Sci 80:170–177

Clemens S (2006) Toxic metal accumulation, responses to exposure and mechanisms of tolerance in plants. Biochemistry 88:1707–1719

Cobbett CS (2000) Phytochelatins and their roles in heavy metal detoxification. Plant Physiol 123:825–832

Cobbett C, Goldsbrough P (2002) Phytochelatin and metallothioneins: roles in heavy metal detoxifixcation and homeostasis. Annu Rev Plant Biol 53:159–182

Fojtová M, Kovařík A (2000) Genotoxic effect of cadmium is associated with apoptotic changes in tobacco cells. Plant Cell Environ 23:531–537

Ghoshroy S, Freedman K, Lartey R, Citovsky V (1998) Inhibition of plant viral systemic infection by non-toxic concentration of cadmium. Plant J 13:591–602

Grant JJ, Yun B-W, Loake GJ (2000) Oxidative burst and cognate redox signaling reported by luciferase imaging: identification of a signal network that functions independently of ethylene, SA and Me-JA but is dependent on MAPKK activity. Plant J 24:569–582

Hall JL (2002) Cellular mechanisms for heavy metal detoxification and tolerance. J Exp Bot 53:1–11

Hare PD, Cress WA, Van Staden J (1997) The involvement of cytokines in plant responses to environmental stress. Plant Growth Regul 23:79–103

He Z, Li J, Zhang MM, Mi M (2005) Different effects of calcium and lanthanum on the expression of phytochelatin synthase gene and cadmium absorption in *Lactuca sativa*. Plant Sci 168:309 318

Hedrich R, Kudla J (2006) Calcium signaling networks channel plant K^+. Cell 125:1221–1223

Hong-Bo S, Li-Ye C, Ming S (2008) Calcium as a versatile plant signal transducer under soil water stress. BioEssays 30:634–641

Honkanen RE (1993) Cantharidin, another natural toxin that inhibits the activity of serine/threonine protein phosphatases types 1 and 2A. FEBS Lett 330:283–286

Jonak C, Nakagami H, Heribert H (2004) Heavy metal stress. Activation of distinct mitogenactivated protein kinase pathways by copper and cadmium. Plant Physiol 136:3276–3283

Kaur N, Gupta AK (2005) Signal transduction pathways under abiotic stresses in plants. Curr Sci 88:1771–1780

Kim H-G, Park K-N, Cho Y-W, Park E-H, Fuchs JA, Lim C-J (2002) Characterization and regulation of glutathione Stransferase gene from Schizosaccharomyces pombe. Biochim Biophys Acta 1520:179–185

Kim K-N, Cheong HY, Gupta R, Luan S (2000) Interaction specificity of Arabidopsis calcineurin B-like calcium sensors and their target kinases. Plant Physiol 124:1844–1853

Knight H (2000) Calcium signaling during abiotic stress in plants. Int Rev Cytol 195:269–325

Kudla J, Xu Q, Harter K, Gruissem W, Luan S (1999) Genes for calcineurin B-like proteins in *Arabidopsis* are differentially regulated by stress signals. Proc Natl Acad Sci USA 96:718–723

Luan S (2004) Protein phosphatases and signaling cascades in higher plants. Plant Physiol 129:908–925

Luan S, Kudla G, Rodriguez-Concepcion M, Yalovsky S, Gruissem W (2002) Calmodulins and calcineurin B–like proteins: calcium sensors for specific signal response coupling in plants. Plant Cell 14:S389–S400

Mackintosh C, Cohen P (1989) Identification of high levels of type 1 and type 2A protein phosphatases in higher plants. Biochem J 262:335–340

Mackintosh C, Coggins J, Cohen P (1991) Plant protein phosphatases: subcellular distribution, detection of protein phosphatase 2 C and identification of protein phosphatases 2A as the major quinate dehydrogenase phosphatase. Biochem J 273:733–738

Maksymiec W (2007) Signaling responses in plants to heavy metal stress. Acta Physiol Plant 29:177–187

Mithöfer A, Schulze B, Boland W (2004) Biotic and heavy metal stress response in plants: evidence for common signals. FEBS Lett 566:1–5

Mithöfer A, Schulze B, Boland W (2005) Biotic and heavy metal stress response in plants: evidence for common signals. FEBS Lett 566:1–5

Mittra B, Ghosh P, Henry SL, Mishra J, Das TK, Ghosh S, Babu CR, Mohanty P (2004) Novel mode of resistance to *Fusarium* infection by a mild dose pre-exposure of cadmium in wheat. Plant Physiol Biochem 42:781–787

Monroy AF, Labbe E, Dhindsa RS (1997) Low temperature perception in plants: effects of cold on protein phosphorylation in cell-free extracts. FEBS Lett 410:206–209

Monroy AF, Sangwan V, Dhindsa RS (1998) Low temperature signal transduction during cold acclimation: protein phosphatase 2A as an early target for cold-inactivation. Plant J 13:653–660

Nakagami N, Pitzschke A, Hirt H (2004) Emerging MAP kinase pathways in plant stress signaling. Trends Plant Sci 10(7):339–346

Navazio L, Mariani P, Sanders D (2001) Mobilization of Ca^{2+} by cyclic ADP-ribose from the endoplasmic reticulum of cauliflower. Plant Physiol 125:2129–2138

Noctor G, Foyer CH (1998) Simultaneous measurement of foliar glutathione, glutamylcysteine and amino acids by high-performance liquid chromatography: comparison with two other assay methods for glutathione. Anal Biochem 264:108–110

Ortiz DF, Ruscitti T, McCue KF, Ow DW (1995) Transport of metal-binding peptides by HMT1, a fission yeast ABC-Type vacuolar membrane protein. J Biol Chem 270:4721–4728

Perfus-Barbeoch N, Leonhardt N, Vavasseur A, Forestier C (2002) Heavy metal toxicity: cadmium permeates through calcium channels and disturbs the plant water status. Plant J 32:539–548

Pinto ETCS, Sigaud-Kutner MAS, Leitao OK, Okamoto D, Morse PC (2003) Heavy metal induced oxidative stress in algae. J Phycol 39:1008–1018

Rauser WE (1999) Structure and function of metal chelators produced by plants. The case of amino acids, organic acids, phytin and methalothioneins. Cell Biochem Biophys 31:1–31

Rauser WE (2000) Roots of maize seedlings retain most of their cadmium through two complexes. J Plant Physiol 156:545–551

Rentel MC, Knight MR (2004) Oxidative stress-induced calcium signaling in Arabidopsis. Plant Physiol 135:1471–1479

Rios-Barrera D, Vega-Segura A, Thibert V, Rodríguez-Zavala JS, Torres-Marquez ME (2009) p38 MAPK as a signal transduction component of heavy metals stress in *Euglena gracilis*. Arch Microbiol 191:47–54

Rivetta A, Negrini N, Cocucci M (1997) Involvement of Ca^{2+} calmodulin in Cd^{2+} toxicity during the early phases of radish (*Raphanus sativus* L.) seed germination. Plant Cell Environ 20:600–608

Sangwan V, Foulds I, Singh J, Dhindsa RS (2001) Cold-activation of *Brassica napus* BN115 promoter is mediated by structural changes in membranes and cytoskeleton, and requires Ca^{2+} influx. Plant J 27:1–12

Sanitá di Toppi L, Gabrielli R (1999) Response to cadmium in higher plants. Environ Exp Bot 41:105–130

Schafer HJ, Haag-Kerwer A, Rausch T (1998) cDNA cloning and expression analysis of genes encoding GSH synthesis in roots of the heavy-metal accumulator *Brassica juncea* L.: evidence for Cd induction of a putative mitochondrial γ-glutamylcysteine synthetase isoform. Plant Mol Biol 37:87–97

Schaller F, Weiler EW (2002) Wound-and mechanical signalling. In: Scheel D, Wasternack C (eds.) Plant signal transduction. Oxford University Press, Oxford, pp 20–44

Shimazakiet K, Goh C-H, Kinoshita T (1999) Involvement of intracellular Ca^{2+} in blue light-dependent proton pumping in guard cell protoplasts from *Vicia faba*. Physiol Plant 105:554–561

Suzuki N (2005) Alleviation by calcium of cadmium-induced root growth inhibition in Arabidopsis seedling. Plant Biotechnol 22:19–25

Tachibana K, Scheuer PJ, Tsukitani Y, Kikuchi H, Van Engen D, Clardy J, Gopichand Y, Schmitz J (1981) Okadaic acid, a cytotoxic polyether from two marine sponges of the genus *Halichondria*. J Am Chem Soc 103:2469–2471

Taylor ATS, Kim J, Low PS (2001) Involvement of mitogen activated protein kinase activation in the signal-transduction pathways of the soya bean oxidative burst. Biochem J 355:795–803

Trewavas A (1999) How plants learn. Proc Natl Acad Sci 96:4216–4218

Vatamaniuk OK, Mari S, Lu YP, Rea PA (2004) Mechanism of heavy metal ion activation of phytochelatin (PC) synthase: blocked thiols are sufficient for PC synthase-catalyzed transpeptidation of glutathione and related thiol peptides. J Biol Chem 275:31451–31459

Vögeli-Lange R, Wagner GJ (1990) Subcellular localization of cadmium and cadmium-binding peptides in tobacco leaves. Plant Physiol 92:1086–1093

Wagner GJ (1993) Accumulation of cadmium in crop plants and it consequences to human health. Adv Agron 51:173–212

Wang HS, Wu JS, Chia JC, Wu YJ, Juang RH (2009) Phytochelatin synthase is regulated by protein phosphorylation at a threonine residue near its catalytic site. Agric Food Chem 57:7348–7355

Wu Y, Kuzma J, Maréchal E, Graeff R, Lee HC, Foster R, Chua NH (1997) Abscisic acid signaling through cyclic ADP-ribose in plants. Science 278:2126–2130

Xiang C, Oliver DJ (1998) Glutathione metabolic genes coordinately respond to heavy metals and jasmonic acid in Arabidopsis. Plant Cell 10:1539–1550

Xing T, Higgins VVJ, Blumwald E (1997) Race-specific elicitors of *Cladosporium fulvum* promote translocation of cytosolic components of NADPH oxidase to the plasma membrane of tomato cells. Plant Cell 9:249–259

Xiong L, Lee H, Ishitani M, Zhu JK (2002) Regulation of osmotic stress responsive gene expression by the LOS6/ABA1 locus in Arabidopsis. J Biol Chem 277:8586–8596

Yang T, Poovaiah BW (2002) Hydrogen peroxide homeostasis: activation of plant catalase by calcium/calmodulin. Proc Natl Acad Sci USA 9:4097–4102

Zenk MH (1996) Heavy metal detoxification in higher plants-a review. Gene 179:21–30

Chapter 11
Microbial Management of Cadmium and Arsenic Metal Contaminants in Soil

Bhoomika Saluja, Abhishek Gupta, and Reeta Goel

Abstract Contamination of soil with heavy metals poses a major environmental and human health problem. Of the various metals, cadmium and arsenic are the two well-known heavy metals. The toxic effects of these metals are due to their abundance while nonbiodegradable nature leads to their concentration buildup in soil. Microbial methods of environment purification and cleanup are promising because of the safety, efficiency, and cost effectiveness. A number of microorganisms including members of Archea, Eukarya, and Bacteria are resistant to cadmium and arsenic and have evolved several defense mechanisms to overcome metal toxicities. The bioremediation of cadmium- and arsenic-contaminated soil involves active microbiological processes, such as biosorption, bioaccumulation, sequestration, and efflux. Furthermore, knowledge of metal ion resistances could provide important insights into environmental processes and help in understanding the basic living processes.

Keywords Arsenic • Cadmium • Bioremediation • Health hazards • Soil contamination

11.1 Introduction

The term "heavy metal" refers to any metallic element that has a relatively high density and is toxic or poisonous at low concentrations. Examples of heavy metals include mercury (Hg), cadmium (Cd), arsenic (As), chromium (Cr), thallium (Tl), and lead (Pb). Some of the heavy metals are essential and are required by the

B. Saluja • A. Gupta • R. Goel (✉)
Department of Microbiology, Govind Ballabh Pant University of Agriculture and Technology,
Pantnagar 263145, Uttarakhand, India
e-mail: rg55@rediffmail.com

M.S. Khan et al. (eds.), *Biomanagement of Metal-Contaminated Soils*,
Environmental Pollution 20, DOI 10.1007/978-94-007-1914-9_11,
© Springer Science+Business Media B.V. 2011

organisms as micro nutrients (e.g., Co, Cr. Ni, Fe, and Zn) and are known as "trace elements" (Bruins et al. 2000). They are involved in redox processes to stabilize molecules through electrostatic interactions, as catalysts in enzymatic reactions, and regulation of osmotic balance (Hussein et al. 2005). On the other hand, some other heavy metals like Cd, Hg, As, Pb, etc., have no biological function and are detrimental to the organisms even at very low concentration They originate from natural sources such as rocks and metalliferous minerals, and anthropogenic inputs from agriculture, metallurgy, energy production, microelectronics, mining, sewage sludge, and waste disposal (Landa 2005; Gilmour and Riedel 2009; Pandey and Pandey 2009a). Soil contamination by heavy metals occurs when the concentration of these elements exceeds the background level in the substratum. A concentration higher than the prescribed limit may lead to the formation of nonspecific complex compounds in the cell, which leads to toxic effects. These atmospherically driven heavy metals have been shown to significantly contaminate soil and vegetables causing a serious risk to human health when plant-based foodstuffs are consumed (Voutsa et al. 1996; Pandey and Pandey 2009b, c).

Microbes have a variety of properties that can bring about changes in metal speciation, toxicity, and mobility. They are intimately associated with the biogeochemical cycling of metals, and associated elements, wherein their activities can result in mobilization and immobilization of metals depending on the mechanism involved and the microenvironment where the organism(s) are located (Gadd 2004, 2007, 2009; Violante et al. 2008; Ehrlich and Newman 2009). The contribution of microbial activities to rock weathering, mineral dissolution, and element cycling is also intimately related to metal movements and microbial strategies for metal transformations (Purvis and Pawlik-Skowronska 2008; Gilmour and Riedel 2009; Uroz et al. 2009). Many microorganisms can absorb and concentrate heavy metals, thereby providing resistance (Burke and Pfister 1986), and, thus, help in removing them from contaminated sites (Roane et al. 2001). The mechanism of heavy metal resistance and its genetic basis, however, varies with the microbe and the metal in question. Therefore, understanding the role of microorganisms in cycling of metals may lead to improved processes employed to detoxify contaminated sites. This chapter deals with the two well-known toxic heavy metals "cadmium and arsenic," their effect on plants and animals, the mechanism behind microbial resistance to these metals, and microbial removal of these metals from contaminated soil.

11.2 Sources of Cadmium and Arsenic in Soil

11.2.1 Cadmium

Cadmium, a highly toxic metal, has been ranked seventh among the top 20 toxins, mainly due to its negative influence on enzymatic system of cell (Al-Kheldhairy et al. 2001). Cadmium can mainly be found in the earth's crust. It always occurs in combination with zinc. Cadmium also exists in industries as an inevitable by-product of zinc, lead, and copper extraction. This metal enters the environment mainly from industrial processes and fertilizers and is transferred to animals and humans through food chain (Wagner 1993). Anthropogenic activities such as industrial waste disposals, fertilizer application, and sewage sludge disposals on land have also led to accumulation of cadmium in soil. The leaching of Cd under certain soil and environmental conditions (Alloway 1990; Naidu et al. 1997) eventually increases its concentration in food crops. The concentration of Cd in soil solution varies significantly with soil properties and nature of management practices. Naturally, a very large amount of Cd is released into the environment, about 25,000 ton a year. About half of this Cd is released into rivers through weathering of rocks and some of it is released into air through forest fires and volcanoes while the rest is released through human activities. Man-made Cd emissions arise from the manufacture, use, and disposal of products intentionally utilizing Cd (e.g., nickel–Cd batteries, Cd alloys, etc.) or from the presence of Cd as a natural but nonfunctional impurity in non-Cd containing products (e.g., fossil fuel, cement, phosphate fertilizers, etc).

11.2.2 Arsenic

Although As has almost exclusively been associated with criminal poisoning for many centuries (Rusyniak et al. 2002), the matter of concern today is its contribution to environmental pollution through man's use of As containing insecticides, herbicides, fungicides, pesticides, wood preservatives, and through mining and burning of coal (Leonard 1991). Thus, anthropogenic use makes As a common inorganic toxicant found at contaminated sites nationwide. Furthermore, mining activities and widespread use of As in the wood preserving industry and in agriculture as a pesticide and herbicide represent a major source of As in the environment (Fig. 11.1).

The common valence states of As in nature include -3, 0, $+3$, and $+5$ (Leonard 1991; Jain and Ali 2000; Oremland et al. 2000). In soils, the most commonly found As forms are inorganic As(III) (arsenite) and As(V) (arsenate) (Cullen and Reimer 1989; Masscheleyn et al. 1991; Pantsar-Kallio and Korpela 2000; Balasoiu et al. 2001). In general, the toxicity of As is dependent on its oxidation state: trivalent As forms are approximately 100 times more toxic than the pentavalent derivatives (Cervantes et al. 1994; Mukhopadhyay et al. 2002; Muller et al. 2003). Methylated species, monomethyl arsenic acid (MMAA), dimethyl arsinic acid (DMAA), and trimethyl arsine oxide (TMAO) have predominantly been found in biomass and have also been detected in soil (Leonard 1991).

It is evident from literature that As(V) functions as an analogue of phosphate (PO_4) and enters the cell through phosphate transport system (*Pit* or *Pst*), short circuiting the life's main energy generation system by inhibiting oxidative phosphorylation. The stable PO_4 anion is replaced with the less stable As(V) anion leading to rapid hydrolysis of high energy bonds in compounds such as ATP, a process that

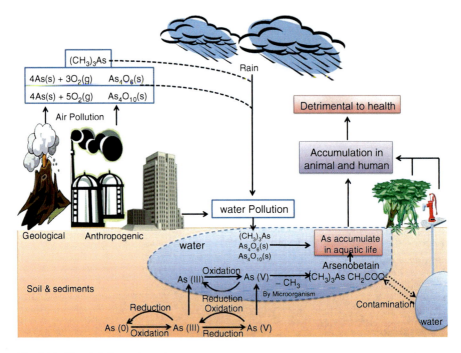

Fig. 11.1 The global arsenic cycle

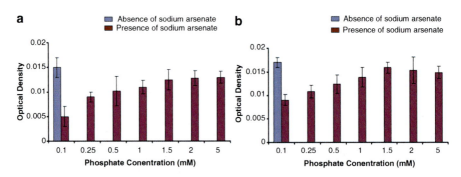

Fig. 11.2 Effects of various phosphate concentrations on the growth of (**a**) *B. cereus* strain, AG27, and (**b**) unidentified strain AGM13 in the presence of 5 mM sodium arsenate compared to growth occurring in the absence of sodium arsenate (Adapted from Gupta 2006)

leads to loss of high energy PO_4 bonds and effectively "uncouples" mitochondrial respiration (Rosenman 2007). Therefore, the effect of PO_4 ions in the medium on As-induced growth/toxicity by varying the PO_4 concentration was documented (Gupta 2006).

For revealing the effect of PO_4-As(V) interaction, the cells of two arsenic resistant bacterial strains, namely, *Bacillus cereus* strain AG27 (AY970345) and AGM13 (unidentified) were transferred to minimal medium containing varying concentrations (0.1, 0.25, 0.5, 1.0, 1.5, 2.0, and 5 mM) of PO_4 along with 5 mM sodium arsenate. The growth of the bacterial cells (strains AG27 and AGM13) after 14 h increased with increasing concentration of PO_4, but this effect was noticeable only up to 1.5 mM of PO_4 concentration and became static thereafter. This finding suggested that PO_4 in the medium could play a protective role for the bacterial cultures in the presence of 5 mM concentration of sodium arsenate (Fig. 11.2).

11.3 Possible Impacts of Cadmium and Arsenic Metal Contaminants

Cadmium and As cycle has broadened as a consequence of human interference, and due to this, large amounts of these metals accumulate in the environment and in living organisms. The mode of toxicity, however, depends upon their chemical forms. They are toxic to humans, animals, and plants, and are the widespread pollutants with a long biological life (Wagner 1993).

11.3.1 Cadmium Toxicity

The toxicity of Cd depends primarily on route of exposure and, hence, varying rates of its absorption have varying health effects. Due to slow elimination, the level of Cd in the body increases over time. In man, chronic exposure to low levels of Cd results in damage to kidneys and has been linked to neoplastic disease and aging. People living near hazardous waste sites or factories that release Cd into the air and people that work in the metal refinery industry breathe in Cd, which damage the lungs and may even cause death. Cadmium may be a catalyst to oxidation reaction, which can generate tissue damage. For example, Cd is reported to increase oxidative stress by acting as a catalyst in the formation of reactive oxygen species (ROS), increasing lipid peroxidation, and depleting glutathione and protein-bound sulfhydryl groups. Among plants, leafy vegetables such as spinach and lettuce are examples of crop species which readily accumulate Cd from enriched soil and would also result in a high dietary intake of this element. When taken up in excess by plants, Cd directly or indirectly inhibits physiological processes such as respiration, photosynthesis, cell elongation, plant–water relationships, and nitrogen metabolisms resulting in poor growth and low biomass. (Chaffei et al. 2004; Rani et al. 2008). Furthermore, after accumulation in the plant tissues, Cd alters the catalytic efficiency of enzymes (Piqueras et al. 1999; Romero-Puertas et al. 1999), damages cellular membranes (Tu and Brouillette 1987), and inhibits root growth.

Table. 11.1 Sources of some heavy metals and their possible hazards

Metal	Sources	Disease
Lead	Mining, coal, automobile, paper dyeing, petrochemicals	Mental retardation, emesis, anorexia, fatigue, anemia, neuritis, palsy
Chromium	Leather tanning, thermal power plant, petroleum refining, textile photography	Bronchial asthma, Allergies
Nickel	Mining, coal, power plant, phosphate fertilizers, automobile electroplating	Dermatitis pneumonia
Mercury	Chloralkali plants, pulp and paper, Antiseptics, fungicides	Minamata disease

11.3.2 *Arsenic Toxicity*

The As cycle has broadened as a consequence of human interference, and due to this, a large amount of As ends up in the environment and in living organisms. Arsenic compounds cause acute and chronic effects in individuals, populations, and communities at concentrations ranging from a few micrograms to milligrams per liter, depending on species, time of exposure, and endpoints measured. Arsenic is highly toxic and ingestion of large doses leads to gastrointestinal symptoms, disturbances of cardiovascular and nervous system functions, and eventually death. It can also cause various health effects, such as irritation of stomach and intestine, decreased production of red and white blood cells, skin changes, and lung irritation. A very high exposure may lead to infertility and miscarriages in women. Drinking water contaminated with As leads to increased risks of cancer in the skin, lungs, bladder, and kidney, as well as other skin changes such as hyperkeratosis and pigmentation changes. Plants absorb As fairly easily, so high ranking concentration may be present in food. It has been reported that As toxicity affects photosynthesis which ultimately results in the reduction of rice growth and yield (Rahman et al. 2007). The concentrations of the dangerous inorganic arsenics that are currently present in surface waters enhance the chances of alteration of genetic material of fishes. This is mainly caused by accumulation of As in the bodies of plant-eating freshwater organisms. These fishes containing eminent amounts of As when eaten by birds, lead to their death due to As poisoning. The overall of evidence indicates that Cd and As can cause clastogenic damage in different cell types with different endpoints in exposed individuals. Moreover, some other heavy metal contaminants and their possible hazards are summarized in Table 11.1.

11.4 Mechanism of Bacterial Resistance to Heavy Metals: An Overview

Microbial resistance to toxic metals is widespread, with frequencies ranging from a few percent in pristine environments to nearly 100% in heavily polluted environments (Silver and Phung 2009). Metals and their compounds interact with microbes

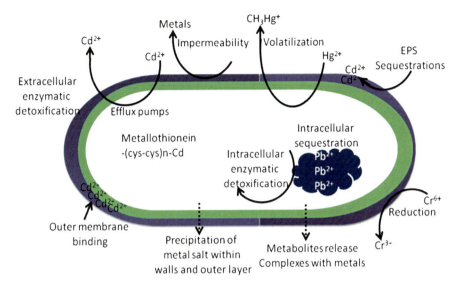

Fig. 11.3 Mechanisms of metal–microbe interaction

in various ways depending on the metal species, organism, and environment. Structural components and metabolic activities of microbes also influence metal speciation and therefore solubility, mobility, bioavailability, and toxicity of metals (Gadd 2004, 2005, 2007) (Fig. 11.3).

Bacterial resistance mechanisms generally involve efflux or enzymatic detoxification of metals (Rosen 2002; Nies 2003; Osman and Cavet 2008; Silver and Phung 2009). It seems that most of the survival mechanisms depend on some changes in metal speciation leading to decreased or increased mobility. These include redox transformations, production of metal-binding peptides and proteins (e.g., metallothioneins, phytochelatins), organic and inorganic precipitation, active transport, efflux, and intracellular compartmentalization. Such metal transformations are central to metal biogeochemistry and emphasize the link between microbial responses and geochemical cycles for metals (Loveley et al. 1991; Gilmour and Riedel 2009). The mechanism adopted by microbes to resist Cd and As toxicity are reviewed and discussed in the following section.

11.4.1 Bacterial Resistance to Cadmium

Three major families of efflux transporters namely, P-type ATPase, CBA transporters, and CDF family transporters are involved in Cd resistance. P-type ATPases span the inner membrane and use energy provided by ATP to pump metal ions from the cytoplasm to periplasm (Rensing et al. 1997). CBA transporters are three-component trans-envelope pumps of Gram-negative bacteria that act as chemiosmotic antiporters (Franke et al. 2003) and cation diffusion facilitator (CDF) family transporters act as chemiosmotic ion-proton exchangers (Anton et al. 1999; Grass et al. 2001). P-type ATPases and CDF transporters export metal ions from the cytoplasm to the periplasm; whereas CBA transporters mainly detoxify periplasmic metal (outer membrane efflux. P-type ATPases and CDF transporters can functionally replace each other but they cannot replace CBA transporter and vice versa (Scherer and Nies 2009).

11.4.1.1 P-type ATPase

P-type ATPases constitute a superfamily of transport proteins that transport ions against the concentration gradient using energy provided by ATP hydrolysis. The term "P-type" refers to the formation of a phosphoenzyme intermediate in the reaction cycle. The energy released by the removal of the PO_4 from ATP is used in the translocation of an ion across biological membranes. Divalent metal efflux ATPases are widespread in both Gram-positive and Gram-negative bacteria (Rensing et al. 1999). In Gram-positive bacteria, the first example of a Cd-exporting P-type ATPase was the CadA pump from *S. aureus* (Nucifora et al. 1989). Cadmium translocating ATPases have been characterized in *Synechocystis* (Thelwell et al. 1998), *P. putida* (Lee et al. 2001; Hu and Zhao 2007) and *Cupriavidus metallidurans* (Legatzki et al. 2003; Scherer and Nies 2009). Two different *cadA* Cd resistance determinants (*cadA1*, first identified in Tn*5422*, and *cadA2*, associated with pLM80) were detected among Cd-resistant *Listeria monocytogenes* strains from turkey processing plants (Mullapudi et al. 2010).

11.4.1.2 CBA Transporters

The CBA transporters are three-component protein complexes that span the whole cell wall of Gram-negative bacteria. The most important component of the transporter is a resistance nodulation cell division (RND) protein that is located in the inner membrane. The RND protein family was first described as a related group of bacterial transport proteins involved in heavy metal resistance (*C. metallidurans*), nodulation (*Mesorhizobium loti*), and cell division (*E. coli*) (Saier et al. 1994). The RND protein is usually accompanied by the membrane fusion protein (MFP) and outer membrane factor (OMF). These three proteins form an efflux protein complex that may export the substrate (ions) from the cytoplasm, the cytoplasmic membrane, or the periplasm across the outer membrane (Nies 1999, 2003). RND-driven export systems are referred to as CBA efflux systems or CBA transporters. This way, they can be distinguished from ABC transport systems and this name also reflects the sequence of the genes in the operon encoding for the components of the transporter complex. In bacteria and archaea, CBA transporters are involved in transport of heavy metals, hydrophobic compounds, nodulation factors, and proteins. By diminishing not only the cytoplasmic concentration of heavy metal cations but additionally the periplasmic concentration, CBA transport systems could remove cations

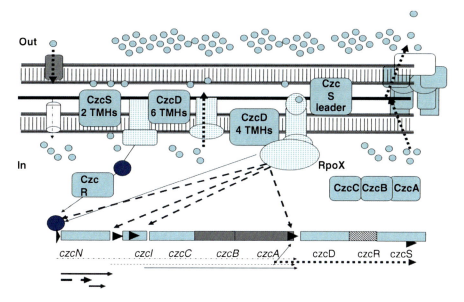

Fig. 11.4 The Czc resistance system of *Ralstonia* sp. CH34

even before they have the opportunity to enter the cell. Moreover, these efflux systems could mediate further export of the cation that had been removed from the cytoplasm by other efflux systems.

The best characterized CBA transporter is the CzcCBA complex from Gramnegative soil bacterium *Ralstonia eutropha* (formerly called *Alcaligenes eutrophus*) strain CH34. The *czc* determinant encodes resistance to (cobalt (Co), zinc (Zn), and cadmium (Cd)) by metal-dependent efflux (Nies et al. 1989b) driven by the proton motive force (Nies 1995). CBA transporters responsible for Zn^{2+} and Cd^{2+} efflux can also be found in *P. aeruginosa* (Hassan et al. 1999) and *P. putida* (Hu and Zhao 2007). The three metal cations (Co, Zn, and Cd), which are taken up into the cell by the fast and unspecific transport system for magnesium ions (Nies et al. 1989a), are actively extruded from the cell by the products of the *czc* resistant determinants (Nies et al. 1989b). The actual efflux protein complex is composed of three subunits: CzcC (outer membrane protein), CzcB (membrane fusion protein), and CzcA (Basic inner membrane transport protein) (Fig. 11.4).

11.4.1.3   CDF Family Transporters

The cation diffusion facilitator family (CDF) comprises of a group of transporters which can catalyze either influx or efflux of heavy metals. Members of the family have been found from both prokaryotes and eukaryotes. CDF family of chemiosmotic efflux systems was first described with the Cd and Zn ions efflux system of *C. metallidurans* (Nies 1992; Anton et al. 1999). CDF proteins are driven by a potassium gradient in addition to the proton motive force. Generally, very little is known about the role of CDF transporters in heavy metal resistance. They provide very low level resistance, but it has been assumed that their main role is to function as a kind of heavy metal buffer for the cell at low cytoplasmic metal concentrations (Anton et al. 1999).

11.4.2   Bacterial Resistance to Arsenic

The best characterized, and probably the most widespread, As resistance system in microorganisms is the "*ars* gene" system. At the basic level, the *ars* system consists of a series of three or more genes coding for a transmembrane pump system and an arsenate reductase. The operon includes: (1) a regulatory gene (*ars*R), (2) a gene coding for an arsenite-specific transmembrane pump (*ars*B), and (3) a gene coding for an arsenate reductase (*ars*C). Arsenic (III) is pumped directly out of the cell by the membrane protein encoded by *ars*B; however, As(V) must first be reduced to As(III) by the soluble arsenate reductase encoded by *ars*C gene. Moreover, *ars*R codes for a repressor protein that regulates *ars* gene expression (Ordonez et al. 2005). In some bacteria, the operon contains other genes: *ars*A that produces an oxyanion-stimulated ATPase (Kaur and Rosen 1994) that couples ATP hydrolysis to the extrusion of arsenicals (and antimonite) through the *ars*B protein; *ars*D that encodes for a regulatory protein capable of controlling the upper level of *ars* expression (Yang et al. 2010). A relatively large number of microorganisms are capable of resisting the toxic effects of arsenic by using methods such as arsenite oxidation (to produce the less toxic arsenate) and minimizing the uptake of arsenic from the environment. For example, *P. stutzeri* strain GIST-BDan2 (EF429003) contain aoxB and aoxR gene, which play an important role in As(III) oxidation to As(V) (Chang et al. 2010). The cell membrane of bacterial cell is a primary site of heavy metal toxicity. Toxic metal ions, including Cu, Co, Ni, Cd, As, and Hg, inhibit plasma membrane ATPase by means of various binding interaction (Ochiai 1987). The above effect leads to an increased permeability of the cell to external material, i.e., adverse effect on membrane integrity and a reduced ability to maintain electrochemical gradient or membrane potential. Therefore, membrane potential and integrity were recorded for *B. cereus* strain AG27 and AGM13 (unidentified) (Gupta 2006) using the fluorescent dyes Bis-oxonal (Ox [DiBAC$_4$] (3)) and Propidium Iodide (PI).

Bis-oxonol is lipophilic, anionic, and accumulates intracellularly producing green fluorescence only when the cytoplasmic membrane is hyperpolarized/depolarized and PI binds to nucleic acids and produces red fluorescence, but cannot cross the intact cytoplasmic membrane, hence can be used to indicate cell membrane integrity. These fluorescent dyes alone or in combination can be used to detect the effect of stress of heavy metals on the cytoplasmic membrane integrity and physiology of bacterial populations (Zhang and Crow 2001). Flow cytometry data for Bis-oxonol for both *Bacillus cereus* strain AG27 and an unidentified strain AGM13 showed

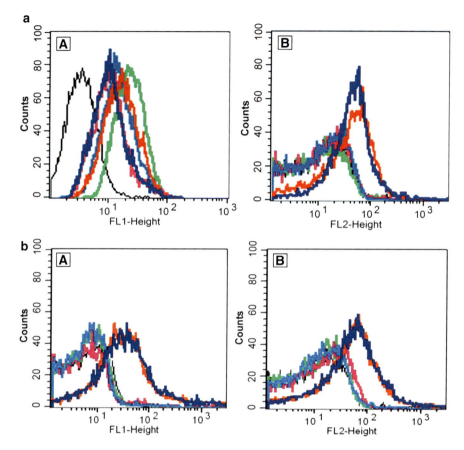

Fig. 11.5 (**a**) Flow Cytometric measurement of changes (*A*) Membrane potential (*B*) Membrane integrity of *Bacillus cereus strain* AG27 in the presence of arsenic after different time intervals. *Green* (30 min), *magenta* (60 min), *sky blue* (90 min), *navy blue* (120 min), and *orange* (150 min), respectively, FL1 and FL2-Fluorescence labels (Adapted from Gupta 2006). (**b**) Flow Cytometric measurement of changes (*A*) Membrane potential (*B*) Membrane integrity of unidentified strain AGM13 in the presence of arsenic after different time intervals. *Green* (30 min), *magenta* (60 min), *sky blue* (90 min), *navy blue* (120 min), *orange* (150 min), respectively, FL1 and FL2-Fluorescence labels (Adapted from Gupta 2006)

significant reduction in membrane potential after addition of sodium arsenate to the cells growing in log phase. The flow cytometric graphs showed a shift in peak toward a higher fluorescence with changing time periods in the presence of As, which revealed the loss in membrane potential (Fig. 11.5a, b).

In case of *Bacillus cereus* (AG27), the fluorescence reached to maximum after 30 min and then decreased suggesting that cells had recovered from initial shock due to addition of As whereas in case of AGM13, the rate of depolarization was more, revealing that the toxic effect was more pronounced in AGM13 as compared to AG27 as Bis-oxonol, which is an important marker (Epps et al. 1994) for determining the change in membrane potential accumulated in the cell leading to increase in fluorescence. Further, the flow cytometric analysis for PI did not show much change in fluorescence in both the strains which clearly indicated that the addition of As disrupted the membrane potential, and thus, the metabolic rate. However, the cells retained the membrane integrity and viability even after addition of As.

11.5 Influence of Microbes on Speciation and Mobility of Arsenic and Cadmium

Microorganisms play an important role in the environmental fate of Cd and As with a multiplicity of mechanisms affecting transformations between toxic and nontoxic forms of Cd and As. The potential of microorganisms to immobilize or volatilize soluble Cd has been explored. Biomass of several bacterial, fungal, and algal species has been evaluated as biosorbents for the removal of soluble Cd from solution. The anionic nature of bacterial cell surface enables them to bind to metal cations through electrostatic interactions. Three strains of thermotolerant polymer-producing bacteria; *Bacillus subtilis* WD90, *B. subtilis* SH29, and *Enterobacter agglomerans* 5M38 were capable of Cd removal by biosorption (Kaewchai and Praseptan 2002). The biosorption of Cd and As by filamentous fungus *Aspergillus clavatus* DESM has been reported (Cernasky et al. 2007). Cadmium-binding proteins have an important role in moderating Cd toxicity in some fungi and bacteria. These have been reported in *P. putida* (Higham et al. 1984). Cadmium-binding metallothioneins have been identified in cyanobacteria (Turner et al. 1996). A metallothionein encoded by *CUP1* gene binds Cd in *Candida glabrata*. Low-molecular-mass carboxylic acids play an important role in chemical attack of minerals, providing protons as well as metal-chelating anions (Burgstaller and Schinner 1993; Jacobs et al. 2002a, b; Huang et al. 2004; Lian et al. 2008a, b). Phytochelatins (PCs); the metal-binding cysteine-rich peptides are enzymatically synthesized in plants and certain fungi from glutathione in response to heavy metal stress. In an attempt to increase the ability of bacterial cells to accumulate heavy metals, the *Arabidopsis thaliana* PC synthase gene (*AtPCS*) was expressed in *E. coli*. When the bacterial cells expressing AtPCS were exposed to metals like Cd or As, cellular metal content was increased 20- and 50-folds, respectively. Thus, the overexpression of PC synthase in bacteria could be a means of improving the metal content of organisms for use in bioremediation (Sauge-Merle et al. 2003). Phytochelatins with good binding affinities for a wide range of heavy metals were also exploited to develop microbial sorbents for Cd removal. Phytochelatin synthase from *Schizosaccharomyces pombe* (SpPCS) was overexpressed in *E. coli*, resulting in PC synthesis and seven times higher Cd accumulation (Kang et al. 2007).

Several microorganisms have been shown to precipitate soluble Cd as insoluble sulfides (Holmes et al. 1997), or carbonates (Cunningham and Lundie 1993). The yeast Cd factor (YCF1, EC 3.6.3.46) mediates accumulation of Cd–glutathione complexes in *Saccharomyces cerevisiae* vacuoles, and metal-binding peptides

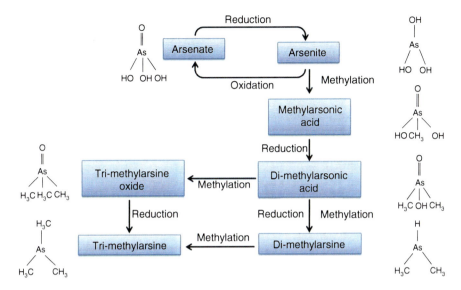

Fig. 11.6 Microbial formation of trimethylarsine from inorganic arsenate

("phytochelatins") sequester Cd in subcellular organelles in *S. pombe* and *C. glabrata* (Perego and Howell 1997). *Pseudomonas aeruginosa* strain KUCd1 exhibiting high Cd accumulation under in vitro aerobic condition has also been reported (Sinha and Mukherjee 2009). Similarly, transmission electron microscopy analysis of *P. aeruginosa* strain 62BN demonstrated intracellular and periplasmic accumulation of Cd (Rani et al. 2009). Cadmium-resistant mutants of *pseudomonas* species NBRI4014 developed through selective enrichment and *P. aeruginosa* MCCB102 also showed bioaccumulation (Zolgharnein et al. 2010). Cytoplasmic (Yoshida et al. 2002) and periplasmic (Naz et al. 2005; Pazirandeh et al. 1998) accumulation of heavy metal ions has also been reported in *E. coli*.

Microbes are also able to detoxify the poisonous As species. They biomethylate inorganic As species to Monomethyl arsonic acid (MMAA) and Dimethyl arsinic acid (DMAA) (Cullen and Reimer 1989). Fungi dominated the microbial screen regarding the production of volatile, garlic-smelling trimethylarsine (Craig et al. 2000), although bacterial and animal tissues also are known to have this potential (Hall et al. 1997). Nonetheless, the conversion of arsenate to MMAA or to DMAA is another possible mechanism for detoxification (Fig. 11.6).

Microbial activity can also result in volatilization of As to gaseous arsines (Gao and Burau 1997). These arsines may travel in air or may oxidize rapidly depending on environmental conditions (Pongratz 1998). *Pseudomonas stutzeri* exhibited a maximum accumulation of 4 mg As g^{-1} (dry weight). Arsenic (V) can be reduced by dissimilatory reduction, where microorganisms utilize As(V) as a terminal electron acceptor for anaerobic respiration. To date, dissimilatory reduction has been observed in several bacteria, such as *Sulfurospirillum barnesii*, *S. arsenophilum*, *Desulfotomaculum auripigmentum*, *Bacillus Asoselenatis*, *B. selenitireducens*,

Crysiogenes arsenatis, Sphingomonas spp., *Pseudomonas* spp. and *Wolinella* spp., *Bacillus* sp. SF-1 (Ahmann et al. 1994; Lovely and Coates 1997; Newman et al. 1998; Stolz and Oremland 1999; Oremland et al. 2000; Macur et al. 2001). In addition, microorganisms may possess As(V) reduction mechanisms that are not coupled to respiration but instead are thought to impart As resistance. The energetics of the oxidation of As(III) to AS(V) suggest that enough energy for growth can be produced through this reaction (Oremland et al. 2002). Since these organisms are almost heterotrophic As(III)-oxidizing bacteria, they require the presence of organic matter for growth. Examples include *A. faecalis, P. arsenitoxidans*, NT-26, *P.* stutzeri strain GIST-BDan2 (EF429003), Pseudomonas strain RS-19 (Phillips and Taylor 1976; Santini et al. 2000; Chang et al. 2010). An As(III) oxidase has been obtained and purified from *A. faecalis* (Anderson et al. 1992; Anderson et al 2002). *Thermus aquaticus* and *T. thermophilus* were also found to rapidly oxidize As(III) to As(V), but they were not able to grow with As(III) as the sole energy source, thus suggesting that the ecological role of As(III) oxidation was detoxification of As (Gihring and Banfield 2001). One microbe in particular, *A. ehrlichii* strain MLHE-1T, can express two completely different physiologies. As an aerobe, it grows heterotrophically, with acetate as the electron donor and carbon source. As an anaerobe, it is a chemolithoautotroph, coupling the oxidation of As(III) to the reduction of nitrate to nitrite (Hoeft et al. 2007). Briefly, microbiological processes can either solubilize metals, thereby increasing their bioavailability and potential toxicity, or immobilize them, and thereby reduce the bioavailability of metals. These biotransformations are thus an important component of biogeochemical cycles of metals and may be exploited in bioremediation of metal-contaminated soils (Gadd 2000; Barkay and Schaefer 2001; Lloyd and Lovley 2001).

11.6 Conclusion

Unlike organic pollutants, that can be mineralized to harmless products, Cd and As cannot be biodegraded, but persist indefinitely, complicating the remediation of contaminated soils. They are released from the earth crust via natural processes and from certain human activities. Environmental levels of Cd and As vary, the concentrations being highest in the air close to industrial sources, in areas with natural geological contamination, and in soils or sediments near contamination sources.

Driven by the realization that large areas of land contaminated with these heavy metals cannot be economically remediated by conventional chemical approaches, microbial bioremediation proved to be the best alternative. Thus, the main strategy employed involves the reduction in bioavailability, mobility, and toxicity of these metals. Biological methods for remediation of Cd and As contaminated soils include detoxification, bioleaching, biosorption, biotransformation, etc. Biomass of several bacterial, fungal, and algal species biosorb Cd and As from contaminated sites and help in their removal. In As-contaminated soils, microbial methylation of inorganic arsenic to water soluble methylated arsenic forms may function as a detoxification method. Furthermore, study of genes responsible for Cd and As resistance may aid in production of genetically modified microorganisms capable of remediation of contaminated soil. Since, microbial transformations of metals are a vital part of natural biosphere processes and can have beneficial as well as detrimental consequences for societal benefit. Therefore, our understanding of this important area of microbiology and its applications needs to be unfolded completely.

References

Ahmann D, Roberts AL, Krumholz LR, Morel FM (1994) Microbe grows by reducing arsenic. Nature 371:750

Al-Kheldhairy AA, Al-Rokayan SA, Al-Minsed FA (2001) Cadmium toxicity and cell stress response. Pak J Biol Sci 4:1046–1049

Alloway BJ (1990) Cadmium. In: Alloway BJ (ed) Heavy metals in soils. Wiley, New York, pp 100–124

Anderson GL, Williams J, Hille R (1992) The purification and characterization of arsenite oxidase from *Alcaligenes faecalis*, a molybdenum containing hydroxylase. J Biol Chem 267:23674–23682

Anderson GL, Ellis PJ, Kuhn P, Hille R (2002) Oxidation of arsenite by *Alcaligenes faecalis*. In: Frankenberger WT Jr (ed.) Environmental chemistry of arsenic. Marcel Dekker, Inc., New York, pp 343–362

Anton A, Grosse C, Reissmann C, Pribyl T, Nies DH (1999) CzcD is a heavy metal ion transporter involved in regulation of heavy metal resistance in *Ralstonia* sp. strain CH34. J Bacteriol 181:6876–6881

Balasoiu C, Zagury G, Deshenes L (2001) Partitioning and speciation of chromium, copper and arsenic in CCA-contaminated soils: influence of soil composition. Sci Total Environ 280:239–255

Barkay T, Schaefer J (2001) Metal and radionuclide bioremediation: issues, considerations and potentials. Curr Opin Microbiol 4:318–323

Bruins MR, Kapil S, Oehme FW (2000) Microbial resistance to metals in the environment. Ecotoxicol Environ Saf 45:198–207

Burgstaller W, Schinner F (1993) Leaching of metals with fungi. J Biotechnol 27:91–116

Burke BE, Pfister RM (1986) Cadmium transport by a Cadmium sensitive and a cadmium resistant strain of *Bacillus subtilis*. Can J Microbiol 32:539–542

Cernasky S, Urik M, Sevc J, Littera P, Hiller E (2007) Biosorption of arsenic and cadmium from aqueous solutions. Afr J Biotechnol 6:1932–1934

Cervantes C, Ji G, Ramirez JL, Silver S (1994) Resistance to arsenic compounds in microorganisms. FEMS Microbiol Rev 15:355–367

Chaffei C, Pageau K, Suzuki A, Gouia H, Ghorbel MH, Masclaux-Daubresse C (2004) Cadmium toxicity induced changes in nitrogen management in *Lycopersicon esculentum* leading to a metabolic safeguard through an amino acid storage strategy. Plant Cell Physiol 45:1681–1693

Chang JS, Yoon IH, Lee JH, Kim KR, An J, Kim KW (2010) Arsenic detoxification potential of aox genes in arsenite-oxidizing bacteria isolated from natural and constructed wetlands in the Republic of Korea. Environ Geochem Health 32:95–105

Craig PJ, Foster SN, Jenkins RO, Miller DP, Ostah N, Smith LM, Marris TA (2000) Occurrence, formation and fate of organoantimony compounds in marine and terrestrial environments. In: Gianguzza A, Pelizzetli E, Sammartano S (eds.) Chemical processes in the environment. Springer, Berlin, pp 265–280

Cullen WR, Reimer KJ (1989) Arsenic speciation in the environment. Chem Rev 89:713–764

Cunningham DP, Lundie LL (1993) Precipitation of cadmium by *Clostridium thermoaceticum*. Appl Environ Microbiol 59:7–14

Ehrlich HL, Newman DK (2009) Geomicrobiology, 5th edn. CRC Press/Taylor and Francis, Boca Raton

Epps DE, Wolfe ML, Groppi V (1994) Characterization of the steady-state and dynamic fluorescence properties of the potential sensitive dye bis-(1,3, dibutylbarbituric acid) trimetheine oxonol, (DiBAC4(3)-), in model system and cells. Chem Phys Lipids 69:137–150

Franke S, Grass G, Rensing C, Nies DH (2003) Molecular analysis of the copper transporting efflux system CusCFBA of *Escherichia coli*. J Bacteriol 185:3804–3812

Gadd GM (2000) Bioremedial potential of microbial mechanisms of metal mobilization and immobilization. Curr Opin Biotechnol 11:271–279

Gadd GM (2004) Microbial influence on metal mobility and application for bioremediation. Geoderma 122:109–119

Gadd GM (2005) Microorganisms in toxic metal polluted soils. In: Buscot F, Varma A (eds.) Microorganisms in soils: roles in genesis and functions. Springer, Berlin, pp 325–356

Gadd GM (2007) Geomycology: biogeochemical transformations of rocks, minerals, metals and radionuclides by fungi, bioweathering and bioremediation. Mycol Res 111:3–49

Gadd GM (2009) Biosorption: critical review of scientific rationale, environmental importance and significance for pollution treatment. J Chem Technol Biotechnol 84:13–28

Gao S, Burau RG (1997) Environmental factors affecting rates of arsine evolution from and mineralization of arsenicals in soil. J Environ Qual 26:753–763

Gihring TM, Banfield JF (2001) Arsenite oxidation and arsenate respiration by a new *Thermus* isolate. FEMS Microbiol Lett 204:335–340

Gilmour C, Riedel G (2009) Biogeochemistry of trace metals and metalloids. In: Likens GE (ed.) Encyclopedia of Inland waters. Elsevier, Amsterdam, pp 7–15

Grass G, Fan B, Rosen BP, Franke S, Nies DH, Rensing C (2001) ZitB (YbgR), a member of the cation diffusion facilitator family, is an additional zinc transporter in *E. coli*. J Bacteriol 183:4664–4667

Gupta A (2006) Diversified arsenic resistant microbial population from industrial and ground water sources and their molecular characterization. Unpublished doctoral dissertation, G.B. Pant University of Agriculture and Technology, Pantnagar

Hall LL, George SE, Kohan MJ, Styblo M, Thomas DJ (1997) In vitro methylation of inorganic arsenic in mouse intestinal cecum. Toxicol Appl Pharmacol 147:101–109

Hassan MT, Van der Lelie D, Springael D, Romling U, Ahmed N, Mergeay M (1999) Identification of a gene cluster, *czr*, involved in cadmium and zinc resistance in *Pseudomonas aeruginosa*. Gene 238:417–425

Higham PD, Sadler PJ, Scawen MJ (1984) Cadmium-resistant *Pseudomonas putida* synthesizes novel cadmium proteins. Science 225:1043–1046. http://www.ncbi.nlm.nih.gov/pubmed/17783048

Hoeft SE, Switzer Blum J, Stolz JF, Tabita FR, Witte B, King GM, Santini JM, Oremland RS (2007) *Alkalilimnicola ehrlichii*, sp. nov. a novel arsenite-oxidizing halophilic gamma proteobacterium capable of chemoautotrophic or heterotrophic growth with nitrate or oxygen as the electron acceptor. Int J Sys Evol Microbiol 57:504–512

Holmes JD, Richardson DJ, Saed S, Evan-Growing R, Russel DA, Sodeau JR (1997) Cadmium specific formation of metal sulphide 'Q-particles' by *Klebsiella pneumoniae*. Microbiology 43:2521–2530

Hu N, Zhao B (2007) Key genes involved in heavy-metal resistance in *Pseudomonas putida* CD2. FEMS Microbiol Lett 267:17–22

Huang PM, Wang MC, Wang MK (2004) Mineral–organic–microbial interactions. In: Hillel D, Rosenzweig C, Powlson DS, Scow KM, Singer MJ, Sparks DL, Hatfield J (eds.) Encyclopedia of soils in the environment. Elsevier, Amsterdam, pp 486–499

Hussein H, Farag S, Kandil K, Moawad H (2005) Tolerance and uptake of heavy metals by pseudomonads. Process Biochem 40:955–961

Jacobs H, Boswell GP, Ritz K, Davidson FA, Gadd GM (2002a) Solubilization of metal phosphates by *Rhizoctonia solani*. Mycol Res 106:1468–1479

Jacobs H, Boswell GP, Ritz K, Davidson FA, Gadd GM (2002b) Solubilization of calcium phosphate as a consequence of carbon translocation by *Rhizoctonia solani*. FEMS Microbiol Ecol 40:65–71

Jain CK, Ali I (2000) Arsenic: occurrence, toxicity and speciation techniques. Water Res 34:4304–4312

Kaewchai S, Praseptan P (2002) Biosorption of heavy metal by thermotolerant polymer producing bacterial cells and the bioflocculant. Songklanakarin J Sci Technol 24:421–430

Kang SH, Singh S, Kim JY, Lee W, Mulchandani A, Chen W (2007) Bacteria metabolically engineered for enhanced phytochelatin production and cadmium accumulation. Appl Environ Microbiol 73:6317–6320

Kaur P, Rosen BP (1994) Identification of the site of a-[^{32}P] ATP adduct formation in the ArsA protein. Biochemistry 33:6456–6461

Landa ER (2005) Microbial biogeochemistry of uranium mill tailings. Adv Appl Microbiol 57:113–130

Lee SW, Glickmann E, Cooksey DA (2001) Chromosomal locus for cadmium resistance in *Pseudomonas putida* consisting of cadmium transporting ATPase and a Mer R family response regulator. Appl Environ Microbiol 67:1437–1444

Legatzki A, Grass G, Anton A, Rensing C, Nies DH (2003) Interplay of the Czcsystem and two P-type ATPases in conferring metal resistance to *Ralstonia metallidurans*. J Bacteriol 185:4354–4361

Leonard A (1991) Arsenic. In: Merian E (ed.) Metals and their compounds in the environment. VCH, Weinheim, pp 751–772

Lian B, Chen Y, Zhu L, Yang R (2008a) Effect of microbial weathering on carbonate rocks. Earth Sci Front 15:90–99

Lian B, Wang B, Pan M, Liu C, Teng HH (2008b) Microbial release of potassium from K-bearing minerals by thermophilic fungus *Aspergillus fumigatus*. Geochim Cosmochim Acta 72:87–98

Lloyd JR, Lovley DR (2001) Microbial detoxification of metals and radionuclides. Curr Opin Biotechnol 12:248–253

Lovely DR, Coates JD (1997) Bioremediation of metal contamination. Curr Opin Biotechnol 8:285–289

Loveley DR, Philips EJP, Gorby YA, Landa ER (1991) Microbial reduction of uranium. Nature 350:413–416

Macur RE, Wheeler JT, McDermott TR, Inskeep WP (2001) Microbial populations associated with the reduction and enhanced mobilization of arsenic in mine tailings. Environ Sci Technol 35:3676–3682

Masscheleyn PH, Delaune RD, Patrick WH (1991) Effect of redox potential and pH on arsenic speciation and solubility in a contaminated soil. Environ Sci Technol 25:1414–1419

Mukhopadhyay R, Rosen, BP, Phung LT, Silver S (2002) Microbial arsenic: from geocycles to genes. FEMS Microbiol Rev 26:311–325

Mullapudi S, Siletzky RM, Kathariou S (2010) Diverse cadmium resistance determinants in *Listeria monocytogenes* isolates from the Turkey processing plant environment. Appl Environ Microbiol 76:627–630

Muller D, Lievremont D, Simeonova DD, Hubert JC, Lett MC (2003) Arsenite oxidase *aox* genes from a metal-resistant beta-proteobacterium. J Bacteriol 185:135–141

Naidu R, Kookana RS, Sumner ME, Harter RD, Tiller KG (1997) Cadmium sorption and transport in variable charge soils: a review. J Environ Qual 26:602–617

Naz N, Young HK, Ahmed N, Gadd GM (2005) Cadmium accumulation and DNA homology with metal resistant genes in sulphate reducing bacteria. Appl Environ Microbiol 71: 4610–4618

Newman DK, Ahmann D, Morel FMM (1998) A brief review of microbial arsenate respiration. Geomicrobiology 15:255–268

Nies DH (1992) Resistance to cadmium, cobalt, zinc, and nickel in microbes. Plasmid 27:17–28

Nies DH (1995) The cobalt, zinc and cadmium efflux system czc ABC from *Alcaligens eutrophus* functions as a cation-proton antiportar in *Escherichia coli*. J Bacteriol 177:2707–2712

Nies DH (1999) Microbial heavy-metal resistance. Appl Microbiol Biotechnol 51:730–750

Nies DH (2003) Efflux-mediated heavy metal resistance in prokaryotes. FEMS Microbiol Rev 27:313–339

Nies DH, Silver S (1995) Ion efflux systems involved in bacterial metal resistance. J Ind Microbiol 14:186–199

Nies A, Nies DH, Silver S (1989a) Cloning and expression of plasmid genes encoding resistance to chromate and cobalt in *Alcaligenes eutrophus* CH34. J Bacteriol 171:5065–5070

Nies DH, Nies A, Chu L, Silver S (1989b) Expression and nucleotide sequence of a plasmid determined divalent cation efflux system from *Alcaligenes eutrophus*. Proc Natl Acad Sci USA 86:7351–7355

Nucifora G, Chu L, Mishra TK, Silver S (1989) Cadmium resistance from Staphylococcus aureus plasmid pI258 CadA gene results from a cadmium efflux-ATPase. Proc Natl Acad Sci USA 86:3544–3548

Ochiai EI (1987) General principals of biochemistry of the elements. Plenum Press, New York

Ordonez E, Letek M, Valbuena N, Gil JA, Mateos LM (2005) Analysis of gene involved in arsenic resistance of *Corynebacterium glutamicum* ATCC 13032. Appl Environ Microbiol 71:6206–6215

Oremland RS, Dowdle PR, Hoeft S, Sharp JO, Schaefer JK, Miller LG, Blum JS, Smith RL, Bloom NS, Wallschlaeger D (2000) Bacterial dissimilatory reduction of arsenate and sulfate in meromictic Mono Lake, California. Geochimica et Cosmochimica Acta 64:3073–3084

Oremland RS, Hoeft SE, Santini JM, Bano N, Hollibaugh RA, Hollibaugh JT (2002) Anaerobic oxidation of arsenite in Mono Lake water and by a facultative, arsenite-oxidizing chemoautotroph, strain MLHE-1. Appl Environ Microbiol 68:4795–4802

Osman D, Cavet JS (2008) Copper homeostasis in bacteria. Adv Appl Microbiol 65:217–247

Pandey J, Pandey U (2009a) Microbial processes at land-water interface and cross-domain causal relationship as influenced by atmospheric deposition of pollutants in three freshwater lakes in India. Lakes Reservoirs Res Manag 13:71–84

Pandey J, Pandey U (2009b) Accumulation of heavy metals in dietary vegetables and cultivated soil horizon in organic farming system in relation to atmospheric deposition in a seasonally dry tropical region of India. Environ Monit Assess 148:61–74

Pandey J, Pandey U (2009c) Atmospheric deposition and heavy metal contamination in an organic farming system in a seasonally dry tropical region of India. J Sustain Agric 33:361–378

Pantsar-Kallio M, Korpela A (2000) Analysis of gaseous arsenic species and stability studies of arsine and trimethylarsine by gas-chromatography-mass spectrometry. Analytica Chimica Acta 410:65–70

Pazirandeh M, Wells BM, Ryan RL (1998) Development of bacterium based heavy metal biosorbants: enhanced uptake of cadmium and mercury by E. coli expressing a metal binding motif. Appl Environ Microbiol 64:4068–4072

Perego P, Howell SB (1997) Molecular mechanisms controlling sensitivity to toxic metal ions in yeast. Toxicol Appl Pharmacol 147:312–318

Phillips SE, Taylor ML (1976) Oxidation of arsenite to arsenate by *Alcaligenes faecalis*. Appl Environ Microbiol 32:392–399

Piqueras A, Olmos E, Martinez-solano IR, Hellin E (1999) Cadmium induced oxidative burst in tobacco BY2 cells: time course, subcellular location and antioxidant response. Free Rad Res 31:33–38

Pongratz R (1998) Arsenic speciation in environmental samples of contaminated soil. Sci Total Environ 224:133–141

Purvis OW, Pawlik-Skowronska B (2008) Lichens and metals. In: Avery SV, Stratford M, van West P (eds.) Stress in yeasts and filamentous fungi. Elsevier, Amsterdam, pp 175–200

Rahman MA, Hasegawa H, Rahman MM, Islam MN, Miah MA, Tasmen A (2007) Effect of arsenic on photosynthesis, growth and yield of five widely cultivated rice (*Oryza sativa* L.) varieties in Bangladesh. Chemosphere 67:1072–1079

Rani A, Shouche Y, Goel R (2008) Declination of copper toxicity in pigeon pea and soil system by growth promoting *Proteus vulgaris* KNP3 strain. Curr Microbiol 57:78–82

Rani A, Shouche Y, Goel R (2009) Comparative assessment of in-situ bioremediation potential of cadmium resistant acidophilic *Pseudomonas putida* 62BN and alkalophilic *Pseudomonas montelli* 97AN strains on soyabean. Int J Biodet Biodegrad 63:62–66

Rensing C, Mitra B, Rosen BP (1997) The zntA gene of Escherichia coli encodes a Zn(II)-translocating P-type ATPase. Proc Natl Acad Sci USA 94:4326–4331

Rensing C, Sun Y, Mitra B, Rosen BP (1999) Pb(II)-translocating P-type ATPases. J Biol Chem 273:32614–32617

Roane TM, Josephon KL, Pepper IL (2001) Dual-bioaugmentation strategy to enhance remediation of co-contaminated soil. Appl Environ Microbiol 67:3208–3215

Romero-Puertas MC, McCarthy I, Sandalio LM, Palma IM, Corpas FI, Gomez M, Del Rio LA (1999) Cadmium toxicity and oxidative metabolism of pea leaf peroxisomes. Free Rad Res 31:525–531

Rosen BP (2002) Transport and detoxification systems for transition metals, heavy metals and metalloids in eukaryotic and prokaryotic microbes. Comp Biochem Physiol 133:689–693

Rosenman K (2007) Occupational heart disease. In: Rom W, Markowitz S (eds,) Environmental and occupational medicine, 4th edn. Lippincott Williams and Wilkins, Hagerstown, p 688

Rusyniak DE, Furbee RB, Kirk MA (2002) Thallium and arsenic poisoning in a small midwestern town. Ann Emerg Med 39:307–311

Saier MH, Tam R, Reizer A, Reizer J (1994) Two novel families of bacterial membrane proteins concerned with nodulation, cell division and transport. Mol Microbiol 11:841–847

Santini JM, Sly LI, Schnagl RD, Macy JM (2000) A new chemolithoautotrophic arsenite-oxidizing bacterium isolated from a gold mine: phylogenetic, physiological, and preliminary biochemical studies. Appl Environ Microbiol 66:92–97

Sauge-Merle S, Cuine S, Carrier P, Lecomte-Pradines C, Luu DT, Peltier G (2003) Enhanced toxic metal accumulation in engineered bacterial cells expressing Arabidopsis thaliana phytochelatin synthase. Appl Environ Microbiol 69:490–494

Scherer J, Nies DH (2009) CzcP is a novel efflux system contributing to transition metal resistance in *Cupriavidus metallidurans* CH34. Mol Microbiol 73:601–621

Silver S, Phung LT (2009) Heavy metals, bacterial resistance. In: Schaechter M (ed.) Encyclopedia of microbiology. Elsevier, Oxford, pp 220–227

Sinha S, Mukherjee SM (2009) *Pseudomonas aeruginosa* KUCd1, a possible candidate for cadmium bioremediation. Braz J Microbiol 40:655–662

Stolz JF, Oremland RS (1999) Bacterial respiration of arsenic and selenium. FEMS Microbiol Rev 23:615–627

Thelwell C, Robinson NJ, Turner-Cavet JS (1998) An SmtB-like repressor from Synechocystis PCC 6803 regulates a zinc exporter. Proc Natl Acad Sci USA 95:10728–10733

Tu SI, Brouillette JN (1987) Metal ion inhibition of cotton root plasma membrane ATPase. Phytochemistry 26:65–69

Turner JS, Glands PD, Samson AC, Robinson JJ (1996) Zn^{2+} sensing by the cyanobacterial metallothionien repressor Smt B: different motifs mediate metal induced protein-DNA dissociation. Nucleic acids Res 24:3714–3721

Uroz S, Calvaruso C, Turpault MP, Frey-Klett P (2009) Mineral weathering by bacteria: ecology, actors and mechanisms. Trends Microbiol 17:378–387

Violante A, Huang PM, Gadd GM (2008) Biophysico-chemical processes of heavy metals and metalloids in soil environments. Wiley, Chichester

Voutsa D, Grimanis A, Samara C (1996) Trace elements in vegetables grown in an industrial area in relation to soil and air particulate matter. Environ Poll 94:325–335

Wagner GJ (1993) Accumulation of cadmium in crop plants and its consequence to human health. Adv Agron 51:173–212

Yang J, Rawat S, Stemmler TL, Rosen BP (2010) Arsenic binding and transfer by the arsD As (III) metallochaperone. Biochemistry 49:3658–3666

Yoshida N, Kato T, Yoshida T, Ogawa K, Yamashita M, Murooka Y (2002) Bacterium based heavy metal biosorbents: enhanced uptake of cadmium by *E. coli* expressing a metallothionein fused to beta-galactosidase. Biotechniques 32:551–558

Zhang S, Crow SA (2001) Toxic effect of Ag (I) and Hg (II) on *Candida albicans* and *C. maltosa*: a flow cytometric evaluation. Appl Environ Biol 67:4030–4035

Zolgharnein H, Karami K, Mazaheri Assadi M, Dadolahi SA (2010) Investigation of heavy metals biosorption on *Pseudomonas aeruginosa* strain MCCB 102 isolated from the Persian gulf. Asian J Biotechnol 2:99–109

Chapter 12
Phytotechnologies: Importance in Remediation of Heavy Metal–Contaminated Soils

Sas-Nowosielska Aleksandra

Abstract Soil contamination by toxic metals is a major problem that has threatened the sustainability of various agro-ecosystem worldwide. Generally, heavy metals are not destructed and, therefore, persist in the environment. The traditional physical and chemical methods applied for metal removal from contaminated sites produce undesirable products and are expensive. The bioremediation methods including phytotechnologies, on the other hand, is an emerging simple and inexpensive in situ technology used for remediating contaminated sites. A comparative analysis of phytoremediation methods for heavy metal–contaminated soils with a special emphasis on the feasibility and applicability to established methods for soil cleaning is presented. Results of the field trials conducted to examine the applicability of technical soil cleaning methods are also highlighted. Phytoextraction when used to clean up polluted soils was found as an efficient method for slightly and medium-contaminated soils. Chemophytostabilization that involved the use of indigenous plant species was identified as the most practical remediation option for pollutant stabilization in soil. The results of studies on the use of phytotechnologies in the utilization of various plant species for direct application in soils contaminated with heavy metals under a wide range of agro-ecological conditions with a view to restore contaminated soils and consequently facilitate plant yields in metal-poisoned soils around the world are discussed.

Keywords Heavy metals • Phytotechnologies • De-contamination

S.-N. Aleksandra (✉)
Institute for Ecology of Industrial Areas, Kossutha 6 Street, Katowice, Poland
e-mail: sas@ietu.katowice.pl

M.S. Khan et al. (eds.), *Biomanagement of Metal-Contaminated Soils*,
Environmental Pollution 20, DOI 10.1007/978-94-007-1914-9_12,
© Springer Science+Business Media B.V. 2011

12.1   Introduction

Heavy metals are released from various industrial sources (like electroplating and metal extractive operations), agrochemicals, and sewage sludge into soil environment (Muchuweti et al. 2006; Marshall et al. 2007; Singh et al. 2010). Heavy metals cannot be destructed and, therefore, persist in soil (Kucharski and Sas-Nowosielska 2001; Tomohito et al. 2010). Once they accumulate beyond permissible limits in soils, heavy metals pose a serious ecological, toxicological, and human health problems, since they are carried into the food web as a result of leaching from agricultural products, polluted soil, waste dumps, or contaminated drinking water (Intawongse and Dean 2006; Martelli et al. 2006; Palmgren et al. 2008). Among various metals, lead and cadmium, for example, enters the body via digestive tract and adversely affect the human health (Hovmond et al. 1983). The concentration of heavy metals, however, varies significantly among different food products like potatoes, cereals, vegetables, meat, dairy products, etc. Of these, vegetables, potatoes, and cereals accumulate substantial amounts of metals. Apart from soil contamination resulting from anthropopression, high natural content of metals in soils is also the cause of problems. It is obvious in the areas contaminated heavily with lead and zinc (Gzyl 1999). Soil protection and rehabilitation of contaminated sites are, therefore, extremely important in order to preserve the structural integrity and fertility of soil.

In addition, agricultural practices are often performed in the areas that are either devastated or are under active industry pressure. And hence, the cultivation of edible or pasture plants in these areas should either be stopped or limited (Kucharski et al. 1994). Other aspect that requires considerable attention is the high cost of transportation of foods to longer distances and the quality of foods that may deteriorate during transportation. To overcome these problems, it is suggested to change the cultivation practice of crops, land use, etc. The change in the pattern of land use may, however, be expensive and might lead to serious social problems like unemployment or requalification of farmers to switch to jobs other than agricultural practices. That is why in the future, agricultural production in polluted environment has to be considered at priority basis. The producers, while working in contaminated areas, however, when properly informed, would be able to reduce the pollutant risk by selecting a suitable plant species, which could also be safe for the consumers. Other procedures, though more expensive and organizationally complicated, are directed toward the improvement of soil quality. The most frequently used methods for improving soil quality includes crop selection, proper agriculture practices, deep plowing, etc. (Table 12.1). Taking these facts into consideration, proper identification and careful assessment of the source, scope, and level of contamination threat are important for developing an efficient and sustainable preventive remedial measure and management of already contaminated sites. In this context, the conventional mechanical and chemical technologies or long-term biological methods for cleaning up the contaminated sites have been suggested and employed. So far as technical measures are concerned, the following methods have been used to some effect: (1) excavation of the contaminated soil, (2) immobilization of the contaminants, and (3) mixing the contaminated

Table 12.1 The most frequently used methods adopted to reduce metal contamination of soils

Action	Description
Phytoremediation	The use of plants to remove metals from soil
Crop selection	An adequate choice of crop, according to individual species accumulation abilities and contamination of soil to provide the consumer with safe food or food products
Good agriculture practices	Maintains a proper pH and a satisfactory level of organic matter and fertility of soil
Deep plowing	Plowing at the level of 40–50 cm to cover the contaminated soil underneath and to expose the clean layer of soil
Top soil replacement	Removal of ca. 20 cm of top soil and its replacement with clean material from some other place
Total soil replacement	Complete removal of soil and replacement with uncontaminated material. The contaminated material is transported to permitted off-site treatment and disposal facilities
Use of binding materials in soil	Introducing various binding materials to the topsoil to bind metals and make them less available to plants
Chemical and electrolytic method, soil washing	Various hard technical soil cleaning methods using electrolysis, chemicals, thermal applications, washing, etc., usually leading to destruction of basic soil properties including soil microflora (side effect)
Placement of clean soil on surface	Uncontaminated soils are applied onto the soil surface. The thickness of the layer applied depends on intended land use
Dilution of contaminated soil by mixing with clean soil	Mixing the contaminated material with clean soil or subsoil in order to reduce the maximum concentrations of contaminants to below the threshold values
Use of site for urban purposes	If any other use of contaminated agricultural land is not feasible, an alternative use of the land should be considered like for urban purposes such as parking, roads, warehouses, etc.
Cultivation of nonedible plants	In order to preserve agricultural practices on contaminated land, nonedible plants might be cultivated, i.e., those for industrial purposes, woods, or biofuels

material with clean soil or subsoil in order to reduce the maximum concentrations of contaminants to below the threshold trigger values.

While considering remediation strategies, users can thus select any one of the methods like immobilization, extraction, or separation. The techniques, which allow stabilization of contaminants in soil using chemical or biological methods, for example, chemophytostabilization, could be very practical, due to its technical simplicity and relatively low cost. The chemicals bind the excess of metals and help to maintain an appropriate pH of soil and may serve as plant nutrient as well. The engineering solutions, for example, soil washing (Anderson 1993; Peijnenburg et al. 2007; Dermont et al. 2008), soil heating (Abramovitch et al. 2003; Jou 2006;

Kucharski et al. 2005b), and electrokinetics method (Faulkner et al. 2005; Altin and Degirmenci 2005) are, however, quite expensive but may be the only practical solution to highly dangerous conditions. So, the search is on to develop efficient and inexpensive remediation method and bioremediation can fulfill this requirement. In a less threatening situations, combination of engineering and/or bioremediation could, however, be the most effective, sustainable, and practical approach. Moreover, the plants, when used in bioremediation technology, also build a dense root mat, which may (1) prevent wind erosion of contaminated material, (2) reduce the leaking of polluted water, and (3) allow metals to accumulate in roots (Berti and Cunningham 2000). The choice of methods, however, depends on the site characteristics, current or intended land use, extent and nature of the pollutants, and available resources.

12.2   Phytoremediation: A Natural Way for Restoration of Polluted Soils

The term phytoremediation (phyto = plant and remediation = correct evil) is the name collectively used for a set of technologies that employ plants to clean up contaminated sites. Broadly, this is an environmentally friendly and visually attractive technology that involves the use of metal-accumulating plants to remove, transfer, or stabilize the contaminants from polluted soils. Plant-based soil remediation systems can be viewed as solar-driven, pump-and-treat systems with an extensive, self-extending uptake network (the root system) that enhances the belowground ecosystem for subsequent productive use. The performance of the phytoremediation technique is, however, affected largely by plant genotypes, speciation, and concentration of metals present in sites to be remediated and action of soil microbes. Based on their ability to remove heavy metals from contaminated sites, plants have been categorized as indicators, accumulators, and excluders.

Application of special amendments has, however, shown the increase (Huang et al. 1997; Salt et al. 1998; Evangelou et al. 2007) or decrease (Vangronsveld et al. 2009) in the availability of metals to plants. The soil amendment like any other emerging new technology has not been fully tested and, therefore, should be checked. The main risk of the method is, however, that it may lead to the contamination of groundwater, which may occur: (1) when dose of amendments applied is too high, (2) if weather conditions are not appropriate, (3) when soil does not react as expected, and (4) when acid rains overlap with soil amendment and increase the mobility of metals in soils and thereby leach to the groundwater. A routine risk assessment procedure is therefore required to determine the impact of pollutants on land users when exposed to pollutants (Paustenbach et al. 1992) and also to see whether this is socially accepted or not. In this regard, the exposure assessment method of Kucharski et al. (1994) might help in predicting the hot spots and to identify the contaminated sites, which could be subjected to cleaning in residential and agricultural areas. The key issue is to determine the intended land use and to decide "how clean is clean" for each specific case. It should further be stressed that

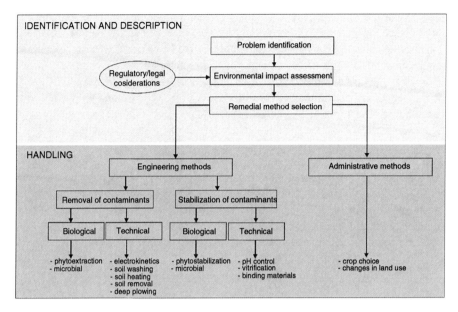

Fig. 12.1 A flow chart showing management of contaminated land

technical activities aimed at safe use of soils must follow the existing legal regulations. When the hazard is already well identified, an appropriate policy has to be followed, which, however, may vary from technical to administrative means. Possibilities of various approaches that can be applied to large areas, agricultural and brownfields, distributed around a decommissioned industrial metal production complex, are presented in Fig. 12.1.

12.2.1 *Advantages and Limitations of Phytotechnologies*

Even though phytoremediation is considered possibly the cleanest and cheapest technology used widely in the remediation of selected polluted sites, this technology has both advantages and disadvantages. For example, phytoremediation has the advantage of the unique and selective uptake abilities of root systems, together with the translocation, bioaccumulation, and contaminant storage/degradative abilities of the whole plants (Nascimento and Xing 2006; January et al. 2008). Moreover, it offers large-scale and on-site treatment of contaminated areas (Vangronsveld et al. 2009; Kucharski et al. 2005a). Thus, the major advantages of this technology include (1) low cost, (2) easy to implement and maintain, (3) far less disruptive to the soil environment, and (4) avoids excavation and is socially acceptable. The disadvantages of this remediation method, on the other hand, are as follows: (1) since root contact with contaminants in this technology is important, the contaminants must be in contact with the root zone of the plants. Therefore, plants should be able to extend roots to the contaminant region. (2) High concentrations of pollutants may inhibit plant growth, which may result in yield losses and poor plant cover. (3) It is time consuming due to the slow growth rate of plants. (4) It can be affected by fluctuating environments. (5) The effect of agronomic practices on this technology is less understood. (6) After remediation, plant biomass needs to be removed. (7) There is danger of spread of pollutants (e.g., metals) into the soil environment even during remediation. Therefore, to overcome these problems and to make phytoremediation a viable and affordable technology, we need to search plants that could grow faster and be able to produce (1) extensive root systems, (2) high biomass, (3) should have lower-level contaminant uptake ability, and (4) be able to accumulate higher amounts of contaminants. Despite the conflicting reports on the use of phytotechnologies and its tremendous potential in the containment of hazardous sites, this technology in clean-up program is still preferred over conventional physico-chemical methods around the world. Some of the commonly applied phytotechnologies and their importance in remediation of heavy metals are described in the following section.

12.2.2 Phytoextraction

Phytoextraction, which involves the use of plants to extract toxic metals from contaminated soils and store them in harvestable tissues, has emerged as a cost-effective, environment friendly cleanup alternative. Among plants, certain plant species called hyperaccumulators are reported to accumulate excessively high concentrations of heavy metals (Szchwartz et al. 2003; McGrath and Zhao 2003; McGrath et al. 2006) at concentrations 10–100 times greater than could be tolerated by many normal plants (Kukier et al. 2004). The most commonly used plants in metals extraction are *Brassica* sp. and *Helianthus* sp. (Sas-Nowosielska et al. 2005; Gupta et al. 2010). Indeed, the use of such plants at larger scale in metal extraction is limited due to lack of good quality seeds, low biomass producing ability of the plants, and harvesting problems. The low metal accumulation ability of plants could, however, be circumvented by the use of plant species capable of producing extensive biomass on contaminated soils. Plants with massive biomass production but poor accumulation rates remove certain amounts of metals, and hence, the process becomes very slow. The addition of chelators to contaminated soil however can enhance metal uptake by plants. This approach is often called as "induced phytoextraction" (Salt et al. 1998). For phytoextraction, site characterization involving description of target contaminants, treatability study (TS), site layout and design, supply and application of amendments, field engineering and metal analysis, and crop disposal (Ensley 2000) should be considered. The other important issues associated with phytoextraction is the selection of plant species, identification of optimal conditions for metal uptake into the aboveground portion of the plants, and assessment of soil whether it supports plant growth during remediation (Sas-Nowosielska et al. 2000) or not. These parameters are evaluated during treatability study, which includes short-term plant uptake investigations, conducted under controlled conditions toward evaluating growth, and metal uptake potential of selected plant species. The TS is generally applied to (1) demonstrate applicability, (2) speed up the process, and (3) decrease the cost of phytoextraction. The purpose of TS is, therefore, broadly to determine (1) whether the soil to be remediated will support the growth of candidate plant species, (2) the type and quantity of amendments needed, and (3) the optimal plant growth period for applying the amendments. The results of growth-chamber studies, however, often fail under field environment due to variation in factors, like humidity, soil structure and heterogeneity, rainfall, and pests; all of which may lead to differing plant response. The TS in general, thus, helps to optimize plant growth and maximize the removal of metals from soils.

Site characterization and TS are conducted sequentially prior to the start of full-scale planting experiments. The traditional TS are conducted under controlled environments in greenhouse conditions. Therefore, the results observed in the laboratory may not be the same under field conditions. The purpose of these activities is, however, to identify the nature and extent of contamination at the target site, and to determine if, and under what conditions, proposed plant species will extract the target contaminants. Furthermore, most of the phytoextraction processes used to assay the inorganic contaminants involve the use of a chelating agent for increasing the bioavailability of the target contaminant (Table 12.2). For example, EDTA, organic acid, and herbicide in a study were found to stimulate the accumulation of lead and cadmium in plants growing in contaminated soil (Sas-Nowosielska et al. 2001). In a follow-up study, about 100 mg of Pb and 10 mg of Cd were extracted from 1 m^2 of medium-contaminated soil in Poland (Sas-Nowosielska 2009). Since EDTA has been reported to exhibit toxicity to soil microbes, its impact on microbial life was investigated in a 4-year field experiment (Galiulin et al. 1998; Galimska-Stypa et al. 2000). No adverse effect of EDTA on soil biological function was, however, detected.

Another test used to determine the efficacy of phytoextraction method is the streamline test (ST), reported by several workers (Korcz et al. 1998; Sas-Nowosielska et al. 2001). This test is based on a combination of lab (TS) and field methods (ST) and is used to assess the applicability of phytoextraction in a given environmental situation. In addition, it provides an early indication of the suitability of the site for the application of phytoremediation technologies.

The concept of the streamline test was based on a geostatistical assumption that an adequately distributed number of soil and plant samples may describe the distribution of metals across an investigated site (Fig. 12.2a, b). The variability of Pb and Cd contents in soil for example was estimated at field scale in phytoextraction experiments (Kucharski et al. 1998, 2002). Based on these findings, it was suggested that two crossing strips covering approximately 20% of the total site surface would be sufficient to represent the entire area for site characterization purposes. To prove this hypothesis, topsoil samples were taken from outside and inside of the strips and were analyzed for metal contents. No significant difference was found among concentrations of Pb, Cd, and Zn in soil collected from inside and outside strips. It was concluded that the ST or TS better reflects the "real world" conditions as compared to the regular treatability study. The phytotoxic effect of heavy metal–contaminated soil

Table 12.2 Evaluation of lead and zinc plant concentration as well as extraction of metals during phytoextraction process in natural conditions

Plant species		Accumulation						d.m.	Phytoextraction					
		Control		amendment I		amendment II			Control		amendment I		amendment II	
		mg kg⁻¹						kg m²	mg m⁻²					
		Pb	Cd	Pb	Cd	Pb	Cd		Pb	Cd	Pb	Cd	Pb	Cd
Brachinia	mean	18.90	8.29	61.60	8.18	56.10	8.86	0.4	8	3	24	3	22	4
	SD	5.80	2.44	33.60	1.50	26.80	1.89							
Corn	mean	9.90	3.15	26.00	3.50	35.10	4.10	0.9	9	3	23	3	31	4
	SD	4.00	0.75	17.80	0.75	5.40	0.88							
Indian	mean	15.00	11.06	153.00	18.65	176.90	14.59	0.7	10	8	107	13	123	10
mustard	SD	1.30	2.25	43.80	8.29	36.40	2.45							

Adapted from Sas-Nowosielska (2009)

Control no additives, *amendment I* 5 mmol EDTA + R*; *amendment II* 5 mmol EDTA + 5 mmol citric acid + R*

*R** herbicide, *SD* standard deviation

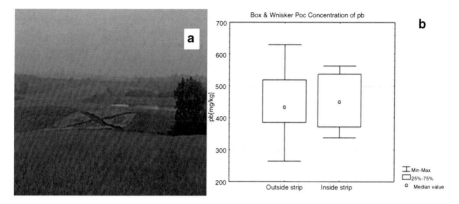

Fig. 12.2 Streamline test (**a**) and statistical evaluation (**b**)

collected from inside and outside the strips was also analyzed. It was observed that the strips reflected the pattern of plant growth at the test site.

Based on these findings, it was suggested that ST can be applied as an alternative to other standard methods used for site characterization and assessment of the toxicity of metals on the biomass-producing potential of plants. This is in contrast to traditional treatability studies where homogenized soil is used. In addition, by employing ST, the potential effectiveness of phytoextraction can be evaluated, and the method can be removed at the early stages of the process, if it is not applicable to test areas.

12.2.2.1 Economics of Phytoextraction

The major problem in any environmental remediation strategy is the cost associated with pollutants removal from derelict sites. To understand it further, 1-year data on the effect of various operational aspects of phytoextraction in order to identify means for reducing the costs was studied. It is unrealistic to expect to extrapolate lab or greenhouse experiments costs to full-scale operation. To know the cost of soil remediation under field environment, a detailed accounting should be maintained for all expenditures connected with cleaning-up activities and costs for all activities like, environmental monitoring, routine agricultural activities such as, planting, fertilization, harvesting, phytoremediation process (e.g., amendment application), contaminated crop disposal, and scientific supervision of the process. In a demonstration project, Kucharski et al. (2002) considered all activities from site characterization through final disposition of contaminated biomass. Moreover, attention was also paid to calculating the costs of the technical processes in order to see how these could be decreased while increasing the effectiveness and safety of the operation. To determine the generic value of each operation, all expenditures connected with the project were recorded and categorized. During the early period of research, the most urgent needs and gaps in knowledge were identified and cost analysis was done.

Table 12.3 Total cost of
phytoextraction process

Step of the process	% of total cost
Field preparation	<1
Fertilizers and plant protection	<1
Chemicals for plant protection	<1
Plant care	<1
Irrigation	<1
Seeds and planting	7
Sampling and monitoring	7
Amendments	70
Contaminated crop disposal	<1
Scientific supervision	15

Adapted from Kucharski et al. (2002)

Fig. 12.3 Amendment application device

Of the total average cost (about 15 US $/m^3/year), cost of amendment was the important factor impacting the cost of phytoextraction (Table 12.3). Based on this finding, it was suggested that the use of reagents should be reduced. As the required amounts of chelating agent (EDTA) were computed stoichiometrically to meet the amounts of metals contained in the soil, investigations focused on amendment distribution in terms of (1) precise orientation to target the plant and (2) location-specific application based on actual metal concentration. For that purpose, a computer-driven device was designed and a prototype was built (Fig. 12.3).

According to such approach, it was possible to reduce the costs of EDTA application by about 20% (Kucharski et al. 2002). In highly contaminated areas, as found in the vicinity of lead and zinc smelter, the performance of phytoextraction process may, however, be highly limited due to shortage of plant species and/or the ability of plants to produce low level of biomass (Sas-Nowosielska et al. 2008). The loamy soil with neutral pH and organic matter content had high level of metals

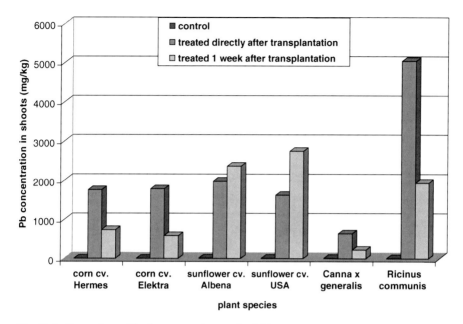

Fig. 12.4 Comparison of lead concentration (mg/kg DW) in plants in relation to time of amendment application to transplanted plants

(Houba et al. 1995; ISO 2008). Accordingly, when sunflower (*Helianhus annuus*) and *Ricinus communis* were grown in such soils, they accumulated higher concentrations of cadmium and lead in shoots (Figs. 12.4 and 12.5) despite optimum care, like, proper fertilizer application and regular watering. These plants also grew poorly in the area contaminated with metals. The content of Pb and Cd in sunflower cultivars was higher when plants were treated 1 week after transplantation. On the contrary, *Ricinus communis* showed higher concentrations of metals when treated directly after transplantation.

These results, thus, suggest that phytoextraction should be employed only for low- or medium-contaminated sites. However, for highly contaminated soil, stabilization seems to be the most appropriate method of remediation. The integral part of phytoremediation strategy is the isolation or disposal of contaminated materials. And hence, when plants are used in various schemes of phytoremediation, the contaminated tissues must be handled carefully and the contaminant be properly sequestered (Sas-Nowosielska et al. 2004). Considering these, we tried to identify the locally available plants and potential procedures for processing the contaminated biomass. Incineration was also included as a treatment process for contaminated plant material since incineration has been found effective and results in dramatic reduction in both mass and volume. However, it must be handled under carefully controlled conditions to avoid redistribution of carefully recovered metals. Incineration was conducted by professional institutions equipped and licensed to handle hazardous wastes. The other option is to incinerate the material in a lead/zinc smelter or cement kilns, using simple technologies such as rotary kilns (the Waelz process) in

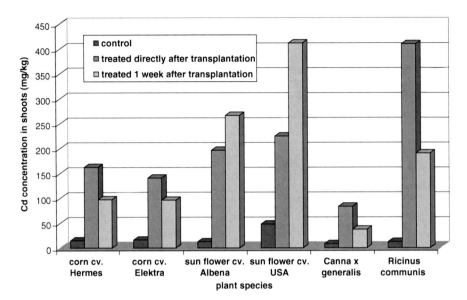

Fig. 12.5 Comparison of cadmium concentration (mg/kg DW) in plants in relation to time of amendment application to transplanted plants

which even heterogeneous material can be accommodated in the process. Modern flue-gas cleaning technology will assure capture of metal-containing dust.

12.2.3 Phytostabilization

Phytostabilization is yet another remediation technology that involves the immobilization of soil pollutants by absorbing and accumulating them through roots or precipitated within the root zone of plants. The use of plants and plant roots in this technology also helps to prevent contaminant migration via wind and water erosion, leaching, and soil dispersion from contaminated sites. In this case, the upper layer of soils is first treated with chemicals (e.g., lime, fertilizers, stabilizers) to adjust the soil pH; fertilized and metal compounds are transformed into non-soluble forms. The next step is to develop a robust plant cover. The techniques, which allow for stabilizing pollutants in soil using chemical or biological methods, could be very practical, considering their technical simplicity and low cost. The chemicals bind the excess of metals and help to maintain pH and required plant nutrition. Various modifications in phytochemostabilization technique have shown promising results in preventing the heavy metals migration and erosion from polluted land (Bidar et al. 2007; Stuczynski et al. 2007; Bes and Mench 2008). The most important features of plant species used in soil phytostabilization should however be that plants be able to (1) tolerate high concentrations of pollutants, (2) develop a dense root mat, (3) accumulate pollutants in aboveground parts, and (4) resist variation in climatic conditions (Vangronsveld et al. 2009). The most ideal plants for stabilization purposes are, however, those plants that retain pollutants in underground parts, do not permit unwanted substances to be grazed by animals, and penetrate the food chain. Based on our investigations, *Agrostis capillaris, Salix viminalis, Festuca rubra, Armoracia lapathifolia,* and *Helianthus tuberosus* cannot be recommended for very high polluted soils, even after chemostabilization processing of soils. Usage of soil amendments (stabilizers) at reasonable levels strongly reduces the availability of the pollutants for plant uptake (Knox et al. 2000). Therefore, in order to identify more ideal plants species, screening should be focused on local vegetation. Especially in extremely polluted sites, the use of indigenous species of plants provides an opportunity to develop a nice soil cover. Polluted site-related species have already proven its importance in surviving under pollution stress. In this context, a better growth was observed for indigenous plant species *Deschampsia caespitosa L., Silene inflate, and Melandrium album* while growing on the soil close to the foundry.

Locally and easily available soil additives that may act as metal stabilizers (e.g., sewage sludge, mixture of dolomite and zeolite (Biodecol), zeolites, lignite, ammonium polyphosphate, or calcium phosphate) may help stabilization process immensely. Of these additives, biodecol and sewage sludge may slightly change the pH and EC of soil while others are reported to reduce the concentration of bioavailable metals (Kucharski et al. 2004). In a study, we determined the bioavailable forms of different metal using stabilization experiment in brownfield, which is located in the vicinity of former nonferrous metal smelter (the Upper Silesia Region, southern part of Poland). The soil of this experimental site was loamy, with neutral pH and had higher organic matter and EC and was highly polluted with Pb (7,679 mg/kg), Zn (9,879 mg/kg), and Cd (427 mg/kg). Of these, 73% Pb, 69% Zn, and 68% Cd were present in bioavailable form while 0.08, 3.52, and 14.5% of Pb, Zn, and Cd, respectively, were in solution form (Sas-Nowosielska 2009).

In addition, despite proper fertilization and irrigation, plants such as *Agrostis capillaries, Salix viminalis* and *Festuca rubra,* grown in highly contaminated soil treated with stabilizers, showed poor growth. Therefore, it was suggested to use the local vegetation that may have a chance to develop soil cover with plants. In this context, plant species like *Silene inflata, Cardaminopsis arenosa,* and *Deschampsia cespitosa* (Warynski ecotype) used separately and as mixture (20% *S. inflata,* 40% *C. arenosa* and 40% *D. cespitosa*) were investigated in stabilization studies. Metal stabilization was enhanced after lime addition and calcium phosphate application (3.8% w/w). Lime was applied to 0–20 cm of soil layer, whereas calcium phosphate was introduced at 10 cm of soil depth. During growing season, natural succession of *S. inflata, C. arenosa,* and *Melandrium album* from the local vegetation was observed, even though only *D. caespitose* was planted (Table 12.4). Metal-accumulating abilities of *D. caespitosa* and *C. Arenosa,* however, differed considerably. Well-known Zn and Cd hyperaccumulator *C. arenosa* was not found suitable for stabilization purposes. In plots treated with additives, *C. arenosa* did not grow while very poor soil coverage was observed for *Deschampsia. Cardaminopsis arenosa,* however, dominated the test site in the absence of additives.

Table 12.4 Characteristic of plant cover and plant communities in field experiment

Additive/plant	First year		Second year	
	Plant communities[a]	Plant cover	Plant communities	Plant cover
No additive/*Deschampsia caespitosa*	*Deschampsia caespitosa* 3	25	*Deschampsia caespitosa* 4	40
	Cardaminopsis arenosa 1		*Silene inflata* 1	
	Silene inflata <1		*Cardaminopsis arenosa* 1	
Calcium phosphate/ *Deschampsia caespitosa*	*Deschampsia caespitosa* 5	90	*Deschampsia caespitosa* 5	100
	Melandrium album <1		*Melandrium album* 1	
	Silene inflata <1		*Silene inflata* <1	
	Cardaminopsis arenosa <1		*Cardaminopsis arenosa* R	

Adapted from Gombert et al. (2004) and Sas-Nowosielska (2009)
[a]The Braun–Blanquet scale was used to estimate the cover of each species in a following scale: 0.5 (<1% cover); *1* (1–5% cover); *2* (6–25% cover); *3* (26–50% cover); *4* (51–75% cover); *5* (76–100% cover); *R* (only several plants were found)

Amendments introduced to the contaminated soil, however, did not change the pH and reduced metal concentrations by several folds in roots and shoots (Table 12.5). Even when additives were applied at a depth of 0–10 cm, the binding effect of metals was found at the depth of 20–40 cm (Sas-Nowosielska 2009). The decrease of cadmium and zinc concentration between the start and the end of experiment was two and threefolds, respectively. Lead content in leachates however, did not change significantly (Fig. 12.6).

12.2.3.1 Advantages

1. In this technology, the mobility of contaminants could be reduced and, hence, the risk associated with them is minimized.
2. No contaminated secondary waste is generated during this process.
3. Compared to other remediation technologies, this is simple to operate and less expensive.
4. This technology may be used in combination with other technology and improves soil fertility.

12.2.3.2 Disadvantages

1. Since the contaminants are left in place, so a regular monitoring of site is needed.
2. At higher concentrations, pollutants impair plant growth unless uptake of pollutants by plants is reduced.
3. If soil additives are used, they are required to be applied consistently so that the efficacy of the method is maintained.

Table 12.5 Metal accumulation (mg kg⁻¹) in *Deschampsia caespitosa* roots and shoots in field experiment

Additive		Pb (mg kg⁻¹)		Cd (mg kg⁻¹)		Zn (mg kg⁻¹)	
		Shoot	Root	Shoot	Root	Shoot	Root
First year	Control	759 ± 215	10,419 ± 1,031	44.4 ± 11.4	972 ± 28.2	1,282 ± 207	8,171 ± 1,149
		(335–1,027)	(8,587–12,045)	(21.6–57.0)	(924.5–1,022)	(903–1,616)	(6,900–10,464)
	Calcium	648 ± 11	387 ± 28	63 ± 1.3	513 ± 29	1,108 ± 35	6,332 ± 164
	phosphate	(635–670)	(356–442)	(62–66)	(459–558)	(1,053–1,173)	(6,015–6,563)
Second year	Control	157 ± 13.9	526 ± 112	47 ± 1.7	1,258 ± 73	1,154 ± 44	2,785 ± 78
		(131–179)	(350–735)	(44–50)	(1,125–1,375)	(1,092–1,238)	(2,635–2,897)
	Calcium	140 ± 10	585 ± 122	19.2 ± 1.8	437 ± 24	754 ± 32	2,992 ± 178
	phosphate	(122–157)	(378–799)	(17–23)	(398–480)	(719–818)	(2,760–3,342)

Adapted from Kucharski et al. (2005a)

Values represent mean of three replicates ± SE; *values in parentheses* indicate minimum and maximum concentrations

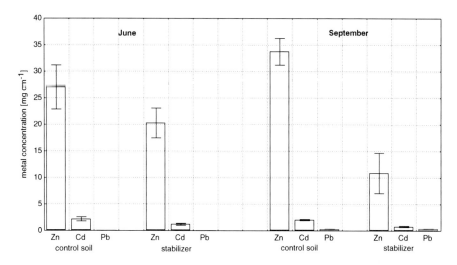

Fig. 12.6 Zinc, cadmium, and lead concentration in leachates after addition of stabilizer (Adapted from Kucharski et al. 2005a)

12.3 Conclusion

The advent of remediation technologies has made it possible to clean up the polluted soil. However, the major constraints while applying remediation approaches have been the side effects of the remediation methods coupled with cost of technology. In the case of improvement of agricultural soil contaminated heavily with metal, the choice of methods is essentially limited to (1) proper land use, (2) "soft" biological methods with all their limitations, and (3) so-called hard technical methods, which are rapid but expensive and accompanied by many unwanted side effects. To overcome such constraints, chemophytostabilization offers an interesting option for restoration of metal-contaminated sites. In this context, the chemical stabilizers change the metal solubility and mobility, whereas plant exudates alter the chemistry and microbiology of the root zone, and may have variable impact on the physico-chemical properties of soils. Plant roots may also prevent contaminant migration and mobilization and play a critical role in reducing the erosion of soil.

Metal stabilizers have been found to diminish bioavailable content of metals in soil and have reduced metal movement down the soil profile. Even though amendments can be applied only onto the upper layer of soil, they may drift to the deeper layers and therefore reduce bioavailable metals, even at the levels of 20–40 cm. Apart from using chemicals in the metal stabilization process, it is also important to generate a plant cover from those growing locally. From such native areas, plants able to accumulate and stabilize heavy metals in their organs should be selected. Other method like phytoextraction is simple but requires an extensive understanding of the technical aspects associated with plant propagation and how environmental factors impact plant development in different ecosystems. Exploitation of biological processes requires flexibility and willingness to modify the planned approach based on site-specific conditions. Single soil pollutants are generally removed in laboratory experiments. Presence of elements as mixtures may however destroy the efficiency of phytoextraction. Therefore, future research on phytotechnologies should focus on identifying plants capable of accumulating higher concentration and more than one metal at a time or on engineering the existing plant species in order to find an ideal plant that could survive well under metal-stressed soils (Jiang et al. 2010).

Acknowledgment The US Department of Energy, EM, State University through Cooperative Agreement # DE-FCZI-95EW55101, EU Fifth Frame Program, and Polish Ministry of Science and Highest Education Poland are greatly acknowledged for funding this project.

References

Abramovitch RA, Chang QL, Hicks E, Sinard J (2003) In situ remediation of soils contaminated with toxic metal ions using microwave energy. Chemosphere 53:1077–1085

Altin A, Degirmenci M (2005) Lead (II) removal from natural soils by enhanced electrokinetic remediation. Sci Total Environ 337:1–10

Anderson WC (1993) Soil washing/soil flushing, vol 1, The innovative site remediation technology series. American Academy of Environmental Engineers, Annapolis

Berti WR, Cunningham SD (2000) Phytostabilization of metals. In: Raskin I, Ensley BD (eds.) Phytoremediation of toxic metals: using plants to clean up the environment. Wiley, New York, pp 71–88

Bes C, Mench M (2008) Remediation of copper-contaminated topsoils from a wood treatment facility using *in situ* stabilization. Environ Pollut 156(3):1128–1138

Bidar G, Garcon G, Dewaele D, Cazier F, Douay F, Shirali P (2007) Behavior of *Trifolium repens* and *Lolium perenne* growing in a heavy metal contaminated field: plant metal concentration and phytotoxicity. Environ Pollut 147:546–553

Dermont G, Bergeron M, Mercier G, Richer-Laflèche M (2008) Soil washing for metal removal: a review of physical/chemical technologies and field applications. J Hazard Mater 152:1–31

Ensley BD (2000) Rationale for use of phytoremediation. In: Raskin I, Ensley BD (eds.) Phytoremediation of toxic metals. Using plants to clean up the environment. Wiley, New York, pp 3–11

Evangelou MWH, Ebel M, Schaeffer A (2007) Chelate assisted phytoextraction of heavy metals from soil. Effect, mechanism, toxicity, and fate of chelating agents. Chemosphere 66:816–823

Faulkner DWS, Hopkinson L, Cundy AB (2005) Elektrokinetic generation of reactive iron-rich barriers in wet sediments: implications for contaminated land management. Mineral Mag 69:749–757

Galimska-Stypa R, Sas-Nowosielska A, Kucharski R, Dushenkov S (2000) Ecological risks caused by application of EDTA to soil. In: Fifth international symposium and exhibition on environmental contamination in central and eastern Europe, Manuscript 754, Prague

Galiulin RV, Bashkin VN, Galiulina RR, Kucharski R, Malkowski E (1998) The impact of phytoextraction effectors on the enzymatic activity of soil contaminated by heavy metals. Agric Chem 2:243–251

Gombert S, Asta J, Seaward MRD (2004) Assessment of lichen diversity by index of atmospheric purity (IAP), index of human impact (IHI) and other environmental factors in an urban area (Grenoble, southeast France). Sci Total Environ 324:183–199

Gupta AK, Mishra RK, Sarita S, Lee B-K (2010) Growth, metal accumulation and yield performance of *Brassica campestris* L. (cv. Pusa Jaikisan) grown on soil amended with tannery sludge/fly ash mixture. Ecol Eng 36:981–991

Gzyl J (1999) Soil protection in central and eastern Europe. J Geochemical Exploration 66:333–337

Houba VJG, Van der Lee JJ, Novozamsky I (1995) Soil analysis procedures, other procedures (Soil and plant analysis, part 5b). Department of Soil Science and Plant Nutrition, Wageningen Agricultural University, Wageningen

Hovmond MF, Tjell JC, Mosbaek H (1983) Plant uptake of air-borne cadmium. Environ Pollut A 30(1):27–38

Huang JW, Chen JJ, Berti WR, Cunningham SD (1997) Phytoremediation of lead contaminated soils: role of synthetic chelates in lead phytoextraction. Environ Sci Technol 31:800–805

Intawongse M, Dean JR (2006) Uptake of heavy metals by vegetable plants grown on contaminated soil and their bioavailability in the human gastro-intestinal tract. Food Addit Contam 23:36–48

ISO (2008) 17402: Soil quality – requirements and guidance for the selection and application of methods for the assessment of bioavailability of contaminants in soil and soil materials. ISO, Geneva

January MC, Cutright TJ, Van Keulen H, Wei R (2008) Hydroponic phytoremediation of Cd, Cr, Ni, As, and Fe: can *Helianthus annuus* hyperaccumulate multiple heavy metals. Chemosphere 70:531–537

Jiang C, Wu QT, Sterckeman T, Schwartz C, Sirguey C (2010) Co-planting can phytoextract similar amounts of cadmium and zinc to mono-cropping from contaminated soils. Ecol Eng 36:391–395

Jou CG (2006) An efficient technology to treat heavy metal-lead-contaminated soil by microwave radiation. J Environ Manag 78:1–4

Knox AS, Seaman J, Adriano DC, Pierzynski G (2000) Chemophytostabilization of metals in contaminated soils. In: Wise DL, Trantolo DJ, Cichon EJ, Inyang HI, Stottmeister U (eds.) Bioremediation of contaminated soils. Marcel Dekker, New York/Basel, pp 811–836

Korcz M, Bronder J, Sas-Nowosielska A, Kucharski R (1998) Analysis of soil Cd and Pb over small areas for phytoremediation purpose. In: Warsaw'98, Fourth international symposium and exhibition on environmental contamination in central and eastern Europe, symposium proceedings, Manuscript 116

Kucharski R, Sas-Nowosielska A (2001) Soil remediation–an alternative to abate human exposure to heavy metals. Int J of Occup Med Environ Health 14(4):385–387

Kucharski R, Marchwińska E, Gzyl J (1994) Agricultural policy in polluted areas. Ecol Eng 3:299–312

Kucharski R, Sas-Nowosielska A, Dushenkov S, Kuperberg JM, Pogrzeba M, Malkowski E (1998) Technology of phytoextraction of lead and cadmium in Poland. Problems and achievements. In: Warsaw'98, Fourth international symposium and exhibition on environmental contamination in central and eastern Europe, symposium proceedings, Manuscript 55

Kucharski R, Sas-Nowosielska A, Malkowski RE (2002) Integrated approach to the remediation of heavy-metal contaminated land. Final report. U.S. Department of Energy/EM 50. Office of science and technology, Germantown

Kucharski R, Sas-Nowosielska A, Malkowski RE (2004) A decision support system to quantify cost/benefit relationship of the use of vegetation in the management of heavy metal polluted soils and dredged sediments. Final report. Alterra, Wageningen

Kucharski R, Sas-Nowosielska A, Małkowski E, Japenga J, Kuperberg JM, Pogrzeba M, Krzyżak J (2005a) The use of indigenous plant species and calcium phosphate for the stabilization of highly metal-polluted sites in southern Poland. Plant Soil 273:291–305

Kucharski R, Zielonka U, Sas-Nowosielska A, Kuperberg JM, Worsztynowicz A, Szdzuj J (2005b) A method of mercury removal from topsoil using low-thermal application. Environ Monit Assess 104:341–351

Kukier U, Peters CA, Chaney RL, Angle JS, Roseberg RJ (2004) The effect of pH on metal accumulation in two *Alyssum* species. J Environ Qual 33:2090–2102

Marshall FM, Holden J, Ghose C, Chisala B, Kapungwe E, Volk J, Agrawal M, Agrawal R, Sharma RK, Singh RP (2007) Contaminated irrigation water and food safety for the urban and peri-urban poor: appropriate measures for monitoring and control from field research in India and Zambia. Inception Report DFID Enkar R8160, SPRU, University of Sussex, Brighton. www.pollutionandfood.net

Martelli A, Rousselet E, Dycke C, Bouron A, Moulis JM (2006) Cadmium toxicity in animal cells by interference with essential metals. Biochimie 88:1807–1814

McGrath SP, Zhao FJ (2003) Phytoextraction of metals and metalloids from contaminated soils. Curr Opin Biotechnol 14:277–282

McGrath SP, Lombi E, Gray CW, Caille N, Dunham SJ, Zhao FJ (2006) Field evaluation of Cd and Zn phytoextraction potential by the hyperaccumulators *Thlaspi caerulescens* and *Arabidopsis halleri*. Environ Pollut 141:115–125

Muchuweti M, Birkett JW, Chinyanga E, Zvauya R, Scrimshaw MD, Lister JN (2006) Heavy metal content of vegetables irrigated with mixtures of wastewater and sewage sludge in Zimbabwe: implication for human health. Agric Ecosys Environ 112:41–48

Nascimento CWA, Xing B (2006) Phytoextraction: a review on enhanced metal availability and plant accumulation. Sci Agric 63:299–311

Palmgren MG, Clemens S, Wiliam LE, Krämer U, Borg S, Schøjrring JK, Sanders D (2008) Zinc biofortification of cereals: problems and solutions. Trends Plant Sci 13:464–473

Paustenbach DJ, Jernigan JD, Bass R, Kalmes R, Scott P (1992) A Proposed approach to regulating contaminated soil: identify safe concentrations for seven of the most frequently encountered exposure scenarios. Regul Toxicol Pharmacol 16:21–56

Peijnenburg W, Zablotskaja M, Vijverb M (2007) Monitoring metals in terrestrial environments within a bioavailability framework and a focus on soil extraction. Ecotoxicol Environ Saf 67:163–179

Salt DE, Smith RD, Raskin I (1998) Phytoremediation. Ann Rev Plant Physiol Plant Mol Biol 49:643–668

Sas-Nowosielska A (2009) Application of phytotechnologies in lead and zinc contaminated areas. Monograph in Polish. Czestochowa University of Technology, Czestochowa. ISBN 978-83-7193-466-7

Sas-Nowosielska A, Kucharski R, Małkowski E, Nowosielski O, Pogrzeba M (2000) Zinc plant toxicity in process of lead and cadmium phytoextraction. Ochr Środ Zas Nat Warszawa 20:35–39

Sas-Nowosielska A, Kucharski R, Korcz M, Kuperberg JM, Małkowski E (2001) Optimizing of land characterization for phytoextraction of heavy metals. In: Gworek B, Mocek A (eds.) Element cycling in the environment, bioaccumulation–toxicity–prevention, Monograph. Institute of Environmental Protection, Warsaw, pp 345–348. ISBN 83-85805-76-11

Sas-Nowosielska A, Kucharski R, Małkowski E, Pogrzeba M, Kuperberg JM, Krynski K (2004) Phytoextraction crop disposal – an unsolved problem. Environ Pollut 128:373–379

Sas-Nowosielska A, Kucharski R, Małkowski E (2005) Feasibility studies for phytoremediation of metal-contaminated soil. In: Margesin R, Schinner F (eds.) Soil biology, manual for soil analysis, vol 5. Springer, Berlin/Heidelberg, pp 161–177

Sas-Nowosielska A, Kucharski R, Pogrzeba M, Małkowski E (2008) Soil remediation scenarios for heavy metal contaminated soil. In: Simeonov L, Sargsyan R (eds.) Soil chemical pollution, risk assessment, remediation and security. The NATO science for peace and security programme. Springer Science, Dordrecht, The Netherlands pp 301–309

Singh A, Sharma RK, Agrawal M, Marshall FM (2010) Health risk assessment of heavy metals via dietary intake of foodstuffs from the wastewater irrigated site of a dry tropical area of India. Food Chem Toxicol 48:611–619

Stuczynski T, Siebielec G, Daniels WL, McCarty GW, Chaney RL (2007) Biological aspect of metal waste reclamation with sewage sludge. J Environ Qual 36:1154–1162

Szchwartz C, Echevaria G, Morel JL (2003) Phytoextraction of cadmium with *Thlaspi caerulescens*. Plant Soil 249:27–35

Tomohito A, Satoru MM, Kaoru A, Yuji M, Tomoyuki M (2010) Heavy metal contamination of agricultural soil and countermeasures in Japan. Paddy Water Environ 8:247–257

Vangronsveld J, Herzig R, Weyens N, Boulet J, Adriaensen K, Ruttens A, Thewys T, Vassilev A, Meers E, Nehnevajova E, van der Lelie D, Mench M (2009) Phytoremediation of contaminated soils and groundwater: lessons from the field. Environ Sci Poll Res 16:765–794

Chapter 13
Chromium Pollution and Bioremediation: An Overview

Nilanjana Das and Lazar Mathew

Abstract Chromium, a steel-gray, lustrous, hard, and brittle metal, occurs in nature in bound forms and has been widely used in various industries. Chromium exists in several oxidation states, of which hexavalent chromium is a priority toxic, mutagenic, and carcinogenic chemical, whereas trivalent form is much less toxic and insoluble. Hexavalent chromium causes various chronic health disorders including organ damage, dermatitis, respiratory impairment, etc. Moreover, the discharge of chromium-containing wastes has also led to the destruction of many agricultural lands and water bodies. Therefore, the remediation of chromium contaminated sites is essentially required to offset the chromium toxicity. Many technologies like land filling, stabilization/solidification, physicochemical extraction, soil washing, and flushing are used to clean up chromium-contaminated soils. None of these techniques are completely accepted because either they do not offer a permanent solution, or they simply immobilize the contaminant or are costly when applied to a large area. Bioremediation involving microorganisms is considered the most promising option in cleaning up the chromium-contaminated environment. Phytoremediation has gained importance in chromium remediation, which can be achieved by phytoextraction, rhizofiltration, and phyto-detoxification. A selective overview of the past achievements and current perspective of chromium remediation technologies reported by different workers using promising microorganisms and plants is given.

Keywords Chromium • Toxic pollutant • Bioremediation • Phytoremediation

N. Das (✉) • L. Mathew
School of Biosciences and Technology, VIT University, Vellore 632014,
Tamil Nadu, India
e-mail: nilanjana00@lycos.com; lazarmathew@rediffmail.com

M.S. Khan et al. (eds.), *Biomanagement of Metal-Contaminated Soils*,
Environmental Pollution 20, DOI 10.1007/978-94-007-1914-9_13,
© Springer Science+Business Media B.V. 2011

13.1   Introduction

Chromium (Cr) is the seventh most abundant element on earth (Nriagu and Pacyna 1988) and a carcinogen to humans and animals. In the environment, some chromium salts do not readily precipitate or become bound to soil components. Therefore, it can move throughout aquifers to contaminate groundwater and other sources of drinking water, which can be hazardous to humans, livestock, and wildlife. The release of chromium waste from many industrial applications such as leather tanning, textile production, electroplating, metallurgy, and petroleum refinery has led to large-scale contamination of land and water (Barlett 1991; Katz and Salem 1994). Consequently, the presence of toxic level of chromium in soils and wastewaters has become a major environmental problem. In air, chromium compounds are present mostly as fine dust particles, which eventually settle over land and water. In aqueous solutions, chromium occurs mainly as trivalent chromium [Cr (III)] and hexavalent chromium [Cr (VI)]. Chromium (III) is an essential micronutrient (Bailar 1997) that is slightly soluble in aqueous solution. It is essential to glucose, lipid, and protein metabolism at concentration below 5 ppm, but can be toxic and mutagenic at large doses (Shen and Wang 1993). By contrast, Cr (VI) is a toxic oxidizing agent with an environmental standard of the order of 0.05 ppm (Dönmez and Kocberber 2005). In the environment, Cr (VI) contamination alters the structure of soil microbial communities (Zhou et al. 2002; Turpeinen et al. 2004). As a result of reduced microbial growth and activities, organic matter accumulates Cr (VI) in soils (Mazierski 1994; Shi et al. 2002).

Several conventional methods of chromium remediation of the impacted soil rely on soil excavation, which is expensive and disruptive. On the other hand, remediation of chromium-contaminated environment through microorganisms (Hasin et al. 2010) and plants (Mangkoedihardjo et al. 2008; Butler et al. 2009) may be the best alternative technology to clean up the chromium-contaminated sites. Compared to conventional techniques, these technologies are eco-friendly and cost effective. Present review reports the current information on how chromium could be removed from soil and wastewaters employing microbes and plants under natural environment.

13.2   General Description, Discovery, and Occurrence of Chromium

Chromium is a hard, steel grey and shiny transition metal that breaks easily. It has a melting point of 1,900°C, a boiling point of 2,642°C, and a density of 7.1 g cm^{-3}. Chromium is a relatively active metal that does not react with water but does react with most of the metals. At room temperature, chromium combines slowly with oxygen to form chromium oxide (Cr_2O_3). The chromium oxide formed acts as a protective layer, preventing the metal from reacting further with oxygen. Chromium was discovered in 1797 by French chemist Louis-Nicolas Vaquelin (1763–1829) in a mineral known as Siberian red lead. The element was named after the Greek word "chroma" meaning "color" because many chromium compounds have a distinctive color, ranging from purple to black to green to orange to yellow (Young 2000). The anthropogenic inputs of chromium have increased rapidly since the industrial revolution (Ayres 1992). Chromium is extensively used in electroplating (as chrome plating), resistant alloys (e.g., stainless steel), leather tanneries, and dye productions (United States Environmental Protection Agency 1998; Ryan et al. 2002). It is reported that every 1,000 kg of "normal soil" contains 200 g Cr, 80 g Ni, 16 g Pb, 0.5 g Hg, and 0.2 g Cd (IOCC 1996). Chromium occurs mainly in three forms. Metallic chromium (Cr [0]) is a steel-grey solid with a high melting point which is used to make steel and other alloys. Chromium metal does not occur naturally; it is produced from chrome ore. Trivalent chromium occurs naturally in rocks, soil, plants, animals, and volcanic emissions. Cr (III) is produced industrially to make metals, metal alloys, and chemical compounds. Hexavalent chromium is produced industrially by heating Cr (III) in the presence of mineral bases and atmospheric oxygen.

13.3 Chemistry of Chromium

Chromium exists in oxidation state ranging from 0 to VI. Only two of them, Cr (III) and Cr (VI), are stable enough to occur in the environment. Cr (II), Cr (IV), and Cr (V) are unstable forms and very little information is available about its hydrolysis (Kotas and Stasicka 2000). The Cr (III) oxidation state is most stable and considerable energy would be required to convert it to lower or higher states. Cr (III) can form hexacoordinate octahedral complexes with a variety of ligands such as water, ammonia, urea, 15 ethylenediamine, and other organic ligands containing O_2, N, or S atoms. The complexation of Cr (III) by ligands other than OH^- increases its solubility when the ligands are in discrete molecules or ionic forms. When donor atoms are bound in a macromolecular system, the Cr (III) complex becomes more or less immobile. If the complexation from these ligands is neglected under redox and pH conditions, which normally are found in natural systems, Cr is removed from the solution as $Cr(OH)_3$ or in the presence of Fe(III), in the form of $(Cr_x, Fe_{1x})(OH)_3$ (where x is the mole fraction of Cr). When the redox potential of Cr (VI)/Cr (III) couple is high, only a few oxidants which are present in natural systems are capable of oxidizing Cr (III) to Cr (VI). Oxidation of Cr (III) by dissolved oxygen without any mediate species has been reported to be negligible, whereas mediation by manganese oxides was found to be the effective oxidation pathway in environmental systems (Kotas and Stasicka 2000).

13.4 Industrial Uses of Chromium

Chromium has wide application in the industries. Metallic chromium is mainly found in alloys such as stainless steel. It is the supreme additive, endowing alloys or metals with new properties, such as a resistance to corrosion, wear, temperature, and decay, as well as strengths, hardness, permanence, hygiene, and color. Trivalent chromium is used in a number of commercial products including dyes, pigments, and salts for leather tanning. Hexavalent chromium is used in industrial processes such as chrome plating (Gomez and Callao 2006). Chromium (II) chloride is used as reducing agent, as a catalyst in organic reactions, and in chromium plating of metals. It is used to reduce alpha-haloketones to parent ketones, epoxides to olefins, and aromatic aldehydes to corresponding alcohols as a reducing agent (Patnaik 2003). Chromium (III) chloride is used for chromium plating and tanning and can serve as textile mordant, waterproofing agent, and catalyst for polymerization of olefins. Chromium (III) sulfate is used as the electrolyte for obtaining pure chromium metal. It is used for chromium plating of other metals for protective and decorative purposes. Other important applications of this compound are (a) as a mordant in the textile industry and leather tanning, (b) to dissolve gelatin, (c) to impart green color to paints, varnishes, inks, and ceramic glazes, and (d) as a catalyst. Chromium (III) oxide is used as pigment or coloring green on glass and fabrics. It is also used in metallurgy as a component of refractory bricks, abrasives, and ceramics and to prepare other chromium salts. Chromium (III) fluoride is used in printing and dyeing woolens, metal polishing, and coloring marbles. Chromium (III) hydroxide trihydrate is used as green pigment, as mordant, as a tanning agent, and as a catalyst. Chromium (VI) oxide is used for chromium plating, copper stripping as an oxidizing agent for conversion of secondary alcohols into ketones as a corrosion inhibitor in purification of oil, and in chromic mixtures for cleaning laboratory glassware (Patnaik 2003).

13.5 Chromium Toxicity

13.5.1 Human Health Risks Associated with Chromium

Exposure to chromium may occur from natural or industrial sources. The reduction/oxidation reactions between Cr (VI) and Cr (III) are thermodynamically possible under physiological conditions. Cr (III) is relatively immobile in the aquatic system due to its low solubility in water. The low solubility retains Cr (III) in the solid phase as colloids or precipitates (Lin 2002). It is known that Cr (III) is essential for the maintenance of effective glucose, lipid, and protein metabolism in mammals (Marques et al. 2000). It is an essential micronutrient in the body and combines with various enzymes to transform sugar, protein, and fat. Cr (III) salts such as chromium polynicotinate and chromium picolinate are used as micronutrients and dietary supplements (Bagchi et al. 2001). Besides this, Cr (III) can stabilize the tertiary structure of proteins and conformation of the RNA and DNA (Zetic et al. 2001). United States Environmental Protection Agency (1999) has classified Cr (III) in Group D, not as carcinogenic to humans. Acute animal tests have shown that Cr (III) have moderate toxicity from oral exposure (ATSDR 1998). No information is available on the reproductive or developmental effects of Cr (III) in humans (United States

Environmental Protection Agency 1999). On the other hand, Marques et al. (2000) reported that Cr (VI) compounds can be toxic for biological systems. Environmental Protection Agency has classified Cr (VI) in Group A, a known human carcinogen (Environmental Protection Agency 1998; WHO 1998). The respiratory tract is the major target organ of Cr (VI) toxicity during acute (short-term) and chronic (long-term) inhalation exposures in humans. Shortness of breath, coughing, and wheezing were reported in acute exposure cases, while perforations and ulcerations of the septum, bronchitis, decreased pulmonary function, pneumonia, and other respiratory effects have been reported for chronic exposure. Human studies have clearly established that chromium (VI) is a human carcinogen, resulting in an increased risk of lung cancer. Animal studies have shown chromium (VI) to cause lung tumors via inhalation exposure. Acute inhalation exposure to very high concentrations of Cr (VI) includes gastrointestinal and neurological effects, while dermal exposure causes skin burns in humans.

13.5.2   Chromium Toxicity to Microbes and Plants

Chromium is a highly toxic nonessential metal for microorganisms and plants. The hexavalent form of the metal, Cr (VI), is considered a more toxic species than the relatively innocuous and less mobile Cr (III) form (Nies 1999). The presence of chromium in the environment has selected microbial and plant variants able to tolerate high levels of chromium compounds. The interactions of bacteria, algae, fungi, and plants with Cr and its compounds have been discussed by Cervantes et al. (2001). The toxic effects of chromium on plant growth and development include alterations in the germination process as well as in the growth of roots, stems, and leaves, which may affect total dry matter production and yield. Sharma et al. (2003), for example, reported that chromium caused visible lesions of interveinal chlorosis in maize (*Zea mays*) while plant physiological processes such as photosynthesis, water relations, and mineral nutrition are reported to be adversely affected by Shanker et al. (2005). Metabolic alterations in plants following chromium exposure could either be due to a direct effect on various enzymes or other metabolites or because of its ability to generate reactive oxygen species (ROS), which may cause oxidative stress. The potential of plants with the capacity to accumulate or to stabilize chromium compounds for bioremediation of chromium contamination has gained interest in recent years.

The diverse Cr-resistance mechanisms displayed by microorganisms, and probably by plants, include biosorption, diminished accumulation, precipitation, reduction of Cr (VI) to Cr (III), and chromate efflux. Some of these systems have been proposed as potential biotechnological tools for the remediation of chromium pollutant. Some of these systems have been proposed as potential biotechnological tools for the remediation of chromium pollutant. The best characterized mechanisms for chromium detoxification/removal from Cr-contaminated environment include the efflux of chromate ions from the cell cytoplasm and reduction of Cr (VI) to Cr (III). Chromate efflux by the ChrA transporter has been established in

Pseudomonas aeruginosa and *Cupriavidus metallidurans* (formerly *Alcaligenes eutrophus*) and consists of an energy-dependent process driven by the membrane potential (Ramírez-Díaz et al. 2008). Mechanisms of hexavalent chromium detoxication by microorganisms and bioremediation application potential have been reviewed by Cheung and Gu (2007).

13.6 Remediation Technologies for Chromium-Contaminated Soil

13.6.1 Conventional Methods

Chromium (VI) contamination of soils results mainly from the discharge of chromium-containing waste and wastewater from ore refining, production of steel and alloys, metal plating, tannery, wood preservation, and pigmentation. In soil environment, hexavalent chromium can be leached into surface water and groundwater because of its high solubility and mobility (Messer et al. 2006). Concentrations of Cr (VI) as low as 0.5 mg L^{-1} in solution and 5 mg kg^{-1} in soil can be toxic to plants (Turner and Rust 1971). Hence, the presence of hexavalent chromium is a significant risk to human health as well as plants when it is released into soil environment.

Conventional technologies for Cr (VI) remediation in soils include physicochemical extraction, land filling, stabilization/solidification, soil washing, flushing, and excavation. However, most of these methods require high energy and large quantities of chemical reagents (Jeyasingh and Philip 2005). Remediation methods that involve the excavation of chromium-contaminated soils are known to generate airborne dust that creates an additional exposure hazard to field workers and general public (Allen et al. 1995; Greenwood and Earnshaw 1997; Macintyre 1992). To address these environmental concerns and health risks, various groups have proposed to facilitate in situ reduction of chromium (Cotton et al. 1999; Greenwood and Earnshaw 1997). Chemical reduction of chromium by these methods is known to decrease the toxicity and bioavailability of this metal drastically. Remediation of chromium in contaminated soil using bioremediation process has been reported by Krishna and Philip (2005). Electrokinetic remediation method for chromium remediation has been reported by Sawadaa et al. (2004). All these methods have certain merits and demerits and are limited to small-scale facilities requiring special monitoring techniques like ion chromatography, chromium speciation, atomic absorption, inductively coupled plasma (ICP), etc.

13.6.1.1 Laser-Induced Breakdown Spectroscopy (LIBS)

Conventional techniques like ICP and atomic absorption used for remediation of soil contaminated with chromium are time consuming and quite expensive. The cost associated with such methods often limits their effectiveness for monitoring the in situ remediation process. Laser-induced Breakdown Spectroscopy (LIBS) on the other hand has been recognized as an advanced technique that is amenable to field applications. Recent advances in hardware components, lasers, detectors, and spectrometers have made LIBS technique more attractive for industrial and environmental analysis (Brouard et al. 2007; Gondal et al. 2009; Haisch et al.1998; Vadillo et al. 2005). It is an emerging technique for rapid and accurate analysis of solid waste. For example, Gondal et al. (2009) used this technique to monitor the remediation process of soil contaminated with chromium. The study was conducted at laboratory scale and the important parameters viz. the laser pulse characteristics (pulse width, energy), the sample homogeneity, and the sampling geometry (distance from the focusing lens to the sample, focal length of the collecting lens, fiber optics, etc.) were optimized to achieve the best limit of detection. The minimum detection limit of the spectrometer for chromium in soil matrix was found to be 2 mg kg^{-1}.

13.6.1.2   Electrokinetic Technique

Electrokinetic remediation is an in situ technique in which a low-level direct current is applied across the soil medium to remove the contaminants. The contaminant transport takes place by two mechanisms (a) electromigration and (b) electroosmosis. Electromigration is the movement of positively and negatively charged ionic species to the corresponding electrodes of opposite sign. Thus, positive ions move toward the cathode and the negative ions move toward the anode. Electroosmosis is the movement of water from the anode to the cathode as a result of dipolar water molecules interacting with double diffuse layer when an electric potential is applied (Lei 2004). Electrolysis reactions take place at the electrodes when a direct electric current is applied across the soil. Prashanth et al. (2009) studied the feasibility of electrokinetic technique for removal of chromium from contaminated soil. The initial concentration of chromium in the soil was 3,100 mg/kg. Experimental results showed that the removal efficiencies of chromium were as high as 72%. Chromium which was in hexavalent state was transported as anion prior to its reduction. Zhang et al. (2010) in a recent study reported that chromium-contaminated soil can be remediated by electrokinetic technique. However, in practical application, Cr (VI) may migrate with water deep into the soil, contaminating previously unpolluted layers. Both horizontal and vertical electric fields were applied simultaneously to improve traditional electrokinetic remediation. Among the three operational modes, 2D crossed mode significantly prevented Cr (VI) from migrating downward and the chromium-contaminated soil was treated effectively.

13.6.1.3   Use of Organic Ligands

Recently, the use of organic ligands has emerged as a reasonable means of enhancing or reducing chromium mobility and considered as a key factor for remediation of chromium-contaminated sites (Deiana et al. 1991; Deng and Stone 1996; Jardine et al. 1999; Johnson et al. 2001; Puzon et al. 2005). In situ stabilization of Cr (VI) in polluted soil using organic ligands has been reported by Kantar et al. (2008). The role of organic ligands such as galacturonic, glucuronic, and alginic acids (main constituents of bacterial exopolymeric substances [EPS]) on Cr (VI) uptake and transport in heterogeneous subsurface media has been investigated. They demonstrated that the addition of galacturonic, glucuronic, and alginic acids to soils enhanced Cr (VI) uptake by soil at pH <7.7 depending on the concentration of the ligand and pH used. Organic ligands have no or little effect on Cr (VI) uptake under highly alkaline pH conditions since the catalytic Cr (VI) reduction decreases with increasing pH. Microorganisms produce EPS for a variety of purposes in response to environmental stresses. Depending upon bacterial strains and metal exposure, quantity and composition of EPS have been shown to vary (Aquino and Stuckey 2004; Guibaud et al. 2005; Priester et al. 2006). In a study with the hydrogen-producing photosynthetic bacteria strain *Rhodopseudomonas acidophila*, Sheng et al. (2005) found that toxic metals such as Cr (VI) and Cd (II) stimulated the production of microbial EPS.

13.6.2 Biological Methods

Biological treatments in recent times have received greater attention for Cr (VI) remediation of contaminated sites because it is an economical and environmentally friendly as compared to conventional technologies. The bioremediation strategy is to convert Cr (VI) into less toxic and less mobile Cr (III). Consequently, Cr (III) is immobilized in the soil matrix (Chai et al. 2009). Many microbes have been reported to reduce Cr (VI) under aerobic and anaerobic conditions (Fulladosa et al. 2006: Guha et al. 2001, 2003; Shen and Wang 1994; Srivastava and Thakur 2006). Bio-reduction of Cr (VI) can be achieved directly as a result of microbial metabolism or indirectly by bacterial metabolite such as H_2S (Michel et al. 2001; Cheung and Gu 2003; Battaglia-Brunet et al. 2002). Desjardin et al. (2002) found that Cr (VI) in soils was reduced by *Streptomyces thermocarboxydus* isolated from the contaminated soil. Bader et al. (1999) studied Cr (VI) reduction in soil by microbial community under aerobic conditions and reported that Cr (VI) was reduced as much as 33% within 21 days. Jeyasingh and Philip (2005) isolated bacterial strains from the contaminated site of Tamil Nadu Chromates and Chemicals Limited (TCCL) premises, Ranipet, Tamil Nadu, India and evaluated Cr (VI) reduction both in aerobic and anaerobic conditions. For maximum Cr (VI) reduction, bacterial concentration of 15 ± 1 mg/g soil (wet/wt), and 50 mg of molasses/g soil, as C source was required. The bioreactor operated at these conditions could reduce entire Cr (VI) (5.6 mg Cr (VI)/g of soil) in 20 days.

Bioremediation of Cr (VI) in contaminated soil using bioreactor-biosorption system was evaluated by Krishna and Philip (2005). Experiments were conducted using different eluents. Leaching of Cr (VI) from the contaminated soil using various eluents showed that desorption was strongly affected by the solution pH. The leaching process was accelerated at alkaline conditions (pH 9). Though, desorption potential of ethylene diamine tetraacetic acid (EDTA) was the maximum among various eluents tried, molasses (5 g/L) could also elute 72% of Cr (VI). Reduction studies of Cr (VI) were carried out under aerobic and facultative anaerobic conditions using the bacterial isolates from contaminated soil. Cr (VI) reduction was moderately higher in aerobic conditions than in facultative anaerobic conditions. The time required for complete Cr (VI) reduction was increased with increase in the initial Cr (VI) concentration. However, specific Cr (VI) reduction was increased with increase in initial Cr (VI) concentration. Sulfates and nitrates did not compete with Cr (VI) for accepting the electrons. A bioreactor was developed by the authors for the detoxification of Cr (VI). Cr (VI) reduction was achieved above 80% in the bioreactor with an initial Cr (VI) concentration of 50 mg/L at an HRT of 8 h. An adsorption column was developed packed with *Ganoderma lucidum* (a wood rooting macrofungus) as the adsorbent for the removal of Cr (III) and excess electron donor from the effluent of the bioreactor. The specific Cr (III) adsorption capacity of *G.lucidum* in the column was found to be 576 mg/g. The new biosystem was found to be a promising alternative for the ex situ bioremediation of Cr (VI) contaminated soils. Recently, Chai et al. (2009) reported the Cr (VI) remediation by indigenous bacteria in soils contaminated by chromium-containing slag. They isolated the bacterial strain, which was identified as *Pannonibacter phragmitetus* sp. by gene sequencing of 16S rRNA. The indigenous strain proved to be a potential strain for chromium remediation in the soils contaminated by chromium-containing slag.

13.6.3 Chromium Remediation in Aqueous Environment

13.6.3.1 Conventional Methods

There are various technologies available to remove Cr (VI) from wastewater such as chemical precipitation (Uysal and Irfan 2007), ion exchange (Jianlong et al. 2000; Rengaraj et al. 2003), membrane separation (Kozlowski and Walkowiak 2002), electro-coagulation (Roundhill and Koch 2002), solvent extraction (Li et al. 2004), reduction (Chen and Hao 1998), reverse osmosis (Li et al. 2004), and adsorption (Baral et al. 2007; Mohan et al. 2005). These technologies have many disadvantages such as incomplete metal removal, high reagent and energy requirements, and generation of toxic sludge or waste products, which require proper disposal without creating any problem to the environment (Aliabadi et al. 2006; Mohan and Pittman 2006).

13.6.3.2 Biological Methods

The use of biological materials including living and nonliving microorganisms for the removal of chromium from wastewaters has gained important credibility during the last few years. Metal uptake by dead cells takes place only by the passive mode known as biosorption. Living cells employ both active and passive modes for chromium uptake. The combination of active and passive mode is called bioaccumulation.

Table 13.1 Biosorbents used in chromium removal

Algae	Fungi	Bacteria	Biowaste materials/ plant parts
Spirogyra	*Aspergillus niger*	*Zoogloea ramigera*	Rice bran
Dunaliella	*R. arrhizus*	*Bacillus* sp.	Wheat bran
Chlorella vulgaris	*P. chrysogenum*	*Aeromonas caviae*	Rice husk
Ecklonia sp.	*P. purpurogenum*	*B. thuringiensis*	Rice straw
Scenedesmus obliquus	*R. nigricans*	*Pantoea* sp.TEM18	Saw dust (Indian Rose wood,
Synechocystis sp.	*Neurospora crassa*	*C. luteola* TEM05	Beech, Rubber wood)
Cladophora crispata	*Lentinus sajor caju*		
Sargassum wightii	Unmethylated yeast		Sugarcane bagasse
Turbinaria sp.	Methylated yeast		Maize corn cob
Nitella pseudoflabellata			Maize bran
			Jatropha oil cake
Phormedium bohneri			Neem sawdust
			Coconut fiber pith
Oscillatoria tenuis			Eucalyptus bark
Ulothrix tenuissima			Pine needles
Oscillatoria nigra			Cactus leaves
Chlamydomonas angulosa			Neem leaf powder
			London leaves

Microorganisms-based technologies compete with both operational and economical terms in existing metal removal treatment systems. Biosorption has several advantages: it does not produce any chemical sludge, is easy to operate, and is very efficient for the removal of pollutants from very dilute solutions. A major advantage of biosorption is that it can be used *in situ* and with proper design, it may not need any industrial process operations and can be integrated with many systems (Tewari et al. 2005). Biosorption is the cost-effective and versatile method when combined with an appropriate step for desorbing the chromium (VI) from adsorbent and avoids the problem of disposal of adsorbent (Kumar et al. 2007). The advantages of biosorption process have prompted to extend the use of various biomaterials with structural, compositional, or chemical characteristics suitable to make this technique viable for removal of hexavalent chromium from the wastewater streams (Alvarez-Ayuso et al. 2007). A number of biosorbents like algae, fungi, bacteria, various plant parts, agricultural by-products, and biowaste materials have been reported (Table 13.1) for the effective removal of Cr (VI) and Cr (III).

Algae as Biosorbent

Remediation by algae known as phycoremediation is considered as a viable option of heavy metal remediation. Algae used for chromium remediation include

Spirogyra (Gupta et al. 2001), *Dunaliella* (Dönmez and Aksu 2002), *Chlorella vulgaris* (Aksu and Acikel 1999, 2000; Aksu et al. 1999; Dönmez et al. 1999), *Ecklonia* sp (Park et al. 2005b). Gupta et al. (2001) studied Cr (VI) biosorption by biomass of filamentous algae *Spirogyra* species. Maximum removal of Cr (VI) was 1.47 g metal/kg dry weight biomass at pH of 2. Two strains of living *Dunaliella* algae were tested as a function of pH, initial metal ion, and salt (NaCl) concentration. The biosorption capacities of both *Dunaliella* strains were obtained at pH 2 in the absence or presence of salt concentration. Both Langmuir and Freundlich models were used to describe Cr (VI) biosorption (Dönmez and Aksu 2002). Dönmez et al. (1999) carried out a similar experiment using dried biomass of three algal species viz. *C. vulgaris, Scenedesmus obliquus,* and *Synechocystis* sp. Optimum Cr (VI) adsorption occurred at pH 2 for all the three algal species. Aksu and Acikel (2000) explored the competitive biosorption of Cu (II) and Cr (VI) on *C. vulgaris* from a binary mixture in a single staged bioreactor at pH 2. Nourbakhsh et al. (1994) investigated Cr (VI) biosorption onto dead biomass from *C. vulgaris* and *Cladophora crispate, etc.* The optimum pH ranged from 1 to 2 for five organisms. Maximum metal uptake was recorded at 25–35°C. In other study, Aravindhan et al. (2004a) utilized the abundant brown seaweed *Sargassum wightii* for chromium removal. The *Sargassum* species showed maximum uptake of 35 mg Cr/g seaweed and the optimum pH was 3.5–3.8. The same group further used brown seaweed (*Turbinaria* sp.), which had been pretreated with sulfuric acid, calcium chloride, and magnesium chloride, to remove chromium from tannery wastewater (Aravindhan et al. 2004b). Protonated seaweeds showed more chromium uptake than seaweed treated with calcium and magnesium. *Turbinaria* exhibited maximum uptake of 31 mgCr/g seaweed at an initial concentration of 1,000 ppm. Park et al. (2005b) utilized protonated brown seaweed *Eclonia* sp. for remediating Cr (VI), which was completely reduced to Cr (III) when wastewater containing Cr (VI) was used with biomass. The converted Cr (III) appeared in the solution or was partly bound to the biomass. The Cr (VI) removal efficiency was always 100% in the pH range of 1–5. The sorption capacity of *Eclonia* sp. was 4.49 mmol Cr (VI)/g. Chromium biosorption by thermally treated biomass of the brown seaweed *Ecklonia* was also studied by Park et al. (2004). Protonated *Eclonia* sp. was also utilized for Cr (III) adsorption (Yun et al. 2001) and it was found to contain at least three types of functional groups. Fourier transform Infrared Spectroscopy (FTIR) showed that the carboxyl group was the chromium binding site within the pH range of 1–5. Cr (III) did not participate in this range. Remediation of Cr (VI)-contaminated waters by *Nitella pseudoflabellata*, a charophyte, has been reported by Gomes and Asaeda (2009). Dwivedi et al. (2010) studied the bioaccumulation potential of green and blue green microalgae growing naturally in selected Cr-contaminated sites in districts Unnao and Kanpur (Uttar Pradesh, India). The maximum accumulation of chromium was shown by *Phormedium bohneri* (8,550 μg/g dw) followed by *Oscillatoria tenuis* (7,354 μg/g dw), *Clamydomonas angulosa* (5,325 μg/g dw), *Ulothrix tenuissima* (4,564 μg/g), and *O. nigra* (1,862 μg/g). All species demonstrated a transfer factor of >10% for chromium.

Fungi as Biosorbent

Various types of fungal and yeast biomasses have been used for the removal and recovery of trivalent and hexavalent chromium from water environment. Both living and nonliving cells possess a remarkable capability for uptake of chromium ions from aqueous phase. Some of the examples are *R. arrhizus* (Sag and Kutsal 1996; Merrin et al. 1998; Prakasham et al. 1999; Priester et al. 2006), *Penicillium chrysogenum* (Deng and Ting 2005), *P. purpurogenum* (Say et al. 2004), dead fungal biomass (Sekhar et al. 1998), *Lentinus sajorcaju* mycelia (Arýca and Bayramoğlu 2005; Bayramoglu et al. 2005), *R. nigricans* (Bai and Abraham 2001, 2002, 2003), *Neurospora* crassa (Tunali et al. 2005), and unmethylated and methylated yeast (Seki et al. 2005). Cr (VI) biosorption by nonliving free and immobilized biomass from *R. arrhizus* was investigated by Prakasham et al. (1999). Adsorption capacities of free *R. arrhizus* biomass were 11 mg/g and immobilized biomass was 8.63 mg/g respectively. Deng and Ting (2005) modified *P. chrysogenum* fungal biomass by grafting polyethylenimine (PEI) onto the biomass surface. The presence of PEI on the biomass surface was verified by FTIR and X-ray photo-electron spectroscopy (XPS) analyses. Cr (VI) biosorption onto untreated, heat-treated, acid- and alkali-treated mycelia of *L. sajorcaju* was investigated by Bayramoglu et al. (2005). The maximum biosorption capacities of the untreated and heat-, acid-, and alkali-treated fungal biomass at pH 2 were 0.36, 0.61,0.48, and 0.51 mmol Cr (VI)/g of dry biomass. Bai and Abraham (2003) investigated Cr (VI) biosorption on immobilized *R. nigricnas*. Five different polymeric matrices, namely, calcium alginate, polyvinyl alcohol (PVA), polyacrylamide, polyisoprene, and polysulfone were employed to entrap finely powdered biomass of *R. nigricnas*. The chromium sorption capacity of all immobilized biomass samples was less than that of native, powdered biomass and followed the order: free biomass > polysulfone entrapped > polyisoprene immobilized > PVA immobilized > calcium alginate entrapped > polyacrylamide at 500 mg/L Cr (VI). Cr (VI) removal was further evaluated by Arýca and Bayramoğlu (2005) using free and CMC immobilized *L. sajorcaju* mycelia. The maximum biosorption capacities of free and immobilized fungus were 18.9 and 32.2 mg/g dry weights, respectively. The highest biosorption was noted at pH 2.

Most of the workers have reported that Cr (VI) adsorption in aqueous phase by dead fungal biomass takes place by "anionic adsorption." In case of *S. cerevisiae*, Zhao and Duncan (1998) observed "partial reduction along with anion adsorption" in packed bed column studies. Park et al. (2005a) suggested that these findings were misinterpreted due to errors in measuring total chromium in aqueous solution. They demonstrated that Cr (VI) was totally reduced to Cr (III) and completely removed from the solution. Complete removal of Cr (VI) by *A. niger, R.oryzae,* and *P. chrysogenum* in 48 h were reported by Park et al. (2005c)

Bacteria as Biosorbent

Numerous studies have identified different bacterial species capable of accumulating metals from aqueous environment. Among bacteria, *Zoogloea ramigera*

(Nourbakhsh et al. 1994), *Bacillus* sp. (Nourbakhsh et al. 2002), *Aeromonas caviae* (Loukidou et al. 2004a, b), *Bacillus thuringiensis* (Sahin and Öztürk 2005), and *Pantoea* sp. (Ozdemir et al. 2004) have been reported as potential chromium remediation agents. *A. caviae,* a Gram-positive bacterium isolated from potable groundwater supplies, removed Cr (VI) maximally (284.4 mg/g) at pH 2.5. Sahin and Öztürk (2005) investigated Cr (VI) biosorption by *B. thuringiensis* var. *thuringiensis* in batch mode. The optimum pH was found to be 2. The equilibrium sorption data were fitted to Langmuir and Freundlich isotherm both. A Gram-negative bacterium *Pantoea* sp. TEM 18 was utilized by Ozdemir et al. (2004) for Cr (VI) removal. Optimum Cr (VI) adsorption occurred at pH 3. Both Langmuir and Freundlich isotherm sorption models were found to be suitable for describing the short-term biosorption of Cr (VI) by *Pantoea* sp. Ozdemir and Baysal (2004) studied the Cr (VI) biosorption by *Chryseomonas luteola* TEM05. The optimum adsorption pH was 4. Rabbani et al. (2005) reported 17 different bacterial strains isolated from Ramsar warm springs in Iran for Cr (III) remediation. The maximum removal (100%) of Cr (III) at 10 ppm concentration occurred at pH 4.

Biowaste Materials as Sorbent

The uses of rice bran and wheat bran as adsorbents have been found less effective as they could remove only 50% toxic chromium (Farajzadeh and Monji 2004; Oliveira et al. 2005). Rice husk in natural form as well as activated rice husk carbon was used for the removal of Cr (VI) and results were also compared with commercial activated carbon and other adsorbents (Bishnoi et al. 2004; Mehrotra and Dwivedi 1988; Srinivasan et al. 1988). The efficiency of activated rice husk was fairly high at pH 2 whereas with activated alumina it was at pH 4. Adsorption increased with increasing dose and time at initial stages and then became somewhat constant due to attainment of equilibrium. Sawdust of Indian rose wood prepared by treatment with formaldehyde and sulphuric acid showed effective removal of chromium Cr (VI) (Garg et al. 2004). Beech sawdust and rubber wood saw dust were also tried for chromium removal (Acar and Malkoc 2004; Karthikeyan et al. 2005). Sugarcane bagasse was used in natural as well as modified form and efficiency for both the forms was compared for the removal of chromium (Gupta and Ali 2004; Krishanani et al. 2004). Most of the studies showed that the chromium biosorption by agricultural waste materials was quite high and varied from 50% to 100%. Mostly, biosorption occurred in acidic range particularly at pH 2 (Garg et al. 2007). Maize bran has been successfully utilized by Hasan et al. (2008) for the removal of Cr (VI) from aqueous solution. The effect of different parameters such as contact time, sorbate concentration, pH of the medium, and temperature were investigated and maximum uptake of Cr (VI) was 312.52 (mg/l) at pH 2, initial Cr (VI) concentration of 200 mg/l, and temperature of 40°C. Effect of pH showed that maize bran was not only removing Cr (VI) from aqueous solution but also reducing toxic Cr (VI) into less toxic Cr (III). The sorption kinetics was tested with pseudo-first order and pseudo-second order reaction and it was found that Cr (VI) uptake process followed the pseudo-second order rate expression. The Langmuir and Freundlich equations for describing sorption equilibrium were applied and the process was found to be well described by Langmuir isotherm. Desorption studies was also carried out and complete desorption of Cr (VI) was found at pH 9.5. The removal of Cr (VI) from aqueous solution by rice straw, agricultural by-product was investigated by Gao et al. (2008). The optimal pH was 2 and Cr (VI) removal rate increased with decreased Cr (VI) concentration and with increased temperature. Decrease in straw particle size led to an increase in Cr (VI) removal. Equilibrium was achieved in about 48 h under standard conditions, and Cr (III), which appeared in the solution, remained stable. This indicated that both reduction and adsorption played a role in the Cr (VI) removal. Isotherm tests showed that equilibrium sorption data were better represented by Langmuir model and the sorption capacity of rice straw was found to be 3.15 mg/g.

Bansal et al. (2009) reported the feasibility of using pre-consumer processing agricultural waste to remove Cr (VI) from synthetic wastewater under different experimental conditions. For this, rice husk, was used after pretreatments (boiling and formaldehyde treatment). Effects of various process parameters viz. pH, adsorbent dose, initial chromium concentration, and contact time were studied in batch systems. Maximum metal removal was observed at pH 2. The efficiencies of boiled and formaldehyde-treated rice husk for Cr (VI) removal were 71% and 76.5%, respectively for dilute solutions at 20 g/l adsorbent dose. Removal of Cr (III) and Cr (VI) from aqueous solution by using different types of sand viz. white, yellow, and red sand from the UAE was reported by Khamis et al. (2009). Adsorption of Cr (VI) on all sands forms was very low at pH 2 (removal <10%) whereas Cr (III) was totally removed at pH 5. Our recent work has proved neem sawdust as the most potential biosorbent for the removal of Cr (VI) from tannery wastewater (Vinodhini and Das 2009, 2010).

13.6.4 Plant Biomass/Plant Parts Used as Biosorbent

Diverse plant parts such as coconut fiber pith, coconut shell fiber, plant bark (*Acacia arabica*, Eucalyptus), pine needles, cactus leaves, neem leaf powder, etc. have been tried and have shown chromium removal efficiency as high as 90–100% at optimum pH (Dakiky et al. 2002; Manju and Anirudhan 1997; Mohan et al. 2006; Sarin and Pant 2006; Venkateswarlu et al. 2007). *Avena monida* (whole plant biomass) showed 90% removal efficiency of Cr (VI) at pH 6 (Gardea-Torresdey et al. 2000). Other plants including *Fagus orientalis* (Acar and Malkoc 2004), *Agave lechuguilla* (Romero-González et al. 2005), *Atriplex canescens* (Sawalha et al. 2005), *Thuja orientalis* (Oguz 2005), *Pinus sylvestris* (Ucun et al. 2002) and *Jatropha curcas* (Yadav et al. 2009) have also been used for chromium remediation. Romero-González et al. (2006) reported the Cr (III) sorption onto *Agave lechuguilla* biomass. The average adsorption capacities calculated from Freundlich (4.7 mg/g) and Langmuir (14.2 mg/g) isotherms showed that *A. lechuguilla* effectively removed chromium (III) in aqueous environment. Cr (III) binding was due to interactions with surface carboxyl groups of the adsorbent's cell tissue. In a similar study, Romero-González et al. (2005) used *Agave lechuguilla* for Cr (VI) adsorption. Cr (VI) binding at pH 2 could be due to either electrostatic attraction to Cr (VI) oxyanions by positively charged ligands such as protonated amines or through reduction of Cr (VI) to Cr (III), subsequently resulting in the binding of Cr (III) to the biomass. Gardea-Torresdey et al. (2000) studied Cr (VI) adsorption and its possible reduction to Cr (III) by agricultural by-products of *Avena monida*. Sawalha et al. (2005) studied chromium adsorption by native, esterified, and hydrolyzed saltbush (*Atriplex canescens*) biomass. X-ray absorption spectroscopy (XAS) studies determined the chromium oxidation state when bound to the biomass. The amounts of chromium adsorbed by saltbush biomass were determined by inductively coupled plasma-optical emission spectroscopy. The percentages of Cr (III) bound by native stems, leaves, and flowers at pH 4 were 98, 97, and 91%, respectively. On the other hand, the Cr (VI) binding by the native stems, leaves, and flowers of the native and hydrolyzed saltbush biomass decreased as pH increased. At pH 2, the stems, leaves, and flowers of native biomass were found to bind 31, 49, and 46%, of Cr (VI), respectively. XAS experiments showed that Cr (VI) was reduced to Cr (III) to some extent by saltbush biomass at both pH 2 and 5. Cr (VI) removal by London plane leaves in aqueous environment was studied by Aoyama (2003). It was found that Cr (VI) did not reduce to Cr (III) and the dominating removal mechanism was adsorption. The total chromium removal by London leaves was almost equal to the amount of Cr (VI) adsorbed.

13.7   Mechanism of Cr (VI) Biosorption

Many studies have claimed that Cr (VI) could be removed from the aqueous phase through an adsorption mechanism, whereby anionic Cr (VI) ion species bind to the positively charged groups of nonliving biomass (Acar and Malkoc 2004; Park et al. 2004; Malkoc and Nuhoglu 2003). According to Park et al. (2005c), Cr (VI) can be removed from an aqueous system by both direct (Mechanism I) and indirect (Mechanism II) mechanisms (Fig. 13.1). In direct reduction, Cr (VI) is directly reduced to Cr (III) in the aqueous phase by contact with the electron-donor groups of the biomass, that is, groups having lower reduction potential values than that of Cr (VI). The indirect reduction consists of three steps: (1) the binding of anionic Cr (VI) ion species to the positively charged groups present on the biomass surface, (2) the reduction of Cr (VI) to Cr (III) by adjacent electron-donor groups, and (3) the release of the Cr (III) ions into the aqueous phase due to electronic repulsion between the positively charged groups and the Cr (III) ions, or the complexation of the Cr (III) with adjacent groups capable of Cr binding. Amino and carboxyl groups take part in direct mechanism. As the pH of the aqueous phase is lowered, the large number of H ions can easily coordinate with the amino and carboxyl groups present on the biomass surface. Thus, low pH makes the biomass surface more positive. The more positive the surface charge of the biomass, the faster the removal rate of Cr (VI)

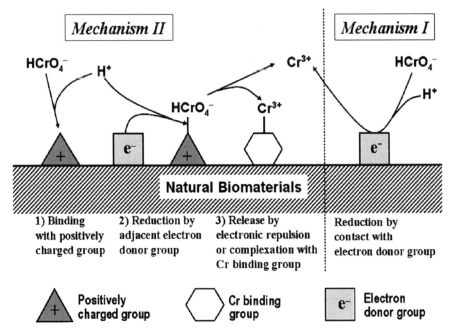

Fig. 13.1 Proposed mechanism of Cr (VI) biosorption by natural biomaterials (Adapted from Park et al. 2005c)

in the aqueous phase, since the binding of anionic Cr (VI) ion species with the positively charged groups is enhanced (Malkoc and Nuhoglu 2003). The low pH also accelerates the reduction reaction in both direct and indirect mechanisms. The solution pH is the most important controlling parameter in the practical use of non-living biomass in the adsorption process. Hence, it is of significance that the pH of wastewaters containing heavy metals is generally very acidic. Meanwhile, if there are a small number of electron-donor groups in the biomass or protons in the aqueous phase, the chromium bound to the biomass can remain in the hexavalent state.

13.8 Phytoremediation

Phytoremediation is an emerging technology that uses plants to remove contaminants from soil and water. The use of plants for remediation of metals offers an attractive alternative, because it is a solar-driven process and can be carried out *in situ* (Salt et al. 1995, 1998). Phytoremediation techniques include: (a) phytotransformation, (b) phytostabilization, (c) phytoextraction or phytoaccumulation, (d) phytodegradation, and (e) rhizofiltration. Phytotransformation involves the uptake of organic contaminants from soil, sediments, and water and subsequently contaminants are transformed to a more stable, less toxic, or less mobile form. Surface

Table 13.2 Overview of phytoremediation applications

Technique	Surface medium	Plant mechanism
Phytotransformation	Surface water, ground water	Plant uptake and degradation of organic compounds
Phytostabilization	Soils, ground water, mine tailing	Precipitation of metals on the root exudates causing less availability of metals
Phytoextraction	Soils	Uptake and concentration of metals through direct uptake into the plant tissue with subsequent removal by the plants
Phytodegradation	Soils, ground water within rhizosphere	Microbial degradation is enhanced in the rhizosphere
Rhizofiltration	Surface water	Uptake of metals into the plant roots
Vegetative cap	Soils	Rainwater is evapotranspirated by plants to prevent leaching of contaminants from disposal sites

water and groundwater are considered as surface medium. Metal chromium can be reduced from hexavalent to trivalent chromium, which is less mobile and noncarcinogenic. Phytostabilization is a technique in which plants reduce the mobility and migration of contaminated soil. Leachable constituents are adsorbed and bound into the plant structure so that they form a stable mass of plant from which the contaminants will not reenter the environment. Phytoextraction is the process used by plants to accumulate contaminants into the roots and aboveground shoots and leaves. This technique is inexpensive and may accumulate low levels of contaminants from a widespread area and the surface medium. Phytodegradation involves the breakdown of contaminants through the rhizosphere microbial activity. The proteins and enzymes released by plants or by soil organisms such as bacteria, yeast, and fungi influence phytodegradation. Plants provide nutrients necessary for the microbes to thrive, while microbes provide a healthier soil environment. Rhizofiltration is a water remediation technique that involves the uptake of contaminants by plant roots. It is used to reduce contamination in natural wetlands and estuary areas. Phytoremediation is well suited for use at very large field sites where other methods of remediation are expensive and not practicable (Eccles 1999). Sometimes, rainwater is evapotranspirated by plants to prevent leaching of contaminants from disposal sites and the phenomenon is known as vegetative cap. An overview of phytoremediation applications is presented in Table 13.2.

Demir and Arisoy (2007) did cost and benefit analysis of biological and chemical removal of hexavalent chromium ions and cost per unit in chemical removal was calculated € 0.24 and the ratio of chrome removal was 99.68%, whereas those of biological removal were € 0.14 and 59.3%. Therefore, it was seen that cost per unit in chemical removal and chrome removal ratio were higher than those of biological removal method. Reports on the phytoextraction of chromium from contaminated soils and sediments are less. Attempts have been made to use promising aquatic plant species (*Scirpus lacustris, Phragmites karka,* and *Bacopa monnieri*) for the phytoextraction of chromium from contaminated tannery sludge (Yadav et al. 2005). The expansion of this research work has promoted phytoremediation to be eco-friendly as well as cost-effective technology (Rai 2009; Rai and Tripathi 2007, 2008).

The disadvantage of the phytoremediation technology is that most of the works have not been scaled up at industrial level. Genetic engineering of biological systems, however, may be considered as potential future prospect for scaling up for phytoremediation (Eapen and D'Souza 2005). Vegetable crops are the extremely important life-supporting materials for humans and other animal species in the developing country like India, since vegetables contain essential components of the diet by the contribution of protein, vitamin, iron, calcium, and other nutrients (Bean et al. 2009; Borah et al. 2009; Rai and Tripathi 2008). These local vegetable products are the basis of human nutrition in many places and of great relevance to human health due to the presence of various antioxidants (Bean et al. 2009). Priority should be given to biological and advanced treatment in order to ameliorate metals concentration especially chromium in treated wastewater used for irrigation during the cultivation of vegetables. There is an urgent need to make sustainable management policies to solve this problem, which is linked with human health in many developing countries.

13.9 Conclusion

Developing health-based cleanup standards and remediation strategies for chromium-contaminated soils and wastewaters seems to be a complex and controversial task. On the basis of the reported works done so far on chromium remediation, it is found that technology based on the use of microorganisms, biowaste materials, or plants may be the best suited and cost-effective technology for remediation of chromium from contaminated sites. In some plants, 99% chromium is adsorbed in the root where it is reduced to Cr (III) species within a short period of time. Microorganisms and plants both can, therefore, serve as highly efficient bioaccumulators of chromium especially in the aqueous environment. The bioremediation technologies are likely to provide an alternative or adjunct to conventional techniques of chromium recovery or removal from polluted sites.

References

Acar FN, Malkoc E (2004) The removal of chromium (VI) from aqueous solutions by *Fagus orientalis* L. Bioresour Technol 94:13–15
Agency for toxic substances and Disease Registry (ATSDR) (1998) Toxicological profile of chromium. US Public Health Service, U.S. Department of Health and Human Services, Atlanta

Aksu Z, Acikel Ü (1999) A single-staged bioseparation process for simultaneous removal of copper (II) and chromium (VI) by using *C. vulgaris*. Process Biochem 34:589–599

Aksu Z, Acikel Ü (2000) Modeling of a single-staged bioseparation process for simultaneous removal of iron (III) and chromium (VI) by using *C. vulgaris*. Biochem Eng J 4:229–238

Aksu Z, Acikel Ü, Kutsal T (1999) Investigation of simultaneous biosorption of copper (II) and chromium (VI) on dried *Chlorella vulgaris* from binary metal mixtures: application of multi-component adsorption isotherms. Sep Sci Technol 34:501–552

Aliabadi M, Morshedzadeh K, Soheyli H (2006) Removal of hexavalent chromium from aqueous solution by lignocellulosic solid wastes. Int J Environ Sci Technol 3:321–325

Allen HE, Huang CP, Bailey GW, Bowers AR (1995) Metals speciation and contamination of soil. Lewis Publishers, Boca Raton

Alvarez-Ayuso E, Garcia-Sanchez A, Querol X (2007) Adsorption of Cr (VI) from synthetic solutions and electroplating wastewaters on amorphous aluminium oxide. J Hazard Mater 142:191–198

Aoyama M (2003) Removal of Cr (VI) from aqueous solution by London plane leaves. J Chem Technol Biotechnol 78:601–604

Aquino SF, Stuckey DC (2004) Soluble microbial products formation in anaerobic chemostats in the presence of toxic compounds. Water Res 38:255–266

Aravindhan R, Madhan B, Rao JR, Nair BU, Ramasami T (2004a) Bioaccumulation of chromium from tannery wastewater: an approach for chrome recovery and reuse. Environ Sci Technol 38:300–306

Aravindhan R, Madhan B, Rao JR, Nair BU (2004b) Recovery and reuse of chromium from tannery wastewaters using *Turbinaria ornata* seaweed. J Chem Technol Biotechnol 79:1251–1258

Arýca MY, Bayramoğlu G (2005) Cr (VI) biosorption from aqueous solutions using free and immobilized biomass of *Lentinus sajorcaju*: preparation and kinetic characterization. Colloids Surf A Physicochem Eng Aspects 253:203–221

Ayres RU (1992) Toxic heavy metals: materials cycle optimization. Proc Natl Acad Sci USA 89:815–820

Bader JL, Gonzalez G, Goodell PC, Ali AS, Pillai SD (1999) Aerobic reduction of hexavalent chromium in soil by indigenous microorganisms. Biorem J 3:201–211

Bagchi D, Bagchi M, Stohs SJ (2001) Chromium (VI)-induced oxidative stress, apoptotic cell death and modulation of p53 tumour suppressor gene. Mol Cell Biochem 222:149–158

Bai RS, Abraham TE (2001) Biosorption of Cr (VI) from aqueous solution by *Rhizopus nigricans*. Bioresource Technol 79:73–81

Bai RS, Abraham TE (2002) Studies on enhancement of Cr (VI) biosorption by chemically modified biomass of *Rhizopus nigricans*. Water Res 36:1224–1236

Bai RS, Abraham TE (2003) Studies on chromium (VI) adsorption– desorption using immobilized fungal biomass. Bioresource Technol 87:17–26

Bailar JC (1997) Chromium. In: Parker SP (ed.) McGraw-Hill encyclopedia of science and technology, vol 3, 8th edn. McGraw- Hill, New York

Bansal M, Garg U, Singh D, Garg VK (2009) Removal of Cr (VI) from aqueous solutions using pre-consumer processing agricultural waste: a case study of rice husk. J Hazard Mater 162:312–320

Baral SS, Das SN, Rath P, Chaudhary GR (2007) Chromium (VI) removal by calcined bauxite. Biochem Eng J 34:69–75

Barlett RJ (1991) Chromium cycling in soils and water: links, gaps and methods. Environ Health Perspect 92:17–24

Battaglia-Brunet F, Foucher S, Denamur A, Ignatiadis I, Michel C, Morin D (2002) Reduction of chromate by fixed films of sulfate-reducing bacteria using hydrogen as an electron source. J Ind Microbiol Biotechnol 28:154–159

Bayramoglu G, Celik G, Yalcin E, Yilmaz M, Arica MY (2005) Modification of surface properties of *Lentinus sajor-caju* mycelia by physical and chemical methods: evaluation of their Cr (VI) removal efficiencies from aqueous medium. J Hazard Mater 119:219–229

Bean H, Schuler C, Leggett RE, Levin RM (2009) Antioxidant levels of common fruits, vegetables, and juices versus protective activity against in vitro ischemia/reperfusion. Int Urol Nephrol 42:409–415

Bishnoi NR, Bajaj M, Sharma N, Gupta A (2004) Adsorption of Cr (VI) on activated rice husk carbon and activated alumina. Bioresource Technol 91:305–307

Borah S, Baurah AM, Das AK, Borah J (2009) Determination of mineral content in commonly consumed leafy vegetables. Food Anal Method 2:226–230

Brouard D, Gravel JFY, Viger ML, Boudreau D (2007) Use of sol-gels as solid matrixes for laser-induced breakdown spectroscopy. Spectrochim Acta B 6:1361–1369

Butler LR, Edwards MR, Farmer R, Greenly KJ, Hensler S, Jenkins SE, Joyce JM, Mann JA, Prentice BM, Puckette AE, Shuford CM, Porter SEG, Rhoten MC (2009) Investigation of the use of *Cucumis sativus* for remediation of chromium from contaminated environmental matrices. An interdisciplinary instrumental analysis project. Chem Educ 86:1095

Cervantes C, Campos-García J, Devars S, Gutiérrez-Corona F, Loza-Tavera H, Torres-Guzmán JC, Moreno-Sánchez R (2001) Interaction of chromium with microorganisms and plants. FEMS Microbiol Rev 25:335–347

Chai L, Huang S, Yang Z, Peng B, Huang Y, Chen Y (2009) Cr (VI) remediation by indigenous bacteria in soils contaminated by chromium containing slag. J Hazard Mater 167:516–522

Chen JMN, Hao OJN (1998) Microbial chromium (VI) reduction, critical reviews. Environ Sci Technol 28:219–251

Cheung KH, Gu JD (2003) Reduction of chromate (CrO_4^{2-}) by an enrichment consortium and an isolate of marine sulfate-reducing bacteria. Chemosphere 52:1523–1529

Cheung KH, Gu JD (2007) Mechanism of hexavalent chromium detoxication by microorganisms and bioremediation application potential: a review. Int Biodeter Biodegr 59:8–15

Cotton FA, Wilkinson G, Murillo CA, Bochmann M (1999) Advanced inorganic chemistry. Wiley, New York

Dakiky M, Khami A, Manassra A, Mereb M (2002) Selective adsorption of chromium (VI) in industrial wastewater using low cost abundantly available adsorbents. Adv Environ Res 6:533–540

Deiana S, Gessa C, Usai M, Piu P, Seeber R (1991) Analytical study of the reduction of chromium (VI) by d-galacturonic acid. Anal Chim Acta 248:301–305

Demir A, Arisoy M (2007) Biological and chemical removal of Cr (VI) from waste water: cost and benefit analysis. J Hazard Mater 147:275–280

Deng B, Stone AT (1996) Surface-catalyzed chromium (VI) reduction: reactivity comparisons of different organic reductants and different oxide surfaces. Environ Sci Technol 30:2486–2494

Deng S, Ting YP (2005) Polyethylenimine-modified fungal biomass as a high capacity biosorbent for Cr(VI) anions: sorption capacity and uptake mechanisms. Environ Sci Technol 39:8490–8496

Desjardin V, Bayard R, Huck N, Manceau A, Gourdon R (2002) Effect of microbial activity on the mobility of chromium in soils. Waste Manage 22:195–200

Dönmez G, Aksu Z (2002) Removal of chromium (VI) from saline wastewaters by *Dunaliella* species. Process Biochem 38:751–762

Dönmez G, Kocberber N (2005) Bioaccumulation of hexavalent chromium by enriched microbial cultures obtained from molasses and NaCl containing media. Process Biochem 40:2493–2498

Dönmez G, Aksu Z, Oztürk A, Kutsal T (1999) A comparative study on heavy metal biosorption characteristics of some algae. Process Biochem 34:885–892

Dwivedi D, Srivastava S, Mishra S, Kumar A, Tripathi RD, Rai UN, Dave R, Tripathi P, Chakraborty D, Trivedi PK (2010) Characterization of native microalgal strains for their chromium bioaccumulation potential: phytoplankton response in polluted habitats. J Hazard Mater 173:95–101

Eapen S, D'Souza DF (2005) Prospects of genetic engineering of plants for phytoremediation of toxic metals. Biotechnol Adv 23:97–114

Eccles H (1999) Treatment of metal-contaminated wastes: why select a biological process? Trends Biotechnol 17:462–465

Farajzadeh MA, Monji AB (2004) Adsorption characteristics of wheat bran towards heavy metal cations. Sep Purif Technol 38:197–207

Fulladosa E, Desjardin V, Murat JC, Gourdon R, Villaescusa I (2006) Cr (VI) reduction into Cr (III) as a mechanism to explain the low sensitivity of *Vibrio scheri* bioassay to detect chromium pollution. Chemosphere 65:644–650

Gao H, Liu Y, Zeng G, Xu W, Li T, Xia W (2008) Characterization of Cr (VI) removal from aqueous solutions by a surplus agricultural waste-rice straw. J Hazard Mater 150:446–452

Gardea-Torresdey JL, Tiemann KJ, Armendariz V, Bess-Oberto L, Chianelli RR, Rios J, Parsons G, Gamez G (2000) Characterization of Cr (VI) binding and reduction to Cr (III) by the agricultural byproducts of *Avena monida* (oat) biomass. J Hazard Mater 80:175–188

Garg VK, Gupta R, Kumar R, Gupta RK (2004) Adsorption of chromium from aqueous solution on treated sawdust. Bioresource Technol 92:79–81

Garg UK, Kaur MP, Garg VK, Sud D (2007) Removal of hexavalent Cr from aqueous solutions by agricultural waste biomass. J Hazard Mater 140:60–68

Gomes PIA, Asaeda T (2009) Phycoremediation of chromium (VI) by *Nitella* and impact of calcium encrustation. J Hazard Mater 166:1332–1338

Gomez V, Callao MP (2006) Chromium determination and speciation since 2000. Trends Anal Chem 25:1006–1015

Gondal MA, Hussain T, Yamani ZH, Baig MA (2009) On-line monitoring of remediation process of chromium polluted soil using LIBS. J Hazard Mater 163:1265–1271

Greenwood NN, Earnshaw A (1997) Chemistry of the elements, 2nd edn. Butterworth, London

Guha H, Jayachandran K, Maurrasse F (2001) Kinetics of chromium (VI) reduction by a type strain *Shewanella alga* under different growth conditions. Environ Pollut 115:209–218

Guha H, Jayachandran K, Maurrasse F (2003) Microbiological reduction of chromium (VI) in presence of pyrolusite-coated sand by *Shewanella alga* Simidu ATCC 55627 in laboratory column experiments. Chemosphere 52:175–183

Guibaud G, Comte S, Bordas F, Dupuy S, Baudu M (2005) Comparison of the complexation potential of extracellular polymeric substances (EPS), extracted from activated sludges and produced by pure bacteria strains, for cadmium, lead and nickel. Chemosphere 59:629–638

Gupta VK, Ali I (2004) Removal of lead and chromium from wastewater using bagasse fly ash: a sugar industry waste. J Colloid Interface Sci 271:321–328

Gupta VK, Shrivastava AK, Jain N (2001) Biosorption of chromium (VI) from aqueous solutions by green algae *spirogyra* species. Water Res 35:4079–4085

Haisch C, Panne U, Niessner R (1998) Combination of an intensified charge coupled device with an echelle spectrograph for analysis of colloidal material by laser induced plasma spectroscopy. Spectrochim Acta B 53:1657–1667

Hasan SH, Singh KK, Prakash O, Talat M, Ho YS (2008) Removal of Cr (VI) from aqueous solutions using agricultural waste maize bran. J Hazard Mater 152:356–365

Hasin A, Gurman SJ, Murphy LM, Perry A, Smith TJ, Gardiner PHE (2010) Remediation of chromium(VI) by a methane-oxidizing bacterium. Environ Sci Technol 44:400–405

IOCC, CAOBISCO (1996) Heavy metals rapport

Jardine PM, Fendorf SE, Mayes MA, Larsen L, Brooks SC, Bailey WB (1999) Fate and transport of hexavalent chromium in undisturbed heterogeneous soil. Environ Sci Technol 33:2939–2944

Jeyasingh J, Philip L (2005) Bioremediation of chromium contaminated soil: optimization of operating parameters under laboratory conditions. J Hazard Mater 118:113–120

Jianlong W, Xinmin Z, Yi Q (2000) Removal of Cr (VI) from aqueous solution by macroporous resin adsorption. J Environ Sci Health A 35:1211–1230

Johnson CR, Hellerich LA, Nikolaidis NP, Gschwend PM (2001) Colloid mobilization in the field using citrate to remediate chromium. Groundwater 39:895–903

Kantar C, Cetin Z, Demiray H (2008) In situ stabilization of chromium (VI) in polluted soils using organic ligands: the role of galacturonic, glucuronic and alginic acids. J Hazard Mater 159:287–293

Karthikeyan T, Rajgopal S, Miranda LR (2005) Chromium (VI) adsorption from aqueous solution by *Hevea brasiliensis* sawdust activated carbon. J Hazard Mater B124:192–199

Katz SA, Salem H (1994) The biological and environmental chemistry of chromium. VCH, New York

Khamis M, Jumean F, Abdo N (2009) Speciation and removal of chromium from aqueous solution by white, yellow and red UAE sand. J Hazard Mater 169:948–952

Kotas J, Stasicka Z (2000) Chromium occurrence in the environment and methods of its speciation. Environ Pollut 107:263 283

Kozlowski CA, Walkowiak W (2002) Removal of chromium (VI) from aqueous solutions by polymer inclusion membranes. Water Res 36:4870–4876

Krishanani KK, Parmila V, Meng X (2004) Detoxification of chromium (VI) in coastal water using lignocellulosic agricultural waste. Water SA 30:541–545

Krishna KR, Philip L (2005) Bioremediation of Cr (VI) in contaminated soils. J Hazard Mater 121:109–117

Kumar KS, Ganesan K, Rao PVS (2007) Phycoremediation of heavy metals by the three-color forms of *Kappaphycus alvarezii*. J Hazard Mater 143:590–592

Lei S (2004) Chromium slag treatment and utilization. Chin J Resour Comp Util 10:5–8 (in Chinese)

Li C, Chen H, Li Z (2004) Adsorptive removal of Cr (VI) by Fe-modified steam exploded wheat straw. Process Biochem 39:541–545

Lin CJ (2002) The chemical transformation of chromium in natural waters – a model study. Water Air Soil Poll 139:137–158

Loukidou MX, Zouboulis AI, Karapantsios TD, Matis KA (2004a) Equilibrium and kinetic modeling of chromium (VI) biosorption by *Aeromonascaviae*. Colloids Surf A Physicochem Engg Aspects 242:93–104

Loukidou MX, Karapantsios TD, Zouboulis AI, Matis KA (2004b) Diffusion kinetic study of chromium(VI) biosorption by *Aeromonas caviae*. Ind Eng Chem Res 43:1748–1755

Macintyre JE (1992) Dictionary of inorganic compounds, 1–3. Chapman & Hall, London

Malkoc E, Nuhoglu Y (2003) The removal of chromium (VI) from synthetic wastewater by *Ulothrix zonata*. Fresenius Environ Bull 4:376–381

Mangkoedihardjo S, Ratnawati R, Alfianti N (2008) Phytoremediation of hexavalent chromium polluted soil using *Pterocarpus indicus* and *Jatropha curcas* L. World Appl Sci J 4:338–342

Manju GN, Anirudhan TS (1997) Use of coconut fiber pith-based pseudo activated carbon for chromium (VI) removal. Ind J Environ Health 4:289–298

Marques MJ, Salvador A, Morales-Rubio A, de la Guardia M (2000) Chromium speciation in liquid matrices: a survey of the literature. Fresenius J Anal Chem 367:601–613

Mazierski J (1994) Effect of chromium (VI) on the growth rate of denitrifying bacteria. Water Res 28:1981–1985

Mehrotra R, Dwivedi NN (1988) Removal of chromium (VI) from water using unconventional materials. J Ind Water Works Assoc 20:323–327

Merrin JS, Sheela R, Saswathi N, Prakasham RS, Ramakrishna SV (1998) Biosorption of chromium (VI) using *Rhizopus arrhizus*. Ind J Exp Biol 36:1052–1055

Messer J, Reynolds M, Stoddard L, Zhitkovich A (2006) Causes of DNA single-strand breaks during reduction of chromate by glutathione in vitro and in cells. Free Radic Biol Med 40:1981–1992

Michel C, Brogan M, Aubert C, Bermuda A, Bruschi M (2001) Enzymatic reduction of chromate: comparative studies using sulfate reducing bacteria—key role of polyheme cytochromes c and hydrogenases. Appl Microbiol Biotechnol 55:95–100

Mohan D, Pittman CU Jr (2006) Activated carbons and low cost adsorbents for remediation of trivalent and hexavalent chromium from water. J Hazard Mater B 137:762–811

Mohan D, Singh KP, Singh VK (2005) Removal of hexavalent chromium from aqueous solution using low-cost activated carbons derived from agricultural waste materials and activated carbon fabric cloth. Ind Eng Chem Res 44:1027–1042

Mohan D, Singh KP, Singh VK (2006) Chromium (III) removal from wastewater using low cost activated carbon derived from agriculture waste material and activated carbon fabric filter. J Hazard Mater B135:280–295

Nies DH (1999) Microbial heavy metal resistance. Appl Microbiol Biotechnol 51:730–750

Nourbakhsh MN, Sag Y, Ozer D, Aksu Z, Kutsal T, Caglar A (1994) A comparative study of various biosorbents for removal of chromium (VI) ions from industrial waste waters. Process Biochem 29:1–5

Nourbakhsh MN, Kilicarslan S, Ilhan S, Ozdag H (2002) Biosorption of Cr (VI), Pb2+ and Cu2+ ions in industrial waste water on Bacillus sp. Chem Eng J 85:351–355

Nriagu JO, Pacyna JM (1988) Quantitative assessment of worldwide contamination of air, water and soils by trace metals. Nature 333:134–139

Oguz E (2005) Adsorption characteristics and the kinetics of the Cr (VI) on the *Thuja orientalis*. Colloids Surf A Physicochem Eng Aspects 252:121–128

Oliveira EA, Montanher SF, Andrade AD, Nobrega JA, Rollemberg MC (2005) Equilibrium studies for the sorption of chromium and nickel from aqueous solutions using raw rice bran. Process Biochem 40:3485–3490

Ozdemir G, Baysal SH (2004) Chromium and aluminum biosorption on *Chryseomonas luteola* TEM05. Appl Microbiol Biotechnol 64:599

Ozdemir G, Ceyhan N, Ozturk T, Akirmak F, Cosar T (2004) Biosorption of chromium (VI), cadmium (II) and copper(II) by *Pantoea* sp. TEM18. Chem Eng J 102:249–253

Park D, Yun YS, Cho HY, Park JM (2004) Chromium biosorption by thermally treated biomass of the brown seaweed. *Ecklonia* sp. Ind Eng Chem Res 43:8226

Park D, Yun YS, Park JM (2005a) Use of dead fungal biomass for the detoxification of hexavalent chromium: screening and kinetics. Process Biochem 40:2559

Park D, Yun YS, Park JM (2005b) Studies on hexavalent chromium biosorption by chemically-treated biomass of *Ecklonia* sp. Chemosphere 60:1356–1364

Park D, Yun YS, Jo JH, Park JM (2005c) Mechanism of hexavalent chromium removal by dead fungal biomass of *Aspergillus niger*. Water Res 39:533

Patnaik P (2003) Handbook of inorganic chemicals. McGraw-Hill, New York

Prakasham RS, Merrie JS, Sheela R, Saswathi N, Ramakrishna SV (1999) Biosorption of chromium (VI) by free and immobilized *Rhizopus arrhizus*. Environ Pollut 104:421–427

Prashanth RB, Brica RM, David BG (2009) Electrokinetic remediation of wood preservative contaminated soil containing copper, chromium and arsenic. J Hazard Mater 162:490–497

Priester JH, Olson SG, Webb SM, Neu MP, Hersman LE, Holden PA (2006) Enhanced exopolymer production and chromium stabilization in *Pseudomonas putida* unsaturated biofilms. Appl Environ Microbiol 72:1988–1996

Puzon GJ, Roberts AG, Kramer DM, Xun L (2005) Formation of soluble organochromium (III) complexes after chromate reduction in the presence of cellular organics. Environ Sci Technol 39:2811–2817

Rabbani M, Ghafourian H, Sadeghi S, Nazeri Y (2005) Biosorption of chromium (III) by new bacterial strain (NRC-BT-2). Int Congr Ser 1276:268–269

Rai PK (2009) Heavy metal phytoremediation from aquatic ecosystems with special reference to macrophytes. Crit Rev Environ Sci Technol 39:697–753

Rai PK, Tripathi BD (2007) Heavy metals removal using nuisance blue green alga *Microcystis* in continuous culture experiment. Environm Sci 4:53–59

Rai PK, Tripathi BD (2008) Heavy metals in industrial wastewater, soil and vegetables in Lohta village, India. Toxicol Environ Chem 90:247–257

Ramírez-Díaz MI, Díaz-Pérez C, Vargas E, Riveros-Rosas H, Campos-García J, Cervantes C (2008) Mechanisms of bacterial resistance to chromium compounds. Biometals 21:321–323

Rengaraj S, Joo CK, Kim Y, Yi J (2003) Kinetics of removal of chromium from water and electronic process wastewater by ion exchange resins: 1200H, 1500H and IRN97H. J Hazard Mater 102:257–275

Romero-González J, Peralta-Videa JR, Rodríguez E, Ramirez SL, Gardea-Torresdey JL (2005) Determination of thermodynamic parameters of Cr (VI) adsorption from aqueous solution onto *Agave lechuguilla* biomass. J Chem Thermodyn 37:243–247

Romero-González J, Peralta-Videa JR, Rodríguez E, Delgado M, Gardea-Torresdey JL (2006) Potential of *Agave lechuguilla* biomass for Cr (III) removal from aqueous solutions: thermodynamic studies. Bioresour Technol 97:178–182

Roundhill DM, Koch HF (2002) Methods and techniques for the selective extraction and recovery of oxo-anions. Chem Soc Rev 31:60–67

Ryan MP, Williams DE, Chater RJ, Hutton BM, McPhail DS (2002) Why stainless steel corrodes? Nature (Lond) 415:770–774

Sag Y, Kutsal Y (1996) Fully competitive biosorption of Cr (VI) and Fe(III) ions from binary metal mixtures by *R. arrhizus*: use of the competitive Langmuir model. Process Biochem 31:573–585

Sahin Y, Öztürk A (2005) Biosorption of chromium (VI) ions from aqueous solution by the bacterium *Bacillus thuringiensis*. Process Biochem 40:1895–1901

Salt DE, Blaylock M, Kumar PBAN, Dushenkov V, Ensley V, Chet D, Raskin I (1995) Phytoremediation: a novel strategy for the removal of toxic elements from the environment using plants. Biotechnology 13:468–474

Salt DE, Smith RD, Raskin I (1998) Phytoremediation. Annu Rev Plant Physiol Plant Mol Biol 49:643–648

Sarin V, Pant KK (2006) Removal of chromium from industrial waste by using eucalyptus bark. Bioresour Technol 97:15–20

Sawadaa A, Mori K, Tanaka S et al (2004) Removal of Cr (VI) from contaminated soil by electro-kinetic remediation. Waste Manag 24:483–490

Sawalha MF, Gardea-Torresdey JL, Parsons JG, Saupe G, Peralta-Videa JR (2005) Determination of adsorption and speciation of chromium species by saltbush (*Atriplex canescens*) biomass using a combination of XAS and ICP–OES. Microchem J 81:122–132

Say R, Yilmaz N, Denali A (2004) Removal of chromium (VI) ions from synthetic solutions by the fungus *Penicillium purpurogenum*. Eng Life Sci 4:276–280

Sekhar KC, Subramanian S, Modak JM, Natarajan KA (1998) Removal of metal ions using an industrial biomass with reference to environmental control. Int J Min Process 53:107–120

Seki H, Suzuki A, Maruyama H (2005) Biosorption of chromium (VI) and arsenic (V) onto methy-lated yeast biomass. J Colloid Interface Sci 281:261–266

Shanker AK, Cervantes C, Loza-Tavera H, Avudainayagam S (2005) Chromium toxicity in plants. Environ Int 31:735–753

Sharma DC, Sharma CP, Tripathi RD (2003) Phytotoxic lesions of chromium in maize. Chemosphere 51:63–68

Shen H, Wang YT (1993) Characterization of enzymatic reduction of hexavalent chromium by *Escherichia coli* ATCC 33456. Appl Environ Microbiol 59:3771–3777

Shen H, Wang YT (1994) Biological reduction of chromium by *E. coli*. J Environ Eng 120:560–572

Sheng GP, Yu H-Q, Yue Z-B (2005) Production of extracellular polymeric substances from *Rhodopseudomonas acidophila* in the presence of toxic substances. Appl Microbiol Biotechnol 69:216–222

Shi W, Becker J, Bischoff M, Turco RF, Konopka AE (2002) Association of microbial community composition and activity with lead, chromium, and hydrocarbon contamination. Appl Environ Microbiol 68:3859–3866

Srinivasan K, Balasubramanian N, Ramakrishnan TV (1988) Studies on chromium removal by rice husk carbon. Ind J Environ Health 30:376–387

Srivastava S, Thakur IS (2006) Evaluation of bioremediation and detoxication potentiality of *Aspergillus niger* for removal of hexavalent chromium in soil microcosm. Soil Biol Biochem 38:1904–1911

Tewari N, Vasudevan P, Guha BK (2005) Study on biosorption of Cr (VI) by *Mucor hiemalis*. Biochem Eng J 23:185–192

Tunali S, Kiran I, Akar T (2005) Chromium(VI) biosorption characteristics of *Neurospora crassa* fungal biomass. Min Eng 18:681–689

Turner MA, Rust RH (1971) Effects of chromium on growth and mineral nutrition of soybeans. Soil Sci Soc Am J 35:755–758

Turpeinen R, Kairesalo T, Haggblom MM (2004) Microbial community structure and activity in arsenic-, chromium- and copper contaminated soils. FEMS Microbiol Ecol 47:39–50

Ucun H, Bayhan YK, Kaya Y, Cakici A, Algur OF (2002) Biosorption of chromium (VI) from aqueous solution by cone biomass of *Pinus sylvestris*. Bioresour Technol 85:155–158

United States Environmental Protection Agency (1998) Toxicological review of hexavalent chromium. National Centre for Environmental Assessment, Office of Research and Development, Washington, DC

United States Environmental Protection Agency (1999) Integrated Risk Information System (IRIS) on chromium (III). National Centre for Environmental Assessment, Office of Research and Development, Washington, DC

Uysal M, Irfan A (2007) Removal of Cr (VI) from industrial wastewaters by adsorption. Part I: determination of optimum condition. J Hazard Mater 149:482–491

Vadillo JM, Garcia CC, Alcantara JF, Laserna JJ (2005) Thermal to plasma transitions and energy thresholds in laser ablated metals monitored by atomic emission/mass spectrometry coincidence analysis. Spec Acta B At Spectrosc 60:948–954

Venkateswarlu P, Venkata Ratnam M, Subba Rao D, Venkateswara Rao M (2007) Removal of chromium from an aqueous solution using *Azadirachta indica* (neem) leaf powder as an adsorbent. Int J Phys Sci 2:188–195

Vinodhini V, Das N (2009) Biowaste materials as sorbents to remove chromium (VI) from aqueous environment- a comparative study. ARPN J Agric Biol Sci 4:19–23

Vinodhini V, Das N (2010) Relevant approach to assess the performance of sawdust as adsorbent of Cr (VI) ions from aqueous solutions. Int J Environ Sci Technol 7:85–92

World Health Organization (1998) Chromium environmental health criteria. WHO, Geneva

Yadav S, Shukla OP, Rai UN (2005) Chromium pollution and bioremediation. Environ News Arch 11:1–4

Yadav SK, Juwarkar AA, Kumar GP, Thawale PR, Singh SK, Chakrabarti T (2009) Bioaccumulation and phyto-translocation of arsenic, chromium and zinc by *Jatropha curcas* L.: impact of dairy sludge and biofertilizer. Bioresour Technol 100:4616–4622

Young RV (2000) World of chemistry. Gale Group, Farmington Hills

Yun YS, Park D, Park JM, Volesky B (2001) Biosorption of trivalent chromium on the brown seaweed biomass. Environ Sci Technol 35:4353–4358

Zetic VG, Steklik-Tomas V, Grba S, Lutilsky L, Kozlek D (2001) Chromium uptake by *Saccharomyces cerevisiae* and isolation of glucose tolerance factor from yeast biomass. J Biosci 26:217–223

Zhang P, Jin C, Zhao Z, Tian G (2010) 2D crossed electric field for electrokinetic remediation of chromium contaminated soil. J Hazard Mater 177:1126–1133

Zhao M, Duncan JR (1998) Column sorption of Cr (VI) from electroplating effluent using formaldehyde cross-linked *Saccharomyces cerevisiae*. Biotechnology 20:603–606

Zhou J, Xia B, Treves DS, Wu LY, Marsh TL, O'Neill RV, Palumbo AV, Tiedje JM (2002) Spatial and resource factors influencing high microbial diversity in soil. Appl Environ Microbiol 68:326–334

Chapter 14
Genotoxicity Assessment of Heavy Metal–Contaminated Soils

**Javed Musarrat, Almas Zaidi, Mohammad Saghir Khan,
Maqsood Ahmad Siddiqui, and Abdulaziz A. Al-Khedhairy**

Abstract The soil environment is a major sink for multitude of chemicals and heavy metals, which inevitably leads to environmental contamination problems. Indeed, a plethora of different types of heavy metals are used and emanated through various industrial activities. Millions of tonnes of trace elements are produced every year from the mines in demands for newer materials. On being discharged into soil, the heavy metals get accumulated and may disturb the soil ecosystem, plant productivity, and also pose threat to human health and environment. Therefore, the establishment of efficient and inexpensive methodology and techniques for identifying and limiting or preventing metal pollution, causing threats to the agricultural production systems and human health, is earnestly required. The possible genotoxic effects of heavy metals on plants and other organisms have been extensively investigated worldwide and sufficiently discussed in this chapter. Also, the development and applications of new biomonitoring methodologies for assessment of soil genotoxicity have been emphasized. The molecular techniques being employed either alone or in combination for detecting the DNA damage induced by heavy

J. Musarrat (✉)
Al-Jeraisy Chair for DNA Research, Department of Zoology, College of Science,
King Saud University, Riyadh 11451, Saudi Arabia

Department of Agricultural Microbiology, Faculty of Agricultural Sciences,
Aligarh Muslim University, Aligarh 202002, Uttar Pradesh, India
e-mail: musarratj1@yahoo.com

A. Zaidi • M.S. Khan
Department of Agricultural Microbiology, Faculty of Agricultural Sciences,
Aligarh Muslim University, Aligarh 202002, Uttar Pradesh, India

M.A. Siddiqui • A.A. Al-Khedhairy
Department of Zoology, College of Science, King Saud University,
Riyadh 11451, Saudi Arabia

M.S. Khan et al. (eds.), *Biomanagement of Metal-Contaminated Soils*,
Environmental Pollution 20, DOI 10.1007/978-94-007-1914-9_14,
© Springer Science+Business Media B.V. 2011

metal–contaminated soils and other potentially genotoxic compounds are adequately elaborated. Indeed, the combination of two techniques leads to the precise and efficient detection and quantification of the sublethal genotoxic effects induced in the plant bioindicators by contaminated soil. Thus, the application of biomonitoring protocols in conjunction with the genotoxic assessment of contaminated soil will be advantageous in effective management of heavy metal–polluted soils.

Keywords Biomonitoring • Biosensor • Genotoxicity • Genotoxic assessment

14.1   Introduction

In agricultural practices, a variety of chemical inputs in the form of fertilizers, pesticides (herbicides, fungicides, and insecticides), or sewage sludge are constantly applied to optimize the crop production. The excessive use of these agrochemicals and other activities such as the burning of fossil fuels, mining, and smelting of metalliferous ores, municipal wastes, and industrial activities adds substantial amounts of heavy metals to soils (Bunger et al. 2007; Devi et al. 2007; Kim et al. 2007; Periyakaruppan et al. 2007; Roos et al. 2008), which cause contamination of the urban and agricultural soils. Metals are notable for their wide environmental dispersion, their tendency to accumulate in selected tissues of the human body, and their overall potential to be toxic even at low level of exposure. Some metals, such as copper and iron, are essential to life and play irreplaceable role. Other metals are xenobiotics; they have no useful role in human physiology and, even worse, as in the case of cadmium (Cd), lead (Pb), arsenic (As), chromium (Cr), nickel (Ni), and mercury (Hg). Even those metals that are essential, however, have the potential to turn harmful at very high levels of exposure. The annual toxicity of all metals mobilized exceeds the combined total toxicity of all radioactive and organic wastes generated every year from all other sources (Nriagu and Pacyna 1988). The continued and excessive discharge of these metals from various sources and their subsequent accumulation in soils pose a significant threat to human health and the environment due to their non-degrading ability. Heavy metals, which often act as genotoxic agents (Panda and Panda 2002; Sarkar et al. 2010), enter the human body through inhalation of dust, ingestion of plants that uptake the metal compounds from soil, and leaching from soil to groundwater and surface water used for drinking purposes. Toxic metal ions enter cells by means of the same uptake processes that move essential micronutrient metal ions. Class A metals (e.g., K, Ca, Mg) preferentially bind with oxygen-rich ligand (e.g., carboxylic groups), class B metals (e.g., Hg, Pb, Pt, Au) preferentially bind with sulfur- and nitrogen-rich ligands (e.g., amino acids), and borderline metals (e.g., Cd, Cu, Zn) show intermediate preferences, with the heavier metals tending toward class B characteristics (Nieboer and Richardson 1980). Heavy metal pollutants have a high bioaccumulation rate and at supra-optimal concentrations affect the human health, microorganisms, soil enzyme activity, and plants (Renella et al. 2005;

Simmons et al. 2005; Garnier et al. 2006; Wang et al. 2007; Unhalekhaka and Kositanont 2009). Pereira et al. (2009) have demonstrated the phytotoxicity and genotoxicity of soils from an abandoned uranium mine area. In plants, some metal compounds have shown the genotoxic effects (Radetski et al. 2004). For instance, in tobacco (*Nicotiana tabacum*), cadmium has reportedly caused cell death through the accumulation of superoxide anions (O_2^-) of mitochondrial origin and membrane peroxidation (Garnier et al. 2006), while in *Allium sativum* and *Vicia faba,* higher concentrations of cadmium induced the lipid peroxidation, resulting in oxidative stress that contributes to the genotoxicity and cytotoxicity of cadmium ions (Ünyayar et al. 2006). Exposure to higher cadmium concentrations has also been found to be carcinogenic, mutagenic, and teratogenic for a large number of animal species (Waalkes 2000, 2003). Cadmium metal and cadmium-containing compounds are known to cause lung cancer, and possibly prostate cancer, or tumors at multiple tissue sites (Mitrov and Chernozemski 1985; Vodenicharska et al. 1992; Tzonevski et al. 1998; Bruning and Chronz 1999; Chernozemski and Shishkov 2001). The studies also revealed induced DNA breaks in human blood lymphocytes with low micromolar concentrations of cadmium metal (Depault et al. 2006). Besides cadmium, other metals like lead, bismuth, indium, silver, and antimony also act as a genotoxicants (Asakura et al. 2009) and binds to the phosphate, deoxyribose, and heterocyclic nitrogenous bases of DNA. Consequently, the integrity of cells gets adversely affected due to systematic loss of altered genetic material through a process often referred as genotoxicity. Therefore, the genomic protection of organisms from the increasing environmental pollution is important for preservation of biodiversity.

The heavy metal and xenobiotics-mediated genotoxicity of soils depends on the bioavailability of contaminants including the physicochemical attributes of the soil and the magnitude of heavy metal contamination. The movement of heavy metals in soils is influenced by pH, particle size distribution, and carbon content of soil (Alloway and Ayres 1993; Wang et al. 2007). Generally, the soils having low pH are more genotoxic (Katnoria et al. 2008). Mostly, the environmental risk assessments of contaminated soils are based on chemical analysis, which reveals the presence of many mutagenic and carcinogenic compounds like heavy metals in soil. However, a major limitation of standard chemical analyses is that many soil genotoxicants are still unknown and most of the soil ecotoxicity data relates to relatively less known compounds. Therefore, there is a need to develop new methods for soil genotoxicity assessment. Bioassays in this regard provide a means of assessing the toxicity of a complex soil mixture without prior knowledge of its chemical composition. This has led to the discovery of different mutagenicity test for soil samples, which include Ames test (Ames et al. 1975; Brooks et al. 1998; Hughes et al. 1998; Monarca et al. 2002), *Tradescantia* micronucleus test (Knasmüller et al. 1998; Cabrera et al. 1999), *Tradescantia* stamen hair mutation (*Trad*-SHM) (Cabrera et al. 1999; Gichner 1999), and *Vicia* root micronucleus assay (Wang 1999; Cotelle et al. 1999). Because of the simplicity and sensitivity, these tests are likely to play an important role in the screening of genotoxic agents, especially for the detection of genotoxic substances from contaminated environments.

14.2 Bioavailability of Contaminants in Soil

There are several well-documented factors, which affect bioavailability of contaminant in soil environment such as the soil pH (Lock and Janssen 2003), redox potential (Rensing and Maier 2003), ionic strength, organic matter (Pardue et al. 1996), type of soil (Lock et al. 2002), clay fraction (Babich and Statzky 1977), water content, oxygen content, temperature and soil organisms, plant roots, invertebrates, etc. Other important factors that may affect bioavailability include the contaminants' physicochemical properties, such as molecular structure, aqueous solubility, polarity, lipophilicity, hydrophobicity, volatility (Reid et al. 2000), speciation of metals (Arnold et al. 2003), mineral form (Davies et al. 2003), mobility, and persistence. Residence time of contaminants referred as "aging" is another factor, which is basically a time-dependent interaction between the contaminant and the soil. Contaminants become sorbed to mineral and organic matter components of soil and trapped in micropores and become biologically inaccessible. The longer a contaminant is in contact with the soil, the more they become associated, reducing bioavailability and consequent potential toxicity (Alexander 2000; Hatzinger and Alexander 1995).

14.3 Genetic Effects of Heavy Metals

Understanding the action and reaction of chemical pollutants is important for preserving the gene pool and management of a healthy ecosystem. The uptake and translocation of heavy metals from soils to plants depends on factors such as the (1) total amount of potentially available elements (intensity factor) and (2) rate of element transfer from solid to liquid phases and to plant roots (Brummer et al. 1986). In the organisms exposed to the genotoxicants, the normal cellular processes are disrupted due to structural modifications in the DNA and influence the cell survival. It has been observed that due to the heavy metal accumulation in soils, the genetic constitution of populations especially the herbaceous or grassy plants is altered (Geburek 2000). Indeed, the heavy metals as genotoxicants affect the synthesis and duplication of DNA and chromosomes both directly or indirectly (Gichner 2003) and cause chromosomal aberrations in plant cells. These effects are influenced greatly by the types and dosage of heavy metals. For example, different barley (*Hordeum vulgare*) cultivars (Tokak and Hamidiye) when grown in nutrient solution under controlled environmental conditions and subjected to increasing concentrations of cadmium (0, 15, 30, 60, and 120 µmol/l) for different time periods exhibited large genotypic variation between barley cultivars. The differential cadmium tolerance observed in the barley cultivars may not be related to uptake or accumulation of cadmium in plants, but is attributed to internal antioxidative mechanisms. In the Cd-sensitive barley cultivar Hamidiye, the high sensitivity is related to oxidative damage due to enhanced production of ROS (Tiryakioglu et al. 2006). Exposure of cadmium, lead, and mercury leads to polyploidy, C-Karyokinesis, chromosome fragmentation, chromosome fusion, micronuclei formation, and nuclear decomposition in beans, garlic (*Allium sativum*) and onions (*Allium cepa*) (Liu et al. 2004). The high concentration of heavy metals in medium, in which plants could not grow normally, affects the sister chromatid exchange (SCE) frequency in root tip cells of *Hordeum vulgare* (Liu et al. 2005). Recently, Yi et al. (2010) demonstrated the cytogenetic effects of aluminum ($AlCl_3$) using *Vicia* cytogenetic tests, which are commonly used to monitor the genotoxicity of environmental pollutants. Significant increase in the frequency of micronuclei (MN) formation and anaphase chromosome aberrations is reported in *Vicia faba* root tips exposed to $AlCl_3$ over a concentration range of 0.01–10 mM for 12 h. The frequency of micronucleated cells is reported to be higher in Al-treated groups at pH 4.5 than that at pH 5.8. The $AlCl_3$ treatment also caused a decrease in the number of mitotic cells in a dose- and pH-dependent manner. The number of cells in each mitotic phase changed in Al-treated samples. Mitotic indices (MI) decreased with the increase of pycnotic cells. Thus, $AlCl_3$ has been classified as clastogenic, genotoxic, and cytotoxic agent in *Vicia* root cells.

The formation of free radicals by genotoxicants can result in the breakage of phosphodiester linkages within the DNA molecule. Genotoxicants can also interfere with normal DNA processing activities such as replication, methylation, and repair, which may result in mutations. Aina et al. (2004) studied the effect of heavy metal stress on the DNA methylation of a metal-sensitive plant, *Trifolium repens,* (L.) and a metal-tolerant plant, *Cannabis sativa,* (L.), and compared the variations in the level of 5-methylcytosine (5mC) in the root DNA of plants grown on soils contaminated with different concentrations of Ni^{2+}, Cd^{2+}, and Cr^{6+} with that of untreated plants, through immunolabeling with a monoclonal antibody. The DNA of hemp control plants has been found to be methylated about three times more than clover DNA. Heavy metals have shown to induce a global dose-dependent decrease of 5mC content, both in hemp and clover that varied between 20% and 40%. Moreover, the changes in methylation pattern of 5′-CCGG-3′ containing sequences were investigated by methylation-sensitive amplification polymorphism (MSAP) technique. Control plants of the same species had a very similar pattern, suggesting that under normal conditions, methylation involves precise sites. Heavy metal–induced DNA methylation changes are mainly the hypomethylation events. These variations are not randomly directed but involve the specific DNA sequences, since the detected polymorphisms have been found to be the same in all the plants analyzed for each treatment. A decline in DNA and RNA content in *Phaseolus vulgaris* under heavy metal stress (Hamid et al. 2010) and submerged aquatic plant (Jana and Choudhuri 1984) has also been reported. The reduced efficiency of DNA synthesis, weaker DNA protection from damaged chromatin proteins (histones), and increased deoxyribonuclease (DNase) activity have been reported in plants exposed to cadmium, copper, chromium, nickel, lead, mercury, and zinc (Prasad and Strzalka 2002). Heavy metals such as copper, nickel, cadmium, and lead are reported to decrease the RNA synthesis and to activate ribonuclease (RNase) activity, leading to further decrease in RNA content (Schmidt 1996).

The heavy metal like cadmium when applied in the form of cadmium chloride on tobacco roots induces significantly higher levels of DNA damage as measured by the cellular comet assay. DNA damage induced by Cd^{2+} in roots of a transgenic catalase-deficient tobacco line (CAT1AS) is reported to be higher than the wild-type tobacco (SR1) roots. While comparing the effects of ethyl methanesulfonate (positive control) and Cd^{2+}, it has been shown that Cd^{2+} does not induce any significant DNA damage in leaf nuclei. Also, the somatic mutations or homologous recombination did not occur in leaves, as measured by the GUS gene reactivation assay. Furthermore, the roots were found accumulating almost 50-fold more cadmium than did the above-ground parts of the tobacco seedlings, as revealed by Inductively Coupled Plasma (ICP) optical emission spectrometry (Gichner et al. 2004). Subsequently, Gichner et al. (2006) cultivated heterozygous tobacco (var. *xanthi*) and potato (*Solanum tuberosum* var. Korela) plants in soil from the site Střimice, which is highly polluted with heavy metals and on non-polluted soil from the recreational site Jezeří, both in North Bohemia, Czech Republic. The total content, the content of bio available, easily mobile, and potentially mobile heavy metals like Cd, Cu, Pb, and Zn in the tested soils, and the accumulation of these metals in the roots and above-ground biomass of test plants have been measured by atomic absorption spectrometry. The data revealed that the average leaf area (tobacco) and plant height (potato) were significantly reduced when these plants were grown in metal-stressed soil. Interestingly, a small but significant increase in DNA damage in nuclei of leaves of both plant species has been observed in plants growing on the polluted soil. The enhanced DNA damage with necrotic or apoptotic DNA fragmentation in heavy metal–stressed tobacco and potato plants leads to growth inhibition and distorted leaves. However, no increase in the frequency of somatic mutations occurred in tobacco plants growing on the polluted soil. The inability of plants to cope with heavy metal stress and to maintain the structural integrity provides an opportunity to test for the genotoxicity of pollutants present in the environment. Therefore, it is desirable to develop and establish new toxicological approaches to evaluate the potential cytotoxic and genotoxic effects of heavy metals. Some of the tests used commonly to assess the genotoxicity of heavy metal–polluted soils are discussed in the following section.

14.4   Assessment of Heavy Metal Genotoxicity

Genotoxicants are usually present in the environment and even at low concentrations can modify or damage the DNA (Fig. 14.1). Qualitative and quantitative assessment of DNA damage is, therefore, an important issue. In this context, various in vitro and in vivo genotoxicity tests designed to detect the substances that induce genetic damage directly or indirectly have been developed. The analytical techniques with sufficient selectivity and sensitivity have been used to detect extremely low levels of DNA damage. The difficulties in the measurements of pollutants in the field and the interpretation of such measurements in terms of bioavailability,

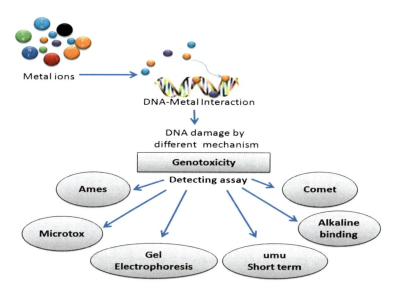

Fig. 14.1 Types of genotoxicity assays commonly used for testing heavy metal contaminated soils

associated with analytical techniques, have, however, generated a strong interest in using bioindicators and biomarkers for detection of damage to DNA. Advances and developments in molecular biology have provided new insight into the assessment of genotoxicity of soils and DNA damage in plants (Conte et al. 1998; Savva 2000; Citterio et al. 2002). Use of bioindicators, for example, in the measurement of contaminants is interesting because it helps in detecting different forms of pollutants, which are hard to measure in the field. Among the bioindicators, plants are considered to be the good bioindicators due to their greater role in food chain transfer and in defining habitat. Moreover, higher plants provide valuable genetic assay systems for screening and monitoring environmental pollutants and have a high sensitivity with few false negatives. They are now recognized as excellent indicators of cytotoxic, cytogenetic, and mutagenic effects of environmental chemicals and can be used to detect mutagens both indoor and outdoor. The higher plant genetic assays are inexpensive and easy to handle, which make them most suitable for use by researchers in developing countries (Grant 1994). Moreover, plants are ethically more acceptable and aesthetically more appealing than animals as sensors of environmental pollution.

14.4.1 Toxicity Assessment Using Multicellular Organisms

Toxicity assessment assays using whole animals or plants have been used for several decades. Several methods have been considered as International Standards Organization (ISO) guidance notes (ISO 1993, 1999, 2004), and as the Organisation for Economic Co-operation and Development (OECD) Chemicals Testing Guidelines (OECD 1984a, b, 2003). They include earthworm acute toxicity, earthworm reproduction, terrestrial plant growth (both monocotyledons and dicotyledons), inhibition of root growth, emergence and growth of higher plants, and effects on invertebrate reproduction and survival. The advantage of these tests is that they are directly relevant to the specific species and represent in situ conditions. These assays do not require a soil extract to be made and, therefore, represent actual pollutant bioavailability to a selected organism in soil over time. The UK Environment Agency assesses these tests on the five "R" criteria: reproducibility, representative, responsiveness, robustness, and relevance. A good example of bioavailability assessment of both organics and metals using multicellular organism toxicity tests is given by Cook et al. (2002). They found that soils with levels of contamination above intervention values, according to chemically based soil criteria, did not generate a toxic response to earthworms' mortality test or seed germination and root elongation, algal growth inhibition, and bacterial luminescence tests. Mammalian tests on rodents, dogs, pigs, etc., are usually used as a surrogate for human risk assessment (Hund-Rinke and Kordel 2003).The overall advantages of using multicellular organisms to test for bioavailability are that the tests are directly relevant to the organism used and can show systemic changes to the reactions. However, the limitations include no direct representation of other organisms, difficulties in data normalization between different field sites, and the time it takes to perform the tests in days/weeks rather than hours. Since, the mammalian tests are further complicated by strict regulations governing animal welfare and the expenses involved, many plant assays are the subject of choice.

Seven higher plant species like, *Allium cepa* (Fiskesjo 1997), *Arabidopsis thaliana, Glycine max, Hordeum vulgaris. Tradescantia paludosa, Vicia faba* (Koppen and Verschaeve 1996), and *Zea mays* were used to detect genotoxicity of chemical agents under the US Environmental Protection Agency (U.S. EPA) Gene-Tox program in the late 1970s. Six bioassays – *Allium* and *Vicia* root tip chromosome breaks, *Tradescantia* chromosome break, *Tradescantia* micronucleus, *Tradescantia*-stamen-hair mutation, and *Arabidopsis*-mutation bioassays – were established from four plant systems that are currently in use for detecting the genotoxicity of environmental agents. Under the Gene-Tox program, the *Crepis capillaris*-chromosome-aberration test was added to the existing six bioassays (Ma et al. 2005). Three of these plant bioassays, the *Allium* root chromosome aberration (AL-RAA) assay, the *Tradescantia* micronucleus (Trad-MCN) assay, and the *Tradescantia* stamen hair (Trad-SHM) mutation assay were validated in 1991 by the International Programme on Chemical Safety (IPCS) under the auspices of the World Health Organization, and the United Nations Environment Programme (UNEP) (Cabrera and Rodriguez 1999). The Tradescantia-Micronucleus (Trad-MCN) bioassay is also recommended and used in the International Program on Plant bioassays (IPPB) under the auspices of the United Nations Environment Programme (UNEP). Using chromosomal damage as indicator of the carcinogenic properties of environmental agents, the Trad-MCN bioassay is a quick and efficient tool for screening carcinogens in gaseous, liquid, and solid forms. Test results can be obtained within 24–48 h after the exposure either on site or in the laboratory. Under the IPPB/UNEP, more than 40 institutes including public health, medical, and cancer research in the major countries of the world are involved in the monitoring task on genotoxicity of polluted air, water, and soil. At the same time, the Trad-MCN can be used at a global scale to detect carcinogens as a preventive measure of cancer (Ma 2001; Kong and Ma 1999). Thus, these plant bioassays have proven to be efficient tests for chemical screening and especially for in situ monitoring for genotoxicity of environmental pollutants. The results from higher plant genetic assays are likely to make a significant contribution in public health protection from hazardous agents that can cause mutations and cancer. Furthermore, plant-based assays applied for toxicity evaluation in the field are likely to reduce animal sacrifices and testing costs.

14.4.2 Toxicity Assessment Using Whole Cell Test System

In the last few decades, an outbreak of microbial/whole cell tests has been noticed for estimating contaminant bioavailability in soil. Genetic engineering and modification techniques have allowed indicator genes to be coupled to genes of specific interest to give a qualitative and quantitative response. The first use of a reporter gene to show a phenotypic response was the Ames test (Ames et al. 1973). Although not directly tested with contaminated soil, it is still suitable to use with soil extracts. The extract used (water/solvent/buffer, etc.) is as important as the biological test chosen when assessing the toxicity of a contaminated soil as the extract will be the factor that determines bioavailability Wegrzyn and Czyz (2003). These mutagenic biosensors tend to be sensitive, reasonably quick to perform (days or hours, not weeks), but are limited by the need for specialist (expensive) equipment and their relevance to other organisms.

The best established microbial test in environmental testing is the Microtox assay (http://www.azurenv.com). A bioluminescent marine bacterium, *Vibrio fischeri*, produces light as a by-product of normal cell functions. Any toxicant inhibits cell functions and proportionally, light emission, allowing toxicity to be quantitatively measured. Microtox itself is still widely used and has been shown to be highly sensitive (Munkittrick et al. 1991). The major advantages include sensitivity in its general toxic response, simplicity and rapidity, robustness, and reproducibility. Major limitations are that it relies on exposure to an extract, which may present difficulties in interpreting results, and is a marine organism and therefore not strictly relevant in its response to soil contamination. The principle behind Microtox, of light emission, has given rise to a huge number of genetically engineered bacteria, yeast, and mammalians cells, which use light as an indication of bioavailability. Microbial biosensors are extensively reviewed by Hansen and Sorensen (2001), Leveau and Lindow (2002), and Belkin (2003). Leveau and Lindow (2002) and Belkin (2003) also discussed similar reporter gene-based systems that use β-galactosidase/lac Z and green fluorescent protein (GFP). The former was originally used in the SOS chromotest (Quillardet et al. 1982), an Ames-like test. The latter is now equally as popular as luminescent-based tests, using fluorescence rather than luminescence. Knight et al. (2004)

presented a yeast/GFP genotoxicity assay used to test a range of pesticides, metals, and solvents. Biosensors that utilize both luminescence and fluorescence, for testing acute and genotoxic threats simultaneously are also available. The most high profile usage of these tools is as health monitors in the International Space Station (Rabbow et al. 2003). Advantages of these light/color-based toxicity indicator tests are ease of assay, speed of assay, versatility, and sensitivity. Microbial and whole cell biosensors may be best employed as initial screening tools for environmental and soil contamination.

14.4.3   Toxicity Assessment Using Subcellular or Molecular Assays

The nucleic acid-based DNA hybridization array (Fredrickson et al. 2001) and reverse transcriptase PCR for monitoring gene expression (Environment Agency 2003), enzyme-based, and antibody- and receptor-based biosensors are in common use for monitoring environmental and food contaminants (Baeumner 2003). The UK Environment Agency has recommended the use of reverse transcriptase PCR to measure gene expression as a tool for looking at thousands of genes at once and their responses to vast arrays of contaminants. Sturzenbaum et al. (1998a, b, 2001) examined changes in gene expression of metallothionein, carboxypeptidase, and other metal-sensitive genes and found transcription levels up to 100-fold greater in exposed organisms, showing that the technique has a high degree of sensitivity. A great advantage of this technique is that organisms that have been directly in contact with contaminated soil can be analyzed and, therefore, bioavailability is the parameter being assessed. A disadvantage is determining what "normal" levels of gene expression are in order to determine whether contamination has had an effect. It is also important to differentiate between "normal stress responses," for example, to drought and those actually related to pollution. Antibody interactions are also highly sensitive and very specific. Immunoaffinity has been adapted from clinical research to quantify environmental pollutants like metals (Chavez-Crooker et al. 2003) and dioxins (Okuyama et al. 2004). Antibodies can be customized and raised against any contaminant and are supplied by various biotechnology companies. The relevance of any antibody assay for bioavailability purposes will be dependent on the soil extraction method adopted. The potential use of this method is to analyze soil DNA to assess in situ bioavailability. Thus, these in vitro molecular assays are very useful as initial bioavailability and toxicity screening tools.

14.5   Genotoxicity Assay for Soil

Soil contaminants are common in industrialized countries, causing widespread contamination directly of soil and indirectly of ground water and food. Among these pollutants, particular attention has been paid to soil mutagens and carcinogens due to their potentially hazardous effects on animal populations and human health. In this context, both physicochemical methods and bioassays (bacteria or plants) have been employed. However, because of the complex chemical nature of soil, standard chemical analyses are limited in their ability to characterize the chemical composition of genotoxicants in soil to assess its potential genotoxicity. On the other hand, the bioassays like *Salmonella* mutation assay has most widely been used for assessing the genotoxicity of toxic substances (Brown et al. 1985; McDaniels et al. 1993; Ehrlichmann et al. 2000; Watanabe and Hirayama 2001). The Ames test (*Salmonella* assay) is a relatively sensitive and specific in vitro test used widely in the screening of mutagenic potential of chemical compounds (Mortelmans and Zeiger 2000; Maron and Ames 1983). The test uses amino acid-dependent strains of *S. typhimurium* and *E. coli*. In the absence of an external histidine source, the cells cannot grow to form colonies. Colony growth is resumed if a reversion of the mutation occurs, allowing the production of histidine to be resumed. Spontaneous reversions occur with each of the strains; mutagenic compounds cause an increase in the number of revertant colonies relative to the background level. A positive test indicates that the chemical might act as a carcinogen (although a number of false-positives and false-negatives are known). It reveals the gene mutation-inducing ability (mutagenicity). Numerous studies all over the world have shown the presence of different mutagenic substances in different soils using Ames test (Smith 1982; Knize et al. 1987; Kool et al. 1989). However, in some studies, Ames test has yielded negative results (Steinkellner et al. 1998). Since 1970s, higher plant bioassays have been recommended for genotoxic evaluation of complex environmental mixtures by Anonymous (1973), Committee 17 of Environmental Mutagen Society (Drake et al. 1975), the World Health Organization (1985), and National Swedish Environmental Protection Board in 1989 (Cabrera et al. 1999). Considering the potential health hazards posed by heavy metals in soil and non-availability of data on content of heavy metals in agricultural soils, Katnoria et al. (2008) conducted a study to evaluate the genotoxic potential of extracts of soil samples collected from different agricultural fields of Amritsar, India, employing Ames and *Allium* root anaphase aberration assay (Al-RAAA). The water soil extract prepared in distilled water (soil: water, 1:2 w/v) was evaporated to dryness and re-dissolved in distilled water. Different concentrations corresponding to 0.25, 0.5, 1, 2, and 2.5 g equivalent of soil per plate have been tested employing *Salmonella typhimurium* TA98 and TA100 strains, with and without in vitro metabolic activation (S9) to detect direct and indirect mutagenic effects. For Al-RAAA, different concentrations of extracts (10, 25, 50, 75, and 100%) have been used for treatment of root tips of *A. cepa*. In situ conditions could be simulated by allowing the onion bulbs to root directly in soil samples contained in small pots. The genotoxic potential of soil samples can be correlated with content of heavy metals like chromium, cobalt, copper, manganese, mercury, nickel, and zinc. The pH, alkalinity, water holding capacity, bulk density, moisture content, nitrates, phosphates, and potassium should also be studied for better correlations with the abiotic factors. Wang et al. (2007) determined the combined effects of cadmium (10 mg/kg soil) and butachlor (5, 10, and 50 mg/kg soil) on enzyme activities and microbial community structure in phaeozem soil. The phosphatase activity is reportedly decreased in soils with 10 mg Cd/kg soil, when used alone while urease activity remained unaffected. However, when the Cd and butachlor were added to soils at 2:1 or 1:2 ratio, the urease and phosphatase activities were decreased. The enzyme activities were, however, greatly improved at 1:5 suggesting that the combined effects of Cd and butachlor on soil urease and phosphatase activities largely depend on the added concentration ratios to soils. Furthermore, the Random Amplification of Polymorphic DNA (RAPD) analysis showed changes in RAPD profiles of different treated samples including the variations in loss of normal bands and appearance of new bands compared with the control soil. The RAPD fingerprints exhibit apparent changes in the number and size of amplified DNA fragments attributed to significant changes in the microbial diversity. Studies also suggest that RAPD analysis in conjunction with other biomarkers such as soil enzyme parameters could prove to be a powerful ecotoxicological tool. Gao et al. (2010) have studied the response of soil enzyme activities, viz. dehydrogenase, phosphatase, and urease in polluted soil using ecological dose model and RAPD in order to determine soil health. They have determined the 50% ecological dose (ED_{50}) values modified by toxicant coefficient from the best-fit model, and studied the determination values from the regression analysis for the three enzyme activities. In general, the elevated heavy metal concentration negatively affected the total population size of bacteria and actinomycetes and enzymatic activity. The dehydrogenase ($ED_{50} = 777$) was the most sensitive soil enzyme, whereas urease activity ($ED_{50} = 2,857$) showed the lowest inhibition. Composite metals or elevated toxicant level resulted in significant disappearing of RAPD bands, and the number of denoting polymorphic bands was greater in combined polluted soils.

The genotoxicity of soil samples collected at six sampling sites in a Slovenian industrial and agricultural region, contaminated by heavy metals and sulfur dioxide (SO_2), has been assessed by Ames test, Comet assay, and preliminary *Tradescantia* micronucleus assay (Lah et al. 2008). Genotoxicity of all six water soil leachates has been proven by the Comet assay on human cell lines; however, no positive results were detected by Ames test. The *Tradescantia* micronucleus assay showed an increase in micronuclei formation for three samples. Of these tests, Comet assay was found as the most sensitive assay, followed by the micronucleus test. The Ames test was not sensitive enough for water soil leachates genotoxicity evaluations where heavy metal contamination is anticipated. In other study, Knasmüller et al. (1998) employed plant bioassays for the detection of genotoxic effects of heavy metal–contaminated soils. In this case, four metal salts, namely $Cr(VI)O_3$, $Cr(III)Cl_3$, $Ni(II)Cl_2$, and $Sb(III)Cl_3$ were tested in MN tests with pollen tetrad cells of *Tradescantia* clone #4430 and in meristematic root tip cells of *Vicia faba*. With Cr^{6+} and Ni^{2+}, obvious dose-dependent effects have been reported, whereas in *Vicia*, negative results were obtained with the four metal salts under all conditions. In order to compare the mutagenic property of the metals, the regression curves (k-values) was calculated, which indicated that the number of MN induced per mM in 100 tetrad cells. The corresponding values for Cr^{6+} and Ni^{2+} were 0.87 and 1.05, respectively. Thus, the *Tradescantia* system has been found to be sensitive toward those metal species, which cause DNA damage in animals and man such as Cr^{6+}, Cd^{2+}, Ni^{2+}, and Zn^{2+}, whereas no clear positive results were obtained with less harmful metal ions such as Cu^{2+}, Cr^{3+}, or Sb^{3+}. Also, the mutagenic effects of four metal-contaminated soils and two types of standardized leachates (pH 4 and pH 7) of these soils were tested in *Tradescantia* and in *Vicia*. Direct exposure of the *Tradescantia* plants in the soils resulted in a drastic increase of the MN frequencies over the background. The lowest effect has been reported with the Slovakian soil containing the Sb and As (4.5-fold increase over the background). However, the induced frequencies with the other soils were 11–15-fold over the control values. Thus, the direct exposure of intact plants is an appropriate method, which enables to detect genotoxic effects of metal-contaminated soils in situ. In a similar study, Kong and Ma (1999) conducted *Allium* root anaphase aberration (Allium-AA), *Tradescantia*-micronucleus (Trad-MCN), and the *Tradescantia* stamen hair mutation (Trad-SHM) tests in soil solutions or shallow well water samples to determine genotoxicity. The results of Allium-AA tests suggested a 2.78–3.01-fold increase in anaphase aberration frequencies in contaminated soil solution samples and well water samples as compared with the negative control. The Trad-MCN tests demonstrated a 1.66–4.75-fold increase of MCN frequencies in contaminated soil solution samples and shallow well water samples as compared with the frequencies of the controls. The Trad-SHM tests exhibited a 2.7–2.86-fold increase of pink mutation events in the contaminated soil solution samples over that of the controls. Control groups of the Allium-AA tests had an average of 0.75/1,000 anaphase figures, and control groups of the Trad-MCN tests had an average of 3.2 MCN/100 tetrads, while control groups of the Trad-SHM tests had an average of 1.4 mutation events/1,000 hairs. The soil solutions of DMSO extracts showed higher genotoxicity than that of distilled water extracts. Among the three plant bioassays tested, the Trad-MCN assay was found to be the most efficient for genotoxicity testing in soil solution (Kong and Ma 1999).

Combining genotoxicity/mutagenicity tests and physico-chemical methodologies can be more useful for determining the potential genotoxic contaminants in soils. For example, the genotoxicity of contaminated soils collected from a highly industrialized area in the Lombardy region, in Northern Italy, was evaluated by employing an integrated chemical/biological approach involving a short-term bacterial mutagenicity test (Ames test), a plant genotoxicity test (*Tradescantia/* micronucleus test), and chemical analyses (Monarca et al. 2002). Soil samples were extracted with water or with organic solvents. Water extracts of soil samples were tested for polycyclic aromatic hydrocarbons (PAH) and heavy metals. The organic solvent extracts were analyzed for their mutagenicity using the Ames mutagenecity and *Tradescantia* genotoxicity tests. The soils with high concentrations of genotoxic PAH and heavy metals showed mutagenic activity with the Ames test and clastogenicity with the Tradescantia/micronucleus test.

14.6 Genotoxicity Assessment Assay for Plants

Generation of DNA damage is considered as an important initial outcome in carcinogenesis. Therefore, the assessment of genotoxins-induced DNA damage and mutations is vital in eco-genotoxicology. In order to detect the various genotoxic effects of compounds, a battery of assays are available. Some of the tests may, however, have limited use because of complicated technical setup or because they are applicable only to a few cell types. Despite limitations, bioassay, which can be applied to various tissues and/or special cell types, is being used due to its sensitivity for detecting low levels of DNA damage, requirement for small numbers of cells per sample, general ease of test performance, the short time needed to complete a study, and its relatively low cost. In recent years, several plant species have been used as bioindicators, and several molecular tests have been developed to evaluate the toxicity of environmental contaminants on vegetal organisms. For example, the genotoxic changes in plants can be detected by the RAPD technique. The RAPD is a simple, reliable, sensitive, and reproducible assay with wide range of DNA damage (e.g., DNA adducts, DNA breakage) and mutations (point mutations and large rearrangements) detecting potential (Savva 1998; Atienzar et al. 2000). Many factors can affect the generation of RAPD profiles. It is, therefore, important that these factors are identified and taken into account while using these assays. In addition, the relevant bands generated in RAPD profile allow to identify some of the molecular events implicated in the genomic instability and, hence, to discover genes involved in the initiation and development of malignancy (Atienzar and Jha 2006). Liu et al. (2007) applied RAPD and other related fingerprinting techniques to detect the genotoxin-induced DNA damage and mutations in rice (*Oryza sativa*) seedlings exposed to varying concentrations (15–60 mg/L) of cadmium. The inhibition in root growth and increase of total soluble protein content in root tips of rice seedlings occurred in a manner similar to those observed for barley (Liu et al. 2005). The RAPD profiles of root tips after cadmium treatment showed modifications in band intensity and gain or loss of bands when compared with control. Thus, DNA polymorphisms detected by RAPD analysis could be used as an investigation tool in environmental toxicology and as a useful biomarker assay for the detection of genotoxic effects of other metals as well. Similarly, Cenkci et al. (2009) used RAPD to detect DNA damage in the roots and leaves of bean (*Phaseolus vulgaris* L.) seedlings exposed to heavy metals like Hg ($HgCl_2$), B (H_3BO_3), Cr ($K_2Cr_2O_7$), and Zn ($ZnSO_4 7H_2O$) at concentrations of 150 and 350 ppm for 7 days. With increasing concentrations, there was a substantial decrease in growth of shoot and root growth, while the contents of Hg, B, Cr, and Zn increased in the roots and leaves at elevated concentration of each heavy metal. During the RAPD analyses, 12 RAPD primers of 60–70% GC content were found to produce unique polymorphic band profiles and were later used to produce a total of 120 bands of 263–3,125 bp in the roots and leaves of untreated and treated seedlings. Polymorphisms became evident as disappearance and/or appearance of DNA bands in 150 and 350 ppm treatments compared with untreated control treatments. The DNA changes in RAPD profiles were more in the roots than in the leaves (Cenkci et al. 2009).

Micronucleus assay is yet another bioassay that has been used for the detection of genotoxic effects of heavy metal ions on plants. Steinkellner et al. (1998) investigated the genotoxic effects of heavy metals As^{3+}, Pb^{2+}, Cd^{2+}, and Zn^{2+} through micronucleus tests with *Tradescantia* pollen mother cells (Trad MCN), and meristematic root tip cells of *Allium cepa* and *Vicia faba* (Allium/Vicia MCN). The order of genotoxicity of metals for three tests was determined as: $As^{3+} > Pb^{2+} > Cd^{2+} > Zn^{2+}$ Cu^{2+}.

In *Tradescantia* experiment, induction of MCN was observed at concentration ranging between 1 and 10 mM, whereas in tests with root tip cells, higher concentrations (10–1,000 mM) were required to show significant effects. Further increase in the concentration of heavy metals reduced root growth, delayed cell division, and showed decreased MCN frequencies. Comparisons by linear regression analyses indicated that the sensitivity of the three bioassays for heavy metals decreases in the order: Trad MCN > Vicia root MCN > Allium root MCN. Moreover, a soil sample which contained high concentrations of the five metals and a control soil were analyzed. Aqueous soil extracts induced only weak effects in Trad MCN tests and no effects in the root tip assays, whereas cultivation of the plants in the soils resulted in a pronounced induction of MCN in the *Tradescantia* system and moderate effects in *Vicia* and *Allium*. Thus, the Trad MCN assay detects the genotoxic effects of heavy metals and can be used for biomonitoring metal-contaminated soils. Also, Sarkar et al. (2010) determined the effect of nickel on shoot regeneration in tissue culture and identified polymorphisms induced in leaf explants exposed to nickel through RAPD. In vitro leaf explants of *Jatropha curcas* were grown in nickel amended Murashige and Skoog (MS) medium at four different concentrations (0, 0.01, 0.1, 1 mM) for 3 weeks. Percent regeneration, number of shoots produced, and genotoxic effects were evaluated by RAPD using leaf explants obtained from the first three treatments following 5 weeks of their subsequent subculture in metal-free MS medium. Percent regeneration decreased with increase in addition of nickel to the medium up to 14 days from 42.31% in control to zero in 1.0 mM. The number of shoot buds scored after 5 weeks was higher in control as compared to all other treatments except in one of the metal-free subculture medium wherein the shoot number was higher in 0.01 mM treatment (mean = 7.80) than control (mean = 7.60). RAPD analysis produced only 5 polymorphic bands (3.225%) out of a total of 155 bands from 18 selected primers. Only three primers OPK-19, OPP-2, and OPN-08 produced polymorphic bands.

14.7   Conclusion

The genotoxicity of contaminated soils originating from industrial sources has been widely studied and reported across the globe. Till date, various chemical and biomonitoring methods for assessment of soil genotoxicity are available. Amongst all, the *Salmonella* mutation assay has been the most commonly and frequently used method. The mutagenicity evaluation of soil helps in identifying the heavily contaminated sites with genotoxic chemicals released from various industrial operations or agricultural practices. These genotoxic substances persist in soils and not only adversely affect the quality of soils but also influence the overall performance of various crops grown in agricultural soil polluted with mutagenic compounds. It is interesting to develop the new bioindicators, biomarkers, and molecular tools for sensitive, rapid, and economical analysis of soil genotoxicity for efficient management and control of bioremediation of contaminated sites for safer environment and increased productivity involving lesser human health hazards.

References

Aina R, Sergio S, Angela S, Massimo L, Alessandra G, Sandra C (2004) Specific hypomethylation of DNA is induced by heavy metals in white clover and industrial hemp. Physiol Plant 121:472–480

Alexander M (2000) Aging, bioavailability, and overestimation of risk from environmental pollutants. Environ Sci Technol 34:4259–4265

Alloway BJ, Ayres DC (1993) Chemical principles of environmental pollution. Blackie Academic and Professional, London

Ames BN, Lee FD, Durston WE (1973) An improved bacterial test system for the detection and classification of mutagens. Proc Natl Acad Sci USA 70:782–786

Ames BN, Kammen HO, Yamasaki E (1975) Hair dyes are mutagenic: identification of a variety of mutagenic ingredients. Proc Natl Acad Sci USA 72:2423–2427

Anonymous (1973) Evalution of gentic risk of environmental chemicals. Royal Swedish academy of Sciences, Ambio, Special Report No.3

Arnold RE, Langdon CJ, Hodson ME, Black S (2003) Development of a methodology to investigate the importance of chemical speciation on the bioavailability of contaminants to *Eisenia andrei*. Pedobiologia 47:633–639

Asakura K, Satoh H, Chiba M, Okamoto M, Serizawa K, Nakano M, Omae K (2009) Genotoxicity studies of heavy metals: lead, bismuth, indium, silver and antimony. J Occup Health 51:498–512

Atienzar FA, Jha AN (2006) The random amplified polymorphic DNA (RAPD) assay and related techniques applied to genotoxicity and carcinogenesis studies: a critical review. Mut Res 613:76–102

Atienzar FA, Cordi B, Donkin ME, Evenden AJ, Jha AN, Depledge MH (2000) Comparison of ultraviolet-induced genotoxicity detected by random amplified polymorphic DNA with chlorophyll fluorescence and growth in a marine macroalgae, *Palmaria palmate*. Aquat Toxicol 50:1–12

Babich H, Stotzky G (1977) Reductions in toxicity of cadmium to microorganisms by clay minerals. Appl Environ Microbiol 33:696–705

Baeumner AJ (2003) Biosensors for environmental pollutants and food contaminants. Anal Bioanal Chem 377:434–445

Belkin S (2003) Microbial whole-cell sensing systems of environmental pollutants. Curr Opin Microbiol 6:206–212

Brooks LR, Hughes TJ, Claxton LD, Austern B, Brenner R, Kremer F (1998) Bioassay-directed fractionation and chemical identification of mutagens in bioremediated soils. Environ Health Perspect 106:1435–1440

Brown KW, Donnelly KC, Thomas JC, Davol P, Scott BR (1985) Mutagenicity of three agricultural soils. Sci Total Environ 41:173–186

Brummer G, Gerth J, Herms U (1986) Heavy metal species, mobility and availability in soils. Z Pflanzenernaehr Bodenkd 149:382–398

Bruning T, Chronz C (1999) Occurrence of urinary tract tumors in miners highly exposed to dinitroluene. YOEM 3:144–149

Bunger J, Krahl J, Munack A, Ruschel Y, Schröder O, Emmert B, Westphal G, Müller M, Hallier E, Brüning T (2007) Strong mutagenic effects of diesel engine emissions using vegetable oil as fuel. Arch Toxicol 81:599–603

Cabrera GL, Rodriguez DM (1999) Genotoxicity of soil from farmland irrigated with wastewater using three plant bioassays. Mut Res 426:211–214

Cenkci S, Yildiz M, Ciğerci IH, Konuk M, Bozdağ A (2009) Toxic chemicals-induced genotoxicity detected by random amplified polymorphic DNA (RAPD) in bean (*Phaseolus vulgaris* L.) seedlings. Chemosphere 76:900–9006

Chavez-Crooker P, Pozo P, Castro H, Dice MS, Boutet I, Tanguy A, Moraga D, Ahearn GA (2003) Cellular localization of calcium, heavy metals, and metallothionein in lobster (*Homarus americanus*) hepatopancreas. Comp Biochem Physiol C Toxicol Pharmacol 136:213–224

Chernozemski I, Shishkov T (2001) Oncology. SIELA-SOFT, Sofia, pp 15–21

Citterio S, Aina R, Labra M, Ghiani A, Fumagalli P, Sgorbati S, Santagostino A (2002) Soil genotoxicity: a new strategy based on biomolecular tools and plants bioindicators. Environ Sci Tech 36:2748–2753

Conte C, Mutti I, Puglisi P, Ferrarini A, Regina GRG, Maestri E, Marmiroli N (1998) DNA fingerprint analysis by PCR based method for monitoring the genotoxic effects of heavy metals pollution. Chemosphere 37:2739–2749

Cook SV, Chu A, Goodman RH (2002) Leachability and toxicity of hydrocarbons, metals and salt contamination from flare pit soil. Water Air Soil Pollut 133:297–314

Cotelle S, Masfaraund JF, Férard JF (1999) Assessment of the genotoxicity of contaminated soil with Allium/Vicia-micronucleus and the *Tradescantia*-micronucleus assays. Mut Res 426:167–171

Davies NA, Hodson ME, Black S (2003) Is the OECD acute worm toxicity test environmentally relevant? The effect of mineral form on calculated lead toxicity. Environ Pollut 121:49–54

Depault F, Cojocaru M, Fortin F, Chakrabarti S, Lemieux N (2006) Genotoxic effects of chromium (VI) and cadmium (II) in human blood lymphocytes using the electron microscopy *in situ* end-labeling (EM-ISEL) assay. Toxicol In vitro 20:513–518

Devi SS, Biswas AR, Biswas RA, Vinayagamoorthy N, Krishnamurthi K, Shinde VM, Hengstler JG, Hermes M, Chakrabarti T (2007) Heavy metal status and oxidative stress in diesel engine tuning workers of central Indian population. J Occup Environ Med 49:1228–1234

Drake JW, Abrahamson S, Crow JF, Hollaender A, Lederberg S (1975) Environmental mutagenic hazards. Science 157:503–514

Ehrlichmann H, Dott W, Eisentraeger A (2000) Assessment of the water-extractable genotoxic potential of soil samples from contaminated sites. Ecotoxicol Environ Saf 46:73–80

Environment Agency (2003) Ecological risk assessment – a public consultation on a framework and methods for assessing harm to ecosystems from contaminants in soil

Fiskesjo G (1997) Allium test for screening chemicals; evaluation of cytological parameters. In: Wang W, Gorsuch JW, Hughes JS (eds) Plants for environmental studies. Lewis Publishers, New York, pp 307–333

Fredrickson HL, Perkins EJ, Bridges TS, Tonucci RJ, Fleming JK, Nagel A, Diedrich K, Mendez-Tenorio A, Doktycz MJ, Beattie KL (2001) Towards environmental toxicogenomics development of a flow through high-density DNA hybridization array and its application to ecotoxicity assessment. Sci Total Environ 274:137–149

Gao Y, Zhou P, Mao L, Zhi Y, Shi WJ (2010) Assessment of effects of heavy metals combined pollution on soil enzyme activities and microbial community structure: modified ecological dose response model and PCR-RAPD. Environ Earth Sci 60:603–612

Garnier L, Simon-Plas F, Thuleau P, Angel JP, Blein JP, Ranjeva R, Montillet JL (2006) Cadmium affects tobacco cells by a series of three waves of reactive oxygen species that contribute to cytotoxicity. Plant Cell Environ 29:1956–1969

Geburek T (2000) Effects of environmental pollution on the genetics of forest trees. In: Zhitkovich A, Young A, Bosheir D, Boyle T (eds) Forest conservation genetics, principle and practice. CABI, Wallingford, pp 135–158

Gichner T (1999) Monitoring the genotoxicity of soil extracts from two heavily polluted sites in Prague using *Tradescantia* staminal hair and micronucleus (MNC) assays. Mut Res 426:163–166

Gichner T (2003) DNA damage induced by indirect and direct acting mutagens in catalase-deficient transgenic tobacco: cellular and acellular comet assays. Mut Res/Genetic Toxicol Environ Mut 535:187–193

Gichner T, Patková Z, Száková J, Demnerová K (2004) Cadmium induces DNA damage in tobacco roots, but no DNA damage, somatic mutations or homologous recombination in tobacco leaves. Mut Res/Genetic Toxicol Environ Mut 559:49–57

Gichner T, Zdeňka P, Jiřina S, Kateřina D (2006) Toxicity and DNA damage in tobacco and potato plants growing on soil polluted with heavy metals. Ecotoxicol Environ Saf 65:420–426

Grant WF (1994) The present status of higher plant bioassays for the detection of environmental mutagens. Mut Res 310:175–185

Hamid N, Bukhari N, Jawaid F (2010) Physiological responses of *Phaseolus vulgaris* to different lead concentrations. Pak J Bot 42:239–246

Hansen LH, Sorensen SJ (2001) The use of whole-cell biosensors to detect and quantify compounds or conditions affecting biological systems. Microb Ecol 42:83–494

Hatzinger PB, Alexander M (1995) Effect of aging of chemicals in soil on their biodegradability and extractability. Environ Sci Technol 29:537–545

Hughes TJ, Claxton LD, Brooks L, Warren S, Brenner R, Kremer F (1998) Genotoxicity of bioremediated soils from the Reilly Tar site, St. Louis Park, Minnesota. Environ Health Perspect 106:1427–1433

Hund-Rinke K, Kordel W (2003) Underlying issues in bioaccessibility and bioavailability: experimental methods. Ecotoxicol Environ Saf 56:52–62

ISO (1993) Soil quality effects of pollutants on earthworms (*Eisenia fetida*). Part 1: determination of acute toxicity using artificial soil substrate. International Standards Organization, Geneva

ISO (1999) Soil quality inhibition of reproduction of Collembola (*Folsomia candida*) by soil pollutants. International Standards Organisation, Geneva

ISO (2004) Soil quality effects of pollutants on Enchytraeidae (*Enchytraeus* sp.). Determination of effects on reproduction and survival. International Standards Organisation, Geneva

Jana S, Chaudhury MA (1984) Synergistic effect of heavy metals pollutants on senescence in submerged aquatic plants. Water Air Soil Pollut 21:351–357

Katnoria JK, Arora S, Nagpal A (2008) Genotoxic potential of agricultural soils of Amritsar. Asian J Sci Res 1:122–129

Kim Y, Park J, Shin YC (2007) Dye-manufacturing workers and bladder cancer in South Korea. Arch Toxicol 81:381–384

Knasmüller S, Gottmann E, Steinkellner H, Fomin A, Pickl C, Paschke A, Göd R, Kundi M (1998) Detection of genotoxic effects of heavy metal contaminated soils with plant bioassays. Mut Res 420:37–48

Knight AW, Keenan PO, Goddard NJ, Fielden PR, Walmsley RM (2004) A yeast-based cytotoxicity and genotoxicity assay for environmental monitoring using novel portable instrumentation. J Environ Monit 6:71–79

Knize MG, Takemoto BT, Lewis PR, Felton GS (1987) The characterization of the mutagenic activity of soil. Mut Res 192:23–30

Kong MS, Ma TH (1999) Genotoxicity of contaminated soil and shallow well water detected by plant bioassays. Mut Res 426:221–228

Kool HJ, Vankreyl CF, Prasad S (1989) Mutagenic activity in groundwater in relation to mobilization of organic mutagens in soil. Sci Total Environ 84:185–199

Koppen G, Verschaeve L (1996) The alkaline comet test on plant cells: a new genotoxicity test for DNA strand breaks in *Vicia faba* root cells. Mut Res 360:193–200

Lah B, Vidic T, Glasencnik E, Cepeljnik T, Gorjanc G, Marinsek-Logar R (2008) Genotoxicity evaluation of water soil leachates by Ames test, comet assay, and preliminary *Tradescantia* micronucleus assay. Environ Monit Assess 139:107–118

Leveau JHJ, Lindow SE (2002) Bioreporters in microbial ecology. Curr Opin Microbiol 5:259–265

Liu D, Jiang W, Gao X (2004) Effects of cadmium on root growth, cell division and nucleoli in root tips of garlic. Physiol Plant 47:79–83

Liu W, Li PJ, Qi XM, Zhou QX, Zheng L, Sun TH, Yang YS (2005) DNA changes in barley (*Hordeum vulgare*) seedlings induced by cadmium pollution using RAPD analysis. Chemosphere 61:158–167

Liu W, Yang YS, Zhou Q, Xie L, Li P, Sun T (2007) Impact assessment of cadmium contamination on rice (*Oryza sativa* L.) seedlings at molecular and population levels using multiple biomarkers. Chemosphere 67:1155–1163

Lock K, Janssen CR (2003) Influence of ageing on zinc bioavailability in soils. Environ Pollut 126:371–374

Lock K, De Schamphelaere KAC, Janssen CR (2002) The effect of lindane on terrestrial invertebrates. Arch Environ Contam Toxicol 42:217–221

Ma TH (2001) *Tradescantia*-micronucleus bioassay for detection of carcinogens. Folia Histochem Cytobiol 39((Suppl) 2):54–55

Ma TH, Cabrera GL, Owens E (2005) Genotoxic agents detected by plant bioassays. Rev Environ Health 20:1–13

Maron DM, Ames NB (1983) Revised methods for the *Salmonella* mutagenicity assay. Mut Res 113:173–215

McDaniels AE, Reyes AL, Wymer LJ, Rankin CC, Stelma GN Jr (1993) Genotoxic activity detected in soils from a hazardous waste site by the Ames test and an SOS colorimetric test. Environ Mol Mutagen 22:115–122

Mitrov G, Chernozemski I (1985) Nutrition and cancer. Medicina i Fizkultura, Sofia, pp 67–98

Monarca S, Feretti D, Zerbini I, Alberti A, Zani C, Resola S, Gelatti U, Nardi G (2002) Soil contamination detected using bacterial and plant mutagenicity tests and chemical analyses. Environ Res 88:64–69

Mortelmans K, Zeiger E (2000) The Ames *Salmonella*/microsome mutagenicity assay. Mut Res 455:29–60

Munkittrick KR, Power EA, Sergy GA (1991) The relative sensitivity of microtox daphnid, rainbow-trout, and fathead minnow acute lethality tests. Environ Toxicol Water Qual 6:35–62

Nieboer E, Richardson DHS (1980) The replacement of the nondescript term heavy metals by a biologically and chemically significant classification of metal ions. Environ Pollut 1:3–26

Nriagu JO, Pacyna JM (1988) Quantitative assessment of world-wide contamination of air, water and soils by trace metals. Nature 333:134–139

OECD (1984a) Chemicals testing guidelines: 207 earthworm, acute toxicity tests. Organization for Economic Co-operation and Development, Paris

OECD (1984b) Chemicals testing guidelines: 208 terrestrial plants, growth test. Organisation for Economic Co-operation and Development, Paris

OECD (2003) Chemicals testing guidelines: 208 seedling emergence and seedling growth test. Organisation for Economic Co-operation and Development, Paris

Okuyama M, Kobayashi N, Takeda W, Anjo T, Matsuki Y, Goto J, Kambegawa A, Hod S (2004) Enzyme-linked immunosorbent assay for monitoring toxic dioxin congeners in milk based on a newly generated monoclonal anti-dioxin antibody. Anal Chem 76:1948–1956

Panda BB, Panda KK (2002) Genotoxicity and mutagenicity of metals in plants. In: Prasad KNV, Strzałka K (eds) Physiology and biochemistry of metal toxicity and tolerance in plants. Kluwer Academic Publishers, Dordrecht, pp 395–414

Pardue JH, Kongara S, Jones WJ (1996) Effect of cadmium on reductive dechlorination of trichloroaniline. Environ Toxicol Chem 15:1083–1088

Pereira R, Marques CR, Silva Ferreira MJ, Neves MFJV, Caetano AL, Antunes SC, Mendo S, Gonçalves F (2009) Phytotoxicity and genotoxicity of soils from an abandoned uranium mine area. Appl Soil Ecol 42:209–220

Periyakaruppan A, Kumar F, Sarkar S, Chidananda SS, Ramesh GT (2007) Uranium induces oxidative stress in lung epithelial cells. Arch Toxicol 81:389–395

Prasad MNV, Strzalka K (2002) Physiology and biochemistry of heavy metal toxicity and tolerance in plants. Kluwer Academic Publishers, Dordrecht

Quillardet P, Huisman O, Dari R, Hofnung M (1982) SOS chromotest, a direct assay of induction of an SOS function in *Escherichia coli* K-12 to measure genotoxicity. Proc Natl Acad Sci USA 79:5971–5975

Rabbow E, Rettberg P, Baumstark-Khan C, Horneck G (2003) The SOS-LUX-LAC-FLUORO toxicity test on the International Space Station (ISS). Adv Space Res 31:1513–1524

Radetski CM, Ferrari B, Cotelle S, Masfaraud JF, Ferard JF (2004) Evaluation of genotoxic, mutagenic and oxidative stress potentials of municipal solid waste incinerator bottom ash leachates. Sci Total Environ 333:204–216

Reid BJ, Stokes JD, Jones KC, Semple KT (2000) Nonexhaustive cyclodextrin-based extraction technique for the evaluation of PAH bioavailability. Environ Sci Technol 34:3174–3179

Renella G, Mench M, Landi L, Nannipieri P (2005) Microbial diversity and hydrolase synthesis in long-term Cd contaminated soils. Soil Biol Biochem 37:133–139

Rensing C, Maier RM (2003) Issues underlying use of biosensors to measure metal bioavailability. Ecotoxicol Environ Saf 56:140–147

Roos PH, Angerer J, Dieter H, Wilhelm M, Wölfle D, Hengstler JG (2008) Perfuorinated compounds (PFC) hit the headlines: meeting report on a satellite symposium of the annual meeting of the German society of toxicology. Arch Toxicol 82:57–59

Sarkar T, Anand KGV, Reddy MP (2010) Effect of nickel on regeneration in *Jatropha curcas* L. and assessment of genotoxicity using RAPD markers. Biometals. doi: 10.1007/s10534-010-9364-7

Savva D (1998) Use of DNA fingerprinting to detect genotoxic effects. Ecotoxicol Environ Saf 41:103–106

Savva D (2000) The use of arbitrarily primed PCR (AP-PCR) fingerprinting detects exposure to genotoxic chemicals. Ecotoxicology 9:341–353

Schmidt W (1996) Influence of chromium (III) on root associated Fe (III) reductase in *Plantago isnceolata* L. J Exp Bot 47:805–810

Simmons RW, Pongsakul P, Saiyasitpanich D, Klinphoklap S (2005) Elevated levels of cadmium and zinc in paddy soils and elevated levels of cadmium in rice grain downstream of zinc mineralized area in Thailand: implications for public health. Environ Geochem Health 27:501–511

Smith JW (1982) Mutagenecity of extracts from agricultural soils in the *Salmonella* microsome test. Environ Mut 4:369–370

Steinkellner H, Mun-Sik K, Helma C, Ecker S, Ma TH, Horak O, Kundi M, Knasmüller S (1998) Genotoxic effects of heavy metals: comparative investigation with plant bioassays. Environ Mol Mutagen 31:183–191

Sturzenbaum SR, Kille P, Morgan AJ (1998a) Heavy metal-induced molecular responses in the earthworm, *Lumbricus rubellus* genetic fingerprinting by directed differential display. Appl Soil Ecol 9:495–500

Sturzenbaum SR, Kille P, Morgan AJ (1998b) The identification, cloning and characterization of earthworm metallothionein. FEBS Lett 431:437–442

Sturzenbaum SR, Cater S, Morgan AJ, Kille P (2001) Earthworm pre- procarboxypeptidase: a copper responsive enzyme. Biometals 14:85–94

Tiryakioglu M, Eker S, Ozkutlu F, Husted S, Cakmak I (2006) Antioxidant defense system and cadmium uptake in barley genotypes differing in cadmium tolerance. J Trace Elem Med Biol 20:181–189

Tzonevski D, Sapundzhiev K, Vodenicharov E (1998) Studying the contents of lead and cadmium in the blood of children up to 15 years old from the region of the non- ferrous-metal works near Plovdiv. Higiena i Zdraveopazvane 2(3):20–22

Unhalekhaka U, Kositanont C (2009) Microbial composition in cadmium contaminated soils around zinc mining area, Thailand. Modern Appl Sci 3:1–8

Ünyayar S, Çelik A, Çekiç FÖ, Güzel A (2006) Cadmium-induced genotoxicity, cytotoxicity and lipid peroxidation in *Allium sativum* and *Vicia faba*. Mutagen 21:77–81

Vodenicharska TZ, Petrov I, Razboinikova F (1992) A study for heavy metal (lead, cadmium, zinc, copper, manganese) contamination among the population from the 3 rd metallurgical base region. Higiena i Zdraveopazvane 2:59–62

Waalkes MP (2000) Cadmium carcinogenesis in review. J Inorg Biochem 79:241–244

Waalkes MP (2003) Cadmium carcinogenesis. Mut Res 533:107–120

Wang H (1999) Clastogenecity of chromium contaminated soil samples evaluated by Vicia root-micronucleus assay. Mut Res 426:147–149

Wang J, Lu Y, Shen G (2007) Combined effects of cadmium and butachlor on soil enzyme activities and microbial community structure. Environ Geol 51:1221–1228

Watanabe T, Hirayama T (2001) Genotoxicity of soil. J Health Sci 47:433–438

Wegrzyn G, Czyz A (2003) Detection of mutagenic pollution of natural environment using microbiological assays. J Appl Microbiol 95:1175–1181

WHO (World Health Organization) (1985) Guide to short-term tests for detecting mutagenic and carcinogenic chemicals. Environ Health Criteria 51:208

Yi M, Yi H, Li H, Wu L (2010) Aluminum induces chromosome aberrations, micronuclei, and cell cycle dysfunction in root cells of *Vicia faba*. Environ Toxicol 25:124–129

Chapter 15
Mobility and Availability of Toxic Metals After Soil Washing with Chelating Agents

Domen Lestan and Metka Udovic

Abstract Remediation techniques for soils polluted with toxic metals can be divided into two main groups: immobilization and soil washing. Immobilization technologies leave metals in soil, but minimize their availability, while soil washing with chelating agents removes metals from soil. Metals in soil are not entirely accessible to chelating agents and, hence, not entirely removed. Residual metals left in the soil after remediation remain in chemically stable species bound to non-labile soil fractions and are considered nonmobile and non-bioavailable and thus non-toxic. However, with the reintroduction of remediated soil into the environment, the soil is exposed to various environmental factors, which could eventually promote or initiate the transition of the residual metals back to more labile forms to re-establish the disturbed equilibrium. Such a shift is likely to increase the toxicity of the residual metals and, consequently, decrease the final efficiency of soil remediation. Different extraction techniques are used to assess metals bioavailability and the efficiency of soil remediation. Reduced bioavailability of contaminants for organisms is most often assessed by established chemical extraction tests. However, do the chemical extraction tests really provide (include) reliable information on the availability of metals for soil fauna? In the present chapter, the effect of biotic and abiotic environmental factors on the mobility and availability of residual metals in soil after remediation is discussed. Furthermore, the benefits of in vivo assessment of soil remediation efficiency by terrestrial organisms is highlighted.

Keywords Bioindicators • Chelating agents • Heavy metal mobility • Soil washing

D. Lestan (✉) • M. Udovic
Department of Agronomy, Biotechnical Faculty, Centre for Soil and Environmental Science,
University of Ljubljana, Jamnikarjeva 101, SI-1000 Ljubljana, Slovenia
e-mail: Domen.Lestan@bf.uni-lj.si

M.S. Khan et al. (eds.), *Biomanagement of Metal-Contaminated Soils*,
Environmental Pollution 20, DOI 10.1007/978-94-007-1914-9_15,
© Springer Science+Business Media B.V. 2011

15.1 Soil Remediation

The contamination of soil with toxic metals has been an unfortunate by-product of industrialization and modern agronomic practices and is ubiquitous. Unlike organic compounds, metals are not degradable in the environment. Contamination of soils with metals can have long-term environmental and health implications. In fact, such soils often present an unacceptable risk to human and ecological health and need to be remediated. Various low-cost, efficient, and environmentally friendly soil treatment technologies are now available for remediation of metal-contaminated sites (Dermont et al. 2008; Lestan et al. 2008). Polluted soil removal and its safe deposition are, however, not always an acceptable approach, because of the high costs involved, the lack of adequate waste disposal facilities, or other reasons. However, appropriate methodologies and technologies have to be carefully selected for each polluted site according to the characteristics of the soil and of the contaminants.

Remediation techniques for metal-contaminated soils can be divided into two main groups (Fig. 15.1): immobilization techniques change the speciation and the fractionation of metals in soil solid phases, thus considerably lowering their mobility and biological availability, without removing them (Guo et al. 2006; Udovic and Lestan 2008); removal technologies, on the other hand, remove metals from the soil and are therefore preferred (Dermont et al.2008). One of the permanent solutions is soil washing/leaching, which involves the separation of metals from soil solid phases by solubilizing them in a washing/leaching solution. The effectiveness of washing/leaching can be increased by adding acids, surfactants, or chelating agents (chelants) to the solution (Griffiths 1995; Peters 1999). Acids dissolve carbonates and other metal-bearing soil fractions and exchange metals from soil colloids. Chelants form a coordinate chemical bond with metals and facilitate their solubilization from the soil solid phases, where the majority of soil metals reside, into the washing/leaching solution. Since acidic solutions could cause deterioration of soil physicochemical properties, use of chelants is considered to be environmentally less disturbing (Neale et al. 1997; Xu and Zhao 2005). Of the various chelants, ethylenediaminetetraacetic acid (EDTA) has been recognized as the most effective synthetic chelating agent in removing metals especially Pb, Cd, Cu, and Zn, from polluted soil (Finzgar and Lestan 2007; Dermont et al. 2008). EDTA has received greater attention in soil remediation because of its relatively low cost compared to other chelants (Chaney et al. 2000; Udovic and Lestan 2007a). However, metals in soil are usually not entirely accessible to chelants (Fig. 15.2), especially in soils rich in organic matter or clay. Therefore, such soils can only be partially removed (Levy et al. 1992). Metals occur in various soil 'pools' of different solubilities with varied chemical characteristics and consequently variable functions (Mulligan et al. 2001; Sabiené and Brazauskiené 2004).

Different approaches are used to describe metal chemical forms in soil. Sequential extraction schemes used to determine metal fractions in soil (Sun et al. 2001) are based on reacting the soil with a series of extracting solutions of increasing strength, causing the release of metals from sorption sites with decreasing availability (Abollino et al. 2006). A six-step sequential extraction, for example, described by Tessier et al. (1979)

Remediation technologies

Reduction of metals availability /mobility:

- Increase of soil pH, liming
- Addition of adsorbents
- Metal precipitation by phosphates
- Vitrification
- Solidification/stabilisation hydraulic binders

Removal of soil metals:

- Removal of contaminated soil fractions
- Phytoextraction
- Electrokinetic remediation
- Soil washing with salt, acid or chelant solutions

Fig. 15.1 Remedial options for soil contaminated with toxic metals

Fig. 15.2 Toxic metals (*bio*)availability stripping concept. Chelating agents cannot extract all toxic metals from soil (*lower curve*), but can remove entirely their (*bio*)available and mobile pool (*upper curve*)

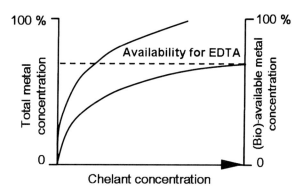

later modified by Lestan et al. (2003), divides metal forms in soil into six fractions. The fraction soluble in soil solution is obtained by extraction of air-dried soil with deionized water. The fraction exchangeable from soil colloids to the soil solution is extracted with 1 M $MgNO_3$, which displaces ions electrostatically bound in the soil matrix. The fraction bound to carbonates is extracted with 1 M ammonoacetate (NH_4OAc) at pH 5, where carbonates (calcite, dolomite) are solubilized and entrapped metals are released. The fraction bound to Fe and Mn oxides is extracted with 0.1 M $NH_2OH \times HCl$ at pH 2, where Fe and Mn oxides are reduced to soluble forms. The fraction bound to organic matter is obtained after oxidizing the organic matter and the soluble sulfides by heating the soil suspension in 0.02 M HNO_3 and 30% H_2O_2 at 85°C, followed by extraction with 1 M NH_4OAc. The last fraction for completion of the mass balance is obtained after the digestion of the remaining soil sample with *aqua regia*. The water soluble, exchangeable, and carbonate bound fractions represent the most available form of metals to organisms (Wen et al. 2004). In general, Pb and Zn form strong bonds with soil solid components. They are retained in the topsoil when the contamination originates from air-borne emissions. Copper is one of the less bioavailable metals in soil (Kabata-Pendias and Pendias 1992) and forms more stable complexes with organic components than other metals (Kizilkaya 2004). Lead is found mostly associated with organic matter and carbonate soil fractions (Li and Thornton 2001). Zinc tends to concentrate in the soil fraction residual after sequential extractions (Rivero et al. 2000; Kabala and Singh 2001). Cadmium, on the other hand, is generally concentrated in exchangeable and carbonate soil fractions or forms weak complexes with soil organic matter. Cadmium is, however, more easily accessible and extractable from the soil than are Pb and Zn (Ramos et al. 1994).

15.2   Efficiency of Soil Washing with Chelating Agents

Fractionation of metals in soil, their availability, mobility, and consequently, the prospect of their removal with soil washing methods and the afterward fate of residual metals are determined by soil properties like texture, content of organic matter, content and type of clay minerals and Al, Fe, and Mn oxides, prevailing physico-chemical conditions in the soil (saturation, aeration, pH, redox potential), and mineralogy of metal contaminants (Levy et al. 1992).

EDTA, the most widely studied chelating agent used in soil remediation is capable of extracting metals from all non-silicate-bound phases in the soil (Ure 1996; Tandy et al. 2004). In a study, Peters and Shem (1992), for example, reported that a maximum of 64 and 19% Pb (compared with initial Pb concentration) was removed with EDTA and nitrilotriacetic acid (NTA) as chelants, respectively, from contaminated soil containing high clay and silt contents. In a similar investigation, Pichtel et al. (2001) reported that various concentrations of EDTA and pyridine-2,6-dicarboxylic acid (PDA) removed up to 58 and 56% of Pb, respectively, from soil material collected from a battery recycling/smelting site. In a follow-up study, Finzgar et al. (2005) reported that using 40 mmol kg^{-1} of [S,S]-ethylenediamine disuccinate (EDDS), 31.1% of total Pb was extracted from vegetable garden soil, rich in organic matter. Borona and Romero (1996) in yet another study extracted Pb-contaminated soil with EDTA and observed that the amount of Pb removed was correlated with the amount of Pb associated with the Fe and Mn oxide and organic matter soil fractions. The decrease of Cu concentration in different fractions in a soil leached with 0.01 M EDTA was in the order of Cu bound to Fe and Mn oxides > Cu bound to organic matter > Cu bound to carbonates > residual Cu (Sun et al. 2001). In a recent study, we also observed differences in the removal efficiency of EDTA when subjected to different soil fractions, as presented in Fig. 15.3 (Udovic and Lestan 2010a, b). Metals were mostly removed from the most labile fractions, as the percentage of the residual metals increased along the fractions. The relatively high percentage of residual metals in the first fraction (data not presented) is caused by the very low metal concentrations, which could probably be due to analytical measurement errors. The residual metals left in soil after remediation remains in chemically stable, nonavailable forms, which are bound to non-labile soil fractions (Nowack et al. 2006; Finzgar and Lestan 2007; Lestan et al. 2008). The final outcome of soil remediation with soil washing/leaching with chelants result in reduced availability of metals for uptake by organisms (metal bioavailability stripping concept) as reported by many workers (Lestan et al. 2008; Pociecha and Lestan 2009).

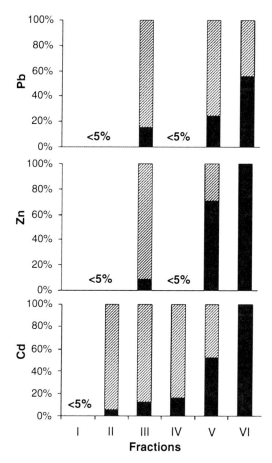

Fig. 15.3 Fractionation of residual (*black*) and removed (*gray*) metals: Pb, Zn, and Cd, in polluted soil after remediation with EDTA leaching. The sum of the residual and the removed respective metal in each fraction is 100%. Only fractions containing >5% of the total metals are represented. I water extractable fraction; II exchangeable fraction; III fraction bound to carbonates; IV fraction bound to Fe and Mn oxides; V organic fraction; and VI residual fraction (Adapted from Udovic and Lestan 2010a)

15.3 Metal Availability and Mobility Before and After Remediation

Since metals are present in soil in various chemical forms (bound to different soil fractions) of different solubilities (Gupta et al. 1996), they are rarely entirely available to organisms (bioavailable) (Arnold et al. 2003). The importance of (bio)availability and mobility of metals beside their total concentration in soil is now largely accepted (Kamnev and van der Lelie 2000; Mulligan et al. 2001). Instead of relying only on the total metal content, assessment of metal bioavailability and mobility are also often used in choosing the appropriate remediation technology as well as to evaluate the final success of soil remediation (Loureiro et al. 2005; Kumpiene et al. 2007).

15.3.1 Chemical Extraction Tests

One-step extraction is used to determine specific chemical forms of metals. The mobility and potential leaching of metals into the groundwater system are frequently assessed using one-step extraction procedure schemes, such as the standardized Toxicity Characteristic Leaching Procedure (TCLP) (US EPA 1995) and soil extraction with 0.05 M EDTA at pH 7.5 (Kosson et al. 2002). Both methods have been developed to simulate the potential leachability of pollutants from soil into the surrounding. The phytoavailability of metals in soil is assayed with different selective extraction methods. The extraction procedure with diethylenetriaminepentaacetic acid (DTPA) was designed by Lindsay and Norvell (1978) to identify the availability of micronutrients for crops growing on near-neutral and calcareous soils (Baker and Senft 1997; Brun et al. 2001). Extracting soil with 0.01 M $CaCl_2$ allows us to determine the phytoavailable share of metals (and nutrients) in soil without disrupting the ionic balance (Novozamsky et al. 1993). More methods are available and choosing the most appropriate one is seldom possible without firstly considering the characteristics of the studied soil.

An important form of exposure to environmental pollutants is the accidental ingestion of soil and inhalation of dust particles by living organisms. It is assumed that children ingest more soil and dust particles than adults due to their mouthing behavior (Davis and Mirick 2006). Several in vitro digestion models based on selected physiological parameters in humans have been developed and used to estimate the oral bioavailability of soil contaminants to humans (Ruby et al.1996; Oomen et al. 2002, 2003). A short overview of the most important aspects to be considered when selecting a suitable test for the assessment of human exposure from ingestion of soil and soil material can be found in the Technical Specification ISO/TS 17924 (2007). The two-step Physiologically Based Extraction Test (PBET) was designed to analyze the oral bioaccessibility of metals in the human stomach, where the pH is low (2.5), and in the small intestine, where the pH is neutral (pH 7). It enables us to detect the amount of metals ready to be absorbed from the intestine into the blood (Wragg and Cave 2002; Dean 2007; Turner and Ip 2007). Therefore, it is preferred to the simplified one-step extraction schemes, where only the stomach phase is considered (ISO/TS 17924, 2007). However, such an analytical approach allows us only to estimate the potential bioavailability of metals in soil, since it does not fully consider the combined effects of different metals and the complex characteristics of the interactions between organisms, soil, and metals (Alvarenga et al. 2008; Sousa et al. 2008). For this reason, a variety of biotests are applied to evaluate the potential risk posed by metals in soil to organisms; when combined with analytical environmental chemistry, they provide more complete and relevant information on the bioavailability of metals in contaminated and remediated soil (Udovic et al. 2009).

15.3.2   *In Vivo Tests with Bioindicators*

When the bioavailability of chemicals and their potential trophic transfer are of interest, the bioassays involving bioindicators must be employed to determine the accumulated contaminants. In this context, some plants and animals, especially invertebrates have already been used for assessing the metal availability in soils (Meier et al.1997). These organisms accumulate metals in proportion to the environmental concentrations and are therefore, used as bioindicators/biomonitors of environmental metal pollution (Heikens et al. 2001; Kennette et al. 2002; Gál et al. 2008; Suthar et al. 2008; Udovic and Lestan 2010c). Among invertebrates, the most popular organisms for metals accumulation research are mollusks, earthworms, crustaceans, insects, myriapods, and arachnids (Hopkin 1989).

Earthworms are often used in soil ecotoxicological tests. While acute tests do not provide an insight into the effects of metals on population dynamics and chronic tests are often time consuming and labor intensive (Loureiro et al. 2005), the avoidance behavior of earthworms is a simple and ecologically relevant measurable endpoint for assessing the effect of metals in soil on earthworm movement (Amorim et al. 2008) as an indicator of soil pollution, also at ecosystem level (Aldaya et al. 2006; Sousa et al. 2008). The presence of chemoreceptors on the prostomium and on the anterior segments and the distribution of tubercles along the body make earthworms highly sensitive to chemicals in their environment, allowing them to avoid unfavorable environments, thanks to their locomotory abilities (Lukkari et al. 2005; Curry and Schmidt 2007). So far, the majority of studies have employed such assays to artificial or natural soils freshly spiked with metals (Hund-Rinke et al. 2005; Langdon et al. 2005; Loureiro et al. 2005; Lukkari and Haimi 2005). It is, however, difficult to extrapolate such results to field situations, where natural soils differ in terms of physical, chemical, and structural properties (Sousa et al.2008). It is therefore, preferable to use natural soils directly, whereby a soil with the same properties as the test soil, but with no contamination, should be chosen as control soil (Aldaya et al. 2006; Amorim et al. 2008; Sousa et al. 2008). In a study, where the feasibility of the standardized two-section vessels earthworm avoidance test (ISO 17512-1, 2008) was tested for assessing the efficiency of soil remediation of Pb-, Zn-, and Cd-polluted soil, groups of 10 *E. fetida* individuals were presented with a choice between the most extensively remediated soil (control soil – leached with four consecutive applications of 40 mmol kg^{-1} EDTA) and test soils (i.e., non-remediated soil and soils remediated using lower EDTA concentrations), avoiding thus differences in soil characteristics, which could result in biased earthworm behavior. However, the earthworms generally did not avoid the tests soils in favor of the control soil (Fig. 15.4).

According to the criteria suggested by Hund-Rinke and Wiechering (2001), no test soil could be classified as toxic with its reduced habitat function, since more than 20% of earthworms were found in these soils. The avoidance test was not sensitive enough to discriminate soil leaching with different EDTA. Some soil properties were taken into consideration for a plausible explanation of the non-avoidance

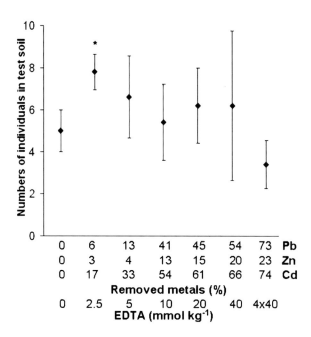

Fig. 15.4 Earthworm (*Eisenia fetida*) avoidance behavior of soils with increasing percentages of removed Pb, Zn, and Cd (remediated with leaching with increasing EDTA concentrations). Asterisk (*) denotes significant non-avoidance behavior of test soil (χ^2-test, $p < 0.05$) (Adapted from Udovic et al. 2009)

behavior, which were pointed out by other authors as important factors affecting earthworms in soil, such as soil pH (Edwards and Bohlen 1996; Edwards 2004) and the presence of sodium due to the usage of disodium-EDTA salt for remediation purposes (Owojori and Reinecke 2009). However, the most probable explanation was found in the metal fractionation pattern in the remediated soil. Most of the Pb, the major metal pollutant, was found bound to organic matter, in which metals are considered to be inaccessible for EDTA (Ure 1996; Tandy et al. 2004). Consequently, while the total Pb concentration decreased after leaching with EDTA, as it was removed from the water soluble, the exchangeable and the carbonate fraction, the share of organically bound Pb was preserved. The same applies for Zn and Cd. Earthworms, however, due to specific routes of exposure, are perhaps susceptible to metals bound to organic matter (Sousa et al. 2008), while having no need to avoid the moderate concentrations of metals in other soil fractions.

Earthworms live in direct contact with the solid and pore-water soil phase and are thus exposed in a manner representative of other soil species (e.g., bacteria, plants, soft-bodied invertebrates) (Spurgeon et al. 2006). However, isopods appear to be the most efficient of them as assimilators of metals. Isopods are omnivorous animals, but they have clear feeding preferences. Many species prefer feeding on decaying leaf litter rather than fresh with high microbial density (Zimmer 2002). Some other species are soil dwelling with differing food preferences. Terrestrial isopod *Porcellio scaber* inhabits a wide range of habitats. They chew dead plants or plant material mixed with soil into small fragments. Terrestrial isopods must have evolved efficient ways of assimilating essential elements from the food, because unlike their marine ancestors, they could no longer obtain them directly from the external medium across the respiratory surfaces (Warburg 1993). They accumulate the highest concentrations of metals such as Zn, Cd, Pb, and Cu so far recorded in any soft tissue of terrestrial animals (Hopkin et al. 1993; Witzel 1998; Vijver et al. 2006). The main metal storage organ in isopods is the hepatopancreas. Metals are bound here to specific low-molecular-weight peptides, or stored in insoluble granules (Hopkin 1989). The metal uptake from food depends on many factors, of which the most important are the bioavailability of metals in the ingested material, gut microflora, the rate of food consumption and pH inside the gut, metal concentration, the duration of exposure, and the combination of factors to which the metals are exposed (Odendaal and Reinecke 2004). Earlier data suggested that isopods accumulate metals for a lifetime, but recent researches provide evidence of metals loss also, once transferred to uncontaminated food sources (Witzel 1998). Nevertheless, a correlation between metal body burden and metal concentrations in food/substratum can be demonstrated (Hopkin 1989). Efficient metal assimilation, the wealth of knowledge on their metal physiology, and ease of handling in the laboratory are the main reasons for choosing terrestrial isopods as experimental animals in studies on soil metal pollution. Because terrestrial isopods accumulate metals from their environment in proportion to their concentration in the soil (Heikens et al. 2001), they appear very suitable as indicators of the metal bioavailable fraction in polluted soil and leaf litter (Paoletti and Hassall 1999; Gál et al. 2008).

So far, the suitability of metal accumulation in earthworms and isopods as a measure of the EDTA remediation efficiency of metal-polluted soil was tested in two studies (Udovic et al. 2009; Udovic and Lestan 2010c). Animals were exposed to non-remediated soil and to soils remediated with increasing EDTA concentrations, keeping thus the soil properties unaltered. In both cases, the gradient of metal removal by increasingly higher EDTA concentration in the leaching solution was generally reflected in the amounts of metal accumulated in the whole animal bodies. However, when the authors used bioaccumulation factors (BAFs) to express the metal accumulation in animals in relation to the total metal concentration in soil as a measure of their bioavailable share, they found the ratio between the non-bioavailable and bioavailable metal share to be constant, even in the most extensively EDTA processed soil (Fig. 15.5). The gradient of metal removal by increasingly higher EDTA concentration of the leaching solution is reflected in decreasing trend of the amounts of metal accumulated by the animals after 14 days of exposure (Fig. 15.5a). However, no such trend was seen when BAFs were calculated (Fig. 15.5b); they indicate a constant ratio between available and nonavailable concentrations of Pb, and even an increasing ratio between available and nonavailable concentrations of Cd. In contrast to the indications given by the chemical extraction tests (sequential extraction, DTPA extraction, TCLP extraction, PBET), leaching with EDTA was unable to reduce the share of Pb, Zn, and Cd in the soil that is bioavailable to *E. fetida* or *P. scaber*. The results substantiate general concern about using chemical extraction methods solely for the assessment of metal availability in non-remediated and remediated soils, since metals that were otherwise unavailable for chemical extractions were available and accumulated by *E. fetida* and *P. scaber*.

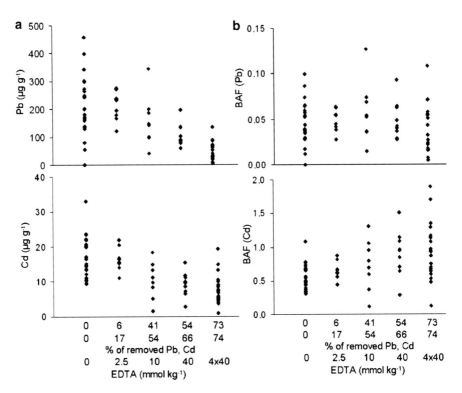

Fig. 15.5 Lead and cadmium concentrations in *Porcellio scaber* (**a**) and respective bioaccumulation factors, BAFs (ratio of metal concentration in the animals to the total soil metal concentration) (**b**) (Adapted from Udovic et al. 2009)

15.4   Aging of Remediated Soil and Toxic Metals Availability and Mobility

With the reintroduction of remediated soil into the environment, we expose the soil to various abiotic and biotic environmental factors (soil ageing factors). Is therefore the reduced mobility and bioavailability of soil residual metals a permanent or only temporal achievement of soil remediation (Fig. 15.6)? Soil is a dynamic natural body and, after remediation, various abiotic (i.e., climatic, hydrological) and biotic soil (microorganisms and fauna) factors could presumably initiate the transition of residual metals from less to more mobile/accessible forms to re-establish the disturbed equilibrium, although the availability of metals in non-remediated contaminated soil is considered to decrease with time (Han et al. 2003). Such a shift would increase the toxicity of the residual metals and consequently decrease and hamper the final efficiency of soil remediation (Fig. 15.7).

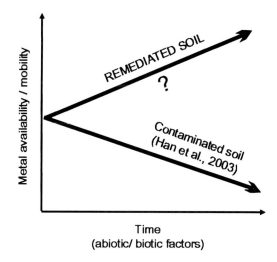

Fig. 15.6 Soil is a dynamic natural body – various environmental factors could initiate the transition of metals residual after remediation back to more available and mobile and therefore more toxic forms

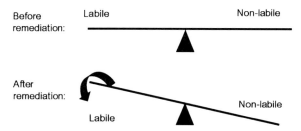

Fig. 15.7 After soil washing, a significant amount of metals remain in the soil. Taking the labile (bio-available, mobile) metal species from the soil could disturb the chemical equilibrium among different species of metals present in soil

15.4.1 Soil Biotic Factors

Earthworms can be considered the most important soil macroorganisms in terms of their impact on soil (Boyle et al. 1997). Already, Aristotle called them the "intestines of the earth." They can affect the environment by stimulating and/or altering the microbial, fungal, and enzymatic activities in soil, by increasing mineral availability and organic matter decomposition and through soil aggregation and cast production along the soil horizons (Tiunov and Scheu 2000; Wen and Wong 2004; Zorn et al. 2005). They play an important role in the transformation of nutrient (C, N, and P) and metal chemical forms, increasing their bioavailability (El Gharmali et al. 2002; Ma et al. 2002; Wen et al. 2004). The scientific literature on the effect of different earthworm species on metal fractionation and availability in contaminated (but not remediated) soil is abundant (Kizilkaya 2004; Udovic and Lestan 2007b), owing to their well-known prevalent importance in the soil processes. For example, Wen et al. (2004) reported an increase of metal concentration (Cr, Co, Ni, Zn, Cu,

Cd, and Pb) in the water soluble, exchangeable, and carbonate soil fraction (the most available form of metals to organisms) due to the presence of earthworms (*Eisenia fetida*). On the other hand, Cheng and Wong (2002) reported that the addition of earthworms, for example, *Pheretima* sp. in soil decreased the concentration of exchangeable Zn and of Zn bound to carbonates, although the significance of the results varied according to the soil type used in the experiment. The variation in results could be explained by differences in soil characteristics and variable physiological and ecological traits of different earthworm species, which, however, are known to be selective consumers (Edwards and Bohlen 1996; Morgan and Morgan 1999). Nevertheless, it has often been reported that the effects of earthworms on soil and on the behavior of different metals in soil varies among ecological categories (determined on similar morpho-ecological characteristics, Bouché 1977) and species (Morgan and Morgan 1988; Udovic et al. 2007). To achieve optimum results that would reflect the conditions in the environment to which metals in soil are exposed is therefore important to use earthworm species actually present in the studied soil. The same concern should also be kept in mind while studying the effect of earthworms on remediated soil. It has been reported that earthworms can rapidly invade remediated soil (Spurgeon and Hopkin 1999; Langdon et al. 2001). The possibility that earthworm activity may raise metal bioavailability is of considerable relevance for the success of soil remediation, especially when the methods that are used (i.e., soil washing, phytoextraction) to remove only part of the metals (presumably labile and bioavailable), or metals even remain in the soil immobilized by the addition of various chemicals (solidification/stabilization). Considering that the estimated annual earthworm cast production ranges between 5 and more than 250 tons ha^{-1} (Bohlen 2002), it is likely that earthworms may considerably affect the residual metals left in soil after remediation.

Studies on the effect of earthworms as key soil biotic environmental factors on other metals like Pb, Zn, and Cd left in soil after remediation with leaching with EDTA indicate that earthworm activity can lead to significant changes in the fractionation, the mobility, and of the bioavailability of residual metals (Udovic and Lestan 2007b, 2010a, b) (Table 15.1). After soil leaching with 20-step leaching, with 2.5 mmol kg^{-1} EDTA used in each step, the pH in the casts of *E. fetida* produced in remediated soil increased compared to the soil itself. Soil pH is a key chemical factor regulating the availability of metals in soil and earthworms are known to actively affect it (Edwards and Bohlen 1996). This increase would enhance the affinity of soil for metals due to the pH-dependent surface-charge density on colloids, thus leading to lower concentrations of metals in the soil solution (Cao al. et 2001; Shan et al. 2002). This, however, was not observed by Udovic and Lestan (2007b) and Udovic et al. (2007), where they reported that the concentrations of Pb, Zn, and Cd in labile (soil solution and exchangeable) fractions in *E. fetida* casts and in the soil itself were not significantly different. The phenomenon suggests the involvement of some metal-chelating metallophores produced by earthworms, or by microorganisms inhabiting their digestive tract, or by other microorganisms present in soil, which otherwise are affected by earthworm activity (Wen et al. 2006).

Table 15.1 Lead and copper mobility assessed by Toxicity Characteristic Leaching Procedure (*TCLP*), stomach and intestine oral bioavailability assayed by Physiologically Based Extraction Test and phytoavailability determined by diethylenediaminepentaacetic acid extraction method in non-processed and processed soil inoculated by three earthworm species before and after remediation with EDTA leaching

	Mobility (mg L^{-1})		Oral bioavailability (mg kg^{-1})				Phytoavailability (mg kg^{-1})	
			Stomach phase		Intestinal phase			
	Before	After	Before	After	Before	After	Before	After
Pb[1]								
Non-processed soil	[a]1.7±0.1	[a]0.3±0.1	[a]1411±269	[a]452±184	[a]66±35	[a]21±11	[a]952±3	[a]78±5
Processed soil								
Eisenia fetida	[a]1.9±0.4	[b]1.7±0.5	[b]646±38	[b]168±43	[b]292±45	[b]50±5	[b]904±2	[b]94±6
Octolasion tyrtaeum	[b]2±0.0	LOQ	[b]782±316	[a]322±127	[b]273±127	[b]59±18	n.d.	n.d.
Cu[2]								
Non-processed soil	n.d.	n.d.	[a]31±2	[a]11±1	[a]38±7	[a]8.7±1	[a]117±2	[a]42.8±2
Processed soil								
Lumbricus terrestris	n.d.	n.d.	[a]38±15	[a]14.5±3.4	[a]39±6	[b]15±5	[a]116±3	[b]48.2±2.6

Adapted from [1]Udovic and Lestan (2007b) and [2]Udovic and Lestan (2010b)

Values indicate mean of three replicates. Mean values (±S.D.) followed by different letters in superscript are significantly different within a column or row at $p \leq 0.05$ according to Duncan test

Earthworms seem to also have a pleasing impact on the mobility of metals in remediated soil, as assessed by Toxicity Characteristic Leaching Procedure (TCLP) (US EPA 1995). While the Pb concentration in the bulk soil leachate decreased due to the leaching process, *E. fetida* annulated the effect of soil remediation (Table 15.1). Again, two earthworm species, the endogeic species *Octolasium tyrtaeum* and the epigeic species *E. fetida* used in that study had variable influence on metal mobility in soil, highlighting the importance of using different test animal species in similar studies (Udovic et al. 2007). From the physiologically based extraction test (PBET) used to assess the biologically accessible metal fraction in soil, it is also evident that the effect of earthworm activity on the bioaccessibility of metals in soil is species specific. While *E. fetida* profoundly increased Pb bioaccessibility up to 5.1 times in remediated soil (Table 15.1), the effect of *Lumbricus rubellus* was not observed (Udovic and Lestan 2007b).

Reports on the effect of earthworm activity on the availability of metals and nutrients for plants (phytoavailability) are, however, conflicting. Several authors reported increased metal uptake in plants due to the presence of earthworms in soil. Ma et al. (2002), for example, reported that soil inoculation with earthworms (*Pheretima guillelmi*) substantially increased Zn and Pb uptake in *Leucaena leucocephala*. Similarly, in a study conducted by Wen et al. (2004), the presence of earthworms (*E. fetida*) in soil increased metals like Cr, Co, Ni, Zn, Cu, Cd, and Pb uptake in wheat (*Triticum aestivum*), but to a variable extent for different soil types. In a similar study, Devilegher and Verstraete (1996) attributed the increased accumulation of Cu and Zn in maize (*Zea mays*) grown in pots containing *L. terrestris* to the incorporation of surface organic matter and the subsequent nutrient (metal) enrichment of casts and soil (nutrient enrichment process) and to the effects of enzyme production and gut-associated microbial activity in the earthworms' gut (gut-associated processes). The potential role of interactions between earthworms and their associated microbes on metal availability in soil is also stressed by Cheng and Wong (2002). Earthworms can affect the soil by excreting organic materials (e.g., amino acids, proteins, and soil-available C), which may form chelates with metals, thus enhancing their availability (Ruiz et al. 2009). Another possible explanation for the increased metal phytoavailability in soil with earthworms present is the release of metals into the soil solution due to the decomposition of soil organic matter subsequent to the feeding behavior of earthworms (El Gharmali et al. 2002; Wen et al. 2004). On the other hand, Liu et al. (2005) concluded that inoculation of sewage sludge with *E. fetida* reduced the uptake of Cu and Cd in Chinese cabbage. Contrarily, *Lumbricus terrestris* promoted a 2.7-fold increase of Cu phytoavailability in polluted soil from a 50-year-old vineyard, regularly treated in the past with copper sulfate ($CuSO_4$, Bordeaux mixture) as fungicide (Udovic and Lestan 2010b). But still, their effect in this study on soil remediation efficiency (soil leaching with 15 mmol kg^{-1} EDTA) was limited and therefore did not hamper its activity. It would, however, be improper to generalize the influence of earthworms, as model soil biotic ageing factors, on metal fractionation, mobility, and phytoavailability in soil, since they are affected by several factors, the most important of which are the soil properties and the earthworm species used (Udovic and Lestan 2007b; Udovic et al. 2007).

The increase/decrease in metal availability in soil depends therefore on the conditions of the bulk soil relative to the conditions of gut of earthworm (Sizmur and Hodson 2009).

15.4.2   *Abiotic Factors*

Literature on the effect of abiotic soil ageing factors on remediated soil is scarcely available. In a study on remediated soil ageing, while investigating the impact of high temperatures on remediated soil, Lacal et al. (2003) observed that simulations can help to a certain extent in predicting the long- or medium-term toxicity of metals in pyritic sludge. While Lock and Janssen (2002) in other experiment exposed artificial soil (OECD) spiked with Zn to four ageing treatments: (1) storage at 20°C, (2) percolation followed by storing at 20°C, (3) alternately heating at 60°C and storing at 20°C, and (4) alternately freezing at −20°C and storing at 20°C. No effect of ageing on Zn speciation and ecotoxicity for enchytraeid worm *Enchytraeus albidus* was detected in this study. However, a simple laboratory scale simulation of a selected abiotic environmental ageing factor, that is, repetitive temperature changes at constant soil moisture regime, has been found as a valuable tool to study, to a certain extent, the fate of residual metals in remediated soil after its reintroduction into the environment (Udovic and Lestan 2009). Repetitive temperature changes lead to a decrease in soil pH, which increased Pb, Zn, and Cd fractionation, affecting thus their availability and mobility, as expected due to the influenced of soil pH on the adsorption–desorption behavior of metals in soil (Basta and Tabatabai 1992; Rieuwerts et al. 1998; Adriano 2001; Cao et al. 2001). Temperature changes are also reported to affect the rates of metal desorption from Fe and Mn oxides and the behavior of the soil organic matter constituents (e.g., humic acids), enhancing metal mobility and solubility by complex formation (Weng et al. 2002). The apparently contradictory results indicate the importance of considering the characteristics of the extraction solutions used to determine metal mobility and availability, as well as the characteristics of each observed element when interpreting the results. In the ageing regime applied to the polluted and remediated soil in a study conducted by Udovic and Lestan (2009) the repetitive cycles of high and low temperatures (105 and −20°C, respectively) had variable effect on mobility of Pb, Zn, and Cd. The mobility of Cd was decreased while the mobility of Pb increased (Table 15.2) and the mobility of Zn remained unaltered. Moreover, the effect of the simulated soil ageing on Pb, Zn, and Cd phytoavailability was contradictory to the results of Pb, Zn, and Cd sequential extraction (fractionation), which showed a significant increase in metal mobility and availability (Udovic and Lestan 2009). The apparent discrepancies and contradictions in the results could possibly be due to (1) variation in the characteristics of the extracting solutions used for analysis, (2) changes in the soil characteristics, and (3) differences in the metal chemical characteristics. A holistic approach, therefore, should be considered while interpreting the results concerning a complex system such as the soil.

Table 15.2 Lead oral bioavailability (PBET), mobility (TCLP), and phytoavailability (DTPA) in aged and non-aged soil, before and after leaching with different EDTA concentrations

	PBET Pb (mg kg⁻¹)		DTPA	TCLP
Before leaching	Stomach phase	Small intestine phase	Pb (mg kg⁻¹)	Pb (mg L⁻¹)
Soil	[a]761 ± 48	[ab]328 ± 64	[a]952 ± 3	[a]1.1 ± 0.0
Aged soil (10% WHC)	[b]550 ± 91	[a]368 ± 27	[b]551 ± 62	[b]1.7 ± 0.1
After leaching 2.5 mmol kg⁻¹ EDTA:				
Soil	[a]914 ± 351	[a]170 ± 78	[a]987 ± 5	[a]1.0 ± 0.0
Aged soil (10% WHC)	[b]459 ± 71	[a]232 ± 24	[b]546 ± 102	[b]1.4 ± 0.1
5.0 mmol kg⁻¹ EDTA:				
Soil	[a]306 ± 36	[a]215 ± 98	[a]901 ± 19	[a]0.8 ± 0.1
Aged soil (10% WHC)	[a]385 ± 14	[b]241 ± 27	[b]483 ± 55	[b]1.3 ± 0.1
10.0 mmol kg⁻¹ EDTA:				
Soil	[a]276 ± 28	[a]127 ± 60	[a]646 ± 46	[a]0.7 ± 0.0
Aged soil (10% WHC)	[ab]378 ± 17	[a]163 ± 43	[b]428 ± 17	[b]1.1 ± 0.1
20.0 mmol kg⁻¹ EDTA:				
Soil	[a]292 ± 77	[a]133 ± 63	[ab]277 ± 13	[a]0.4 ± 0.0
Aged soil (10% WHC)	[a]214 ± 19	[a]116 ± 34	[a]322 ± 75	[b]0.6 ± 0.0
40.0 mmol kg⁻¹ EDTA:				
Soil	[a]238 ± 90	[a]40 ± 24	[a]212 ± 1	[a]0.4 ± 0.0
Aged soil (10% WHC)	[a]192 ± 16	[b]93 ± 4	[a]239 ± 8	[a]0.4 ± 0.0
4 × 40.0 mmol kg⁻¹ EDTA:				
Soil	[a]95 ± 4	[a]54 ± 27	[a]78 ± 5	LOQ
Aged soil (10% WHC)	[a]108 ± 13	[a]50 ± 13	[b]158 ± 22	LOQ

Adapted from Udovic and Lestan (2009)

Values indicate mean of three replicates. Mean values ± S.D. followed by different letters in superscript are significantly different within a column or row at $p \leq 0.05$ according to Duncan test. LOQ indicates below limit of quantification

15.5 Conclusion

Soil washing remediation techniques usually remove only the labile metal species from the soil, leaving the residual ones in less available/mobile forms. Thus, washing of soils contaminated with toxic metals is effective but rarely removes metals completely from soil. More labile and thus bioavailable forms of metals are expected to be removed first (selectively), while metals strongly bound to solid soil fractions and chemically less available to chelants remain in soil even after remediation (metal bioavailability stripping concept). Considering that different chemical forms of metals are present in soil, also metal (bio)availability, mobility, and fractionation, beside the total metal concentration, should therefore be considered when assessing soil pollution, when choosing the most suitable remediation technique and when assessing the effect of remediation. Studies have shown that the effect of remediation and the bioavailability of metals could be misjudged using chemical extraction tests solely. Combined results of in vitro chemical extraction tests and in vivo accumulation and/or avoidance tests with representative indicator animal species should be considered for a more holistic and relevant picture of the soil pollution and the availability stripping of metals after soil remediation. For this purpose, bioaccumulation tests with *P. scaber* and *E. fetida* could be used as a sensitive supplement to chemical extractions in assessing the efficiency of remediation and the metal fraction bioavailable to soil fauna.

The purpose of soil remediation actions is to reintroduce less-polluted soil with improved characteristics into the environment. Studies have shown that the soil remediation achievement seems not to be permanent, but it changes in time due to the effect of abiotic and biotic environmental soil ageing factors, diminishing the effectivity of the remediation. Post-remediation fate of the residual metals left in soil should be therefore monitored to properly assess the effect of remediation in time.

References

Abollino O, Giacomino A, Malandrino M, Mentasti E, Aceto M, Barberis R (2006) Assessment of metal availability in a contaminated soil by sequential extraction. Water Air Soil Pollut 137:315–338

Adriano DC (2001) Trace elements in terrestrial environments: biogeochemistry, bioavailability and risks of metals. Springer, New York

Aldaya MM, Lors CL, Salmon S, Ponge JF (2006) Avoidance bio-assays may help to test the ecological significance of soil pollution. Environ Pollut 140:173–180

Alvarenga P, Palma P, Gonçalves AP, Fernandes RM, de Varennes A, Vallini G, Duarte E, Cunha-Queda AC (2008) Evaluation of tests to assess the quality of mine-contaminated soils. Environ Geochem Health 30:95–99

Amorim MJB, Novais S, Römbke J, Soares AMVM (2008) Avoidance test with *Enchytraeus albidus* (Enchytraeidae): effects of different exposure time and soil properties. Environ Pollut 155:112–116

Arnold RE, Hodson ME, Black S, Davies NA (2003) The influence of mineral solubility and soil solution concentration on the toxicity of copper to *Eisenia fetida* Savigny. Pedobiol 47:622–632

Baker DE, Senft JP (1997) Copper. In: Alloway BJ (ed.) Heavy metals in soils, 2nd edn. Chapman & Hall, Suffolk, pp 179–205

Basta NT, Tabatabai MA (1992) Effect of cropping systems on adsorption of metals by soils: II. Eff pH Soil Sci 153:195–204

Bohlen PJ (2002) Earthworms. In: Lal R (ed.) Encyclopedia of soil science, 1st edn. Marcel Dekker, New York, pp 370–373

Borona A, Romero F (1996) Fractionation of lead in soils and its influence on the extractive cleaning with EDTA. Environ Technol 17:63–70

Bouché MB (1977) Stratégies lombriciennes. In: Lohm U, Persson T (eds.) Soil organisms as components of ecosystems. Ecol Bull 25:122–132

Boyle KE, Curry JP, Farrell EP (1997) Influence of earthworms on soil properties and grass production in reclaimed cutover peat. Biol Fertil Soil 25:20–26

Brun LA, Maillet J, Hinsinger P, Pépin M (2001) Evaluation of copper availability to plants in copper-contaminated vineyard soils. Environ Pollut 111:293–302

Cao X, Chen Y, Wang X, Deng X (2001) Effects of redox potential and pH value on the release of rare elements from soil. Chemosphere 44:655–661

Chaney RL, Brown SL, Li YM, Angle JS, Stuczynski TI, Daniels WL, Henry CL, Siebielec G, Malik M, Ryan JA, Compton H (2000) Progress in risk assessment for soil metals, and in-situ remediation and phytoextraction of metals from hazardous contaminated soils (Paper presented at the US-EPA's conference Phytoremediation). State of the Science, Boston

Cheng J, Wong MH (2002) Effects of earthworms on Zn fractionation in soils. Biol Fertil Soil 36;72–78

Curry JP, Schmidt O (2007) The feeding ecology of earthworms – a review. Pedobiol 50:463–477

Davis S, Mirick DK (2006) Soil ingestion in children and adults in the same family. J Expo Sci Environ Epidemiol 16:3–75

Dean JR (2007) Bioavailability, bioaccessibility and mobility of environmental contaminants, 1st edn. Wiley, Chichester

Dermont G, Bergeron M, Mercier G, Richter-Lafléche M (2008) Soil washing for metal removal: a review of physical(chemical technologies and field applications. J Hazard Mater 152:1–31

Devliegher W, Verstraete W (1996) *Lumbricus terrestris* in a soil sore experiment: effects of nutrient-enrichment processes (NEP) and gut-associated processes (GAP) on the availability of plant nutrients and heavy metals. Soil Biol Biochem 28:489–496

Edwards CA (2004) Earthworm ecology, 2nd edn. CRC Press, Boca Raton

Edwards CA, Bohlen PJ (1996) Biology and ecology of earthworms, 3rd edn. Chapman & Hall, London

El Gharmali A, Rada A, El Meray M, Nejmeddine A (2002) Study of the effect of earthworm *Lumbricus terrestris* on the speciation of heavy metals in soils. Environ Technol 23: 775–780

Finzgar N, Lestan D (2007) Multi-step leaching of Pb and Zn contaminated soils with EDTA. Chemosphere 66:824–832

Finzgar N, Kos B, Lestan D (2005) Heap leaching of lead contaminated soil using biodegradable chelator [S, S]-ethylenediamine disuccinate. Environ Technol 26:553–560

Gál J, Markiewicz-Patkowska J, Hursthouse A, Tatner P (2008) Metal uptake by woodlice in urban soils. Ecotoxicol Environ Saf 69:139–149

Griffiths RA (1995) Soil washing technology and practice. J Hazard Mater 40:175–189

Guo G, Zhou Q, Ma LQ (2006) Availability and assessment of fixing additives for the in situ remediation of heavy metal contaminated soils: a review. Environ Monit Assess 116:513–528

Gupta SK, Vollmer MK, Krebs R (1996) The importance of mobile, mobilisable and pseudo total heavy metal fractions in soil for three-level risk assessment and risk management. The Sci Environ 178:11–20

Han FX, Banin A, Kingery WL, Triplett GB, Zhou LX, Zheng SJ, Ding WX (2003) New approach to studies of heavy metal redistribution in soil. Adv Environ Res 8:113–120

Heikens A, Pejinenburg WJGM, Hendriks AJ (2001) Bioaccumulation of heavy metals in terrestrial invertebrates. Environ Pollut 113:385–393

Hopkin SP (1989) Ecophysiology of metals in terrestrial invertebrates, 1st edn. Elsevier Applied Science Publishers Ltd, London

Hopkin SP, Jones DT, Dietrich D (1993) The isopod *Porcellio scaber* as a monitor of the bioavailability of metals in terrestrial ecosystems: towards a global 'woodlouse watch' scheme. Sci Total Environ 134:357–365

Hund-Rinke K, Wiechering H (2001) Earthworm avoidance test for soil assessments. An alternative for acute and reproduction tests. J Soil Sediment 1:15–20

Hund-Rinke K, Lindermann M, Simon M (2005) Experiences with novel approaches in earthworm testing alternatives. J Soil Sediment 5:233–239

ISO/TS 17924 (2007) Soil quality – Assessment of human exposure from ingestion of soil and soil material – Guidance on the application and selection of physiologically based extraction methods for the estimation of the human bioaccessibility/bioavailability of metals in soil. International Organization for Standardization, Genéve

ISO 17512-1 (2008) Soil quality – avoidance test for testing the quality of soils and the toxicity of chemicals. Part I: test with earthworms (Eisenia fetida and Eisenia andrei). International Organization for Standardization, Genéve

Kabala C, Singh BR (2001) Fractionation and mobility of copper, lead and zinc in soil profiles in the vicinity of a copper smelter. J Environ Qual 30:485–492

Kabata-Pendias A, Pendias H (1992) Trace elements in soils and plants. CRC Press, Boca Raton

Kamnev AA, van der Lelie D (2000) Chemical and biological parameters as tools to evaluate and improve heavy metal phytoremediation. Biosci Rep 20:239–258

Kennette D, Hendershot W, Tomlin A, Suavé S (2002) Uptake of trace metals by the earthworm *Lumbricus terrestris* L. in urban contaminated soils. Appl Soil Ecol 19:191–198

Kizilkaya R (2004) Cu and Zn accumulation in earthworm *Lumbricus terrestris* L. in sewage sludge amended soil and fractions of Cu and Zn in casts and surrounding soil. Ecol Eng 22:141–151

Kosson DS, van der Sloot HA, Sanchez F, Garrabrants AC (2002) An integrated framework for evaluating leaching in waste management and utilization of secondary materials. Environ Eng Sci 19:159–204

Kumpiene J, Lagerkvist A, Maurice C (2007) Stabilization of Pb- and Cu-contaminated soil using coal fly ash and peat. Environ Pollut 145:365–373

Lacal J, da Silva MP, García R, Sevilla MT, Procopio JR, Hernández L (2003) Study of fractionation and potential mobility of metal in sludge from pyrite mining and affected river sediments: changes in mobility over time and use of artificial ageing as a tool in environmental impact assessment. Environ Pollut 124:291–309

Langdon CJ, Piearce TG, Meharg AA, Semple KT (2001) Survival and behavior of the earthworms Lumbricus rubellus and Dendrodrilus rubidus from arsenate-contaminated and non-contaminated sites. Soil Biol Biochem 33:1239–1244

Langdon CJ, Hodson ME, Arnold RE, Black S (2005) Survival, Pb uptake and behaviour of three species of earthworm in Pb treated soils determined using an OECD-style toxicity test and a soil avoidance test. Environ Pollut 138:368–375

Lestan D, Grcman H, Zupan M, Bacac N (2003) Relationship of soil properties to fractionation of Pb and Zn in soil and their uptake into *Plantago lanceolata*. Soil Sedim Contam 12:507–522

Lestan D, Luo C, Li X (2008) The use of chelating agents in the remediation of metal-contaminated soils: a review. Environ Pollut 153:3–13

Levy DB, Barbarick KA, Siemer EG, Sommers LE (1992) Distribution and partitioning of trace metals in contaminated soil near Leadsville, Colorado. J Environ Qual 21:185–195

Li X, Thornton I (2001) Chemical partitioning of trace and major elements in soils contaminated by mining and smelting activities. Appl Geochem 16:1693–1706

Lindsay WL, Norvell WA (1978) Development of a DTPA test for zinc, iron, manganese and copper. Soil Sci Soc Am J 42:421–428

Liu X, Hu C, Zhang S (2005) Effects of earthworm activity on fertility and heavy metal bioavailability in sewage sludge. Environ Int 31:874–879

Lock K, Janssen CR (2002) The effect of ageing on the toxicity of zinc for the potworm *Enchytraeus albidus*. Environ Pollut 116:289–292

Loureiro S, Soares AMVM, Nogueira AJA (2005) Terrestrial avoidance behaviour tests as a screening tool to assess soil contamination. Environ Pollut 138:121–131

Lukkari T, Haimi J (2005) Avoidance of Cu- and Zn-contaminated soil by three ecologically different earthworm species. Ecotoxicol Environ Saf 62:35–41

Lukkari T, Aatsinki M, Väisänen A, Haimi J (2005) Toxicity of copper and zinc assessed with three different earthworm tests. Appl Soil Ecol 30:133–146

Ma Y, Dickinson NM, Wong MH (2002) Toxicity of Pb/Zn mine tailings to the earthworm *Pheretima* and the effects of burrowing on metal availability. Biol Fertil Soil 36:79–86

Meier JR, Cheng LW, Jacobs S, Torsela J, Meckes MC, Smith MK (1997) Use of plant and earthworm bioassays to evaluate remediation of soil from a site contaminated with polychlorinated biphenyls. Environ Toxicol Chem 16:928–938

Morgan JE, Morgan AJ (1988) Earthworms as biological monitors of cadmium, copper, lead and zinc in metalliferous soils. Environ Pollut 54:123–138

Morgan JE, Morgan AJ (1999) The accumulation of metals (Cd, Cu, Pb Zn and Ca) by two ecologically contrasting earthworm species (*Lumbricus rubellus* and *Aporrectodea caliginosa*): implications for ecotoxicological testing. Appl Soil Ecol 13:9–20

Mulligan CN, Yong RN, Gibbs BF (2001) Remediation technologies for metal contaminated soils and groundwater: an evaluation. Eng Geol 60:193–207

Neale CN, Bricka RM, Chao AC (1997) Evaluating acids and chelating agents for removing heavy metals from contaminated soils. Environ Prog 16:274–280

Novozamsky I, Lexmond Th M, Houba VJG (1993) A single extraction procedure of soil evaluation of uptake of some heavy metals by plants. Int J Environ Anal Chem 51:47–58

Nowack B, Schulin R, Robinson BH (2006) Critical assessment of chelant metal phytoextraction. Environ Sci Technol 40:5225–5232

Odendaal JP, Reinecke AJ (2004) Evidence of metal interaction in the bioaccumulation of cadmium and zinc in *Porcellio laevis* (Isopoda) after exposure to individual and mixed metals. Water Air Soil Pollut 156:1–4

Oomen AG, Hack A, Minekus M, Zeijdner A, Cornelis C, Schoeters W, van der Wiele T, Wragg J, Rompelberg CJM, Sips AJAM, van Wijnen JH (2002) Comparison of five *in vitro* digestion models to study the bio-accessibility of soil contaminants. Environ Sci Technol 36: 3326–3334

Oomen AG, Rompelberg CJM, Bruil MA, Dobbe CJG, Pereboom DPKH, Sips AJAM (2003) Development of an *in vitro* digestion model for estimating the bio-accessibility of soil contaminants. Arch Environ Contam Toxicol 44:281–287

Owojori OJ, Reinecke AJ (2009) Avoidance behaviour of two eco-physiologically different earthworms (*Eisenia fetida* and *Aporrectodea caliginosa*) in natural and artificial saline soils. Chemosphere 75:279–283

Paoletti MG, Hassall M (1999) Woodlice (Isopoda: Oniscidea): their potential for assessing sustainability and use as bioindicators. Agr Ecosyst Environ 74:157–165

Peters RW (1999) Chelant extraction of heavy metals from contaminated soils. J Hazard Mater 66:151–210

Peters RW, Shem L (1992) Use of chelating agents for remediation of heavy metal contaminated soil. In: Vandegrift GE, Reed DT, Tasker IR (eds.) Environmental remediation removing organic and metal Ion pollutants. American Chemical Society, Washington, pp 70–84

Pichtel J, Vine B, Kuula-Vaisanen P, Niskanen P (2001) Lead extraction from soils as affected by lead chemical and mineral forms. Environ Eng Sci 18:91–98

Pociecha M, Lestan D (2009) EDTA leaching of Cu contaminated soil using electrochemical treatment of the washimg solution. J Hazard Mater 165:533–539

Ramos L, Hernandez LM, Gonzales MJ (1994) Sequential fractionation of copper, lead, cadmium and zinc in soil from or near Donana national park. J Environ Qual 23:50–57

Rieuwerts JS, Thornton ME, Farago ME, Ashmore MR (1998) Factors influencing metal bioavailability in soils: preliminary investigations for the development of a critical loads approach for metals. Chem Speciat Bioavailab 10:61–75

Rivero VC, Masedo MD, De la Villa RV (2000) Effect of soil properties on zinc retention in agricultural soils. Agrochimica 43:46–54

Ruby MV, Davis A, Schoof R, Eberle S, Sellstone CM (1996) Estimation of lead and arsenic bioavailability using a physiologically based extraction test. Environ Sci Technol 30:422–430

Ruiz E, Rodriguez L, Alonso-Azcárate J (2009) Effects of earthworms on metal uptake of heavy metals from polluted mine soils by different crop plants. Chemosphere 75:1035–1041

Sabienë N, Brazauskienë DM (2004) Determination of heavy metal mobile forms by different extraction methods. Ekologija 1:36–41

Shan XQ, Lian J, Wen B (2002) Effect of organic acids on adsorption and desorption of rare earth elements. Chemosphere 47:701–710

Sizmur T, Hodson ME (2009) Do earthworms impact metal mobility and availability in soil? – A review. Environ Pollut 157:1981–1989

Sousa A, Pereira R, Antunes SC, Cachada A, Pereira E, Duarte AC, Gonçalves F (2008) Validation of avoidance assays for the screening assessment of soils under different anthropogenic disturbances. Ecotoxicol Environ Saf 71:661–670

Spurgeon DJ, Hopkin SP (1999) Tolerance to zinc in populations of the earthworm *Lumbricus rubellus* from uncontaminated and metal-contaminated ecosystems. Arch Environ Contam Toxicol 37:332–337

Spurgeon DJ, Lofts S, Hankard PH, Toal M, McLellan D, Fishwick S, Svedsen C (2006) Effect of pH on metal speciation and resulting metal uptake and toxicity for earthworms. Environ Toxicol Chem 25:788–796

Sun B, Zhao FJ, Lombi E, McGrath SP (2001) Leaching of heavy metals from contaminated soils using EDTA. Environ Pollut 113:111–120

Suthar S, Singh S, Dhawan S (2008) Earthworms as bioindicators of metals (Zn, Fe, Mn, Cu, Pb and Cd) in soils: is metal bioaccumulation affected by their ecological category? Ecol Eng 32:99–107

Tandy S, Bossart K, Mueller R, Ritschel J, Hauser L, Schulin R, Nowack B (2004) Extraction of heavy metals from soils using biodegradable chelating agents. Environ Sci Technol 41:7851–7856

Tessier A, Campbell PGC, Bisson M (1979) Sequential extraction procedure for the speciation of particulate trace metals. Anal Chem 51:844–851

Tiunov AV, Scheu S (2000) Microbial biomass, bio-volume and respiration in *Lumbricus terrestris* L. cast material of different age. Soil Biol Biochem 32:265–275

Turner A, Ip KH (2007) Bio-accessibility of metals in dust from the indoor environment: application of a physiologically based extraction test. Environ Sci Technol 41:7851–7856

Udovic M, Lestan D (2007a) EDTA leaching of Cu contaminated soils using ozone/UV for treatment and reuse of washing solution in a closed loop. Water Air Soil Pollut 181:319–327

Udovic M, Lestan D (2007b) The effect of earthworms on the fractionation and bioavailability of heavy metals before and after soil remediation. Environ Pollut 148:663–668

Udovic M, Lestan D (2008) Remediation of soil from a former zinc smelter area with stabilization with cement. Acta Agr Slovenica 91:283–295

Udovic M, Lestan D (2009) Pb, Zn and Cd mobility, availability and fractionation in aged soil remediated by EDTA leaching. Chemosphere 74:1367–1373

Udovic M, Lestan D (2010a) Redistribution of residual Pb, Zn and Cd in soil remediated with EDTA leaching and exposed to earthworms (*Eisenia fetida*). Environ Technol 31:655–669

Udovic M, Lestan D (2010b) Fractionation and bioavailability of Cu in soil remediated by EDTA leaching and processed by earthworms (*Lumbricus terrestris* L.). Environ Sci Pollut Res 17:561–570

Udovic M, Lestan D (2010c) *Eisenia fetida* avoidance behavior as a tool for assessing the efficiency of remediation of Pb, Zn and Cd polluted soil. Environ Pollut 158:2766–2772

Udovic M, Plavc Z, Lestan D (2007) The effect of earthworms on the fractionation, mobility and bioavailability of Pb, Zn and Cd before and after soil leaching with EDTA. Chemosphere 70:126–134

Udovic M, Drobne D, Lestan D (2009) Bioaccumulation in *Porcellio scaber* (Crustacea, Isopoda) as a measure of the EDTA remediation efficiency of metal-polluted soil. Environ Pollut 157:2822–2829

Ure AM (1996) Single extraction schemes for soil analysis and related applications. Sci Total Environ 178:3–10

Us EPA (1995) Test methods for evaluation of solid waste, vol IA. Laboratory Manual Physical/Chemical Methods, SW 86, 40 CFR Parts 403 and 503. GPO, Washington, DC

Vijver MG, Vink JPM, Jager T, van Straalen NM, Wolterbeek HT, van Gestel CAM (2006) Kinetics of Zn and Cd accumulation in the isopod *Porcellio scaber* exposed to contaminated soil and/or food. Soil Biol Biochem 38:1554–1563

Warburg MR (1993) Evolutionary biology of land isopods, 1st edn. Springer, Berlin

Wen JHC, Wong MH (2004) Effects of earthworm activity and P-solubilizing bacteria on P availability in soil. J Plant Nutr Soil Sci 167:209–213

Wen B, Hu X, Liu Y, Wang W, Feng M, Shan X (2004) The role of earthworms (*Eisenia fetida*) on influencing bioavailability of heavy metals in soils. Biol Fertil Soil 40:181–187

Wen B, Liu Y, Hu X, Shan X (2006) Effect of earthworms (*Eisenia fetida*) on the fractionation and bioavailability of rare earth elements in nine Chinese soils. Chemosphere 63:1179–1186

Weng L, Temminghoff EJM, Lofts S, Tipping E, VanRiemsdijk WH (2002) Complexation with dissolved organic matter and solubility control of heavy metals in a sandy soil. Environ Sci Technol 36:4804–4810

Witzel B (1998) Uptake, storage and loss of cadmium and lead in the woodlouse *Porcellio scaber* (Crustacea, Isopoda). Water Air Soil Pollut 108:51–68

Wragg J, Cave N (2002) *In-vitro* methods for the measurement of the oral bioaccessibility of selected metals and metalloids in soils: a critical review. R&D Technical Report P5-062/TR/01. Environment Agency, Bristol

Xu Y, Zhao D (2005) Removal of copper from contaminated soil by use of poly(aminodiamine) dendrimeres. Environ Sci Technol 39:2369–2375

Zimmer M (2002) Nutrition in terrestrial isopods (Isopoda: Oniscidea): an evolutionary-ecological approach. Biol Rev 77:455–493

Zorn MI, Van Gestel CAM, Eijsackers H (2005) The effect of two endogeic earthworm species on zinc distribution and availability in artificial soil columns. Soil Biol Biochem 37:917–925

Chapter 16
Microalga-Mediated Bioremediation of Heavy Metal–Contaminated Surface Waters

Cristina M. Monteiro, Paula M.L. Castro, and F. Xavier Malcata

Abstract Heavy metal contamination of aquatic ecosystems has been increasing in recent times, owing to disposal of such pollutants in effluents. Their presence to excessive levels leads to serious health problems in living organisms, since they cannot be mineralized to completely innocuous forms. The physicochemical methods, such as chemical precipitation, chemical oxidation/reduction, or ion exchange, employed to recover such pollutants are either expensive or not efficient, especially when heavy metals are present at very low concentrations. Microorganisms in general, and microalgae in particular, have been recognized as suitable vectors for detoxification and have emerged as a potential low-cost alternative to physico-chemical treatments. Uptake of metals by living microalgae occurs in two steps: one takes place rapidly and is essentially independent of cell metabolism – "adsorption" onto the cell surface. The other one is lengthy and relies on cell metabolism – "absorption" or "intracellular uptake." Nonviable cells have also been successfully used in metal removal from contaminated sites. Removal of heavy metals by microalgal biomass is affected by a number of environmental factors. The intrinsic and extrinsic factors affecting uptake of metals by microalgae and how these micro-organisms can be helpful in removing metals from polluted environments are hereby reviewed and highlighted.

Keywords Bioaccumulation • Biosorption • Metal speciation • Microalgae

---

C.M. Monteiro • P.M.L. Castro
CBQF/Escola Superior de Biotecnologia, Universidade Católica Portuguesa,
Rua Dr. António Bernardino de Almeida, P-4200-072 Porto, Portugal

F.X. Malcata (✉)
ISMAI – Instituto Superior da Maia, Avenida Carlos Oliveira Campos, Castelo da Maia,
Avioso S. Pedro P-4475-690, Portugal

CIMAR/CIIMAR – Centro Interdisciplinar de Investigação Marinha e Ambiental,
Rua dos Bragas no 289, P-4050-123 Porto, Portugal
e-mail: fmalcata@ismai.pt

M.S. Khan et al. (eds.), *Biomanagement of Metal-Contaminated Soils*,
Environmental Pollution 20, DOI 10.1007/978-94-007-1914-9_16,
© Springer Science+Business Media B.V. 2011

16.1 Heavy Metals in the Environment

The term "heavy metal" is collectively applied to a group of metals (and metal-like elements) with density greater than 5 g/cm^3 and atomic number above 20. Such metals are associated with environmental pollution and biological toxicity issues (Jjemba 2004). Due to their persistence in the environment and toxicity to living organisms, heavy metals are among the most dangerous and widely studied contaminants (Harte et al. 1991). Heavy metals are natural constituents of the Earth lithosphere and hydrosphere, so nonnegligible background concentrations thereof are expected in soils, sediments and waters, and even in living organisms. However, the wide range and intensive applications of heavy metals in industrial processes (Table 16.1) have caused a dramatic increase in their concentration in the environment relative to their normal background counterpart. They affect negatively both terrestrial and aquatic ecosystems after they are disposed off as untreated effluents and as solid residues, and before they can be eliminated (Alloway and Ayres 1997).

From a biological point of view, heavy metals can be sorted out according to their environmental impact and toxicity. Some of those elements, viz. Fe, Cu, Ni, Co, and Zn, are labeled as essential, because they are required by most living organisms, at minute concentrations, for regular growth and maintenance. They are indeed part of biological structures such as cell membranes and enzyme prosthetic groups, which play crucial roles in key metabolic processes. However, excessive levels of those metals are toxic to most prokaryotic and eukaryotic organisms. On the other hand, metals like As, Cd, Pb, and Hg do not play any known biochemical role, so they are nonessential. These metals are, however, known to cause severe biological damages even at very low concentrations (Kaplan 2004).

Table 16.1 Major industrial uses of the most common heavy metals

Metal	Industrial use
As	Medical uses, insecticides, pigments, paints, electronic devices
Cd	Galvanization, pigments, batteries, metal plating, smelting, paints, pesticides, polymer and plastic stabilization, fertilizers
Cr	Metallurgy, galvanization, paints, wood conservation, chemical industry
Cu	Electrical industry
Hg	Insecticides, metallurgical and pharmaceutical industries, plastic production catalysis, batteries
Ni	Metallurgy, batteries, galvanization
Pb	Batteries, fuels, pigments, paints
Zn	Galvanization, pigments, batteries, smelting, paints, metal plating, agricultural products, fertilizers and pesticides, sewage sludge, fossil fuel combustion, metallurgy, polymer stabilizers

Adapted from Harte et al. (1991) and Alloway and Ayres (1997)

16.2   Physicochemical and Biological Recovery of Heavy Metals

After the industrial revolution, human activities have significantly increased the level of contamination of soils and water bodies by heavy metals, used either deliberately for agricultural and industrial purposes, or accidentally through the mishandling of chemicals. Owing to their nonbiodegradability and consequent bioaccumulation throughout the trophic chain, heavy metals represent a serious threat to all kinds of inhabiting organisms. Therefore, alleviation of the heavy metal burden of industrial wastewaters is important before their discharge into waterways (Mehta and Gaur 2005). Obviously, this should be accomplished right at the source of emissions, i.e., before they enter the ecosystem. Controlling heavy metal discharges and eventually preventing toxic heavy metals from entering surface waters have accordingly become a challenge. To address such problems, several downstream physicochemical approaches exists, besides biological ones (as detailed below), which can be applied to recover heavy metals from aqueous solutions, or from aqueous solutions that soak soils; however, all of them remediate rather than prevent.

16.2.1   Physicochemical Methods

The so-called "best treatment technologies" are a number of physicochemical methods including membrane filtration, adsorption, ion exchange, reverse osmosis, chemical precipitation, chemical oxidation/reduction, coagulation/flocculation, or solvent extraction, which have classically been employed for stripping toxic metals from wastewaters (Eccles 1999; Volesky 2001). However, these methods have disadvantages, like incomplete metal removal, high reagent or energy requirements, and generation of toxic sludge or other heavy metal-containing waste products that may sometimes be more toxic than their parent ones. Hence, additional disposal methods are required. Furthermore, they are often expensive, especially when the heavy metal concentrations are low (e.g., 10–100 mg/L) and inefficient, because a too large volume reduction of effluents is intended, so a limited use in large-scale in situ operations will typically result (Mehta and Gaur 2005).

Extensive studies have been undertaken in recent years aimed at finding alternative and economically feasible technologies for detoxification of heavy metal-contaminated effluents. Such studies have focused mainly on screening living entities, and parts thereof, for their intrinsic capacity to overcome (at least) some of the limitations of physicochemical treatments (Ngah and Hanafiah 2008). Among these, microorganisms, in particular, are considered intrinsically more efficient in bioaccumulating heavy metals when exposed to low concentrations in their surrounding aqueous environment. These are briefly discussed in the following section.

16.2.2   Biological Methods

The biological methods have advantages like reduced requirement for chemicals, low operating costs, eco-friendliness (as no toxic sludge results), and efficiency at low levels of contamination. They also offer possibilities for metal recovery and biosorbent regeneration afterward (Srivastava and Majumder 2008). Of the various biological methods, biosorption is indeed an effective alternative to conventional methods for decontaminating liquid effluents loaded with heavy metals. For a biosorbent to be economical and suitable for large-scale operation, it should be abundant in nature (or released as a by-product from bioprocessing) and should not require pre-processing (Arief et al. 2008). A great deal of interest has recently arisen toward using various kinds of readily available and inexpensive biomass of several microorganisms and microalgae, in particular for removal of heavy metals. Microalgae are used in bioremediation of metal-contaminated sites due to (1) their ability to tolerate those metals, (2) their high yields of recovery per unit mass, and (3) their high specific outer area coupled with a cell wall loaded with ionizable groups (Malik 2004).

16.3   Microalga-Mediated Recovery of Heavy Metals

Microalgae are eukaryotic, unicellular, photoautotrophic organisms that are abundant in natural aquatic (and bordering) environments. They can adapt to a wide range of conditions, including moist soil, fresh and marine habitats, as well as industrial and domestic effluent dumping sites. They are primary producers in the food chain, so they are the most basic trophic support level. Because of their small cell size, microalgae exhibit a large surface area-to-volume ratio, which is readily available for contact with the surrounding environment; and their functionally rich cell wall groups can easily interact with cations in solution. Although they may spontaneously serve as vehicle to introduce and transfer heavy metal cations along the food chain to higher trophic levels, their ability to remove metals from polluted aquatic sites may be advantageous in bioremediation strategies.

16.3.1   Removal Capacity of Microalgae

Biosorption of heavy metals is a complex phenomenon, and accumulation of heavy metals by microalgae is typically considered as a two-stage process: (1) an initial rapid (passive) removal of metals by the cell and (2) a much slower one that occurs inside the cell. During passive removal, heavy metal ions are adsorbed onto functional groups present on the cell surface by electrostatic interactions, which, however, differ in their affinity and specificity for metal binding. This is a non-metabolic, rapid, and essentially reversible process, occurring in both living and nonliving cells. It includes physical adsorption, ion exchange, chemisorption, complexation, chelation, entrapment in the structural polysaccharide network, and diffusion through the cell wall and membrane (Muñoz et al. 2006; Sud et al. 2008). The second phase is essentially a metabolism-dependent process, involving transport of metal ions across the cell membrane barrier and subsequent accumulation inside the cell, with posterior binding to intracellular compounds and/or organelle containment. This metal uptake process is much slower and usually irreversible, and occurs in living cells only. All heavy metal species are hydrophilic, so their transport through the partially lipophilic biological membrane surrounding the cell is mediated by specific proteins (Worms et al. 2006). However, transport of metal ions may also occur through facilitated diffusion, owing to a metal-induced increase in permeability of the cell membrane (Wang and Chen 2006).

The capacity of biomaterials to adsorb heavy metals depends on the composition of their cellular surface, coupled with the chemical composition of the outer solution undergoing treatment. Hence, a rational choice of the most adequate biosorbent for metal decontamination of a specific water stream demands a priori knowledge of the target metals and their concentration. Microalgae are especially suitable as biosorbents, due to their availability (in almost unlimited amounts) in seas and oceans, and their high sorption uptake capacity. These capacities are, on average, higher than those claimed for other biological sorbents or from physicochemical sources (Table 16.2). Therefore, microalgal biomass appears to be an economically feasible and technologically efficient alternative to existing physicochemical methods of heavy metal removal and recovery from wastewaters (Romera et al. 2006).

16.3.2   Toxicity and Tolerance Mechanisms

Toxicity of heavy metals occurs when homeostasis of microalgal cells fails to eliminate, metabolize, or store them in innocuous forms. Such mechanisms depend on the organism itself, including its current biological phase and metabolic state (Torres 1997). Typically, microalgae are sensitive to heavy metals, so they can be used to advantage as biological sensors to detect potential toxic effects thereof. Growth has indeed been used as a key indicator of the toxicity of heavy metals to microalgae. The toxicity impairs the proper functioning of various physiological and biochemical processes in microalgae, e.g., it disrupts photosynthesis or nutrient uptake; these processes can be easily monitored in the laboratory (Carr et al. 1998; Arunakumara and Xuecheng 2008). Growth inhibition in microalgae is a direct function of the amount of heavy metal ions bound to the cell surface or taken up intracellularly (Tripathi and Gaur 2006; Monteiro et al. 2011c). Morlon et al. (2005), for example, investigated the cellular growth and intracellular concentrations of selenium in the unicellular green alga *Chlamydomonas reinhardtii* and concluded that toxicity is mainly linked to its intracellular accumulation.

Table 16.2 Comparison of metal removal capacity of living microalgae, bacteria, and fungi, and physicochemical sorbents

Biosorbent	Heavy metal ion	Operating conditions		Maximum removal yield (mg/g)	Reference
		pH	Initial concentration (mg/l)		
Microalgae					
C. vulgaris	Cd^{2+}	4.0	25–150	58.4	Aksu and Dönmez (2006)
Chlamydomonas reinhardtii	Cd^{2+}	2.0–7.0	20–400	77.62	Tüzün et al. (2005)
Scenedesmus obliquus	Zn^{2+}	—	10–75	836.5	Monteiro et al. (2011a)
S. obliquus	Cd^{2+}	—	0.05–1	11.4	Monteiro et al. (2009b)
Desmodesmus pleiomorphus	Zn^{2+}	—	1–30	360.2	Monteiro et al. (2009a)
D. pleiomorphus	Cd^{2+}	—	0.5–5	61.2	Monteiro et al. (2010)
Macroalgae					
Cladophora fascicularis	Cu^{2+}	5.0	12.7–254.2	70.54	Deng et al. (2007b)
Ulva reticulata	Zn^{2+}	5.5	1,500	125.5	Senthilkumar et al. (2006)
Ascophyllum nodosum	Cd^{2+}	1.0–6.0	10–150	87.7	Romera et al. (2007)
	Zn^{2+}			42.0	
Asparagopsis armata	Cd^{2+}	1.0–6.0	10–150	32.3	Romera et al. (2007)
	Zn^{2+}			21.6	
Chondrus crispus	Cd^{2+}	1.0–6.0	10–150	75.2	Romera et al. (2007)
	Zn^{2+}			45.7	
Fucus spiralis	Cd^{2+}	1.0–6.0	10–150	114.9	Romera et al. (2007)
	Zn^{2+}			53.2	

	Metal ion	pH	Concentration	Capacity	Reference
Bacteria					
Pseudomonas putida	Zn²⁺	5.0	279.4	26.1	Chen et al. (2005)
Escherichia coli (supported on kaolin)	Cd²⁺	7.2	97	15.5 (nonviable)	Quintelas et al. (2009)
	Ni²⁺		101	10.3	
Bacillus sp. (ATS-1)	Pb²⁺	3.0	250	6.9	Tunali et al. (2006)
	Cu²⁺	5.0	200	92.27	
Bacillus jeotgali	Cd²⁺	4.0–7.0	35	16.25	Ruiz et al. (2008)
	Zn²⁺		75	37.3	
Staphylococcus xylosus	Cd²⁺	6.0	10–1,000	105.2	Ziagova et al. (2007)
Fungi					
Rhizopus arrhizus	Pb²⁺	5.0	25–200	250	Sağ et al. (2000)
	Cu²⁺	4.0		50.42 (nonviable)	
	Ni²⁺	5.0		31.71 (nonviable)	
R. arrhizus	Cd²⁺	5.0–6.0	22–394	30.86 (nonviable)	Yin et al. (1999)
Phanerochaete chrysosporium	Cd²⁺	5.0	10–500	65.23	Iqbal et al. (2007)
Mucor rouxii	Zn²⁺	5.0	10	71.36	Yan and Viraraghavan (2003)
	Cd²⁺	5.0	10	7.75	
P. simplicissimum	Cd²⁺	5.0	200	8.46	Fan et al. (2008)
	Cd²⁺	5.0	250	52.50 (nonviable)	
	Zn²⁺	5.0	250	65.60 (nonviable)	
	Pb²⁺	5.0		76.90 (nonviable)	
Ion exchange resins					
Amberlite 200	Cd²⁺	4.8	2249.6	202.46	Vaughan et al. (2001)
Dowex 50 W	Cd²⁺	5.0	112.48	134.97	An et al. (2001)

There are a number of symptoms of heavy metal toxicity to microalgal cells reported by various researchers (Kagalou et al. 2002; Rangsayatorn et al. 2002). These include (1) decrease in nutrient uptake, (2) displacement and/or substitution of essential metal ions in biomolecules, which may lead to modifications of, and constraints upon activity (e.g., urease, acid phosphatase, and ATPase), (3) blockage in functioning of biologically important molecules (e.g., enzymes and transport systems for essential nutrients), (4) disruption of protein structure and membrane integrity, (5) reduction of growth and photosynthetic activity, and (6) stimulation of free radical and reactive oxygen species (ROS) generation. To overcome these negative effects, microalgae have developed several intra- and extracellular resistance or tolerance mechanisms that render heavy metals to (almost) harmless forms.

Like other microorganisms, the microalgal cells can exhibit resistance to toxicity by adsorbing heavy metals onto cell-associated materials and/or cell wall components (Costa and França 2003; la Rocca et al. 2009; Monteiro et al. 2011a), secreting metal-binding organic compounds (Levy et al. 2008), or reducing the rate of uptake (Ahuja et al. 2001). However, metal resistance may also be attained by microalgae by taking advantage of their ability to cope with high intracellular amounts of heavy metals. In addition, detoxification of heavy metals may be achieved via binding of metals to specific intracellular compounds and/or transport to specific cellular compartments (Pawlik-Skowrońska 2003), or even through efflux of the heavy metals back into solution (Monteiro et al. 2009a). One of the most common mechanisms underlying intracellular heavy metal detoxification in microalgae is the formation of metal-binding peptides or proteins, namely, class III metallothioneins (MT) or phytochelatins (PC), as observed in the marine microalga *Tetraselmis suecica* and the freshwater green alga *Scenedesmus vacuolatus* (Pérez-Rama et al. 2001; Faucheur et al. 2005).

Phytochelatins are low-molecular-weight, intracellular, metal binding polypeptides produced by microalgae on exposure to increased metal concentrations in their environment. Phytochelatins are rich in cysteine, capable of chelating metallic ions through their thiol group (−SH), and have the general amino acid structure $(\gamma\text{-Glu-Cys})_n\text{-Gly}$, with n usually varying from 2 to 11; the chain length is a characteristic of each microalgal species coupled with the metal inducer (Perales-Vela et al. 2006). The typical structure of a PC is depicted in Fig. 16.1. Phytochelatin synthesis may be induced by numerous heavy metals, like Ag, Au, Cd, Cu, Hg, Pb, and Zn; of these, Cd^{2+} appears to be the most potent activator, followed by Pb^{2+}, Zn^{2+}, and Cu^{2+}. The synthesis of PC was first reported by Stokes et al. (1977) in *Scenedesmus acutiformis*, and several subsequent studies established its role in metal detoxification. For example, Pawlik-Skowrońska et al. (2004) found induction of synthesis and accumulation of PC when *Stichococcus bacillaris* was exposed to As^{3+}, whereas Morelli and Scarano (2001) observed its synthesis in *Phaeodactylum tricornutum* following exposure to Cd^{2+}, Pb^{2+}, and Zn^{2+}.

After the metal enters the cell cytosol, it is complexed and inactivated at once, thus avoiding any inhibitory effect that might result from eventual binding to active catalytic sites or structural proteins; note that immobilized metals are less toxic than free ions. The metal/PC complex ends up in the vacuoles of the microalga cells, thus

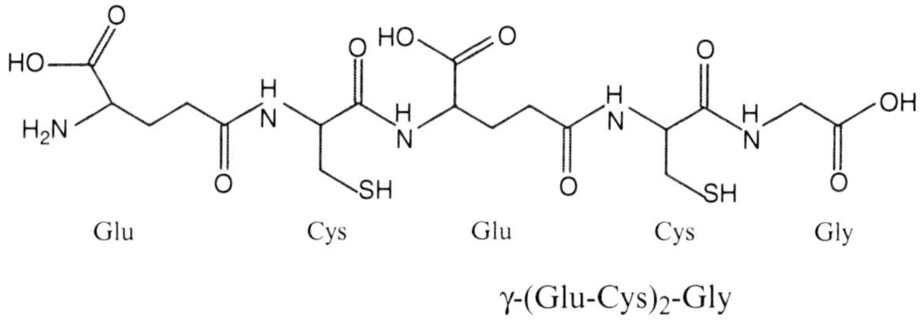

$$\gamma\text{-}(Glu\text{-}Cys)_2\text{-}Gly$$

Fig. 16.1 General structure of phytochelatin containing two γ-Glu-Cys subunits

facilitating appropriate control of the cytoplasmic concentration of heavy metal ions, while neutralizing their potentially toxic effects; hence, some species and ecotypes can live in the presence of otherwise toxic metal concentrations, which would be lethal for many other species or populations. Although PC play an important role in heavy metal detoxification, and their existence is a unique characteristic of microalgae, alternative mechanisms can be followed. The most commonly accepted hypotheses of cell protection exhibited by microalgae are listed in Table 16.3.

16.3.3   Cell/Heavy Metal Interactions

The microalgal cell wall is the first barrier for heavy metal cation uptake; it has indeed the capacity to bind such ions via its negatively charged moieties (García-Ríos et al. 2007). Attempts to localize metal cations bound onto microalga cell walls have been carried out by electron microscopy and X-ray energy dispersive analysis. The evidence available suggests that sites available for heavy metal sorption are present on the surface of those cells (Kaduková and Virčíková 2005; Doshi et al. 2007b). In a study, la Rocca et al. (2009) reported that most (i.e., above 98%) of Cd remained outside *Koliella antarctica* cells, bound to the components of their cell wall. Hence, adsorption via ion exchange appears to be the major mechanism for heavy metal uptake, and as much as 90% of the total metal has been found adsorbed on microalgal cells (Mehta et al. 2002; Monteiro et al. 2010). However, a few reports suggest that metabolic uptake of heavy metals may be more important than adsorption or, at least, identically important (Pérez-Rama et al. 2002; Wilde et al. 2006). For instance, Monteiro et al. (2009a) claimed that the microalga *Desmodesmus pleiomorphus* removed higher amounts of Zn by intracellular incorporation than adsorption onto the cell wall, between 3 and 7 days of exposure to a supernatant concentration of 1 mg/l. The cell walls of microalgae consist mainly of polysaccharides, proteins, and lipids; these offer several functional moieties (e.g., carboxyl, hydroxyl, phosphate, amino, and sulphydryl). These functional groups confer a net negative charge to the cell surface, and concomitantly a high binding affinity for heavy metal cations (Deng et al. 2007b; Volesky 2007; Gupta and Rastogi 2008). Since heavy metals

Table 16.3 Most widely accepted mechanisms of tolerance and/or resistance of microalgae to heavy metals

Mechanism	Description	Metals involved	Example of microalgae	Reference
Adsorption	Binding to, or complexation of heavy metals with cell surface components or extracellular polymers (e.g., polysaccharides, proteins, peptides, small organic acids) capable of precipitating them, which reduces bioavailability and prevents entry of metal into cells.	Zn	*Scenedesmus obliquus*	Monteiro et al. (2011a)
Exclusion	Reduced uptake of heavy metals, via changes in membrane permeability.	Cu	*Dunaliella tertiolecta*	Levy et al. (2008)
Efflux	Export of heavy metals from cell back into solution by active transport.	Zn	*Desmodesmus pleiomorphus*	Monteiro et al. (2009a)
Compartmentalization	Intracellular binding to, or sequestration of heavy metals by complexing agents (e.g., phytochelatins), or accumulation in specific cellular compartments (e.g., polyphosphate bodies, or vacuoles) – thus keeping essential functions apart from being exposed.	Cu	*Tetraselmis* sp.	Levy et al. (2008)
Volatilization	Chemical transformation of heavy metal into less toxic form (e.g., volatile product).	Hg	*Chlorella emersonii*	Wilkinson et al. (1990)

Compiled from Jjemba (2004), Bertrand and Poirier (2005), Mehta and Gaur (2005), Worms et al. (2006), and Levy et al. (2008)

in aqueous media are usually in a cationic form, they tend to adsorb onto the cell surface via counterion interactions; however, Mehta and Gaur (2005) claimed complexation and microprecipitation further to ion exchange, even though the latter dominates.

If ion exchange is present, then a somewhat competitive process for binding between cations with the same charge should occur. For instance, la Rocca et al. (2009) found that binding of Cd^{2+} to *K. antarctica* occurred together with a decrease in Ca^{2+} concentration in the culture, as an obvious outcome of competition for extracellular binding sites by divalent species. Furthermore, Ahuja et al. (1999) claimed that biosorption of Zn^{2+} by *Oscillatoria anguistissima* was accompanied by release of Mg^{2+}. However, the complexity in composition of the microalgal cell surface makes it possible that various mechanisms operate simultaneously, yet to varying degrees of importance, depending on the microalga species and the prevailing environmental conditions. When the extracellular concentration of heavy metal ions is considerably higher than its intracellular counterpart (as is usually the case of interest for bioremediation), the binding groups on the surface may aid in transporting those cations across the cell membrane into the cytoplasm, where they can eventually become compartmentalized in distinct subcellular organelles (Franklin et al. 2002).

16.3.4 *Factors Affecting Sorption Capacity of Microalgae*

Irrespective of the nature of cell/metal interactions, sorption of heavy metals by microalgae is affected by several factors, which include inorganic (e.g., temperature, pH, metal concentration and speciation, and presence of other metals) and biological factors (e.g., biomass of either living or dead cells, or possibility of its reuse).

16.3.4.1 Temperature

The reports on the effect of temperature upon sorption of heavy metals by microalgae are conflicting. Aksu (2002), for instance, observed that the extent of Ni^{2+} adsorbed onto dry biomass of *Chlorella vulgaris* increased with increasing temperature, from a maximum of 48.1 at 15°C to 60.2 mg/g at 45°C, and spanning a range of 50–250 mg/l of initial metal concentrations. These results suggest an endothermic process. Conversely, Gupta and Rastogi (2008) observed that the metal sorption by microalgae is an exothermic process, and that the metal uptake capacity of algae decreases with rising temperature. For instance, adsorption of Cd^{2+} by *Oedogonium* sp. decreased from 88.9 to 80.4 mg/g, when temperature increased from 25°C to 45°C. There are, however, other reports which suggest that temperature does not have any effect on metal sorption (Ahuja et al. 1999; Rangsayatorn et al. 2002).

16.3.4.2 pH

pH is another important environmental variable that plays a major role in adsorption of heavy metals by microalgal cells. Therefore, efforts have been directed to find pH optima, in order to maximize the extent of heavy metal removal by algal cells (Rangsayatorn et al. 2002). The pH dependence of heavy metal uptake is closely related to the acid–base properties of various functional groups present on the microalga cell surface. Since the majority of binding groups are acidic in nature, their availability as charged moieties is affected by environmental pH (Sheng et al. 2007). For instance, at low pH, the functional groups are associated with H^+ ions, which hamper binding of positively charged metal ions because of repulsion forces. On the other hand, at higher pH, the functional sites become deprotonated, so their net negative charges decrease. As a result, the functional groups of microalgal cells bind heavy metal cations to higher and higher extents (Al-Rub et al. 2004). Several reports suggest that the sorption of heavy metals by microalgal cells is increased by pH increases. For example, Ahuja et al. (1999) and Gupta and Rastogi (2008) described an increase in Zn^{2+} and Cd^{2+} removal by *O. anguistissima* and *Oedogonium* sp., respectively, when pH was increased up to 5. Similarly, an increase in Zn^{2+} and Cd^{2+} removal by *S. obliquus* was reported when pH increased up to 6 (Monteiro et al. 2011a) and 7 (Monteiro et al. 2009b), respectively. In a similar study, Han et al. (2006) reported an increase in Cr^{3+} removal by *Chlorella miniata* following increase in solution pH. The work described by Cain et al. (2008) showed as well a maximum uptake of Hg^{2+} by (the related cyanobacterium) *Spirulina platensis* at pH 6. However, under alkaline conditions, precipitation tends to occur for most heavy metals, which reduces their bioavailability and subsequent toxicity. Therefore, no bioremoval occurs at alkaline pH, probably due to dominance of plain inorganic process.

16.3.4.3 Supernatant Metal Concentration

The rate and extent of removal of heavy metals by microalgae also depend on the concentration and type of metals in solution: the sorption degree increases with increase in metal concentration, but eventually reaches saturation (Omar 2002) as predicted by classical adsorption isotherms. La Rocca et al. (2009) accordingly described an increase in the amount of Cd uptake by *K. antarctica* with increasing metal concentration in the growth medium. Bayramoğlu and Arıca (2009) also reported that the adsorption of Cu^{2+}, Zn^{2+}, and Ni^{2+} by immobilized *Scenedesmus quadricauda* increased as the initial concentration of metal ions increased in the medium, with maximum adsorption capacities of 75.6, 55.2, and 30.4 mg/g, respectively.

16.3.4.4 Metal Speciation

Bioavailability and toxicity of metals depend on their speciation in aquatic environments. The possibility of heavy metal cations to bind onto microalgae depends also on their form and charge, which, in turn, is chiefly determined by pH. Heavy metals in wastewaters occur often in a variety of chemical forms, e.g., free ions, complexes with inorganic/organic ligands, and adsorbates on particulate phases; however, the former are those that bind the furthest to microalgae, and thus the most toxic form. Rodea-Palomares et al. (2009) studied the correlation between toxicity of Cd, Zn, Hg, and Cu (which is connected to the amount of metal removed by the biosorbent) and the predicted metal free-ion concentration in solution using the freshwater cyanobacterium *Anabaena* CPB4337 as model. Although the toxicity does in general correlate with the free metal ion concentration, it was interestingly found that low amounts of PO_4^{3-} and CO_3^{2-} increased metal toxicity. Consequently, they concluded that this effect could not be related only to significant changes in metal speciation, but might be attributed to a modulating effect of these anions on uptaken metal toxicity.

16.3.4.5 Presence of Other Metals

The sorption of a desired heavy metal to microalgal biomass is significantly affected by occurrence of other metals in solution. Presence of a similar solute typically inhibits sorption of the desired metal, due to repulsive interactions between them and competition for the adsorption sites located on the cell surface (Arief et al. 2008). In practice, many industrial wastewaters contain high levels of more than one heavy metal. For example, mixtures of Cr, Ni, Cd, and Zn are found in effluents of electroplating operations (Volesky 2001). Unlike accumulation of single species of heavy metal ions by microalgal biomass, little attention has been paid to multi-metal systems. However, examination of the effects of heavy metal cations in various combinations is more representative of common environmental problems than are single metal studies, because a multiplicity of metals interfere with physiological and biochemical processes in a much more complex manner than their single metal counterparts. The underlying mechanisms of multi-metal ion uptake by microalgal biomass are accordingly not trivial. In general, such a mixture can exhibit three types of relationships: (1) synergism/cooperation, when the effect of the mixture is greater than the sum of the individual effects of the constituents; (2) antagonism, when the effect of the mixture is smaller than the sum of the individual effects of the constituents; and (3) no interaction, when the effect of the mixture is essentially similar to the sum of the individual effects of the constituents (Aksu and Dönmez 2006). Senthilkumar et al. (2006) observed a decrease in Zn uptake by *U. reticulata* when the concentration of Ca^{2+} and Mg^{2+} increased in solution. The reduced heavy metal uptake in the presence of light metals with similar charge has indeed been attributed to competition for cellular binding sites, or else to precipitation (or complexation) by Ca and/or Mg carbonates, hydrogenocarbonates, or hydroxides (Mehta and Gaur 2005). Furthermore, Hg^{2+} and Pb^{2+} were reported to reduce the amount of Zn^{2+} adsorbed to *Aphanothece halophytica* (Incharoensakdi and Kitjaharn 2002), whereas Fraile et al. (2005) demonstrated that presence of Cd^{2+} decreased uptake of Zn^{2+} in a competitive manner.

Removal studies pertaining to multi-metal systems have in fact typically unfolded competitive interaction amongst metals for binding onto adsorption sites. Mehta et al. (2000) have provided evidence for mutual interference of Cu^{2+} and Ni^{2+} onto *C. vulgaris*, whereas Aksu and Dönmez (2006) described competitive adsorption of Cd^{2+} and Ni^{2+} by the same species. Likewise, Monteiro et al. (2011b) detected competition for cell surface binding sites in either *S. obliquus* or *D. pleiomorphus* cells, when Zn^{2+} and Cd^{2+} were simultaneously present in solution, with a consequent decrease in the overall metal uptake. On the other hand, Cain et al. (2008) found that the presence of dissolved Co^{2+}, Ni^{2+}, and Fe^{3+} played a synergistic role upon Hg^{2+} uptake by *S. platensis*. In order to overcome this mutual interference, multi-metal solutions are best bioremediated by resorting to higher biomass concentrations (Terry and Stone 2002).

16.3.4.6 Biomass Concentration

The amounts of heavy metals recovered from a solution are obviously affected, in a more or less proportional fashion, by the concentration of biomass. The increased level of metal removed at higher biomass concentration could thus be simply due to a greater availability of total binding sites, even though the amount adsorbed per unit mass will tend to decrease (Fraile et al. 2005). Although increasing biomass means more adsorption sites available, a decrease of metal removal is often observed at biomass levels above a given threshold; this may be explained by partial aggregation of biomass, a cooperative process that reduces the effective surface area available for sorption, besides the average distance between the adsorption sites available (Muñoz et al. 2006). Ahuja et al. (1999) showed that increasing the biomass concentration from 0.04 to 0.2 g/l decreased the metal binding per unit cell mass. Gong et al. (2005) also reported a marked reduction in Pb^{2+} uptake by *Spirulina maxima* from 121 to 21 mg/g, when the biomass concentration was raised from 0.1 to 20 g/l. Excessively high levels of metal ions adsorbed also contribute to unfavorable electrostatic interactions between binding sites and between cells, and so lead to less efficient mixing at high biomass concentration that permits concentration gradient build-up (Fraile et al. 2005).

Living Versus Dead Biomass

Both viable and inactivated microalgal biomass have been used as sorbent material in metal removal from contaminated sites (Kaduková and Virčíková 2005; Doshi et al. 2007a). Using nonliving biomass, Lodi et al. (2008) employed re-hydrated *S. platensis* as biosorbent material for Cr^{3+} removal and observed that 95% Cr^{3+} was removed using a biomass dose of 3 g/l. On the other hand, Folgar et al. (2008) reported that living *Dunaliella salina* could remove only 11.3% Cd of the total metal used (5 mg/l, through a 96-h exposure). Heavy metal removal using living biomass assures a more quantitative removal, which often combines precipitation, adsorption, and bioaccumulation. However, viable biomass is sensitive to the chemical composition of the effluent being treated, and to operating conditions such as temperature and pH (as discussed earlier). Therefore, it is not appropriate for wastewaters that have too high heavy metal concentrations, or contain other toxic impurities (Sánchez et al. 1999). In addition, the metal recovery may also be limited due to the complex-forming abilities of extracellular metabolites secreted by living cells. Removal of metals by inactivated biomass, in contrast, entails a passive process only, in which cations predominantly adsorb onto the functional groups of the cellular surface. This process parallels synthetic sorbents, so it is poorly selective; nevertheless, dead biomass may, under some circumstances, provide a higher capacity for heavy metal uptake than viable biomass (Özer et al. 2000). In addition, it is a rapid and reversible phenomenon, which allows regeneration (and reuse) of the biomaterial in multiple sorption/desorption cycles. Finally, nonliving cells do not need nutrients and are much less affected by the physicochemical characteristics of the supernatant of heavy metal solutions. Use of inactivated biomass can also hold a great interest owing to the large variety and low cost of that biological material; however, it should not be used when biological change in the valence of the heavy metal is required for effective removal.

The method used to inactivate microalgal cells may also influence their heavy metal sorption capacity. Inactivation by heat, for example, may indeed cause partial decay of structural components of microalgae, which may decrease the number of binding sites suitable for interaction with metal cations relative to those in living cells (Costa and França 1998; Vannela and Verma 2006). Monteiro et al. (2009b, 2010) reported that, following heat inactivation, the microalgal biomass entertained lower amounts of Cd removal than its living form, for both *S. obliquus* and *D. pleiomorphus*. Furthermore, Katırcıoğlu et al. (2008) compared the removal capacity of *Oscillatoria* sp. H1 as living and heat-inactivated biomass immobilized on Ca-alginate; maximum biosorption capacities were 32.2 and 27.5 mg/g, so it was concluded that living biomass is more efficient in removing Cd from solution. However, Kaduková and Virčíková (2005) experienced an opposite trend: a higher capacity of dead cells resulted upon thermal processing. Therefore, living biomass should be preferred to inactivated one whenever the heavy metals are not at detrimental concentrations, and continuous generation of fresh adsorbent is intended.

Biomass Regeneration and Reuse

In order to make microalga-mediated biosorption processes successful on the industrial scale, regeneration of the biosorbent for repeated use is important as this keeps processing costs down. Additionally, it is also important to obtain heavy metal(s) originally extracted from the liquid phase in a more concentrated (and convenient) form than the original one, for recovery and reuse afterwards (Chojnacka et al. 2005). When sorbed on microalgal biomass, heavy metals can be desorbed by a suitable eluant or desorbing solution, which in turn allows the ready reuse of biomass in multiple sorption–desorption cycles. However, selection of a desorbing agent depends on the desorption efficiency and the persistence of biosorption capacity of adsorbing materials. Furthermore, the desorbing agent should not cause irreversible physical or chemical changes, or damage to the biomass for that matter. One of the most commonly employed methods of heavy metal desorption from microalgal biomass relies on a swing of pH. The lowering pH of the loaded biomass suspension causes displacement of heavy metal cations back to solution, by protons concomitantly gained by the binding sites. Several organic and inorganic acids and bases, as well as salts and metal chelators have accordingly been tested for their metal desorbing ability. Vannela and Verma (2006) claimed that the elution efficiency was maximum in the case of inorganic acids, followed by inorganic salts, chelating agents, and organic acids (in this order), with recoveries above 90%, except for the latter that could only reach ca. 80%. Chojnacka et al. (2005) investigated desorption of Cr^{3+}, Cd^{2+}, and Cu^{2+} from *Spirulina* sp. using 0.1 M EDTA, 0.1 M HNO_3 or deionized water, and found nitric acid as the most convenient desorbing agent, with efficiencies ranging between 90% and 98%, and without hampering the biosorption capacity of the biomaterial. Rangsayatorn et al. (2004) tested the reusability of *S. platensis* TISTR 8217 biomass immobilized on alginate and silica gel, up to five cycles of adsorption and desorption of Cd^{2+}, using 0.1 M HCl as desorbent. A significant loss of 26% in adsorption capacity resulted after the first cycle. Surprisingly, the Cd^{2+} adsorption capacity remained essentially constant from the second cycle onward.

Although HCl has a high capacity to desorb heavy metals, studies have shown that it decreases the metal sorption ability of biosorbents when applied in sequential cycles, likely due to damage to metal binding sites, including hydrolysis of surface polysaccharides (Chu et al. 1997). Cain et al. (2008) examined the regeneration of biomass over 4 sorption/desorption cycles, and unfolded a dramatic decrease in Hg removal by *S. platensis* after the second cycle when HCl was used as desorbing agent. This decrease could be attributed to partial biomass loss, coupled with acid-induced cell damage resulting during the regeneration cycles. On the other hand, Al-Rub et al. (2004) were able to reuse immobilized *C. vulgaris* biomass up to 3 sorption/desorption cycles using 0.1 M HCl, and noticed that the efficiency of Ni^{2+} removal was improved after the first cycle but leveled off thereafter.

16.3.5 Application of Microalgal Biomass in Bioremediation Processes

Recently, there has been an increasing interest in using biological processes for heavy metal removal/recovery from contaminated environments. This is primarily due to the fact that technologies encompassing naturally occurring biological entities possess numerous advantages, such as low cost, and can be applied even to low contamination levels, over other classical physicochemical approaches. Microalgae in this context have successfully been used owing to their remarkable ability to take up and accumulate heavy metals from their surrounding environment. Several cases of success have indeed been reported. For example, inactivated microalgal biomass in the form of biotraps (algaSORB®) has been used as a commercial adsorbent material for removal of heavy metals from industrial effluents. One of its major advantages is that the heavy metals adsorbed onto the cell surface can be recovered afterwards, so the material can be reused. Another important feature is that high concentrations of common ions do not interfere with sorption of the target heavy metal cations. Another successful approach to remove heavy metals via living microalgal biomass entailed a reactor containing immobilized cells, BIOALGA (Travieso et al. 2002). Using this bioreactor with *S. obliquus*, a maximum removal of 94.5% of Co was achieved by 10 days of exposure to solutions originally containing 3,000 μg/l. Nevertheless, other microalgal species tested have shown that the efficiency of removal depends not only on the species, but also on the specific heavy metal to be removed (Radway et al. 2001).

16.4 Conclusion

The ability of microalgae to sorb high concentrations of heavy metals makes them suitable condidates for efficient and commercially feasible in wastewater bioremediation strategies. However, such a goal usually demands concerted and educated research efforts for identification of microalgal species that would perform better under different ecosystems. In this context, some microalgal species isolated from long-term, metal-contaminated sites have developed a much higher capacity to accumulate heavy metals than those isolated from non-contaminated locations (Wong et al. 2000). Therefore, understanding the resistance/tolerance and uptake mechanism(s) of microalgae when present naturally in the contaminated environment or exposed intentionally to heavy metals is crucial, so as to provide rationally improved vehicles for metal removal from the environment, as part of strategies that are tailor-made for each contaminated site.

Acknowledgments This work was partially supported by Fundação para a Ciência e Tecnologia (Portugal) and Fundo Social Europeu (III Quadro Comunitário de Apoio), via a Ph.D. research fellowship granted to author C. M. Monteiro (ref. SFRH/BD/9332/2002) and supervised by author F. X. Malcata. Author F. X. Malcata acknowledges free access to literature databases made available by CBQF.

References

Ahuja P, Gupta R, Saxena RK (1999) Zn^{2+} biosorption by *Oscillatoria anguistissima*. Process Biochem 34:77–85

Ahuja P, Mohapatra H, Saxena RK, Gupta R (2001) Reduced uptake as a mechanism of zinc tolerance in *Oscillatoria anguistissima*. Curr Microbiol 43:305–310

Aksu Z (2002) Determination of the equilibrium, kinetic and thermodynamic parameters of the batch biosorption of nickel(II) ions onto *Chlorella vulgaris*. Process Biochem 38:89–99

Aksu Z, Dönmez G (2006) Binary biosorption of cadmium(II) and nickel(II) onto dried *Chlorella vulgaris*: co-ion effect on mono-component isotherm parameters. Process Biochem 41:860–868

Alloway BJ, Ayres DC (1997) Inorganic pollutants–heavy metals. In: Alloway BJ, Ayres DC (eds.) Chemical principles of environmental pollution. Blackie Academic and Professional Publishing/Chapman and Hall, London, pp 190–217

Al-Rub FA, El-Naas MH, Benyahia F, Ashour I (2004) Biosorption of nickel on blank alginate beads, free and immobilized algal cells. Process Biochem 39:1767–1773

An HK, Park BY, Kim DS (2001) Crab shell for the removal of heavy metals from aqueous solution. Water Res 35:3551–3556

Arief VO, Trilestari K, Sunarso J, Indraswati N, Ismadji S (2008) Recent progress on biosorption of heavy metals from liquids using low cost biosorbents: characterization, biosorption parameters and mechanism studies. Clean 36:937–962

Arunakumara KKIU, Xuecheng Z (2008) Heavy metal bioaccumulation and toxicity with special reference to microalgae. J Ocean Univ China 7:60–64

Bayramoğlu G, Arıca MY (2009) Construction of a hybrid biosorbent using *Scenedesmus quadricauda* and Ca-alginate for biosorption of Cu(II), Zn(II) and Ni(II): kinetics and equilibrium studies. Biores Technol 100:186–193

Bertrand M, Poirier I (2005) Photosynthetic organisms and excess of metals. Photosynthetica 43:345–353

Cain A, Vannela R, Woo LK (2008) Cyanobacteria as a biosorbent for mercuric ion. Biores Technol 14:6578–6586

Carr HP, Cariño FA, Yang MS, Wong MH (1998) Characterization of the cadmium-binding capacity of *Chlorella vulgaris*. Bull Environ Contam Toxicol 60:433–440

Chen XC, Wang YP, Lin Q, Shi JY, Wu WX, Chen YX (2005) Biosorption of copper(II) and zinc(II) from aqueous solution by *Pseudomonas putida* CZ1. Colloids Surf B: Biointerfaces 46:101–107

Chojnacka K, Chojnacki A, Górecka H (2005) Biosorption of Cr^{3+}, Cd^{2+}, and Cu^{2+} ions by blue-green alga *Spirulina* sp.: kinetics, equilibrium and the mechanism of the process. Chemosphere 59:75–84

Chu AU, Hashim KH, Phang SM, Samuel VB (1997) Biosorption of cadmium by algal biomass: adsorption and desorption characteristics. Water Sci Technol 35:115–122

Costa ACA, França FP (1998) The behaviour of the microalgae *Tetraselmis chuii* in cadmium-contaminated solutions. Aquacult Int 6:57–66

Costa ACA, França FP (2003) Cadmium interaction with microalgal cells, cyanobacterial cells, and seaweeds; toxicology and biotechnological potential for wastewater treatment. Mar Biotechnol 5:149–156

Deng L, Zhu X, Wang X, Su Y, Su H (2007b) Biosorption of copper(II) from aqueous solutions by green alga *Cladophora fascicularis*. Biodegradation 18:393–402

Doshi H, Ray A, Kothari IL (2007a) Biosorption of cadmium by live and dead *Spirulina*: IR spectroscopic, kinetics, and SEM studies. Curr Microbiol 54:213–218

Doshi H, Ray A, Kothari IL (2007b) Bioremediation potential of live and dead *Spirulina*: spectroscopic, kinetics and SEM studies. Biotechnol Bioeng 96:1051–1063

Eccles H (1999) Treatment of metal-contaminated wastes: why select a biological process? Trends Biotechnol 17:462–465

Fan T, Liu Y, Feng B, Zeng G, Yang C, Zhou M, Zhou H, Tan Z, Wang X (2008) Biosorption of cadmium(II), zinc(II) and lead(II) by *Penicillium simplicissimum*: isotherms, kinetics and thermodynamics. J Hazard Mater 160:655–661

Faucheur SL, Behra R, Sigg L (2005) Phytochelatin induction, cadmium accumulation, and algal sensitivity to free cadmium ion in *Scenedesmus vacuolatus*. Environ Toxicol Chem 24:1731–1737

Folgar S, Torres E, Pérez-Rama M, Cid A, Herrero C, Abalde J (2008) *Dunaliella salina* as marine microalga highly tolerant to but a poor remover of cadmium. J Hazard Mater 165:486–493

Fraile A, Penche S, González F, Blázquez ML, Muñoz JA, Ballester A (2005) Biosorption of copper, zinc, cadmium and nickel by *Chlorella vulgaris*. Chem Ecol 21:61–75

Franklin NM, Stauber JL, Apte SC, Lim RP (2002) Effect of initial cell density on the bioavailability and toxicity of copper in microalgal bioassays. Environ Toxicol Chem 21:742–751

García-Ríos V, Freile-Pelegrín Y, Robledo D, Mendoza-Cózatl D, Moreno-Sánchez R, Gold-Bouchot G (2007) Cell wall composition affects Cd^{2+} accumulation and intracellular thiol peptides in marine red algae. Aquat Toxicol 81:65–72

Gong R, Ding Y, Liu H, Chen Q, Liu Z (2005) Lead biosorption and desorption by intact and pretreated *Spirulina maxima* biomass. Chemosphere 58:125–130

Gupta VK, Rastogi A (2008) Equilibrium and kinetic modelling of cadmium(II) biosorption by nonliving algal biomass *Oedogonium* sp. from aqueous phase. J Hazard Mater 153:759–766

Han X, Wong YS, Tam NFY (2006) Surface complexation mechanism and modelling of Cr(III) biosorption by a microalgal isolate, *Chlorella miniata*. J Colloid Interface Sci 303:365–371

Harte J, Holdren C, Schneider R, Shirley C (1991) A guide to commonly encountered toxics. In: Harte J, Holdren C, Schneider R, Shirley C (eds.) Toxics A to Z – a guide to everyday pollution hazards. University of California Press, Berkeley, pp. 244–247, 436–438

Incharoensakdi A, Kitjaharn P (2002) Zinc biosorption from aqueous solution by a halotolerant cyanobacterium *Aphanothece halophytica*. Curr Microbiol 45:261–264

Iqbal M, Saeed A, Zafar SI (2007) Hybrid biosorbent: an innovative matrix to enhance the biosorption of Cd(II) from aqueous solution. J Hazard Mater 148:47–55

Jjemba PK (2004) Interaction of metals and metalloids with microorganisms in the environment. In: Jjemba PK (ed.) Environmental microbiology – principles and applications. Science Publishers, Enfield, pp 257–270

Kaduková J, Virčíková E (2005) Comparison of differences between copper bioaccumulation and biosorption. Environ Int 31:227–232

Kagalou I, Beza P, Perdikaris C, Petridis D (2002) Effects of copper and lead on microalgae (*Isochrysis galbana*) growth. Fres Environ Bull 11:233–236

Kaplan D (2004) Water pollution and bioremediation by microalgae – absorption and adsorption of heavy metals by microalgae. In: Richmond A (ed.) Handbook of microalgal culture – biotechnology and applied phycology. Blackwell Publishing, Ames, pp 439–447

Katırcıŏglu H, Aslım B, Türker AR, Atıcı T, Beyatlı Y (2008) Removal of cadmium(II) ion from aqueous system by dry biomass, immobilized live and heat-inactivated *Oscillatoria* sp. H1 isolated from freshwater (Mogan Lake). Biores Technol 99:4185–4191

la Rocca N, Andreoli C, Giacometti GM, Rascio N, Moro I (2009) Responses of the Antarctic microalga *Koliella antarctica* (Trebouxiophyceae, Chlorophyta) to cadmium concentration. Photosynthetica 47:471–479

Levy JL, Angel BM, Stauber JL, Poon WL, Simpson SL, Cheng SH, Jolley DF (2008) Uptake and internalisation of copper by three marine microalgae: comparison of copper-sensitive and copper-tolerant species. Aquat Toxicol 89:82–93

Lodi A, Soletto D, Solisio C, Converti A (2008) Chromium(III) removal by *Spirulina platensis* biomass. Chem Eng J 136:151–155

Malik A (2004) Metal bioremediation through growing cells. Environ Int 30:261–278

Mehta SK, Gaur JP (2005) Use of algae for removing heavy metal ions from wastewater: progress and prospects. Crit Rev Biotechnol 25:113–152

Mehta SK, Tripathi BN, Gaur JP (2000) Influence of pH, temperature, culture age and cations on adsorption and uptake of Ni by *Chlorella vulgaris*. Eur J Protistol 36:443–450

Mehta SK, Singh A, Gaur JP (2002) Kinetics of adsorption and uptake of Cu^{2+} by *Chlorella vulgaris*: influence of pH, temperature, culture age and cations. J Environ Sci Health A 37:399–414

Monteiro CM, Castro PML, Malcata FX (2009a) Use of the microalga *Scenedesmus obliquus* to remove cadmium cations from aqueous solutions. World J Microbiol Biotechnol 25:1573–1578

Monteiro CM, Marques APGC, Castro PML, Malcata FX (2009b) Characterization of *Desmodesmus pleiomorphus* isolated from a heavy metal-contaminated site: biosorption of zinc. Biodegradation 20:629–641

Monteiro CM, Castro PML, Malcata FX (2011a) Biosorption of zinc ions from aqueous solution by the microalga *Scenedesmus obliquus*. Environ Chem Lett 9:169–176

Monteiro CM, Castro PML, Malcata FX (2010) Cadmium removal by two strains of *Desmodesmus pleiomorphus* cells. Water Air Soil Poll 208:17–27

Monteiro CM, Castro PML, Malcata FX (2011b) Capacity of simultaneous removal of zinc and cadmium from contaminated media, by two microalgae isolated from a polluted site. Environ Chem Lett DOI 10.1007/s10311-011-0311-9

Monteiro CM, Fonseca SC, Castro PML, Malcata FX (2011c) Toxicity of cadmium and zinc on two microalgae, *Scenedesmus obliquus* and *Desmodesmus pleiomorphus*, from Northern Portugal. J Appl Phycol 23:97–103

Morelli E, Scarano G (2001) Synthesis and stability of phytochelatins induced by cadmium and lead in the marine diatom *Phaeodactylum tricornutum*. Mar Environ Res 52:383–395

Morlon H, Fortin C, Adam C, Garnier-Laplace J (2005) Cellular quotas and induced toxicity of selenite in the unicellular green alga *Chlamydomonas reinhardtii*. Radioprotection 40:101–106

Muñoz R, Alvarez MT, Muñoz A, Terrazas E, Guieysse B, Mattiasson B (2006) Sequential removal of heavy metal ions and organic pollutants using an algal-bacterial consortium. Chemosphere 63:903–911

Ngah WSW, Hanafiah MAKM (2008) Removal of metal ions from wastewater by chemically modified plant wastes as adsorbents: a review. Biores Technol 99:3935–3948

Omar HH (2002) Bioremoval of zinc ions by *Scenedesmus obliquus* and *Scenedesmus quadricauda* and its effect on growth and metabolism. Int Biodeter Biodegrad 50:95–100

Özer D, Özer A, Dursun G (2000) Investigation of zinc(II) adsorption on *Cladophora crispata* in a two-staged reactor. J Chem Technol Biotechnol 75:410–416

Pawlik-Skowrońska B (2003) Resistance, accumulation and allocation of zinc in two ecotypes of the green alga *Stigeoclonium tenue* Kütz. coming from habitats of different heavy metal concentrations. Aquat Bot 75:189–198

Pawlik-Skowrońska B, Pirszel J, Kalinowska R, Skowroński T (2004) Arsenic availability, toxicity and direct role of GSH and phytochelatins in As detoxification in the green alga *Stichococcus bacillaris*. Aquat Toxicol 70:201–212

Perales-Vela HV, Peña-Castro JM, Cañizares-Villanueva RO (2006) Heavy metal detoxification in eukaryotic microalgae. Chemosphere 64:1–10

Pérez-Rama M, López CH, Alonso JA, Vaamonde ET (2001) Class III metalothioneins in response to cadmium toxicity in the marine microalga *Tetraselmis suecica* (Kylin) Butch. Environ Toxicol Chem 20:2061–2066

Pérez-Rama M, Alonso JA, López CH, Vaamonde ET (2002) Cadmium removal by living cells of the marine microalga *Tetraselmis suecica*. Biores Technol 84:265–270

Quintelas C, Rocha Z, Silva B, Fonseca B, Figueiredo H, Tavares T (2009) Removal of Cd(II), Cr(VI), Fe(III) and Ni(II) from aqueous solutions by an *E. coli* biofilm supported on kaolin. Chem Eng J 149:319–324

Radway JC, Wilde EW, Whitaker MJ, Weissman JC (2001) Screening of algal strains for metal removal capabilities. J Appl Phycol 13:451–455

Rangsayatorn N, Upatham ES, Kruatrachue M, Pokethitiyook P, Lanza GR (2002) Phytoremediation potential of *Spirulina* (Arthrospira) *platensis*: biosorption and toxicity studies of cadmium. Environ Poll 119:45–53

Rangsayatorn N, Pokethitiyook P, Upatham ES, Lanza GR (2004) Cadmium biosorption by cells of *Spirulina platensis* TISTR 8217 immobilized in alginate and silica gel. Environ Int 30:57–63

Rodea-Palomares I, González-García C, Leganés F, Fernández Piñas (2009) Effect of pH, EDTA, and anions on heavy metal toxicity toward a bioluminescent cyanobacterial bioreporter. Arch Environ Contam Toxicol 57:477–487

Romera E, González F, Ballester A, Blázquez ML, Muñoz JA (2006) Biosorption with algae: a statistical review. Crit Rev Biotechnol 26:223–235

Romera E, González F, Ballester A, Blázquez ML, Muñoz JA (2007) Comparative study of biosorption of heavy metals using different types of algae. Biores Technol 98:3344–3353

Ruiz CG, Tirado VR, Gil BG (2008) Cadmium and zinc removal from aqueous solutions by *Bacillus jeotgali*: pH, salinity and temperature effects. Biores Technol 99:3864–3870

Sağ Y, Kaya A, Kutsal T (2000) Biosorption of lead(II), nickel(II) and copper(II) on *Rhizopus arrhizus* from binary and ternary metal mixtures. Sep Sci Technol 35:2601–2617

Sánchez A, Ballester A, Blázquez ML, González F, Muñoz J, Hammaini A (1999) Biosorption of copper and zinc by *Cymodocea nodosa*. FEMS Microbiol Rev 23:527–536

Senthilkumar R, Vijayaraghavan K, Thilakavathi M, Iyer PVR, Velan M (2006) Seaweeds for the remediation of wastewaters contaminated with zinc(II) ions. J Hazard Mat B 136:791–799

Sheng PX, Ting Y-P, Chen JP (2007) Biosorption of heavy metal ions (Pb, Cu and Cd) from aqueous solutions by the marine algae *Sargassum* sp. in single- and multiple-metal systems. Ind Eng Chem Res 46:2438–2444

Srivastava NK, Majumder CB (2008) Novel biofiltration methods for the treatment of heavy metals from industrial wastewater. J Hazard Mat 151:1–8

Stokes PM, Maler T, Riordan JR (1977) A low molecular weight copper-binding protein in a copper tolerant strain *Scenedesmus acutiformis*. In: Hemphil DD (ed.) Trace substances in environmental health. University of Missouri Press, Columbia, pp 146–154

Sud D, Mahajan G, Kaur MP (2008) Agricultural waste material as potential adsorbent for sequestering heavy metal ions from aqueous solutions – a review. Biores Technol 99:6017–6027

Terry PA, Stone W (2002) Biosorption of cadmium and copper contaminated water by *Scenedesmus abundans*. Chemosphere 47:249–255

Torres JE (1997) Toxicidad del cádmio sobre la diatomacea marina *Phaeodactylum tricornutum* Bohlin. Mecanismo de detoxificación. Sc. D. thesis, Faculdad de Ciências, A Coruña

Travieso L, Pellón A, Benítez F, Sánchez E, Borja R, O'Farrill N, Weiland P (2002) BIOALGA reactor: preliminary studies for heavy metals removal. Biochem Eng J 12:87–91

Tripathi BN, Gaur JP (2006) Physiological behaviour of *Scenedesmus* sp. during exposure to elevated levels of Cu and Zn and after withdrawal of metal stress. Protoplasma 229:1–9

Tunali S, Çabuk A, Akar T (2006) Removal of lead and copper ions from aqueous solutions by bacterial strain isolated from soil. Chem Eng J 115:203–211

Tüzün İ, Bayramoğlu G, Yalçın E, Başaran G, Çelik G, Arıca MY (2005) Equilibrium and kinetic studies on biosorption of Hg(II), Cd(II) and Pb(II) ions onto microalgae *Chlamydomonas reinhardtii*. J Environ Manag 77:85–92

Vannela R, Verma SK (2006) Co^{2+}, Cu^{2+}, and Zn^{2+} accumulation by cyanobacterium *Spirulina platensis*. Biotechnol Prog 22:1282–1293

Vaughan T, Seo CW, Marshall WE (2001) Removal of selected metal ions from aqueous solution using modified corncobs. Biores Technol 78:133–139

Volesky B (2001) Detoxification of metal-bearing effluents: biosorption for the next century. Hydrometallurgy 59:203–216

Volesky B (2007) Biosorption and me. Water Res 41:4017–4029

Wang J, Chen C (2006) Biosorption of heavy metals by *Saccharomyces cerevisiae*: a review. Biotechnol Adv 24:427–451

Wilde KL, Stauber JL, Markich SJ, Franklin NM, Brown PL (2006) The effect of pH on the uptake and toxicity of copper and zinc in a tropical freshwater alga (*Chlorella* sp.). Arch Environ Contam Toxicol 51:174–185

Wilkinson SC, Goulding KH, Robinson PK (1990) Mercury removal by immobilized algae in batch culture systems. J Appl Phycol 2:223–230

Wong JPK, Wong YS, Tam NFY (2000) Nickel biosorption by two *Chlorella* species, *C. vulgaris* (a commercial species) and *C. miniata* (a local isolate). Biores Technol 73:133–137

Worms I, Simon DF, Hassler CS, Wilkinson KJ (2006) Bioavailability of trace metals to aquatic microorganisms: importance of chemical, biological and physical processes on biouptake. Biochimie 88:1721–1731

Yan G, Viraraghavan T (2003) Heavy-metal removal from aqueous solution by fungus *Mucor rouxii*. Water Res 37:4486–4496

Yin P, Yu Q, Jin B, Ling Z (1999) Biosorption removal of cadmium from aqueous solution by using pretreated fungal biomass cultured from starch wastewater. Water Res 33:1960–1963

Ziagova M, Dimitriadis G, Aslanidou D, Papaioannou X, Tzannetaki EL, Liakopoulou-Kyriakides M (2007) Comparative study of Cd(II) and Cr(VI) biosorption on *Staphylococcus xylosus* and *Pseudomonas* sp. in single and binary mixtures. Biores Technol 98:2859–2865

Chapter 17
Decontamination of Radioactive-Contaminated Soils: Current Perspective

Mamdoh F. Abdel-Sabour

Abstract Radionuclides exist in the environment naturally and, in more recent times, have been added by nuclear power and weapons. The carcinogenic nature and long half-lives of many radionuclides make them a potential threat to human health. Moreover, there is an increasing trend of uranium accumulating in soils due to a number of deliberate or wrong practices. Also, the contamination of land by naturally occurring radionuclides from "non-nuclear" industries include uranium mining and milling, metal or coal mining, radium and thorium factories, and the processing of materials containing technologically enhanced levels of natural radioactivity. As a consequence, there would be a risk for ecosystems, agro-systems, and health. It is suggested that knowledge of the mechanisms that control the behavior of such heavy metals must be improved and be used for risk assessment and proposition of remediation treatments. Phytoremediation has been used to extract radionuclides and other pollutants from contaminated sites. The accuracy and success of these applications depend on an understanding of the processes involved in plant uptake of radionuclides. The recent advances in uranium removal from contaminated soils, using either chemical and/or biological techniques (such as hyperaccumulator plants, or high biomass crop species after soil treatment with chelating compounds) are reviewed and discussed.

Keywords Hyperaccumulator • Phytoremediation • NORM • Concentration ratios

M.F. Abdel-Sabour (✉)
Nuclear Research Center, Atomic Energy Authority, P.O. 13759, Cairo, Egypt
e-mail: freemfs73@yahoo.com

M.S. Khan et al. (eds.), *Biomanagement of Metal-Contaminated Soils*,
Environmental Pollution 20, DOI 10.1007/978-94-007-1914-9_17,
© Springer Science+Business Media B.V. 2011

17.1 Introduction

At many hazardous waste sites requiring cleanup, the contaminated soil, groundwater, and/or wastewater contain a mixture of contaminants, often at widely varying concentrations. These include salts, organics, heavy metals, trace elements, and radioactive compounds. The simultaneous cleanup of multiple contaminants using conventional chemical and thermal methods is both technically difficult and expensive. These methods also destroy the biotic component of soils.

Naturally occurring radionuclides are found in most ores and natural resources. The levels at which they are found depend upon the nature of ore or resource in which it is present and can vary from very low levels up to a few percent. The processing of these naturally occurring radioactive materials (NORM) can lead to the enhancement of the concentrations of the radionuclides either within the products, or in the wastes from the processes. The radionuclides which are of most interest are ^{235}U, ^{238}U, and ^{232}Th because they can undergo a series of radioactive decays (Fig. 17.1) and give rise to daughters which may also be found in NORM.

Industries which utilize NORM include uranium mining and milling, metal mining and smelting, phosphate ore processing, coal mining and fossil fuel power production, oil and gas drilling, rare earth extracting and processing, titanium oxide industry, zirconium and ceramic industries, building materials, and application of radium and thorium. These are all long-established activities. Wastes from these industries have built up over the years and a recent survey of Europe has found many sites which have long since been abandoned and where ownership is not known (Lambers et al. 1999).

The use of depleted uranium (DU, ^{238}U) as ammunition is currently a major topic for discussion. Depleted uranium is the main by-product from the processing of nuclear fuel (^{235}U). It is considered to be less radioactive than natural uranium, but despite this, there is still a serious hazard due to the alpha-radiation that is emitted (Lamas et al. 2002). Uranium like other heavy metals is a threat to both health and the environment because of its pronounced toxicity. Significant amounts of uranium have been released in the last decade with armor piercing ammunition that was

^{238}U $\rightarrow$ ^{234}Th $\rightarrow$ ^{234}Pa $\rightarrow$ ^{234}U $\rightarrow$ ^{230}Th $\rightarrow$ ^{226}Ra $\rightarrow$ ^{222}Rn $\rightarrow$ ^{218}Po $\rightarrow$ ^{214}Po $\rightarrow$
^{214}Bi $\rightarrow$ ^{210}Pb $\rightarrow$ ^{210}Bi $\rightarrow$ ^{210}Po $\rightarrow$ ^{206}Pb

^{235}U $\rightarrow$ ^{231}Th $\rightarrow$ ^{231}Pa $\rightarrow$ ^{227}Ac $\rightarrow$ ^{227}Th $\rightarrow$ ^{223}Ra $\rightarrow$ ^{219}Rn $\rightarrow$ ^{215}Po $\rightarrow$ ^{211}Pb $\rightarrow$
^{211}Bi
$\rightarrow$ ^{211}Pb $\rightarrow$ ^{207}Pb

^{232}Th $\rightarrow$ ^{228}Ra $\rightarrow$ ^{228}Ac $\rightarrow$ ^{228}Th $\rightarrow$ ^{224}Ra $\rightarrow$ ^{220}Rn $\rightarrow$ ^{216}Po $\rightarrow$ ^{212}Pb $\rightarrow$ ^{212}Bi $\rightarrow$
^{208}Pb

Fig. 17.1 Decay series for ^{238}U, ^{235}U, and ^{232}Th; environmentally significant radionuclides are shown in bold

manufactured from DU, not only during major conflicts but also on numerous military shooting ranges all over the world (Bosnia, Kosovo, Afghanistan, Iraq, Lebanon, and several Arab countries) (Sansone et al. 2001). A field study, organized, coordinated, and conducted under the responsibility of the United Nations Environment Programme (UNEP), took place in Kosovo, Serbia in November 2000 to evaluate the level of DU released into the environment by the use of DU ammunition during the 1999 conflict (UNEP programme in Balkan, 2000 and 2001). During this field mission, the Italian National Environmental Protection Agency (ANPA) collected water, soil, lichen, and tree bark samples from different sites. The samples were analyzed by alpha-spectroscopy and in some cases by inductively coupled plasma-source mass spectrometry (ICP-MS). The $^{234}U/^{238}U$ and $^{235}U/^{238}U$ activity concentration ratios were used to distinguish natural from anthropogenic uranium. They indicated that all water samples had very low concentrations of uranium (much below the average concentration of drinking water in Europe). However, the surface soil samples showed a very large variability in uranium activity concentration, that ranged from 20 Bq kg^{-1} (environmental natural uranium) to 2.3×10^5 Bq kg^{-1} (18,000 mg kg^{-1} of depleted uranium), with concentrations above environmental levels always due to DU. The uranium isotope measurements refer to soil samples collected at places where DU ammunition had been fired; this variability indicates that the impact of DU ammunitions is very site specific, reflecting both the physical conditions at the time of the impact of the DU ammunition and any physical and chemical alteration which occurred since then. This finding is in agreement with Flues et al. (2002) who investigated 52 soil samples in the vicinity of a coal-fired power plant (CFPP) in Figueira (Brazil). The radionuclide concentration for the uranium and thorium series in soils ranged from <9 to 282 Bq kg^{-1}. The range of 40 K concentration in soils varied from <59 to 412 Bq kg^{-1}. The CFPP (10 MWe) has been operating for 35 years and caused a small increment in natural radionuclide concentration in the surroundings. This technologically enhanced natural radioactivity (TENR) was mainly due to the uranium series (^{234}Th, ^{226}Ra, and ^{210}Pb) and was observable within the first kilometer from the power plant. The CFPP influence was only observed in the 0–25 cm soil horizon. The soil properties prevent the radionuclides of the ^{238}U-series from reaching deeper soil profiles. The same behavior was observed for 40 K as well. No influence was observed for ^{232}Th, which was found in low concentrations in the coal. The results of Sansone et al. (2001) on tree barks and lichens indicated the presence of DU in all cases, showing their usefulness as sensitive qualitative bioindicators for the presence of DU dusts or aerosols formed at the time the DU ammunition had hit a hard target.

17.2   Phytoremediation Technology

Phytoremediation, an emerging cleanup technology for contaminated soils, groundwater, and wastewater, is both low tech and low cost. Phytoremediation is the engineered use of green plants, including grasses, forbs, and woody species, to

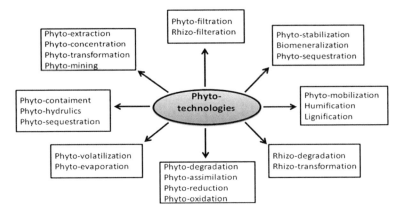

Fig. 17.2 Phytoremediation technologies

remove, contain, or render harmless such environmental contaminants as heavy metals, trace elements, organic compounds, and radioactive compounds in soil or water (Raskin et al. 1997; Salt et al. 1998). This definition includes all plant-influenced biological, chemical, and physical processes that aid in the uptake, sequestration, degradation, and metabolism of contaminants, either by plants or by the free-living organisms that constitute the plant's rhizosphere (Baker et al. 1995; McGrath 1998). Phytoremediation takes advantage of the unique and selective uptake capabilities of plant root systems, together with the translocation, bioaccumulation, and contaminant storage/degradation abilities of the entire plant body. Plant-based soil remediation systems (Fig. 17.2) can be viewed as biological, solar-driven, pump-and-treat systems with an extensive, self-extending uptake network (the root system) that enhances the below-ground ecosystem for subsequent reductive use (Wenger et al. 2002; Liphadzi et al. 2003; Dickinson and Pulford 2005).

Examples of simpler phytoremediation systems (Fig. 17.3) that have been used for years are constructed engineered wetlands, often using cattails to treat acid mine drainage or municipal sewage (Kadlec 1995; Kadlec and Knight 1996).

Phytoremediation of a site contaminated with heavy metals and/or radionuclides involves "farming" the soil with selected plants to "biomine" the inorganic contaminants, which are concentrated in the plant biomass (Ross 1994; Salt et al. 1995). For soils contaminated with toxic organics, the approach is similar, but the plant may take up or assist in the degradation of the organic contaminant (Schnoor et al. 1995). Several sequential crops of hyperaccumulating plants could possibly reduce soil concentrations of toxic inorganics or organics to the extent that residual concentrations would be environmentally acceptable and no longer considered hazardous. The potential also exists for degrading the hazardous organic component of mixed contamination, thus reducing the waste (which may be sequestered in plant biomass) to a more manageable radioactive one.

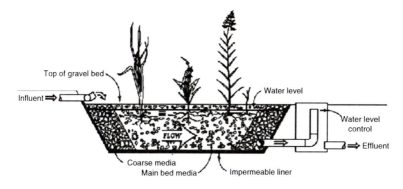

Fig. 17.3 Engineered basin wet land (GBT)

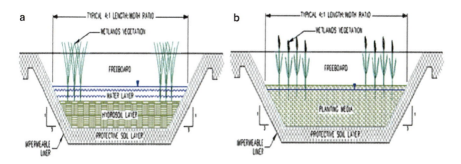

Fig. 17.4 Surface (**a**) and subsurface (**b**) flow of effluent (Okurut et al. 1999)

For treating contaminated wastewater, the phytoremediation plants are grown in a bed of inert granular substrate, such as sand or pea gravel, using hydroponic or aeroponic techniques (Fig. 17.4). The wastewater, supplemented with nutrients if necessary, trickles through this bed, which is ramified with plant roots that function as a biological filter and a contaminant uptake system. An added advantage of phytoremediation of wastewater is the considerable volume reduction attained through evapotranspiration (Hinchman and Negri 1994; Fritioff and Greger 2003; Aksorn and Visoottiviseth 2004). Some of the aquatic plants used in bioremediation of trace elements are listed in Table 17.1 (Prasad 2001a, b, 2004a, b, 2006a, b, 2007; Prasad and Freitas 2003; Prasad et al. 2001, 2006; Williams 2002). In appropriate situations, phytoremediation can be an alternative to the much harsher remediation technologies of incineration, thermal vaporization, solvent washing, or other soil washing techniques, which essentially destroy the biological component of the soil and can drastically alter its chemical and physical characteristics as well, creating a relatively nonviable solid waste. Phytoremediation actually benefits the soil, leaving an improved, functional, soil ecosystem at costs estimated at approximately one tenth of those currently adopted technologies.

Table 17.1 Aquatic plants for and biomonitoring of toxic trace elements used in a wide range of toxicity bioassays

Plant species	Metal
Azolla fililiculioides	Cr, Ni, Zn, Fe, Cu, Pb
A. pinnata	Cd, Cr, Zn
Bacopa monnieri	Hg, Cr, Cu, Cd
Carex juncell	Cu, Pb, Zn, Co, Ni, Cr, Mo, U
Carex rostrata	Cu, Pb, Zn, Co, Ni, Cr, Mo, U
Carex sp.	Cd, Fe, Pb, Mn
Ceratophyllum demersum	Cd, Cu, Cr, Pb, Hg, Fe, Mn, Zn, Ni, Co, and radionuclides
Cyperus eragrostis	Cd, Cu, Pb, Zn
Distichlis spicata	Cd, Fe, Pb, Mn
Elodea densa	Hg, methyl-Hg
E. nuttallia	Cu
E. sptangulare	Hg, Pb, Cd, Cu, and Fe
Eichhornia crassipes	As, Cd, Co, Cr, Cu, Al, Ni, Pb, Zn, Hg, P, Pt, Pd, Os, Ru, Ir, Rh
Elodea canadensis	Cu, Pb, Cd, Zn, Cr, Ni
Eriocaulon septangulare	Hg, Pb, Cd, Fe
Euryale ferox	Cd, Cr, Pb, Cu
Hydrilla verticillata	Hg, Fe, Ni, Hg, Pb
Hygrophila onogaria	Hg, methyl-Hg
Isoetes lacustris	Cu, Pb
Lemna minor	Mn, Pb, Ba, B, Cd, Cu, Cr, Ni, Se, Zn, Fe
L. trisulca	Cu, Cd
L. gibba	Cu, Cd
L. palustris	Zn, Cu, Fe, Hg
L. paucicostata	Cd, Zn, EDTA, Cu, Ca
L. perpusilla	Cd
L. polyrrhiza	Cd
L. valdivinia	Cd, Cu
Littorella uniflora	Cu, Pb
Ludwigia natans	Hg, methyl-Hg
Lysimachia nummularia	Hg, methyl-Hg
Myriophyllum spicatum	Cd, Cu, Zn, Pb, Ni, Cr
M. alterniflorum	Cu, Pb
M. exalbescens	Zn, Pb
M. aquaticum	Zn, Cu, Fe, Hg, Cd, Pb
Melilotus indica	Se
Mentha aquatic	Cd, Zn, Cu, Fe, Hg
Najas marina	Cd, Fe, Pb, Mn
Nasturtium officinale	Cd
Nuphar lutea	Cu, Ni, Cr, Co, Zn, Mn, Pb, Cd, Hg, Fe
N. variegatum	Cu, Zn
Nymphaea alba	Ni, Cr, Co, Zn, Mn, Pb, Cd, Cu, Hg, Fe
Nymphoides germinate	Cd, Cu, Pb, Zn

(continued)

Table 17.1 (continued)

Plant species	Metal
Potamogeton attenuatum	Cd, Cu, Pb, Zn
P. communis	Ni, Cr, Co, Zn, Mn, Pb, Cd, Cu, Hg, Fe
P. crispus	Cu, Pb, Mn, Fe, Cd
P. filiformis	Cd, Fe, Pb, Mn
P. lapathifoilum	Cd, Cu, Pb, Zn
P. orientalis	Cd, Cu, Pb, Zn
P. pectinatus	Mn, Pb, Cd, Cu, Cr, Zn, Ni, As, Se
P. perfoliatus	Cu, Pb, Cd, Zn, Ni, Cr
P. richardsonii	Cd, Cr, Cu, Ni, Zn, Pb
P. subsessiles	Cd, Cu, Pb, Zn
Phragmites karka	Cr
Pistia stratoites	Cu, Al, Cr, P, Hg
Ranunculus aquatilis	Mn, Pb, Cd, Fe, Pb
R. baudotii	Cd, Cu, Cr, Zn, Ni, Pb
Ruppia maritime	Mn, Pb, Cd, Pb, Fe, Se
Salvinia acutes	Mn, Pb
S. maritimus	Cd, Fe, Pb, Mn
S. natans	Pb, Cr
S. undulate	Pb
S. molesta	Hg
Scapania uliginosa	B, Ba, Cd, Co, Cr, Cu, Li, Mn, Mo, Ni, Pb, Sr, V, Zn
Schoenoplectus lacustris	Ni, Cr, Co, Zn, Mn, Pb, Cd, Cu, Hg, Fe
Scirpus lacustris	Cr
Spirodela polyrhiza	Cr
Typha domingensis	Cd, Cu, Pb, Zn
T latifolia	Ni, Cr, Co, Zn, Mn, Pb, Cd, Cu, Hg, Fe
Vallisneria americana	Cd, Cr, Cu, Ni, Pb, Zn
V. spiralis	Hg
Wolffia globosa	Cd, Cr

Compiled from Prasad (2001a, b, 2004 a, b, 2007), Prasad and Freitas (2003) and Prasad et al. (2001, 2006)

17.3 Higher Plants as Indicators of Uranium Occurrence in Soil

Leaves of nine different plant species (terrestrial moss, *Hylocomium splendens,* and *Pleurozium schreberi*; and seven species of vascular plants: blueberry, *Vaccinium myrtillus*; cowberry, *Vaccinium vitis-idaea*; crowberry, *Empetrum nigrum*; birch, Betula pubescens; willow, Salix spp.; pine, *Pinus sylvestris,* and spruce, *Picea abies*) have been collected from up to nine catchments spread over a 1,500,000 km^2 area in Northern Europe (Reimann et al. 2001a). Soil samples were taken from the O- and C-horizon at each sample site. All samples were analyzed for 38 elements (Ag, Al, As, B, Ba, Be, Bi, Ca, Cd, Co, Cr, Cu, Fe, Hg, K, Li, Mg, Mn, Mo, Na, Ni, P, Pb, Rb, S, Sb, Sc, Se, Si, Sn, Sr, Th, Tl, U, V, Y, Zn, and Zr) by ICP-MS, ICP-AES or CV-AAS (for Hg-analysis) techniques. The data showed that the concentrations of some elements like Cd, V, Co, Pb, Ba, and Y vary significantly between different plants. Other elements, for example, Rb, S, Cu, K, Ca, P, and Mg, showed surprisingly similar levels in all plants. Each group of plants including moss, shrubs, deciduous, and conifers shows a common behavior for some elements. Each plant accumulates or excludes some selected elements. Compared to the C-horizon, a number of elements (S, K, B, Ca, P, and Mn) are clearly enriched in plants. The plant:O-horizon and O-horizon:C-horizon ratios show that some elements are accumulated in the O-horizon (e.g., Pb, Bi, As, Ag, Sb). Airborne organic material attached to the leaves can thus result in high values of these elements without any pollution source. In other study, Reimann et al. (2001b) collected additional soil samples from the O-horizon and the C-horizon at each plant sample site. One of the nine catchments was located directly adjacent (5–10 km S) to the nickel smelter and refinery at Monchegorsk, Kola Peninsula, Russia. The high levels of pollution at this site are reflected in the chemical composition of all plant leaves. However, it appears that each plant enriches (or excludes) different elements. Elements emitted at trace levels, such as Ag, As, and Bi, are relatively much more enriched in most plants than the major pollutants Ni, Cu, and Co.

The potential of using higher plants as indicators of uranium distribution in soil was studied at a site in Germany where uranium concentrations ranged from 5 to 1,500 mug/g soil and reached a maximum of 1,860 µg/kg in soil water (Steubing et al. 1993). Results indicated that *Sambucus nigra* was the best indicator of uranium contamination whereas chemical analysis of its leaves provided more detailed information regarding uranium distribution than soil analyses. The plants not only indicate the location of mineralization but also the migration pathway of U-containing soil water. They indicated that adsorption of contaminated water was the main source of the U accumulation in the different plant organs. Elemental composition of soil, herbaceous and woody plant species, and the muscle and liver tissue of two common small mammal species were determined in a wetland ecosystem contaminated with Ni and U from nuclear target processing activities at the Savannah River Site, Aiken, SC (Punshon et al. 2003). Species studied were black willow (*Salix nigra* L.), rushes (*Juncus effusus* L.), marsh rice rat (*Oryzomys palustris*), and cotton rat (*Sigmodon hispidus*). Two mature trees were sampled around the perimeter of the former de facto settling basin, and transect lines sampling rushes and trapping small mammals were laid across the wetland area, close to a wooden spillway that previously enclosed the pond. Nickel and U concentrations were elevated to contaminant levels; with a total concentration of 1,065 (±54) mg kg^{-1} U and 526.7 (±18.3) mg kg^{-1} Ni within the soil. Transfer of contaminants into woody and herbaceous plant tissues was higher for Ni than for U, which appeared to remain bound to the outside of root tissues, with very little (0.03 ± 0.001 mg kg^{-1}) U detectable within the leaf tissues. This indicated a lower bioavailability of U than the co-contaminant Ni. Trees sampled from the drier margins of the pond area contained more Ni within their leaf tissues than the rushes sampled from the wetter floodplain area, with leaf tissues concentrations of 75.5 mg kg^{-1} Ni. Transfer factors of contaminants indicated that U bioavailability is negligible in this wetland ecosystem.

17.3.1 Hyperaccumulator of Uranium

It is known that natural hyperaccumulators do not use rhizosphere acidification to enhance their metal uptake. Recently, it has been found that some natural hyperaccumulators (e.g., *Thlaspi caerulescens*) proliferate their roots positively in patches of high metal availability. In contrast, non-accumulators actively avoid these areas, and this is one of the mechanisms by which hyperaccumulators absorb more metals when grown in the same soil. However, there are few studies on the exudation and persistence of natural chelating compounds by these plants. It is thought that rhizosphere microorganisms are not important for the hyperaccumulation of metals from soil. Applications of chelates have been shown to induce large accumulations of metals like Pb, U, and Au in the shoots of non-hyperaccumulators, by increasing metal solubility and root-to-shoot translocation. The efficiency of metal uptake does vary with soil properties, and a full understanding of the relative importance of mass flow and diffusion in the presence and absence of artificial chelates is not available. To successfully manipulate and optimize future phytoextraction technologies, it is argued that a fully combined understanding of soil supply and plant uptake is needed (McGrath et al. 2002).

Shahandeh and Hossner (2002a, b) evaluated 34 plant species for uranium accumulation from U-contaminated soil. There was a significant difference in U accumulation among plant species. They indicated that sunflower (*Helianthus annuus*) and Indian mustard (*Brassica juncea*) accumulated more U than other plant species. Sunflower and Indian mustard were selected as potential U accumulators for further study in one U mine tailing soil and eight cultivated soils (pH range 4.7 to 8.1) contaminated with different rates (100–600 mg U(VI) kg^{-1}) of uranyl-nitrate ($UO_2(NO_3)_2.6H_2O$). Uranium fractions of contaminated soils [(exchangeable, carbonate, manganese (Mn), iron (Fe) oxides bond, organic bond, and residual)] were determined periodically over an 8-week incubation period. Uranium accumulated mainly in the roots (6,200 mg U kg^{-1}). The highest concentration of U in shoots of plant species was 102 mg kg^{-1}. Plant performance was affected by U contamination rates, especially in calcareous soils. Plants grown in soils with high carbonate–U fractions accumulated the most U in shoots and roots. The lowest plant U occurred in clayey acid soils with high Fe, Mn, and organic U-fractions. They concluded that the effectiveness of U remediation of soils by plants was strongly influenced by soil type and its properties, which determine the tolerance and accumulation of U in plants. Some of the metals hyperaccumulators are presented in Fig. 17.5.

In a similar investigation, Dreesen and Cokal (1984) assessed the uptake of contaminants occurring in chemical waste burial sites using different plant species such as *Atriplex canescens*, *Kochia scoparia*, barley, lucerne, and *Melilotus officinalis* growing on uranium mill tailings materials. There were significant differences among plant species in terms of the nutrients and contaminants in aerial organs. Of the tested plant species, barley contained higher levels of U and much higher levels of Si than the other species while lucerne had higher levels of Al, Ba, Co, and V and *M. officinalis* had higher levels of Ba and V than barley.

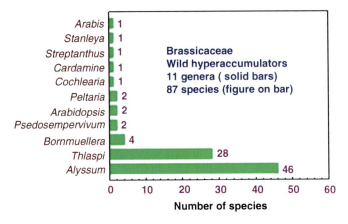

Fig. 17.5 Wild Brassicaceae with 11 genera and 87 metals hyperaccumulator species

Furthermore, significant differences in radionuclide concentrations among crop species (squash were generally higher than beans or sweetcorn) and plant parts (nonedible tissues were generally higher than edible tissues) were observed (Fresquez et al. 1998). They reported that the maximum net positive committed effective dose equivalent of beans, sweet corn, and squash in equal proportions was 74 mrem/year (740 µS/year). This upper bound dose was below the International Commission on Radiological Protection permissible dose limit of 100 mrem/year (1,000 µS/year) from all pathways and corresponds to a risk of an excess cancer fatality of 3.7×10^{-5} (37 in a million), below the US Environmental Protection Agency's guideline of 10^{-4} (US-EPA 2005).

Carrots, squash, and Sudan grass were irrigated with groundwater amended with manganese, molybdenum, selenium, and uranium stock solutions to simulate a range of concentrations found at ten inactive uranium ore milling sites to determine plant tissue levels after a 90-day growth period in sand in a greenhouse experiment (Baumgartner et al. 2000). It was evident from this study that except for squash response to uranium, all plants had increased levels of each metal, some even to unacceptable levels. Squash, on the hand, however, did not accumulate uranium at any dose tested. Similar results were reported by Lotfy (2010), which indicated that sunflower and cotton shoots accumulated the highest U content among the five tested plant species (sunflower, cotton, pankium, napier grass, and squash), irrespective of soil type. Shoot concentrations of U were as high as 69.9 Bq kg^{-1} dry matter of sunflower, followed by cotton and napier grass, panikum then squash with a range of U between 4.2 and 69.9 Bq kg^{-1} dry matter in case of the alluvium soil. However, in the loamy sandy soil, sunflower U-shoots were > cotton > penakium > napier grass > Squash with a lower order of magnitude, which could be explained by the lower U content in sandy soil compared to the alluvial soil.

Activity concentrations and plant/soil concentration ratios (CRs) of ^{239,240}Pu, ^{241}Am, ^{244}Cm, ^{232}Th, and ^{238}U were determined for three vegetable crops grown on an exposed, contaminated lake bed of a former reactor cooling reservoir in South

Carolina, USA (Whicker et al. 1999). The crops were turnip greens and tubers (cv. white globe), bush beans (*Phaseolus vulgaris*), and husks and kernels of sweet corn cv. Silver Queen. Although all plots were fertilized, some received K_2SO_4, while others received no K_2SO_4. The K_2SO_4 fertilizer treatment generally lowered the activity concentrations for [241]Am, [244]Cm, [232]Th, and [238]U, but differences were statistically significant for [241]Am and [244]Cm only. Highly significant differences occurred in activity concentrations among actinides and among crops. In general, turnip greens exhibited the highest uptake for each of the actinides measured, while corn kernels had the least. For turnip greens, geometric mean CRs ranged from 2.3×10^{-3} for [239,240]Pu to 5.3×10^{-2} for [241]Am (no K_2SO_4 fertilizer). For corn kernels, geometric mean CRs ranged from 2.1×10^{-5} for [239,240]Pu and [232]Th to 1.5×10^{-3} for [244]Cm (no K fertilizer). In general, CRs across all crops for the actinides were in the order: [244]Cm > [241]Am > [238]U > [232]Th > [239,240]Pu. They calculated the lifetime health risks from consuming crops contaminated with anthropogenic actinides, which were similar to the risks from naturally occurring actinides in the same crops (total 2×10^{-6}); however, these risks were only 0.3% of that from consuming the same crops contaminated with [137]Cs.

17.4 Metals in Soils and Food Chain

The movement of both essential and non-essential trace elements through agricultural ecosystems and food chains is complex. Such elements as As, B, Cd, Cr, Cu, Hg, Ni, Pb, Se, U, V, and Zn are generally present in soils in low concentrations but concentrations may be elevated because of natural processes and human activities, such as fossil fuel combustion, mining, smelting, sludge amendment to soil, fertilizer application, and agricultural practices. Although a significant effort has been expended over the past 40 years to evaluate and quantify the transfer of trace elements from soils to plants, more attention needs to be given to mechanisms within the soil and plant systems, which influence their solubility, chemical speciation, mobility, and uptake by and transport in plants (Banuelos and Ajwa 1999). The prediction of movement of trace elements in the agricultural ecosystem must be partially based on understanding the soil and plant processes governing chemical form and the uptake and behavior of trace elements within plants.

Sparingly soluble contaminants are less likely to affect human health through food chain transfers, such as plant uptake or passage through animal-based foods, because mobility in these pathways is limited by solubility (Sheppard and Evenden 1992). Direct ingestion or inhalation of contaminated soil becomes the dominant pathway. However, both of these can be selective processes. Clay-sized particles carry the bulk of the sparingly soluble contaminants, and mechanisms that selectively remove and accumulate clay from the bulk soil also concentrate the contaminants. Erosion is another process that selectively removes clays. Sheppard and Evenden (1992) examined the degree of clay and contaminant-concentration enrichment that could occur by these processes, using U, Th, and Pb as representative contaminants and using clayey and loamy soil. They indicated that soil erosion by water in natural rainfall events caused concentration enrichments up to sevenfold, and enrichments varied with characteristics of the erosion events. Adhesion to skin gave modest enrichments of 1.3-fold in these soils, but up to 10-fold in sandy soils. Adhesion to plant leaves, where there was no root contact with contaminated soil, gave leaf concentration comparable to situations where the roots contacted the contaminated soil. Clearly, adhesion to leaves is an important component of plant accumulation of sparingly soluble contaminants. The bioreduction and immobilization of soluble U(VI) to insoluble U(IV) minerals are a promising strategy for the remediation of uranium-contaminated soil and groundwater. While a mechanistic description is not fully resolved, it appears humic materials could interrupt electron transport to U(VI). The results of Lenhart et al. (2000) suggested that humic materials could potentially decrease U(VI) reduction under certain conditions. Furthermore, humic materials could prevent U(IV) precipitation and thus facilitate the transport of U(IV)–humic complexes.

Soils contaminated with heavy metals or radionuclides at concentrations above regulatory limits pose an environmental and human health risk (Elless et al. 1997). Whereas regulatory limits are only concerned with the "extent" of the contamination, knowledge of the "nature" of the contamination (e.g., oxidation state and mineralogy of the contaminant, particulate vs. adsorbed form, etc.) is necessary for developing optimal treatment strategies. Mineralogical identification of the contaminants provides important information concerning the nature of the contamination because once the mineral form is known, its properties can then be determined from geochemical data. A new density-fractionation technique was used to concentrate U particulates from U-contaminated soils. Results from neutron-activation analysis of each density fraction showed that the U had been concentrated (up to 11-fold) in the heavier fractions. Mineralogical analyses of the density fractions of these soils using x-ray diffraction, scanning-electron microscopy, and an electron microprobe showed the predominance of an *autunite* [$Ca(UO_2)_2(PO_4)_2.10-12\ H_2O$]-like mineral with lesser amounts of *uraninite* (UO_2) and *coffinite* ($USiO_4$) as the U-bearing minerals in these soils. The presence of reduced forms of U in these soils suggests that the optimal remediation strategy requires treatment with an oxidizing agent in addition to a carbonate-based leaching to solubilize and remove U from these soils.

The [238]U and [232]Th concentrations in soil and various foods obtained in high natural radiation areas in China were determined for estimating the internal radiation doses caused by these radionuclides (Yukawa et al. 1999). Several analytical methods were evaluated for their applicability and quality assurance. The accuracy and precision of ICP-MS is considerably better for determining trace elements like U and Th in fine powder samples. The estimated annual effective dose is 0.302 μ Sv/year for [238]U and 1.86 μ Sv/year for [232]Th in the high natural radiation area, and 0.0101 μ Sv/year for [238]U and 0.177 μ Sv/year for [232]Th in the control area.

Soil samples were collected around a coal-fired power plant from 81 different locations in Hungary (Papp et al. 2002). Brown coal, unusually rich in uranium, is burnt in this plant that lies inside the confines of a small industrial town and has been operational since 1943. Activity concentrations of the radionuclides [238]U,

^{226}Ra, ^{232}Th, ^{137}Cs, and ^{40}K were determined in the samples. Considerably elevated concentrations of ^{238}U and ^{226}Ra were found in most samples collected within the inhabited area. Concentrations of ^{238}U and ^{226}Ra in soil decreased regularly with increasing depth at many locations, which can be explained by fly ash fallout. Concentrations of ^{238}U and ^{226}Ra in the top (0–5 cm depth) layer of soil in public areas inside the town are 4.7 times higher, on average, than those in the uncontaminated deeper layers, which mean there are approximately 108 Bq kg^{-1} surplus activity concentrations above the geological background. A high emanation rate of ^{222}Rn from the contaminated soil layers and significant disequilibrium between ^{238}U and ^{226}Ra activities in some kinds of samples have been found.

Accumulation of ^{226}Ra into different plant species from contaminated soils was measured further on site within the area of an uranium mill (Soudek et al. 2004). While the ^{226}Ra activity concentration in soil on site ranged from 7.12 to 25.60 Bq.g^{-1} (1 SD < ±10%), in the plant species tested, it ranged from 0.66 to 5.70 Bq.g^{-1} (1 SD < ±10%). Their results proved that the ^{226}Ra accumulation was rather different for the tested higher plant species. They suggested that some of tested plants could be applied for effective large-scale and long-time decrease of ^{226}Ra activity concentration in highly contaminated soils. Moreover, using selected plant species could be considered for biomonitoring. This finding can be helpful in selecting plant species able to extract ^{226}Ra and/or in phytomonitoring within the areas of uranium facilities.

17.5   Remediation Options

A wide variety of remediation technologies are available. Techniques most suited to these particular sites are those which are well established, require little maintenance, and are known to be able to deal with wastes containing radionuclides which arise from the ^{235}U, ^{238}U decay chains. Remediation technologies may be divided into five major categories (Angle et al. 2001; Gisbert et al. 2003; Adriano et al. 2004; Kamal et al. 2004), they are as follows: (1) Removal of source, where the contaminated material is collected and removed to a more secure location. (2) Containment, where barriers are installed between contaminated and uncontaminated media to prevent the migration of contaminants, i.e., capping and sub-surface barriers. Solidification/stabilization (S/S) can be done in situ or ex situ on excavated materials by processing at a staging area either on-site or off-site. Solidification refers also to techniques that encapsulate hazardous waste into a solid material of high structural integrity. (3) Immobilization, where materials are added to the contaminated medium, in order to bind the contaminants and reduce their mobility, i.e., cement-based solidification and chemical immobilization. Contaminated soils can be treated in situ or ex situ to reduce the pollutants and thereby their toxicity and mobility. The redox potential (Eh) depends on the availability of oxygen in soils, water, and sediments, and upon biochemical reactions by which microorganisms extract oxygen for respiration. Redox conditions influence the mobility of metals in two different ways. Firstly, the valence of certain metals changes. For example, under reducing conditions, Fe^{3+} is transformed to Fe^{2+} and, similarly, the valence of manganese and arsenic is subject to direct changes. Since the reduced ions are more soluble, increased concentrations of these metals have been observed in reducing environments such as groundwaters and sediment solutions. Under reducing conditions, sulfate reduction will take place: for example, in sediments, lead sulfide with a low solubility is formed. On the other hand, an increase in the redox potential will cause lead sulfide to become unstable, with a subsequent rise in dissolved lead concentrations (McCutcheon and Schnoor 2003). (4) Separation, where the contaminating radionuclides are separated from the bulk of the material, i.e., soil washing, flotation, and chemical/solvent extraction. Separation can be carried out both in situ and ex situ. The fundamental strategy of soil washing is to extract unwanted contaminants from soil through washing or leaching the soil with liquids, generally aqueous solutions. The contaminant must be separated from the soil matrix and transferred to the washing solution and then the washing solution must be extracted from the soil. (5) Phytoremediation. Phytoremediation is the use of green plants to remove pollutants from the environment or render them harmless. "Current engineering-based technologies used to clean up soils—like the removal of contaminated topsoil for storage in landfills—are very costly"; "green" technology uses plants to "vacuum" heavy metals from the soil through their roots. Certain plant species, known as metal hyperaccumulators, have the ability to extract elements from the soil and concentrate them in the easily harvested plant stems, shoots, and leaves. These plant tissues can be collected, reduced in volume, and stored for later use. While acting as vacuum cleaners, the unique plants must be able to tolerate and survive high levels of heavy metals and radionuclides in soils.

Phytoremediation can be used as part of a treatment train when time constraints require other methods to be employed to achieve a remediation goal in a short period of time. This usually occurs when high contaminant concentrations in sensitive areas (i.e., near drinking water sources) require quick reduction. A series of remediation efforts may be undertaken to reduce the concentrations to an acceptable level before applying phytoremediation as the last "polishing step" to remediate and contain low-level concentrations. Phytoremediation can also be applied in conjunction with other technologies to achieve a treatment goal. The natural solar-powered pumping of deep-rooted trees may need to be coupled with traditional pump-and-treat systems to maintain treatment rates during the less effective growing months of the winter season. Vegetation may also be planted around site perimeters and "hot spots" to maintain hydraulic control and prevent contamination migration, while traditional methods are applied to remediate the source.

17.5.1 Uranium Bioremediation

Depending on solution chemistry, U(VI) often exists as mobile anionic uranyl–carbonate complexes (Langmuir 1978; Grenthe et al. 1992). Biological reduction of soluble U(VI) to a sparingly soluble form of U(IV) (e.g., uraninite UIVO2(s)) has been proposed as a remediation strategy (Lovley 1993). A variety of dissimilatory metal-reducing bacteria (DMRB), like the genus Geobacter (an obligate anerobe) and Shewanella (a facultative anaerobe), are the most extensively studied and sulfate-reducing bacteria, like Thermo-desulfo-bacteria, the Nitrospirae, and the gram-positive Peptococcaceae—for instance, Thermodesulfovibrio and Desulfotomaculum. There is also a genus of Archaea known to be capable of sulfate reduction; Archaeoglobus can catalyze this reaction under anoxic conditions (e.g., Truex et al. 1997; Spear et al. 1999; Liu et al. 2002; Dyer 2003).

The impact of humic materials on the bioreduction of soluble U(VI) is not well understood. For example, if a DMRB preferentially uses humic materials instead of U(VI) as its electron acceptor, then U(VI) bioreduction could be inhibited. However, if the humic materials act as effective electron shuttles, then no inhibition would be observed and, depending on the different reaction rates and the solution chemistry, enhancement may occur (Gu and Chen 2003; Gu et al. 2005). Another possibility is that humic materials may complex U(VI) (Moulin et al. 1992; Higgo et al. 1993; Lenhart et al. 2000), decrease bioavailability, and inhibit bioreduction. Finally, humic materials may also complex U(IV) (Li et al. 1980; Zeh et al. 1997), which could interfere with U(IV) precipitation and facilitate U(IV) transport.

17.5.2 Phytoremediation of Uranium-Contaminated Soils

Radionuclides can effectively be removed from contaminated sites using phytoremediation technologies. In this context, an environmentally friendly and cost-effective uptake of radionuclides by root systems from contaminated soils and/or surface waters has shown promising results. Several organic as well as inorganic agents can effectively and specifically increase solubility and, therefore, accumulation of heavy metals by several plant species (Schmidt 2003). Metal hyperaccumulators should exhibit the following traits: (1) possess highly efficient root uptake, (2) enhanced root-to-shoot transport, and (3) hypertolerance of metal(s), involving internal complexation and compartmentation (McGrath and Zhao 2003). Crops like willow (*Salix viminalis* L.), Indian mustard Czern.], corn (*Zea mays* L.) and sunflower show high tolerance to heavy metals and are, therefore, to a certain extent able to use the surpluses that originate from soil manipulation. Both good biomass yields and, particularly, metal hyperaccumulation (naturally or enhanced) are required in order to make phytoextraction efficient over relatively short time periods.

Uranium concentrations can be strongly increased by applying citric acid (Garbisu and Alkorta 2001). The addition of chelating agents to the soil can also bring metals into solution through desorption of sorbed species, dissolution of Fe and Mn oxides, and dissolution of precipitated compounds (Norwell 1984). These complexes can greatly alter the reactivity of the metal ion. They can alter the oxidation–reduction properties of transition-metal ions, such as iron (Fe) and manganese (Mn), and therefore increase or decrease the reactivity of these systems (Evangelou et al. 2006). Moreover, for large-scale applications, agricultural measures as placement of agents, dosage splitting, the kind and amount of agents applied, and the soil properties are important factors governing plant growth, heavy metal concentrations, and leaching rates. Effective prevention of leaching, breeding of new plant material, and use of the contaminated biomass (e.g., as biofuels) will be crucial for the acceptance and the economic breakthrough of enhanced phytoextraction. Therefore, it is emphasized that the ability to hyperaccumulate metals should be demonstrated on real field contaminated soils. Phytoremediation of uranium contaminated soil has been hampered by a lack of information relating U speciation to plant uptake. For example, Ebbs et al. (2001) investigated U uptake by plants in order to show how the phytoextraction of U from contaminated soil could be improved. Using speciation modeling and hydroponic experiments, they concluded that the uranyl ($UO_2 2+$) cation is the chemical species of U most readily accumulated in plant shoots. A subsequent soil incubation experiment examined the solubilization of U from contaminated soil by synthetic chelates and organic acids. The results of the hydroponic and soil experiments were then integrated in a study that grew red beets in U-contaminated soils amended with citric acid or HEDTA. Citric acid was again a highly effective amendment, increasing shoot U content by 14-fold compared to controls. In another work, Ebbs et al. (1998) studied U-uptake and translocation by plants using a computer speciation model to develop a nutrient culture system that provided U as a single predominant species in solution. A hydroponic uptake study determined that at pH 5, the uranyl (UO_2^{2+}) cation was more readily taken up and translocated by peas (*Pisum sativum*) than the hydroxyl and carbonate–U complexes present in the solution at pH 6 and 8, respectively. A subsequent experiment tested the extent to which various monocot and dicot species take up and translocate the uranyl cation. Of the species screened, tepary bean (*Phaseolus acutifolius*) and red beet (*Beta vulgaris*) were the species showing the greatest accumulation of U. The initial characterization of U-uptake by peas suggested that in the field, a soil pH of <5.5 would be required in order to provide U in the most plant-available form. A pot study using U-contaminated soil was therefore conducted to assess the extent to which two soil amendments, HEDTA and citric acid, were capable of acidifying the soil, increasing U solubility, and enhancing U-uptake by red beet. Of these two amendments, only citric acid proved effective, decreasing the soil pH to 5 and increasing U accumulation by a factor of 14. The results of this pot trial provide a basis for the development of an effective phytoremediation strategy for U-contaminated soils. However, applications of synthetic chelators such as EDTA can lead to a substantially increased risk of leaching of metals to groundwater. This environmental risk is likely to limit the usefulness of chelator-induced phytoextraction. One way to deal with this risk is to use hydrological barriers. However, due to the costs of construction of hydrological barriers, it would probably be simpler and quicker to flush metals out of the soils using chelators, without growing plants. Such operations require that the chelators to be used are cheap and easily degradable in soil; meeting both of these criteria is not easy.

It is worth to mention here that bioconcentration factors obtained from studies using hydroponic culture, sand culture, or even soils spiked with soluble metals do not give a realistic measure of how the plants will perform on field contaminated soils, where metals are usually much less bioavailable. Hydroponic culture or metal spiking experiments are useful for investigating mechanisms of metal uptake and tolerance, but often the results cannot be extrapolated to the field. Metal tolerance is also important, because metal-sensitive plants are not likely to establish and produce large biomass on contaminated soils. Non-hyperaccumulators may achieve an apparently large bioconcentration factor under conditions of metal toxicity, when growth has been severely inhibited. Vandenhove et al. (2001) in a study investigated the potential of ryegrass (*Lolium perenne* cv. Melvina), Indian mustard, and redroot pigweed (*Amarathus retroflexus*) to phytoextract uranium (U) from a sandy soil contaminated at low levels in the greenhouse experiment. Two soils were tested: a control soil (317 Bq ^{238}U kg^{-1}) and the same soil washed with bicarbonate (69 Bq ^{238}U kg^{-1}). The annual removal of the soil activity with the biomass was less than 0.1%. The addition of citric acid (25 mmol kg^{-1}) one week before the harvest increased U-uptake up to 500-fold. With a ryegrass and mustard, yield of 15,000 kg ha^{-1} and 10,000 kg ha^{-1}, respectively, up to 3.5 and 4.6% of the soil activity could annually be removed with the biomass. With a desired activity reduction level of 1.5 and 5 for the bicarbonate washed and control soil, respectively, it would take 10–50 years to attain the release limit. A linear relationship between the plant ^{238}U concentration and the ^{238}U concentration in the soil solution of the control, bicarbonate-washed, or citric acid-treated soil points to the importance of the soil solution activity concentration in determining U-uptake and hence to the importance of solubilizing agents to increase plant uptake. However, they indicated that citric acid addition decreased dry matter accumulation in all plants and inhibited growth of ryegrass. On the contrary, Lotfy (2010) indicated that increasing citric acid rate of application up to 20 mmol/kg resulted in a relative increase in whole plant dry weight, which could be attributed to the release of micro and macro nutrient.

Huang et al. (1998) suggested that key to the success of U phytoextraction is to increase soil U availability to plants. Some organic acids can be added to soils to increase U desorption from soil to soil solution and to trigger a rapid U accumulation in plants. Of the organic acids (acetic acid, citric acid, and malic acid) tested, citric acid was the most effective in enhancing U accumulation in plants. Shoot U concentrations of *Brassica juncea* and *Brassica chinensis* grown in a U-contaminated soil (total soil U, 750 mg/kg) increased from <5 to >5 000 mg/kg in citric acid-treated soils. Using this U hyperaccumulation technique, U accumulation in shoots of selected plant species grown in two U-contaminated soils (total soil U, 280 and 750 mg/kg) can be increased by more than 1,000-fold within a few days. The results suggest that U phytoextraction may provide an environmentally friendly alternative for the cleanup of U-contaminated soils. Lotfy (2010) indicated that U-uptake varies with soil/plant interaction, chelate, rate, and chemical forms (citric acid and EDTA). Chelate addition enhanced U accumulation in sunflower shoots and roots significantly. For example, the addition of citric acid at 20 mmol/kg increased sunflower shoot U concentration from 23.6 to 42.6 mg/kg and from 18.2 to 24.3 mg/kg in roots. It is worth to mention that the nature of the contaminant (recalcitrance, persistence, bioavailability, etc.) is crucial when developing effective phytoremediation strategies for a given site. High contaminant concentrations may limit phytoremediation as a treatment option due to phytotoxicity or the impracticality of using such a slow remediation method. Additionally, the physical location of the contaminant will determine the efficiency of the treatment. Due to plant root limitations, phytoremediation of soils and sediments is typically employed for contaminants in the near surface environment within the root zone. For groundwater treatment, phytoremediation is limited to unconfined aquifer where the water table and the contaminant are both within reach of plant roots (either in direct contact or via transpiration). It can be deduced that no single application of phytoremediation is appropriate for all sites. Rather, a prescription must be made based on a thorough site assessment. Phytoremediation may be the sole solution to a remediation project in instances where time to completion is not a pressing issue. While phytoremediation may not be a stand alone solution to all hazardous waste sites, it can certainly be used as part of a treatment train for site remediation either during peak growing seasons or as a polishing step to clean up the last remaining "hard-to-get" low concentrations.

17.6   Conclusion

Despite problems, phytoremediation is still considered an effective technology, which, however, requires acceptance at commercial scale. Several reports indicated that this technology has received greater acceptance for chlorinated solvents and metals while just starting to gain acceptance within the explosives and pesticides domains. Continued bench-scale studies are needed to determine plant toxicities, degradation pathways, and contaminant fates, and the resulting field scale applications are necessary to provide proof the technology works in order for phytoremediation to be fully accepted by the industry.

References

Adriano DC, Wenzel WW, Vangronsveld J, Bolan NS (2004) Role of assisted natural remediation in environmental cleanup. Geoderma 122:121–142

Aksorn E, Visoottiviseth P (2004) Selection of suitable emergent plants for removal of arsenic from arsenic contaminated water. Sci Asia 30:105–113

Angle JS, Chaney RL, Baker AJM, Li Y, Reeves R, Volk V, Roseberg R, Brewer E, Burke S, Nelkin J (2001) Developing commercial phyto-extraction technologies: practical considerations. S Afr J Sci 97:619–623

Baker AJM, McGrath SP, Sidoli CMD, Reeves RD (1995) The potential for heavy metal decontamination. Mining Environ Manage 3:12–14

Banuelos GS, Ajwa HA (1999) Trace elements in soils and plants: an overview. J Environ Sci Health A Toxic Hazard Subst Environ Eng 34:951–974

Baumgartner DJ, Glenn EP, Kuehl RO, Thompson TL, Artiola JF, Menke SE, Saar RA, Moss GS, Algharaibeh MA (2000) Plant uptake response to metals and nitrate in simulated uranium mill tailings contaminated groundwater. Water Air Soil Pollut 118:115–129

Dickinson NM, Pulford ID (2005) Cadmium phytoextraction using short-rotation coppice Salix: the evidence trail. Environ Int 31:609–613

Dreesen DR, Cokal EJ (1984) Plant uptake assay t determine bioavailability of inorganic contaminants. Water Air Soil Pollut 22:85–93

Dyer DB (2003) A field guide to bacteria. Comstock Publishing Associates/Cornell University Press, Ithaca/London

Ebbs SD, Brady D, Kochian L (1998) Role of uranium speciation in the uptake and translocation of uranium by plants. J Exp Bot 49:1183–1190

Ebbs S, Brady D, Norvell W, Kochian L (2001) Uranium speciation, plant uptake and phytoremediation. Prac Period Hazard Toxic Radioact Waste Manag 5:130–135

Elless MP, Timpson ME, Lee SY (1997) Concentration of uranium particulates from soils using a novel density-separation technique. Soil Sci Soc Am J 61:626–631

Evangelou MWH, Ebel M, Schnffer A (2006) Evaluation of the effect of small organic acids on phytoextraction of Cu and Pb from soil with tobacco Nicotiana tabacum. Chemosphere 63:996–1004

Flues M, Moraes V, Mazzilli BP (2002) The influence of a coal-fired power plant operation on radionuclide concentrations in soil. J Environ Radioact 63:285–294

Fresquez PR, Armstrong DR, Mullen MA, Naranjo L Jr (1998) The uptake of radionuclides by beans, squash, and corn growing in contaminated alluvial soils at Los Alamos National Laboratory. J Environ Sci Health B: Pest Food Contam Agric Wastes 33:99–122

Fritioff Å, Greger M (2003) Aquatic and terrestrial plant species with potential to remove heavy metals from storm water. Int J Phytorem 5:211–224

Garbisu C, Alkorta I (2001) Phytoextraction: a cost effective plant-based technology for the removal of metals from the environment. Biores Technol 77:229–236

Gisbert C, Ross R, De Haro A, Walker DJ, Bernal MP, Serrano R, Navarro-Avino J (2003) A plant genetically modified that accumulates Pb is especially promising for phytoremediation. Biochem Biophy Res Comm 303:440–445

Grenthe I, Fuger J, Konings RJM, Lemire RJ, Muller AB, Cregu CNT, Wanner H (1992) Chemical thermodynamics of uranium. North Holland, Amsterdam

Gu B, Chen J (2003) Enhanced microbial reduction of Cr (VI) and U(VI) by different natural organic matter fractions. Geochim Cosmochim Acta 67:3575–3582

Gu B, Yan H, Zhou P, Watson DB, Park M, Istok J (2005) Natural humics impact uranium bioreduction and oxidation. Environ Sci Technol 39:5268

Higgo J, Kinniburgh D, Smith B, Tippin E et al (1993) Complexation of cobalt^{2+}, Nickel^2 uranyl and calcium^{2+} by humic substances in groundwaters. Radiochim Acta 61

Hinchman R, Negri C (1994) The grass can be cleaner on the other side of the Fence. Logos Argonne Nat Lab 12:8–11

Huang JW, Blaylock MJ, Kapulnik Y, Ensley BD (1998) Phytoremediation of uranium-contaminated soils: role of organic acids in triggering uranium hyperaccumulation in plants. J Environ Sci Technol 32:13

Kadlec RH (1995) Overview: Surface flow constructed wetlands. Water Sci Technol 32:1–12

Kadlec RH, Knight RL (eds.) (1996) Treatment wetlands. Lewis Publishers, Boca Raton, 893p

Kamal M, Ghaly AE, Mahmood N, Côté R (2004) Phytoaccumulation of heavy metals by aquatic plants. Environ Int 29:1029–1039

Lamas M, Fleckenstein J, Schroetter S, Sparovek RM, Schnug E, Kalra YP (2002) Determination of uranium by means of ICP-QMS. Comm Soil Sci Plant Anal 33:3469–3479

Lambers B, Jackson D, Vandenhove H, Hedemann Jensen P, Smith AD, Bousher A (1999) A common approach to restoration of sites contaminated with enhanced levels of naturally occurring radionuclides. In: Thorne MC (ed.) Proceedings of the international symposium organised by the society for radiological protection, Southport, pp 99, 14–18 June 1999, ISBN 0-7058-1784-9

Langmuir D (1978) Uranium solution-mineral equilibria at low temperatures with applications to sedimentary ore deposits. Geochim Cosmochim Acta 42:547

Lenhart JJ, Cabaniss SE, MacCarthy P, Honeyman BD (2000) Uranium(VI) complexation with citric, humic and fulvic acids. Radiochim Acta 88:345

Li WC, Victor DM, Chakrabarti L (1980) Effect of pH and uranium concentration on interaction of uranium (VI) and uranium(IV) with organic ligands in aqueous solutions. Anal Chem 52:520

Liphadzi MS, Kirkham MB, Mankin KR, Paulsen GM (2003) EDTA-assisted heavy-metal uptake by poplar and sunflower grown at a long-term sewage-sludge farm. Plant Soil 257:171–182

Liu C, Gorby YA, Zachara JM, Fredrickso JK, Brown CF (2002) Reduction kinetics of Fe(III), Co(III), U(VI) Cr(VI) and Tc(VII) in cultures of dissimilatory metal-reducing bacteria. Biotechnol Bioeng 80:637–670

Lotfy SM (2010) Decontamination of soil polluted with heavy metals using plants as determined by nuclear technique. Ph.D. thesis, Soil Science Department, Faculty of Agriculture, Zagazig University, Zagazig

Lovley DR (1993) Dissimilatory metal reduction. Annu Rev Microbiol 47:263

McCutcheon SC, Schnoor JL (eds.) (2003) Phytoremediation – transformation and control of contaminants. Wiley Interscience, Hoboken, pp 985

McGrath SP (1998) Phytoextraction for soil remediation. In: Brooks RR (ed.) Plants that hyperaccumulate heavy metals: their role in archeology, microbiology, mineral exploration, phytomining and phytoremediation. CAB International, Wallingford/Oxon/New York, pp 261–287

McGrath SP, Zhao FJ (2003) Curr Opin Biotechnol 14:277–282

McGrath SP, Zhao FJ, Lombi E, Powlson DS, Bateman GL, Davies-KG, Gaunt-JL, Hirsch-PR, Barlow PW (2002) Plant and rhizosphere processes involved in phytoremediation of metal-contaminated soils. Interactions in the root environment:- an integrated approach. In: Proceedings of the millennium conference on rhizosphere interactions, IACR Rothamsted, UK, 10–12 Apr 2001.2002, 207–214

Moulin V, Tits J, Quaounian G (1992) Actinide speciation in the presence of humic substances in natural water conditions. Radiochim Acta 58:179

Norwell WA (1984) Comparision of chelating agents as extractants for metals in diverse soil materials. Soil Sci Soc Am J 48:1285–1292

Okurut TO, Rijs GBJ, van Bruggen JJA (1999) Design and performance of experimental constructed wetlands in Uganda, planted with *Cyperus papyrus* and *Phragmites mauritianus*. Water Sci Technol 40:265–263

Papp Z, Dezso Z, Daroczy S (2002) Significant radioactive contamination of soil around a coal-fired thermal power plant. J Environ Radioact 59:191–205

Prasad MNV (2001a) Metals in the environment: analysis by biodiversity. Marcel Dekker, New York, pp 504

Prasad MNV (2001b) Bioremediation Potential of Amaranthaceae. In: A.Leeson, EA Foote, MK Banks, VS Magar (eds.) Phytoremediation, wetlands, and sediments, vol 6:165–172. Proceedings of the 6th international In Situ and On-Site bioremediation symposium, Battelle Press, Columbus

Prasad MNV (2004a) Phytoremediation of metals in the environment for sustainable development. Proc Indian Natl Sci Acad 70:71–98

Prasad MNV (ed.) (2004b) Heavy metal stress in plants: from biomolecules to ecosystems. Springer, Heidelberg, pp 462

Prasad MNV (2006a) Stabilization, remediation and integrated management of metal-contaminated ecosystems by grasses (Poaceae). In: Prasad MNV, Sajwan KS, Naidu R (eds.) Trace elements in the environment: biogeochemistry, biotechnology and bioremediation. CRC Press/Taylor & Francis, Boca Raton, pp 405–424

Prasad MNV (2006b) Sunflower (*Helianthus annuus* L.)- a potential crop for environmental industry.In: 1st international symposium on sunflower industrial uses, Faculty of Agriculture, Udine, Italy, 11–13 Sept 2006

Prasad MNV (2007) Aquatic plants for phytotechnology. In: Singh SN, Tripathi RD (eds.) Environmental bioremediation technologies. Springer, Berlin, pp 257–274

Prasad MNV, Freitas H (2003) Metal hyperaccumulation in plants– Biodiversity prospecting for phytoremediation technology. Electronic J Biotechnol 6:275–321

Prasad MNV, Greger M, Smith BN (2001) Aquatic macrophytes. In: Prasad MNV (ed.) Metals in the environment: analysis by biodiversity. Marcel Dekker, New York, p 259

Prasad MNV, Greger M, Aravind P (2006) Biogeochemical cycling of trace elements by aquatic and wetland plants: relevance to phytoremediation. In: Prasad MNV, Sajwan KS, Naidu R (eds.) Trace elements in the environment: biogeochemistry, Biotechnology and Bioremediation. CRC Press/Taylor & Francis, Boca Raton, pp 451–482

Punshon T, Gaines KF, Jenkins RA Jr (2003) Bioavailability and trophic transfer of sediment-bound Ni and U in a southeastern wetland system. Arch Environ Contam Toxicol 44:30–35

Raskin I, Kumar PBNA, Dushenkov S, Salt DE (1997) Bio-concentration of heavy metals by plants. Environ Biotechnol 5:285–290

Reimann C, Koller F, Frengstad B, Kashulina G, Niskavaara H, Englmaier P (2001a) Comparison of the element composition in several plant species and their substrate from a 1500 000-km2 area in Northern Europe. Sci Total Environ 278:87–112

Reimann C, Koller F, Kashulina G, Niskavaara H, Englmaier P (2001b) Influence of extreme pollution on the inorganic chemical composition of some plants. Environ Poll 115:239–252

Ross S (ed.) (1994) Toxic metals in soil-plant systems. Wiley, New York

Salt DE, Blaylock M, Kumar NPBA, Dushenkov V, Ensley BD, Chet I, Raskin I (1995) Phytoremediation – a novel strategy for the removal of toxic metals from the environment using plants. Biotechnol 13:468–474

Salt DE, Smith RD, Raskin I (1998) Phytoremediation. Annu Rev Plant Physiol Plant Mol Biol 49:643–668

Sansone U, Danesi PR, Barbizzi S, Belli M, Campbell M, Gaudino S, Jia G, Ocone R, Pati A, Rosamilia S, Stellato L (2001) Radioecological survey at selected sites hit by depleted uranium ammunition during the 1999 Kosovo conflict. Sci Total Environ 281:23–25

Schmidt U (2003) The effect of chemical soil manipulation on mobility, plant accumulation, and leaching of heavy metals. J Environ Qual 32:1939–1954

Schnoor JL, Licht LL, McCutcheon SC, Wolfe NL, Carreira LH (1995) Phytoremediation of organic and nutrient contaminants. Environ Sci Technol 29:318A–323A

Shahandeh H, Hossner LR (2002a) A role of soil properties in phytoaccumulation of uranium. Water Air Soil Pollut 141:165–180

Shahandeh H, Hossner LR (2002b) Enhancement of uranium phytoaccumulation from contaminated soils. Soil Sci 167:269–280

Sheppard SC, Evenden WG (1992) Concentration enrichment of sparingly soluble contaminants (U, Th and Pb) by erosion and by soil adhesion to plants and skin. J Environ Geochem Health 14:121–131

Soudek P, Podracká E, Vágner M, Van kT, Pet!ík P, Tykva R (2004) ^{226}Ra uptake from soils into different plant species. J Radioanalytical Nucl Chem 262:187–189

Spear JR, Figueroa LA, Honeyman BD (1999) Modeling the removal of uranium U(VI) from aqueous solutions in the presence of sulfate reducing bacteria. Environ Sci Technol 33:2667

Steubing L, Haneke J, Markert B (1993) Higher plants as indicators of uranium occurrence in soil. In: Markert B (ed.) Plants as biomonitors: indicators for heavy metals in the terrestrial environment. VCH, Weinheim/New York, pp 155–165

Truex MJ, Peyton BM, Valentine NB, Gorby YA (1997) Kinetics of U(VI) reduction by a dissimilatory Fe(III)-reducing bacterium under non-growth conditions. Biotechnol Bioeng 55:490

UNEP (2000) NATO confirms to the UN use of depleted uranium during the Kosovo Conflict. Press Release, 21 Mar 2000

UNEP (2001) Depleted uranium in Kosovo, post-conflict environmental assessment. In: UNEP scientific team mission to Kosovo" (5th-19th November 2000). United Nations Environment Programme, Geneva, March 2001

US-EPA/630/P-03/001FMarch (2005) Guidelines for Carcinogen Risk Assessment. Risk Assessment Forum, U.S. Environmental Protection Agency, Washington

Vandenhove H, van M H, van S W, van Hees M, van Winckel S (2001) Feasibility of phytoextraction to clean up low-level uranium-contaminated soil. Int J Phytoremed 3:301–320

Wenger K, Gupta SK, Furrer G, Schulin R (2002) Zinc extraction potential of two common crop plants, *Nicotiana tabacum* and *Zea mays*. Plant Soil 242:217–225

Whicker FW, Hinton TG, Orlandini KA, Clark SB (1999) Uptake of natural and anthropogenic actinides in vegetable crops grown on a contaminated lake bed. J Environ Radioact 45:1–12

Williams JB (2002) Phytoremediation in wetland ecosystems: progress, problems, and potential. Crit Rev Plant Sci 21:607–635

Yukawa M, Watanabe Y, Nishimura Y, Guo Y, Yongru Z, Lu H, Zhan W, Wei L, Tao Z, Rossbach M (1999) Determination of U and Th in soil and plants obtained from a high natural radiation area in China using ICP-MS and gamma-counting. J Anal Chem 363:760–766

Zeh P, Czerwinski KR, Kim JI (1997) Speciation of uranium in Gorleben groundwaters. Radiochim Actav 37:76

Chapter 18
Transgenic Approaches to Improve Phytoremediation of Heavy Metal Polluted Soils

Pavel Kotrba, Martina Mackova, and Tomas Macek

Abstract Use of plants to remediate soil contaminated with heavy metals has received an increasing attention during the last decade. Bioremediation using living plant species, referred to as phytoremediation, covers several different strategies, of which bioremediation employs phytoextraction, rhizofiltration, phytostabilization, and phytovolatilization. High efficiency, low cost, and easy operation make phytoremediation an important alternative to current physicochemical methods. Although, a number of metal-hyperaccumulating plant species have been identified, they have little significance in direct application because of their slow growth, low biomass, and intense interaction with a specific habitat. The phytoremediation potential of plants with well-established agricultural properties and high-biomass yield can be substantially improved by genetic manipulations. The transgenic approaches involve implementation of heterologous metal transporters, centrally important in metal uptake, compartmentalization and/or translocation to organs, improved production of intracellular metal-detoxifying chelators, and (over)production of novel enzymes. Efforts are also being directed to obtain better molecular insights into metallomics and physiology of hyperaccumulating plants, which is

P. Kotrba (✉) • M. Mackova
Department of Biochemistry and Microbiology, Institute of Chemical Technology, Prague, Technická 5, Prague 166 28, Czech Republic
e-mail: pavel.kotrba@vscht.cz

T. Macek
Department of Biochemistry and Microbiology, Institute of Chemical Technology, Prague, Technická 5, Prague 166 28, Czech Republic

IOCB & ICT Joint laboratory, Institute of Organic Chemistry and Biochemistry, Academy of Sciences of the Czech Republic, Flemingovo sq. 2, 166 10 Prague, Czech Republic

M.S. Khan et al. (eds.), *Biomanagement of Metal-Contaminated Soils*,
Environmental Pollution 20, DOI 10.1007/978-94-007-1914-9_18,
© Springer Science+Business Media B.V. 2011

likely to provide candidate genes suitable for phytoremediation. Although substantial progress has been made, further efforts require interdisciplinary approach and, more so, field trials are needed to assess the risk of genetic pollution and underlying economics. Here, we discuss the evidence supporting suitability and prospects of transgenic approaches in phytoremediation of heavy metal-contaminated soils.

Keywords Bioremediation • Decontamination • Genetic engineering • Phytoextraction • Phytovolatilization

18.1 Introduction

Heavy metals, due to their elemental non-degradable nature, when released into the environment pose serious risks to health and ecology. Conventional physicochemical methods for remediation of metal-polluted soils involve chemical extraction, electrolysis, separation of high-metal soil particles by size, or immobilization of metallic species in the soil *in situ* by vitrification or chemical precipitation (Iskandar and Adriano 1997; Page and Page 2002). The high costs (both capital and operational), poor efficiency to remove metals at low concentrations, and significant alteration in physicochemical properties of the soils following application are some of the disadvantages of these processes. Phytoremediation, the use of plants to clean up sites with shallow, low to moderate levels of inorganic or organic contaminants, on the other hand, has however, gained increasing attention recently (Eapen et al. 2007; Macek et al. 2008; Doty, 2008; Vangronsveld et al. 2009; Kotrba et al. 2009; Aken et al. 2010). Phytoremediation is both a growing science and a growing eco-friendly industry. This technique can be used along with or, in some cases, in place of physicochemical cleanup methods. Several estimates on the costs for remediating contaminated sites have shown that plants could, in many cases, do that same job as a group of engineers for one tenth of the cost. Use of plants for decontamination have some beneficial features: solar-energy-driven production of high biomass; plants can be sown, watered, and harvested with relatively low input; the capacity to reduce the spread of pollutants through water and wind erosion; storage of the harvested plants as hazardous waste is seldom required and when needed is less demanding than traditional disposal techniques; it has public acceptance and is an aesthetically pleasant method. Several mechanisms may be involved in the direct and indirect action of phytoremediation at metal-contaminated sites (for detail see Chap. 3).

An inherent capacity to efficiently accumulate metals in harvestable aboveground tissues is particularly important in phytoextraction (Chaney 1983) approach. Several studies indicated that heavy metals could be divided into three categories based on their propensity to be translocated to plant shoots: Mn, Zn, Cd, and Mo are readily translocated to the shoots; Ni, Co, and Cu, are intermediate; and Cr, Pb, and Hg are translocated to the lowest extent (Alloway 1995). The natural capacity of some plant species to accumulate heavy metals at large quantities (Table 18.1) has

Table 18.1 Some examples of heavy metal-hyperaccumulating plants

Species	Maximum metal concentration (mg kg^{-1} dry weight basis)					Location
	Ni	Zn	Pb	Co	Cu	
Aeollanthus subacaulis var. Linearisa				5,176	13,700	Central Africa
Agrostis tenuis			13,490			UK
Allyssum sp.	1,280–29,400					Europe, Turkey, Japan, USA
Anisopappus davyi				2,889	3,504	Central Africa
Arabidopsis halleri		13,620	2,740			Europe
Ariadne shafer	13,070–22,360					Cuba
Berkheya coddii	11,600					South Africa
Bornmuellera sp.	11,400–31,200					Greece, Albania, Turkey
Bulbostylis pseudoperennis				2,127	7,783	Central Africa
Buxus sp.	1,320–25,420					Cuba
Dichapetalum gelonioides		30,000				Sumatra, Mindanao, Sabah
Festuca ovina			11,750			UK
Haumaniastrum robertii				10,230	2,070	Central Africa
Homalium sp.	1,160–14,500					New Caledonia
Leucocroton sp.	2,260–27,240					Cuba
Pandiaka metallorum				2,131	6,270	Central Africa
Peltaria emarginata	34,400					Greece
Pentacalia sp.	16,600					Cuba
Phyllanthus sp.	1,090–60,170					New Caledonia, Cuba, Philippines, Indonesia
Rumex acetosa		11,000	5,450			Northern hemisphere
Streptanthus polygaloides	14,800					USA
Thlaspi sp.	2,000–31 000	15,300–43,710	1,210–8,200			Europe, Turkey, Albania, Syria, Iraq
Vigna dolomitica					3,000	Central Africa
Viola calaminaria		10,000				Belgium, Germany

Adapted from Reeves (2006)

sparked the interest of plant physiologists, ecologists, and evolutionary biologists for over 50 years. The term hyperaccumulation, referring to abnormal levels of Ni in the tree *Sebertia acuminata*, was introduced by Jaffre et al. (1976). The plants are considered as hyperaccumulating if they are capable to accumulate at least 100 times higher concentrations of a particular element than other species growing over an underlying substrate with the same characteristics (Brooks 1998). Specifically, the currently accepted concentration limit in shoot tissues of hyperaccumulators on a dry-weight basis is 0.1 wt.% for most metals, except, for zinc (1 wt.%), cadmium (0.01 wt.%), or gold (0.0001 wt.%) (Baker et al. 2000). About 360 plant species worldwide are known to act as Ni hyperaccumulators (Reeves 2006). The plant families most strongly represented are the *Brassicaceae, Euphorbiaceae, Asteraceae, Flacourtiaceae, Buxaceae,* and *Rubiaceae.* Since the discovery of zinc accumulation in certain *Viola* sp. (*Violaceae*) and *Thlaspi* sp. (*Brasisaceae*) in the nineteenth century, other Zn-hyperaccumulating species capable of hyperaccumulating more than 10,000 mgZn kg^{-1}, were described, notably *Arabidopsis halleri.* This plant has colonized calamine soils, which are highly contaminated with Zn, Cd, and Pb as a consequence of industrial activities. In addition, some populations have been reported to contain more than 100 µg g^{-1} of dry biomass of Cd in their leaves. Only few additional plant species, such as *Thlaspi caerulescens* and members of *Salix* genus (Dickinson and Pulford 2005), have been shown to accumulate more than 100 mg of Cd kg^{-1} into their tissue. Lead shows relatively low mobility in soils and into vegetation, which typically contains less than 10 mg of Pb kg^{-1}. Several hyperaccumulating species of *Brassiceae*, *Poaceae,* and *Polygonaceae* families have been reported to contain above 1,000 mg Pb kg^{-1} in shoots. Concentrations of Co and Cu in plants range between 0.03–2 and 5–25 mg kg^{-1}, respectively. The tupelo or black gum of the southeastern United States (*Nyssa sylvatica*) is remarkable in being able to accumulate as much as 845 mg of Co kg^{-1} from pristine soils (Reeves 2006). Extensive screening of many sites of mining and smelting activity throughout Zaire after plant and soil sample collections and analysis, identified 30 hyperaccumulators of Co and 32 of Cu, with 12 species common to both (Table 18.1). The phytoremediation potential of most known hyperaccumulating species is, however, currently rather low because of their slow growth, low biomass, and often tight association with a specific habitat and lack of good agronomic characteristics (Cunnigham et al. 1995; Chaney et al. 2005).

18.2 Ideal Phytoremediation and Genetically Modified Plants

The ideal phytoremediation plants should possess the following characteristics: (1) capacity to tolerate and accumulate metals, (2) ability to produce high and fast-growing biomass, (3) widely distributed highly branched root system, (4) repulsive to herbivores to avoid the escape of accumulated metals to the food chain, (5) must have a wide geographic distribution and be easy to cultivate, and (6) be relatively easy to harvest. In non-hyperaccumulating plants, factors limiting the phytoextraction performance include, limited root uptake and little root-to-shoot translocation of accumulated heavy metals. Chemically enhanced phytoextraction has been shown to overcome these problems (Blaylock et al. 1997; LeCooper et al. 1999). Common crop plants with high biomass can be triggered to accumulate high amount of metals by enhancing the mobility of metal from the roots to the green parts of the plant by adding mobilizing agents when the crop had reached its maximum biomass. Though, this approach results in decontamination of soil, but involves chemical intervention to the soil, thereby causing secondary pollution. In addition, efforts are being made for the genetic manipulation of plants in order to improve their phytoremediation performance. Accumulated knowledge and continuing efforts toward deciphering physiological mechanisms and of the cognate genetic determinants underlying metal accumulation and tolerance provide solid basis for selection of suitable genes to be (over)expressed in high-biomass plants of well-established agriculture (see Sect. 18.3). Some of the high-biomass metallophytes with well-established genetic manipulation procedures eligible for future exploitation include *Brassica juncea*, sunflower (*Helianthus annuus*), yellow poplar (*Liliodendron tulipifera*), and shrub tobacco (*Nicotiana glaucum*) (Eapen and D'Sousa 2005).

Unlike transgenic crops, the issues such as food safety, allergenicity, and labeling (Kok et al. 2008) are not relevant when genetically modified (GM) plants are considered for use in phytoremediation. However, an improved tolerance to toxic metals implemented through genetic engineering would provide GM plants with a selective advantage at contaminated sites, for example, with acquired metallotolerance. Thus, the main risk concerns the gene flow from cultivated plants to wild relatives via cross-pollination. Potential changes in biological diversity due to invasion of privileged GM plants and the effects of GM plants on related soil microorganisms, herbivores, and other organisms along the food chain must be also taken into account. Some risk assessment methods suggest that the danger of entry of metals to food chains through GM accumulator would be low in most cases, because such plants would be in isolated industrial regions, rather than in countryside. The threat of uncontrolled pollination and crossing with the relatives and spreading of seeds could be avoided, if GM plants are harvested before flowering (Linacre et al. 2003). In addition to "physical" barriers, various genetic methods are available that may restrict transgenic flow in a self-maintaining manner. One approach is targeting the heterologous gene into chloroplasts, since chloroplast DNA is maternally inherited, its transmission via pollen occurs rarely (Davison 2005). Use of plastid-specific promoters is desirable to minimize the risk of transfer of a functional heterologous gene to the nucleus, though such danger is only hypothetical. A suitable technique restricting the spread of GM plants by seeds is based on poison/antidote idea and employs lethal ribonuclease barnase of *Bacillus amyloliquefaciens* as poison and protein barstar as antidote (Kuvshikov et al. 2001). To implement poison/antidote pathway, the GM plant is also transformed with the barnase and barstar genes. The barnase gene is controlled by the promoter, which is only active at the time of seed-pod development. Expression of barstar gene is regulated by heat-shock promoter. Correct seed development and germination are possible only when the barstar is produced due to the controlled heating of developing seeds to 40°C. Such conditions are unlikely in the field, making the germination of progeny likely to fail there.

Use of antibiotic resistance genes as a simple method to select for a transformation event is often criticized, although the risk of horizontal antibiotic-resistance transfer from GM plant is essentially negligible (Bennett et al. 2004). The more realistic threat is, however, the mobilization of genes and elements proximal to the gene for antibiotic resistance, which is always also the heterologous gene-of-interest. As genetic determinants of antibiotic resistance are widely distributed in the environment, a potential mechanism of horizontal transfer involving homologous recombination exists. Construction of GM plant to be released into the environment should thus avoid the use of antibiotic resistance markers and employ some of novel markers and screening strategies. Best solution to this problem is the precise deletion of marker gene from a chromosome employing the bacteriophage *cre-lox* or yeast *FLP-FRT* recombination system (Zuo et al. 2002; Gilbertson 2003). This strategy would then render transgenes containing only those heterologous genes, which are to be employed for the phytoremediation job.

18.3   Improving Phytoremediation of Metals Through Genetic Engineering

Prerequisite to the efficient accumulation of metal is its mobilization from soil, efficient metal uptake mechanism, cellular capability to maintain homeostasis of essential metals, and competence to detoxify (over)accumulated metal species (Clemens et al. 2002). As many metallic species exert their toxic effect by induction of reactive oxygen species (ROS) and other free radicals, their elimination is another challenge faced by the cell (Foyer and Noctor 2005).

18.3.1   Molecular Mechanisms of Metal Uptake and Targets to Genetic Manipulations

Essential heavy metals are required by plants for the activity of numerous metalloenzymes and proteins. Some heavy metal ions, are essential for specific metabolic process, but may impair biological equilibrium when over accumulated. Tight control and regulation of essential metal accumulation are thus of central importance, both at organism and cellular level. Uptake of non-essential metals employs the same mechanisms as adopted by essential metals. Unless detoxified, non-essential metal ions may exert their toxic effect at virtually any tissue and cellular concentration. The property of metallophytes to accumulate heavy metal ions in large quantities from metalliferous soils is a consequence of their adaptation. These plants choose accumulation-detoxification pathway, rather than restriction of metal ion entry, which can be regarded to as another adaptation strategy (Callahan et al. 2006). Accordingly, metal (hyper) accumulation requires complex alterations in the plant metal homeostasis network. Though the reason why some plants have evolved the hyperaccumulation phenotype is not clear, it has been suggested that accumulated metals execute some kind of defense function, poisoning plant tissues for herbivores and pathogens (Boyd 2007).

The actual bioavailability of metal ions in soil is limited, because of their presence in mineral form, formation of hydrous oxides at pH >5, and strong binding to soil components like humic and fulvic acids. In order to solubilize metals for uptake, plants need to interact with the rhizosphere soil. To this end, plant can decrease pH within rhizosphere by H^+ excretion and produce various organic chelators (root exudates), such as carboxylates or phytosiderophores from the mugineic acid family (Fig. 18.1). The soil microflora can increase the bioavailability of metals by several mechanisms, involving excretion of H^+ and carboxylic (e.g., citrate) ligands and redox conversion to mobile forms (Gadd 2007, 2010). Metabolic activities of some microorganisms may, in turn, result in immobilization of metallic species in soil by such mechanisms as organic precipitation with oxalates, inorganic precipitation with carbonates, phosphates or hydroxides, redox immobilization, sorption at cell walls and associated polymeric substances, and bioaccumulation. Following mobilization, the initial contact of the metal ion with root cell involves its adsorption at the cell wall via ion-exchange and chelatation at cellulose, hemicellulose, pectin, and some minor polymers. The transport of heavy metal ions across the plant plasma membrane (Fig. 18.1) is likely to take place through secondary transporters of cation diffusion facilitator (CDF) and natural resistance-associated macrophage protein (Nramp) families. There is a growing number of studies on the plant metal transporters of different families involved in root uptake, metal translocation to other organs, and/or sequestration of metals in organelles (Krämer et al. 2007; Krämer 2010). A common feature underlying the interactions of heavy metal with the components of a biological system is relatively high reactivity of metal ions, mostly due to their ability to form coordination and covalent complexes. In some cases, heavy metal ions also trigger formation of free radicals. Only minute proportions of heavy metals, if any, are thus during their passage through the plant body present as free hydrated ions. Specialized ligands ensure functional deposition of the metal ion in the binding centers of metalloproteins and are intimately involved in management of the storage metal pool. It should be noted that metal–ligand complexes are primary substrates for transporters active in metal translocation to organs or compartmentalization in organelles.

The cysteine-rich metallothioneins (MTs) are intracellular ligands capable of tight coordination of heavy metal ions via cysteine residues shared along the peptide sequence in Cys-X-Cys or Cys-Cys motifs (X represents any amino acid). Peptides of MT family have been identified in plants, animals, eukaryotic microorganisms, and certain prokaryotes. Most of plant MTs consist of about 63–85 amino acids with two terminal cysteine-rich domains separated by a central region without any cys residues (Freisinger 2008). The plant MTs play a role in the homeostasis of essential heavy metals and the transcription of their genes is controlled by signals instrumental during germination, organ development, and senescence (Kotrba et al. 1999; Cobbett and Goldsbrough 2002; Clemens 2006). Mammalian and certain

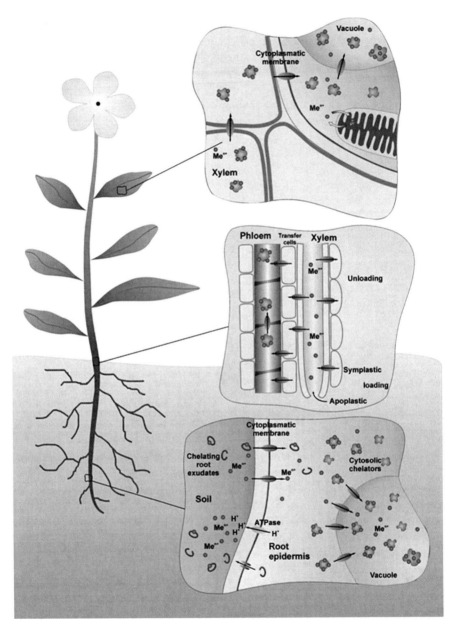

Fig. 18.1 Molecular events proposed for the (hyper)accumulation and detoxification of metals in plants. *Bottom panel*: Mobilization of metals in the rhizosphere via acidic or chelating root exudates. *Middle panel*: Root-to-shoot translocation of metals, either as hydrated ions or metal–ligand complexes occurring via the xylem. *Upper panel*: Translocated metals reaching the leaf apoplast are then captured in different cell types, moving cell-to-cell through plasmodesmata

fungal MTs are, besides their role in homeostasis, responsible for intracellular binding of toxic heavy metal ions. Sequestration of metal species by MTs is the principal mechanisms sustaining tolerance to a particular heavy metal ion in these organisms (Vasák 2005). In yeast S*accharomyces cerevisiae*, 12 cysteine residues of CUP1, a 53 amino-acid MT variant, form eight binding centers for monovalent, and four binding centers for divalent heavy metal ions.

Intracellular detoxification of most heavy metal ions by plants and certain yeasts relies on phytochelatins (PCs). These peptides of general structure $(\gamma\text{-Glu-Cys})_n X$ (PCn; $n = 2-11$; X represents Gly, Ser, β-Ala, Glu, Gln, or no residue) tightly sequester multiple metal ions in metal–thiolate complexes, rendering them inactive in cellular processes. Low-molecular-weight (2–4 kDa) metal–PC complexes, formed in cytosol, could further be transported to vacuoles, which serve as cellular sink for toxic metal species (Fig. 18.1). In this compartment, an inorganic sulfide ion may be incorporated to convert the complex to immobile 6 to 9 kDa high-molecular-weight complex of metal sulfide crystallites covered with PC (Cobbett and Goldsbrough 2002; Clemens 2006). Alternatively, the acidic vacuolar sap may promote dissociation of metal ions from PCs and metals are then complexed there by carboxylic acids or phytate. Phytochelatins can be synthesized upon exposure to heavy metal ions, as well as to some metalloids. The biosynthesis of PCs via transpeptidation reaction from glutathione (γ-glutamylcysteinylglycine, GSH) or its homologues (*iso*-PCs) is catalyzed by the constitutive PC synthase (PCS) in a metal-dependent manner. Glutathione (GSH) and its homologues also act as a fundamental antioxidant molecule. Glutathione directly eliminates reactive oxygen radicals induced by heavy metals in cells (Schutzendubel and Polle 2002) and provides reducing equivalents in the ascorbate-glutathione antioxidation cycle to maintain redox homeostasis (Foyer and Noctor 2005). In yeast, *S. cerevisiae* (Li et al. 1997), and in some ectomycorrhizal fungi (Bellion et al. 2006) that do not produce PCs, cellular detoxification of Cd^{2+} depends upon exclusion of the metal into vacuoles. The metal transport is then effective on the bis(glutathionato)Cd complex. The differential Cd^{2+} stress-dependent expression of homologues of the respective yeast vacuolar ATP-dependent ABC-type transporter YCF1 has been reported in *A. thaliana* (Bovet et al. 2005). In this plant, GSH also appears to play a role in Cd^{2+} sequestration in the mitochondria and bis(glutathionato)Cd conjugates, transported via ABC transporter AtAMT3 into cytoplasm, become substrates for PC synthesis (Kim et al. 2006). The glutathionato-metal or metalloid conjugates also seem to be involved in root-to-shoot translocation of Cd^{2+} in *Brassica napus* (Mendoza-Cózatl et al. 2008) and Hg^{2+} in *A. thaliana* (Li et al. 2006).

Translocation of metal to aboveground organs involves its passage from root symplast to xylem apoplast (Fig. 18.1). Here, we refer to the symplast-apoplast concept, which considers that all the cells of a higher plants are connected, forming symplast. Continuous semipermeable membrane then separates the symplast from the apoplast, the nonliving parts of the plant tissue (cell walls, xylem, and intercellular space). The passage of metal ions into xylem occurs via specific membrane transporters and is generally tightly regulated. While chelatation of heavy metal ions by PCs or MTs is thought to route predominantly to root sequestration, other ligands, such as citrate and nicotianamine, target metals to xylem sap via primary transport mediated by transporters of P-type ATPase family (Krämer et al. 2007). Nevertheless, the possible redistribution of metal–PC complexes within the plant body via a symplasmic or apoplasmic passage, initiated by their export via ATP-binding cassette (ABC) transporters, has been reported (Bovet et al. 2005). The notion that PCs can be involved in long-distance metal transport via symplasmic passage is further supported by the high PC content, compared to xylem, four times higher Cd^{2+} levels in the phloem sap of Cd^{2+}-exposed rapeseed B. *napus* (Mendoza-Cózatl et al. 2008).

Heavy metal ions (or complexes) that reach the apoplast of leaves are unloaded from xylem sap by the transporters of P-type ATPase into the leaf symplast, where they are complexed by MTs or PCs. Intracellular ligands, metallochaperones, and organellar transporters ensure delivery of metals to metal-requiring proteins and maintenance of cytosolic metal levels within physiological ranges. Toxic non-essential metals as well as excess essential metals are sequestered in leaf cell vacuoles. It should be noted that less metabolically active trichomes show pronounced capacity to store accumulated metals (Clemens et al. 2002), thereby indicating that these cell types play an important role in storage and detoxification of heavy metals. The above paragraphs define the targets for genetic modifications of plants directed toward the improved phytoextraction of metals from soils and sediments. These lay in such pathways as follows: (1) Mobilization and uptake of metal from the soil. (2) Competence of metal translocation to shoots via symplast or xylem (apoplast), including efficiency of xylem loading. (3) Distribution to aboveground organs and tissues. (4) Sequestration within tissue cells. (5) Expulsion of accumulated metal to less metabolically active cells. Removal of Hg^{2+} (Sect. 18.3.4) as well as of some metalloids (Sect. 18.4) from contaminated soil by phytovolatilization could be achieved on implementation of enzyme activities promoting plants. (6) Capacity to convert metals to volatile species for phytovolatilization. Although deposition of heavy metals in roots is not desirable in phytoextraction strategy, improved metallotolerance in such organ could be of importance during phytostabilization of contaminated soils. Therefore, some efforts should also be directed to improve root sequestration by metal-complex formation and deposition in vacuoles.

18.3.2   Modifications in Metal Transport Across Plasma Membrane

The transport of essential metals or alkali cations across plasma membranes by means of primary and secondary active transporters is of central importance in the metal homeostasis network in all organisms. Relatively broad substrate specificity of transporters makes them a promising tool to improve toxic metal uptake for phytoremediation. The N. *tabacum* plasma membrane transporter NtCBP4 (calmodulin-binding protein), for example, is structurally similar to vertebrate and invertebrate K^+ and to non-selective cation channels (Arazi et al. 1999). Overproduction of entire NtCBP4 or its C-terminal part (calmodulin-binding and putative cyclic nucleotide-binding domains) in *N. tabacum* resulted in 20% increased uptake and translocation of Pb^{2+} to shoots, reflected in the higher sensitivity of transgenes compared to wild type (WT) controls (Sunkar et al. 2000). While 50% inhibition in elongation of WT roots was observed in hydroponic solutions containing 900 μM Pb^{2+}, growth of NtCBP4 roots was inhibited by 50% at 600 μM Pb^{2+}. Intriguingly, same transgenes showed improved tolerance to elevated Ni^{2+} concentrations, which was apparently due to NtCBP4-promoted Ni^{2+}-exclusion by yet unidentified mechanism. For example, as compared to WT controls, NtCBP4 plants produced roots, 70% longer, in media with 100 μM Ni^{2+} and showed 60% reduction in Ni uptake by shoots.

The overexpression of hypothetical plant iron transporters of the Nramp family in model plants *A. thaliana* or *N. tabacum* has been primarily conducted to assess their function in Fe homeostasis (Curie et al. 2000). It was also found that although the overproduction of intrinsic AtNramp3 in *A. thaliana* markedly increased sensitivity of the transgene to Cd^{2+}, this phenotype was not accompanied by increase in Cd^{2+} accumulation (Thomine et al. 2000). The feasibility of using bacterial metal transporters in plants was first demonstrated in *A. thaliana* transformed with *zntA* coding for Zn^{2+}, Cd^{2+}, and Pb^{2+} P1-ATPase responsible for the metal-efflux-based metalloresistance of *E. coli* (Lee et al. 2003). In transformed *A. thaliana,* localized ZntA on plasma membrane reduced the Cd^{2+} accumulation in protoplasts by promoting release of preloaded Cd^{2+}. Overall, ZntA improved the tolerance of the ZntA plants when exposed to 70 μM Cd^{2+} and 700 μM Pb^{2+} (1.8 and 2.6 higher biomass yields, respectively, compared to WT controls) and shoots of transgenics grown at these concentrations showed 70% and 54% decreased content of Cd^{2+} and Pb^{2+} respectively, a desirable feature for crop plants to be safer from heavy metal contamination.

The widespread bacterial Hg^{2+} resistance mechanism, based on the import of Hg^{2+} into cytoplasm and its subsequent reduction to metallic mercury (see also Sects. 18.3.3 and 18.3.4), involves MerC as one of the plasma membrane transporters for the Hg^{2+} uptake step (Silver and Phung 2005). In a model experiment with *merC*-expressing *A. thaliana*, the leaves when excised and submerged into a solution containing 100 μM Hg^{2+} showed more than threefold increased rate of foliar Hg^{2+} accumulation as compared to WT controls (Sasaki et al. 2006). However, MerC Arabidopsis seedlings also acquired a Hg^{2+} hypersensitive phenotype.

18.3.3   *Modifications for Improved Metallotolerance*

18.3.3.1   Improving Phytochelatin and Glutathione Production

Phytochelatins (PCs) are synthesized by phytochelatin synthase (PCS) enzyme from glutathione or its homologues. The constitutive overexpression of TaPCS1, a PCS from wheat, in shrub tobacco *N. glauca* substantially increased its tolerance to Pb^{2+} and Cd^{2+} (Gisbert et al. 2003) and greatly improved accumulation of Cu^{2+}, Zn^{2+},

Pb^{2+}, and Cd^{2+} in shoots (Martínez et al. 2006). The overexpressed gene conferred up to 36 and 9 times more Pb^{2+} and Cd^{2+} accumulation respectively, in shoots of the transgenic line NgTP1 under hydroponic conditions, reflected in the increased accumulation of these metals from mining soil (Table 18.2). The natural capability of fast-growing *N. glauca* to grow at contaminated sites and to accumulate heavy metals (Barazani et al. 2004), as well as its wide geographic distribution, high biomass, deep roots, and resistance to herbivores, makes this plant a promising candidate in phytoremediation efforts.

An original approach to modulate the heavy metal accumulation capability in leguminous plants by engineering root-associated rhizobia was employed by Ike et al. (2007). Rhizobia establish a symbiotic relationship with leguminous plants and form nitrogen fixing nodule that contains more than 10^8 bacterial progenies. When PCS gene *AtPCS1* from *A. thaliana* along with a genetic fusion of four mammalian MT-coding sequences were expressed in *Mesorhizobium huakuii* subsp. *rengei* (strain B3)*, the natural capability of the bacterium to accumulate Cd^{2+} from media containing 30 μM Cd^{2+} increased by 25-fold. The colonization of leguminous milkvetch *Astragalus sinicum* with the B3 strain in rice-paddy soil containing 1 mg of Cd kg^{-1} promoted Cd^{2+} uptake in roots, but not in nodules, by three times. The contribution of these free-living modified rhizobia to the collection of Cd^{2+} in the soil and the subsequent chemotaxis-mediated transport of accumulated metal for uptake in the legume's roots seems possible. Although the enhanced Cd^{2+} accumulation phenotype of the roots was not accompanied by an increased metal translocation to the shoots, such a strategy is likely to be useful in the rhizofiltration or transient phytostabilization of heavy metals in soil.

Although the overexpression of intrinsic *AtPCS1* in *A. thaliana* resulted in 25 times higher levels of the transcript and up to a twofold increased production of PCs, *AtPCS1*-transformed lines paradoxically showed hypersensitivity to Cd^{2+} and Zn^{2+} (Table 18.2) (Lee et al. 2003). Such a phenotype could be attributed to a non-physiological decrease in the intracellular GSH pool due to the synthesis of supraoptimal levels of PCs. Since GSH molecule is involved in many aspects of the plant response to heavy metal ions, many efforts have been directed toward engineering its biosynthesis pathway. Attempts to increase GSH production in plants, by the implementation of enzyme activities involved in its synthesis and recycling, have aimed mainly at the promotion of increased PC levels under metal stress. GSH is synthesized from its constituent amino acids in two sequential, ATP-dependent enzymatic reactions catalyzed by γ-glutamylcysteine synthetase (γ-ECS) and glutathione synthetase (GS), respectively. Constitutive overproduction of the *E. coli gshI* gene and targeting of encoded γ-ECS in plastids in *B. juncea* increased GSH levels in hydroponically grown transformants threefolds (Zhu et al. 1999a). Consequently, the PC2 levels of shoots and PC2, PC3, and PC4 levels in roots of γ-ECS *B. juncea* stressed at 200 μM Cd^{2+} increased, compared to WT plants, by 30%, which resulted in higher Cd^{2+} tolerance and accumulation in shoots (Table 18.2). The effect of cytosolic overexpression of *gshII* encoding GS on Cd^{2+} tolerance and accumulation from a hydroponic solution was less pronounced, although transformed plants stressed at 100 μM Cd^{2+} had 2.3 and 1.7 times higher PC2 compared to WT control (Zhu et al. 1999b).

Table 18.2 Properties of genetically modified plants overproducing enzymes involved in phytochelatine and glutathione pathways

Overexpressed activity	GM plant	Transformed gene	Phenotype as compared to WT controls	Reference
Phytochelatin synthase	N. glauca	TaPCS1 of T. Aestivum	Shoots of transformed line NgTP1 accumulated from polluted soil[a], respectively, 6.0, 3.3, 4.8, 18.2, and 2.6 times more Pb, Cd, Zn, Cu, and Ni.	Martínez et al. (2006) and Gisbert et al. (2003)
	A. thaliana	AtPCS1 of A. Thaliana	4.5 and 3 times reduced root elongation on media with 85 µM Cd^{2+} and 500 µM Zn^{2+}, respectively.	Lee et al. (2003)
Phytochelatin synthase plus glutathione synthetase	A. thaliana	AsPCS1 of A. sativum plus GSH1 of S. Cerevisiae	Accumulation of Cd from media with 30 ppm Cd^{2+} increased ten times.	Guo et al. (2008)
Glutathione synthetase	A. thaliana	GSH1 of S. Cerevisiae	Accumulation of Cd from media with 30 ppm Cd^{2+} increased four times.	Guo et al. (2008)
	B. juncea	gshII of E. Coli	When grown on polluted soil[b], shoots showed 1.5 higher Cd and Zn levels. On medium with 200 µM Cd^{2+} roots 1.5 times roots. By 20% enhanced Cd^{2+} accumulation from media with 50 µM Cd^{2+}.	Bennett et al. (2003) and Zhu et al. (1999b)
γ-Glutamylcysteine synthetase	B. juncea	gshI of E. Coli	When grown on polluted soil[b], shoots showed 1.5, 2.0, 2.0 and 3.1 times higher Cd, Zn, Cu, and Pb levels, respectively. By 90% higher shoot Cd levels when grown in media with 50 µM Cd^{2+} and 2.1 times longer roots in media with 200 µM Cd^{2+}.	Bennett et al. (2003) and Zhu et al. (1999a)
	Populus	gshI of E. Coli	Cd accumulation from soil containing 225 ppm Cd higher by 2.5 and 3 times when ECS in cytosol and chloroplasts, respectively.	Koprivova et al. (2002)
Cysteine synthase	N. tabacum	OAS-TL of S. oleracea	On medium with 300 µM Cd^{2+}: 2.5 times higher biomass, 2.8 times longer roots. On medium with 500 µM Ni^{2+} 1.3 times higher biomass, 4.2 times longer roots. When grown in media with 100 µM Cd^{2+}, the Cd levels 1.4 times increased in shoots and four times reduced in roots.	Kawashima et al. (2004)

(continued)

Table 18.2 (continued)

Overexpressed activity	GM plant	Transformed gene	Phenotype as compared to WT controls	Reference
ATP sulfurylase	B. juncea	APS1 of A. thaliana	GM plantlets showed increased tolerance (as root elongation) to Cd^{2+} (2.2 times), Cu^{2+} (by 30%), Hg^{2+} (by 20%) in complex metal(loid) media[c]. From media[d] increased accumulation of Cd^{2+} (1.9 times), VO_4^{3-} (2.5 times), CrO_4^{2-} (1.5 times), WO_4^{2-} (1.7 times), MoO_4^{2-} (1.4 times) in shoots.	Wangeline et al. (2004)

[a]The metal content of soil was 2345 ppm Pb, 0.3 ppm Cd, 1826 ppm Zn, 35 ppm Cu, and 27 ppm Ni

[b]640 ppm Cu, 2400 ppm Pb, 2700 ppm Mn, 7800 ppm Zn, 3.6 ppm Cd

[c]20 ppm Cd^{2+}, 5 ppm Cu^{2+}, 8 ppm Hg^{2+}, 40 ppm Zn^{2+}, 4 ppm AsO_4^{3-}, or 25 ppm AsO_2^-

[d]705 ppm Cd^{2+}, 7 ppm VO_4^{3-}, 4.5 ppm CrO_4^{2-}, 50 ppm WO_4^{2-}, or 40 ppm $Mo_7O_{24}^{6-}$

Benett et al. (2003) further demonstrated that overexpression of *gshI* and *gshII* can indeed multiply the natural potential of *B. juncea* for phytoextraction from polluted soils (Table 18.2). Similar results have also been obtained in hybrid poplar (*Populus tremula* × *P. alba*), in which overproduction of *E. coli* γ-ECS enhanced foliar GSH content two to fourfolds (Arisi et al. 1997) and promoted accumulation of Cd^{2+}, but not of Zn^{2+}, in young leaves (Table 18.2) (Koprivova et al. 2002; Bittsánszky et al. 2005). The significance of glutathione reductase (GR) in Cd^{2+} accumulation and tolerance was recorded in transgenic *B. juncea* overproducing the GR of *E. coli* in the cytosol and plastids (Pilon-Smits et al. 2000). Only plastidic GR overproduction, improving natural GR levels 20- to 50-fold, doubled GSH levels in roots. In contrast to the WT control, the plastidic transformants showed no chlorosis when treated with 100 μM Cd^{2+}; however, the shoot Cd^{2+} accumulation was only a half of that of control WT plants. The overproduction of PCs followed by an exhaustion of the GSH pool in *A. thaliana,* however, had a negative impact on the ability of transgenes to tolerate and accumulate Cd^{2+} (Lee et al. 2003; Li et al. 2004). This phenotype was converted to the tolerant and accumulating on expression of yeast *GSH1*-encoded GS in *A. thaliana* lineages overproducing AsPCS1 of garlic *Allium sativum* (Guo et al. 2008).

The sulfur assimilatory mechanism and subsequent production of the antioxidant and PC precursor GSH in plants are known to be highly induced by heavy metal exposure. In the respective pathways, the overall rate of GSH biosynthesis and the capacity to maintain an elevated GSH pool is limited by the activity of cysteine synthase (*O*-acetylserine [thiol] lyase, OAS-TL), which substitutes the acetate of *O*-acetyl-L-serine (OAS) with sulfide (Barroso et al. 1995; Meyer and Fricker 2002). Indeed, constitutive overexpression of *Atcys-3A* encoding intrinsic OAS-TL in *A. thaliana* increased intracellular cysteine and GSH levels, allowing transgenes to survive at 400 μM Cd^{2+} stress (Domínguez-Solís et al. 2004). Over a 14-day period, OAS-TL Arabidopsis accumulated 72% more metal than WT control plants from a medium containing 250 μM Cd^{2+}, the highest Cd^{2+} content being detected in the trichomes. Kawashima et al. (2004) reported a substantial improvement in Cd^{2+} and Ni^{2+} tolerance in *N. tabacum* overproducing OAS-TL from spinach (*Spinacia oleracea*). The authors also determined the Cd^{2+} accumulation potential of the best performing transgenic line and found that the Cd^{2+} concentration was reduced in root but slightly increased in shoots compared to the WT control (Table 18.2), indicating the onset of promoted metal translocation. Moreover, due to highly improved biomass yields on media with 100 μM Cd^{2+}, shoots of a 3-week-old transgenic plant accumulated 2.8 times higher amount of metal than shoots of a WT plant.

Improved supply of *O*-acetyl-L-serine (OAS) to the OAS-TL enzyme has also been shown as an effective method to increase the rate and yield of GSH synthesis. OAS synthesis from L-serine and acetyl-Co-A is catalyzed by serin-*O*-acetyltransferase (SAT). Overproduction of mitochondrial SAT encoded by *TgSATm* of *Thlaspi goesingense* promoted accumulation of GSH in leaves of *A. thaliana*, providing increased tolerance to Ni^{2+}, Co^{2+}, Zn^{2+}, and Cd^{2+}, attributed mainly to the acquired advantage of an improved antioxidative defense potential (Freeman and Salt 2007). In cysteine biosynthesis, inorganic sulfate after uptake is activated by ATP sulfurylase to form adenosine phosphosulfate (APS), which is subsequently reduced to free sulfide by APS reductase. While measuring the effect of ATP sulfurylase overproduction on the accumulation of 12 metal and metalloid cations and oxyanions, Wangeline et al. (2004) observed that the expression of the *APS1* gene of *A. thaliana* in *B. juncea* seedlings markedly contributed to both tolerance and accumulation of certain metal and metalloid species (Table 18.2). Although they did not pinpoint the mechanisms behind the observed phenotypes, it seems likely that the oxyanions MoO_4^{2-}, CrO_4^{2-}, WO_4^{2-} could be, as are sulfate analogues (Leustek, 1996), accumulated via sulfate permease, upregulated on virtual sulfate starvation caused by the removal of free sulfate by the overexpressed enzyme. The higher tolerance and accumulation of cations could be attributed to the ATP sulfurylase-promoted increase in GSH levels (Pilon-Smits et al. 1999). In this study, transgenic APS1 *B. juncea* exhibited a two times increase in ATP sulfurylase activity and GSH contents, both in roots and shoots.

18.3.3.2 Promoting Sequestration of Metals in Vacuoles

Subcellular sequestration of metal ions may, besides chemical complexation via thiol-containing biomolecules, involve ion (or complex) transport into vacuoles as the final metabolically inactive sink. Manipulation of vacuolar exchange activity in *N. tabacum* by the overproduction of the metal ion/H^+ antiporters CAX2 and CAX4 (calcium exchanger 2 and 4) of *A. thaliana* provided transgenic plants the ability to efficiently detoxify Cd^{2+}, Zn^{2+}, and Mn^{2+} (Hirchi et al. 2000; Korenkov et al. 2007a,b). The CAX2 or CAX4 plants showed an improved uptake of metal ions in the roots but not in shoots, which accumulated 70–80% less metals than the roots. However, the net metal uptake was due to the acquired metal tolerance and markedly improved aboveground biomass yields (Table 18.3). Similarly, overexpression of the intrinsic vacuolar ZAT Zn^{2+} transporter (homologous to bacterial metal ion/H^+ ion exchangers of the cation diffusion facilitator CDF family [Silver and Phung 2005] increased the Zn^{2+} tolerance and accumulation in roots of *A. thaliana* (van der Zaal et al. 1999). In other organism like in yeast *S. cerevisiae*, Cd^{2+} is detoxified by transport of cytosolic (glutathione)$_2$Cd complex to vacuoles by ABC-type YCF1 transporter (Li et al. 1997). Accordingly, heterologous expression of *YCF1* gene in *A. thaliana* rendered transgene with an enhanced tolerance to Cd^{2+} and Pb^{2+} (Song et al. 2003). Quite surprisingly, the YCF1 plant also efficiently translocated these metals to shoots (Table 18.3).

Although the overexpression of the mammalian *hMRP1* gene, encoding a different type of the ABC-type multidrug resistance-associated transporter, did not alter Cd^{2+} accumulation in the organs of *N. tabacum*, transgenes showed improved Cd^{2+} tolerance compared to WT controls, manifested by the continuous growth of transgene plantlets, reduced chlorosis, and a 25% faster root elongation on media containing 100–240-μM Cd^{2+} (Yazaki et al. 2006). Mammalian ATP-binding cassette (ABC) transporters involved in the multidrug resistance of cancer cells can efflux cytotoxic compounds that show a wide variety of chemical structures and biological

Table 18.3 Properties of genetically modified plants overproducing vacuolar metal transporters

Expressed transporter	GM plant	Transformed gene	Phenotype as compared to WT controls	Reference
Metal ion/H+ ion exchangers	N. tabacum	CAX2 of A. thaliana	Amount of metal accumulated per plant growing on media with 3 μM Cd^{2+}, 500 μM Mn^{2+}, and 150 μM Zn^{2+} was higher 3.4, 2.3, and 1.9 times, respectively.	Korenkov et al. (2007b)
	N. tabacum	CAX4 of A. thaliana	Amount of metal accumulated per plant growing on media with 3 μM Cd^{2+}, 500 μM Mn^{2+}, and 150 μM Zn^{2+} was higher 2.4, 2.8, and 2.2 times, respectively.	Korenkov et al. (2007b)
	A. thaliana	ZAT of A. thaliana	Six times longer roots and 2.3 times higher Zn levels in roots when grown in hydroponic solution with 200 μM Zn^{2+}.	van der Zaal et al. (1999)
ABC transporter	A. thaliana	YCF1 of S. cerevisiae	2.2 and 1.8 times higher biomass when grown on media with 60 μM Cd^{2+} and 900 μM Pb^{2+}, respectively. Shoots accumulated 1.5 times higher metal levels from media with 70 μM Cd^{2+} or 750 μM Pb^{2+}.	Song et al. (2003)

activities. Human multidrug resistance-associated protein (hMRP1) is one of the most intensively studied ABC transporters and many substrates have been identified, including both organic and inorganic compounds (Zhou et al. 2008). Interestingly, in mammals, members of the MRP family are found in plasma membrane, while in *N. tabacum*, hMRP1 is localized in vacuolar tonoplasts. Besides detoxification of Cd^{2+}, presumably transported to vacuoles as glutathione-Cd conjugate, hMRP1 also conferred vacuolar uptake and resistance to model organic xenobiotic daunorubicin, an anthracycline-type DNA-intercalating drug, suggesting that MRP transporters could be beneficial in constructing plants for the remediation of a complex polluted environment.

18.3.3.3 Modifications with Heterologous Metal Ligands

Overproduction of recombinant MTs to enhance metalloresistance and to support metal accumulation in plants has been the first strategy considered for the construction

Table 18.4 Properties of genetically modified plants overproducing metallothioneins

GM plant	Transformed gene	Phenotype as compared to WT controls	Reference
N. tabacum	MT-I of Mus musculus	Cd^{2+} tolerance enhanced from 10 to 200 μM.	Pan et al. (1994)
	MT-II of Homo sapiens	Accumulation from soil containing 0.2 mg of Cd kg^{-1} reduced by 73%. Tolerated up to 100 μM Cd^{2+} at seedling stage.	de Borne et al. (1998), Misra and Gedamu (1989)
	CUP1 of S. cerevisiae	2–3 times higher Cu^{2+} accumulation from soil containing 1645 mg of Cu kg^{-1}.	Thomas et al. (2003)
	HisCUP1 (recombinant hexahistidine vision to CUP1)	GM plant tolerated up to 16.2 mg of Cd kg^{-1} in sandy soil. By 75–90% higher Cd^{2+} accumulation from sandy soil with 0.2 mg of Cd kg^{-1} and from humus soil with 0.4 mg of Cd kg^{-1}.	Pavlikova et al. (2004) and Macek et al. (2002)
B. oleracea	CUP1 of S. cerevisiae	Cd^{2+} tolerance enhanced from 25 μM to 400 μM in hydroponic medium.	Hasegawa et al. (1997)

of plants suitable in phytoremediation process. This approach, applied in several laboratories, has resulted in different phenotypes (Table 18.4). Although the constitutive expression of genes encoding mouse MT-1, human hMT-1A and h-MT-II, Chinese hamster MT-II, and yeast CUP1 in tobacco, cabbage, *Brassica oleracea,* and *A. thaliana* markedly enhanced Cd^{2+} resistance, the transgenic plants showed a 20–70% reduction in metal accumulation in the shoots (Eapen a D'Souza 2005). On the other hand, production of CUP1 significantly promoted the accumulation of Cu^{2+}, but not of Cd^{2+}, in leaves of *N. tabaccum* (Thomas et al. 2003) (Table 18.4). Increased Cu^{2+} accumulation was also reported for roots of *A. thaliana,* overexpressing the plant MT gene *PsMTA* of pea *Pisum sativum* (Evans et al. 1992). An improved phytoextraction potential for Cd^{2+} from sandy and humus soils was acquired by *N. tabacum* on overexpression of HisCUP1 (Table 18.4), the CUP1 additionally modified with an N-terminal hexahistidine (His) extension (Macek et al. 2002; Pavlikova et al. 2004). The HisCUP produced at levels reaching 10% of cellular cysteine-rich peptides involving glutathione and PCs (Křížková et al. 2007) in transgenic lines also improved Cd^{2+} tolerance. Periplasmic protein MerP is a component of bacterial Hg^{2+} resistance, which is responsible for funneling metal ions to the uptake transporters MerT, MerC, or MerF (Silver and Phung 2005). When overproduced in *A. thaliana,* MerP got localized in the cell membrane and vesicles of plant cells (Hsieh et al. 2009). Unlike the WT control, MerP plants germinated on media with 12.5 μM Hg^{2+} and accumulated 5.35 μg Hg^{2+}/g of fresh seedling weight.

18.3.4   Modifications for Phytovolatilization of Mercury

Phytovolatilization of Hg^{2+} and organomercurial compounds ($R-Hg^+$) involves the accumulation of metal species in GM plant cells and their subsequent conversion to volatile metalic Hg^0, which can be liberated to atmosphere through leaf evaporation. To this end, genetic determinants of widespread bacterial resistance to Hg^{2+} and $R-Hg^+$ are employed, which involve *merA* encoding mercuric ion reductase, which converts Hg^{2+} to non-toxic volatile metalic Hg^0, and *merB* coding for organomercurial lyase, liberating Hg^{2+} from $R-Hg^+$ (Silver and Phung 2005). The main advantage of phytovolatilization is the removal of Hg^{2+} from a site without the need for plant harvesting and disposal. Although, there could be some skepticism regarding the safety of such strategy, safety assessment studies on mercury phytovolatilization have indicated that the advantage of wide dispersion and dilution in the atmosphere and eventually to other environment components outweigh the potential risks (Lin et al. 2000; Moreno et al. 2005). Overexpression of *merA, merB,* or a combination of both, in *A. thaliana* (Bizily et al. 1999, 2003; Yang et al. 2003), *N. tabacum* (He et al. 2001; Ruiz et al. 2003), rice (Heaton et al. 2003), saltmarsh cordgrass *Spartia alterniflora* (Czakó et al. 2006), yellow poplar *L. tulipifera* (Rugh et al. 1998), and cottonwood *Populus deltoides* (Che et al. 2003; Lyyra et al. 2007), resulted in Hg^{2+} and $R-Hg^+$ tolerant phenotypes (Table 18.5). To achieve efficient volatization of mercury, use of modified versions of *merA* optimized for plant codon preferences (*merApe9* and *merA18*) were shown instrumental in achieving efficient production of MerA and pronounced mercury volatization in *A. thaliana*, *N. tabacum*, and *L. tulipifera*. While cytoplasmatic MerB allowed *A. thaliana* plants to grow at fivefolds higher methyl mercury concentrations compared to WT controls, the additional expression of *merApe9* further improved tolerance by a factor of 10 and promoted efficient phenyl mercury removal and Hg^0 volatization from a model solution (Bizily et al. 2000). More than a tenfold higher volatization rate was further achieved by the targeting of MerB in the endoplasmatic reticulum (ER) of *merA/merB* double transformant (Bizily et al. 2003). The likely reason was that ER-localizing MerB exhibited more than a 20 times higher specific activity than in MerB plants with cytoplasmic MerB.

18.4   Plants Genetically Modified for Improved Phytoremediation of Metalloids

Some of genetic modifications conducted so far to enhance PCs and glutathione production have also resulted in improved uptake and/or detoxification of arsenic species. Phytochelatins seems likely to play a role in As detoxification, as arsenite forms tight As(III)-tris-thiolate complexes (Pickering et al. 2000). The tris(glutathionato) As(III) conjugate is a metalloid form for long-distance symplasmic transport in *A. thaliana* (Li et al. 2006). For example, overproduction of PCs in *B. juncea* and

Table 18.5 Properties of GM plants engineered for mercury volatization

Overexpressed activity	GM plant	Transformed gene	Phenotype as compared to WT controls	Reference
Mercuric reductase	A. thaliana	Optimized merA of E. coli (merApe9)	2.5 times higher rate of Hg volatization from hydroponic medium with 5 μM Hg^{2+} (rate of 5 μg Hg^0 [g FW]$^{-1}$ min^{-1}). Germinating on media with 50–100 μM Hg^{2+} (25 μM Hg^{2+} is lethal to germination of WT seeds).	Rugh et al. (1996)
	N. tabacum	Optimized merA of E. coli (merApe9)	Increased rate of Hg volatization from solution with 25 μM Hg^{2+}: nine times by roots (23.8 μg Hg^0 [g FW]$^{-1}$ min^{-1}), 8 times by leaves (6.9 μg Hg^0 g^{-1} min^{-1}), and 5 times by stem (4.1 μg Hg^0 g^{-1} min^{-1}). Germinating on media with 100–350 μM Hg^{2+} (50 μM Hg^{2+} is lethal to germination of WT seeds).	He et al. (2001)
	L. tulipifera	Optimized merA of E. coli (merA18)	Hg volatization from hydroponic media with 10 μM Hg^{2+} improved ten times (rate of 1.2 μg Hg^0 [g FW]$^{-1}$ day^{-1}). Germinating on media with 50 μM Hg^{2+} (25 μM Hg^{2+} is lethal to germination of WT seeds).	Rugh et al. (1998)
Organomercurial lyase plus mercuric reductase	N. tabacum (chloroplast)	merB and merA of E. coli	Doubled biomass yield with seedlings grown on medium with 400 μM phenyl-Hg$^+$.	Ruiz et al. (2003)
		Optimized merB of E. coli (merBpe) and merApe9	Seedlings volatized Hg from solution with 25 μM phenyl-Hg$^+$ at rate of 60 ng Hg^0 [g FW]$^{-1}$ min^{-1}. Germinating on media with 10 μM CH$_3$-Hg$^+$ (1 μM CH$_3$-Hg$^+$ is lethal to germination of WT seeds).	Bizily et al. (2000)
Organomercurial lyase plus mercuric reductase	S. alterniflora	Optimized merB of E. coli (merBpe) and merApe9	Callus culture can tolerate 500 μM Hg^{2+} and 100 μM phenyl-Hg$^+$ (225 μM Hg^{2+} or 50 μM phenyl-Hg$^+$ are lethal to WT callus).	Czakó et al. (2006)
	A. thaliana (MerB to endoplasmatic reticulum)	merB of E. coli plus merApe9	Seedlings volatized Hg from solution with 25 μM phenyl-Hg$^+$ at rate of 760 ng Hg^0 [g FW]$^{-1}$ min^{-1}.	Bizily et al. (2003)

A. thaliana transformed by PCS gene (*AtPCS1*) resulted in a markedly improved resistance to As(V) in media (Li et al. 2004; Gasic and Korban 2007). Simultaneous overproduction of yeast *GSH1*-encoded glutathione synthase and phytochelatin synthase AsPCS1 of *A. sativum* in transgenic *A. thaliana* improved not only arsenite and arsenate tolerance, but also their accumulation (Guo et al. 2008). The toxic nature of arsenate is attributed to its chemical similarity with phosphate, promoting its uptake by the roots via the essential phosphate pathway (Ullrich-Eberius et al. 1989). Dhankher et al. (2002) thus combined two bacterial genes – *gsh1*-encoding-ECS with arsenate reductase gene *arsC* in *A. thaliana* – resulting in a transgene that showed substantially greater tolerance to As(V) and accumulation of As oxyanions in shoots (predominantly as $[glutathione]_3As[III]$) than did the control WT and/or *gsh1*-only-transformed plants (Table 18.6). Most plants appear to have high levels of endogenous root arsenate reductase and arsenate conversion to arsenite sequestered in the roots as As(III)-thiol may prevent the translocation of arsenic species to aboveground tissues (Ramirez-Solis et al. 2004). As such phenotype is not suitable for phytoextraction, Dhankher et al. (2006) more recently employed the RNA interference approach to reduce the arsenate reductase ARS2 activity in *A. thaliana* by 98%. The *ARS2*-knockdown lines retained the ability to grow in a medium with 100 µM arsenate and accumulated 10–16-fold more arsenic species in shoots and up to 40% less in roots than WT controls.

Natural Se hyperaccumulating plants use selenocysteine methyltransferase (SMT) to diminish the misincorporation of selenocysteine (SeCys) and selenomethionine (SeMet) into proteins by decreasing their intracellular levels via a conversion to non-protein amino acid methylselenocysteine (MetSeCys) (Neuhierl et al. 1999). The overexpression of the SMT of Se hyperaccumulating milkvetch *Astragalus bisulcatus* in *A. thaliana* and *B. juncea* (LeDuc et al. 2004) substantially improved the tolerance of transformants to selenate and selenite (Table 18.6). Overall, Se accumulation in shoots was better pronounced with SMT *B. juncea*, which exhibited a threefold higher content of foliar MetSeCys than the WT control. Se accumulation in shoots of *B. juncea* was further promoted on additional implementation of ATP sulfurylase (LeDuc et al. 2006). As MetSeCys can be converted *in planta* to volatile dimethylselenide (DMSe) or dimethyldiselenide (DMDSe), the respective pathway was recognized as an attractive target for genetic modification. Suitability of Se phytovolatization approach is supported by the fact that DMSe has been reported to be 500–700 times less toxic than selenate and selenite in soil (Wilber 1980). As a consequence of increased MetSeCys supply to metabolom of SMT *B. juncea*, this plant had higher DMSe and DMDSe contents than control WT plants (Table 18.6) (LeDuc et al. 2004; Bañuelos et al. 2007). The rate-limiting step in the DMSe formation pathway, the conversion of SeCys to selenocystathionine, is catalyzed by the cystathione-γ-synthase (CGS), which is an enzyme of the physiological methionine pathway with cystathione and homocysteine intermediates (Terry et al. 2000; Van Huysen et al. 2003). The constitutive expression of *CGS1* of *A. thalina* in *B. juncea* resulted in increased DMSe formation and evaporation (Van Huysen et al. 2003). This was accompanied with decreased accumulation of Se species both in roots (by up to 70%) and in shoots (by up to 40%) and thus with improved selenite tolerance.

Table 18.6 GM plants engineered for improved phytoremediation of As and Se

Overexpressed activity	GM plant	Transformed gene	Phenotype as compared to WT controls	Reference
Phytochelatin synthase	A. thaliana	AtPCS1 of A. thaliana	Grown in hydroponic solution with 250–300 µM AsO_4^{3-} produced up to 100 times higher biomass.	Li et al. (2004)
	B. juncea	AtPCS1 of A. thaliana	1.4 times longer roots on media with 500 µM AsO_4^{3-}.	Gasic and Korban (2007)
Phytochelatin synthase plus glutathione synthetase	A. thaliana	AsPCS1 of A. sativum plus GSH1 of S. cerevisiae	Three and four times increased accumulation of As from media with 28 ppm AsO_4^{3-} and AsO_2^-, respectively. On media with 150 µM AsO_4^{3-} or 50 µM AsO_2^- roots 2 times longer.	Guo et al. (2008)
Glutathione synthetase	A. thaliana	GSH1 of S. cerevisiae	2.5 and 4.4 times increased accumulation of As from media with 28 ppm AsO_4^{3-} and AsO_2^-, respectively. No effect on AsO_4^{3-} and AsO_2^- tolerance.	Guo et al. (2008)
Arsenate reductase plus glutathione synthetase	A. thaliana	arsC and gshI of E. coli	Three times higher As accumulation from medium with 125 µM AsO_4^{3-}. Biomass yield from medium with 200 µM AsO_4^{3-} six times higher.	Dhankher et al. (2002)
Cysteine synthase	N. tabacum	OAS-TL of S. oleracea	On medium with 250 µM SeO_4^{2-} 1.5 times higher biomass, 1.8 times longer roots.	Kawashima et al. (2004)
Selenocysteine methyltransferase	B. juncea	SMT of A. bisulcatus	No phytotoxicity of 25 µM SeO_3^{2-} in medium (growth of WT reduced by 97%). Se accumulation from media with 200 µM SeO_4^{2-} and 100 µM SeO_3^{2-} increased four and two times, respectively. DMSe and DMDSe volatization from media with 200 µM SeO_4^{2-} increased three times (10% of Se evaporated as DMDSe).	LeDuc et al. (2004)
ATP sulfurylase plus selenocysteine methyltransferase	B. juncea	APS1 of A. thaliana plus SMT of A. bisulcatu	Se accumulation from media with 200 µM SeO_4^{2-} increased nine times (six times compared to single-transformed APS1 plant).	LeDuc et al. (2006)
cystathione-γ-synthase	B. juncea	CGS1 of A. thaliana	Doubled rate of Se volatization from media with 40 µM SeO_3^{2-} (up to 0.23 g Se [g FW]$^{-1}$ day^{-1}) or 40 µM SeO_4^{2-} (up to 0.3 g Se [g FW]$^{-1}$ day^{-1}).	Van Huysen et al. (2003)

Several lines of evidence suggest the possibility of extending the phytovolatization concept to remediation of arsenic pollution. Many bacteria, fungi, mammals, and some plants employ As(III) methylase (*S*-adenosylmethionine-dependent methyltransferase) to convert arsenite to the gaseous trimethylarsine (TMA) by the mechanism involving cycles of oxidative arsenite methylation and reduction of methyl-arsenate to methyl-arsenite intermediates (Norton et al. 2008; Zhu et al. 2008). Qin et al. (2006, 2009) showed that recombinant production of As(III) methylase from bacterium *Rhodopseudomonas palustris* and the plant-related eukaryotic alga *Cyanidioschyzon merolae* in *E. coli* resulted in production of TMA and improved tolerance to arsenic. These results indicate that (over)expression of single As(III) methylase gene would be sufficient to engineer plants to efficiently produce TMA, which can be volatized from the leaf surface.

18.5 Conclusion

Currently, the approaches employed to develop genetically modified plants suitable for phytoremediation include (a) increasing the number of metal transporters along with modulation of the specificity of the metal uptake system, (b) enhancing intracellular ligand production and the efficiency of metal targeting into vacuoles to keep accumulated metal or metalloid in a safe form without disturbing cellular processes, and (c) biochemical transformation of metal or metalloid to their volatile forms. A substantial progress has been achieved, which has helped to improve the suitability of heterologous and/or promoted intrinsic gene expression for the development of plants useful in phytoremediation. It is generally accepted that hyperaccumulators, when well understood, can be good sources of genes for phytoremediation. One of the major limitations in current efforts addressing phytoextraction of metals and metalloids is, however, the lack of detailed information on the molecular factors governing their translocation. Recent advances in the identification and functional evaluation of metal transporters in model plants *A. thaliana* and *N. tabacum*, and an understanding of the mechanisms and regulation of transport events in (hyper)accumulators *Arabidopsis harlei*, *T. caerulescens* and *B. juncea*, thus offer great promise for the manipulation of suitable plants. Long-distance metal transport would further promote repressed metal deposition in roots and creation of artificial metal sinks in shoots. To this end, specifically decreasing transport into root vacuoles and the expression of engineered cell-wall proteins with high-affinity binding sites for metal deposition in the apoplast of aboveground tissues could be instrumental.

Successful phytoremediation of metal pollution may further involve promoting mobilization of metals in soils and sediments. Increased attention should thus be devoted to modifications that enhance the capacity of plants to secrete metal-complexing exudates such as phytosiderophores and organic acids into the rhizosphere and implementation of the cognate metal-complex transport mechanism. Conversion of immobile metals to their bioavailable forms in soils is largely dependent on the activity of soil microflora, especially in the rhizosphere. An understanding of the complex plant–microbe interactions in the rhizosphere would, thus, further allow for the constructions of GM plants and their microbial symbionts to promote the mobilization of metal species of interest. Genetically modified plants may induce remediation of metal and metalloid polluted soils with obvious benefits, yet some would question their techno-economic perspective and environmental safety. The potential of GM plants should be demonstrated in field phytoremediation trials, some of which have emerged in the last few years (Bañuelos et al. 2005, 2007; Van Huysen et al. 2004). The ecological impact and underlying economics of phytoremediation with transgenics should be carefully evaluated and weighted against known disadvantages of conventional remediation techniques or risks of having the recalcitrant heavy metal or metalloid species in our environment.

Acknowledgements Respective research in the authors' laboratories is supported by the Czech Ministry of Education, Youth and Sports (1M06030) and the Czech Science Foundation (P504/11/0484).

References

Aken BV, Correa PA, Schnoor JL (2010) Phytoremediation of polychlorinated biphenyls: new trends and promises. Environ Sci Technol 44:2767–2776

Alloway BJ (1995) Heavy metals in soils, 2nd edn, Blackie. The University of Reading, Glasgow

Arazi T, Sunkar R, Kaplan B, Fromm H (1999) A tobacco plasma membrane calmodulin-binding transporter confers Ni^{2+} tolerance and Pb^{2+} hypersensitivity in transgenic plants. Plant J 20:171–182

Arisi AC, Noctor G, Foyer CH, Jouanin L (1997) Modification of thiol contents in poplars (*Populus tremula* x *P. alba*) overexpressing enzymes involved in glutathione synthesis. Planta 203: 362–372

Baker A, McGrath S, Reeves R, Smith J (2000) Metal hyperaccumulator plants: a review of the ecology and physiology of a biological resource for phytoremediation of metal polluted soils. In: Terry N, Bañuelos GS (eds.) Phytoremediation of contaminated soil and water. CRC Press, Boca Raton, pp 85–107

Bañuelos G, Terry N, Leduc DL, Pilon-Smits EAH, Mackey B (2005) Field trial of transgenic Indian mustard plants shows enhanced phytoremediation of selenium-contaminated sediment. Environ Sci Technol 39:1771–1777

Bañuelos G, LeDuc DL, Pilon-Smits EAH, Terry N (2007) Transgenic Indian mustard overexpressing selenocysteine lyase or selenocysteine methyltransferase exhibit enhanced potential for selenium phytoremediation under field conditions. Environ Sci Technol 41:599–605

Barazani O, Sathiyamoorthy P, Manandhar U, Vulkan R, Golan-Goldhirsh A (2004) Heavy metal accumulation by *Nicotiana glauca* Graham in a solid waste disposal site. Chemosphere 54:867–872

Barroso C, Vega J, Gotor C (1995) A new member of the cytosolic O-acetylserine(thiol)lyase gene family in *Arabidopsis thaliana*. FEBS Lett 363:1–5

Bellion M, Courbot M, Jacob C, Blaudez D, Chalot M (2006) Extracellular and cellular mechanisms sustaining metal tolerance in ectomycorrhizal fungi. FEMS Microbiol Lett 254:173–181

Bennett LE, Burkhead JL, Hale KL, Terry N, Pilon M, Pilon-Smits EAH (2003) Analysis of transgenic Indian mustard plants for phytoremediation of metal-contaminated mine tailings. J Environ Qual 32:432–440

Bennett PM, Livesey CT, Nathwani D, Reeves DS, Saunders JR, Wise R (2004) An assessment of the risks associated with the use of antibiotic resistance genes in genetically modified plants:

report of the working party of the British society for antimicrobial chemotherapy. J Antimicrob Chemother 53:418–431

Bittsánszky A, Kömives T, Gullner G, Gyulai G, Kiss J, Heszky L, Radimsky L, Rennenberg H (2005) Ability of transgenic poplars with elevated glutathione content to tolerate Zn²⁺ stress. Environ Int 31:251–254

Bizily SP, Rugh CL, Summers AO, Meagher RB (1999) Phytoremediation of methylmercury pollution: merB expression in *Arabidopsis thaliana* confers resistance to organomercurials. Proc Natl Acad Sci USA 96:6808–6813

Bizily SP, Rugh CL, Meagher RB (2000) Phytodetoxification of hazardous organomercurials by genetically engineered plants. Nat Biotechnol 18:213–217

Bizily SP, Kim T, Kandasamy MK, Meagher RB (2003) Subcellular targeting of methylmercury lyase enhances its specific activity for organic mercury detoxification in plants. Plant Physiol 131:463–471

Blaylock MJ, Salt DE, Dushenkov S, Zakharova O, Gussman C, Kapulnik Y, Ensley BD, Raskin I (1997) Enhanced accumulation of Pb in Indian mustard by soil-applied chelating agents. Environ Sci Technol 31:860–865

Bovet L, Feller U, Martinoia E (2005) Possible involvement of plant ABC transporters in cadmium detoxification: a cDNA sub-microarray approach. Environ Int 31:263–267

Boyd R (2007) The defense hypothesis of elemental hyperaccumulation: status, challenges and new directions. Plant Soil 293:153–176

Brooks RR (1998) General introduction. In: Brooks RR (ed.) Plants that hyperaccumulate heavy metals. CABI, Wallingford, pp 1–14

Callahan DL, Baker AJ, Kolev SD, Wedd AG (2006) Metal ion ligands in hyperaccumulating plants. J Biol Inorg Chem 11:2–12

Chaney RL (1983) Plant uptake of inorganic waste constituents. In: Parr JF, Marsh PB, Kla JM (eds.) Land treatment of razardous wastes. Noyes Data Corporation, Park Ridge, pp 50–77

Chaney RL, Angle JS, McIntosh MS, Reeves RD, Li YM, Brewer EP, Chen K-Y, Rosenberg RJ, Perner H, Synkowski EC, Broadhurst CL, Wang S, Baker AJM (2005) Using hyperaccumulator plants to phytoextract soil Ni and Cd. Z Naturforsch 60:190–198

Che D, Meagher RB, Heaton ACP, Lima A, Rugh CL, Merkle SA (2003) Expression of mercuric ion reductase in eastern cottonwood (*Populus deltoides*) confers mercuric ion reduction and resistance. Plant Biotechnol J 1:311–319

Clemens S (2006) Toxic metal accumulation, responses to exposure and mechanisms of tolerance in plants. Biochimie 88:1707–1719

Clemens S, Palmgren M, Krämer U (2002) A long way ahead: understanding and engineering plant metal accumulation. Trends Plant Sci 7:309–315

Cobbett C, Goldsbrough P (2002) Phytochelatins and metallothioneins: roles in heavy metal detoxification and homeostasis. Ann Rev Plant Biol 53:159–182

Cunningham SC, Berti WR, Huang JW (1995) Phytoremediation of contaminated soils. Trends Biotechnol 13:393–397

Curie C, Alonso JM, Le Jean M, Ecker JR, Briat JF (2000) Involvement of Nramp1 from *Arabidopsis thaliana* in iron transport. Biochem J 347:7497–7455

Czakó M, Feng X, He Y, Liang D, Márton L (2006) Transgenic *Spartina alterniflora* for phytoremediation. Environ Geochem Health 28:103–110

Davison J (2005) Risk mitigation of genetically modified bacteria and plants designed for bioremediation. J Ind Microbiol Biotechnol 32:639–650

de Borne FD, Elmayan T, de Roton C, de Hys L, Tepfer M (1998) Cadmium partitioning in transgenic tobacco plants expressing mammalian metallothionein gene. Mol Breed 4:83–90

Dhankher OP, Li Y, Rosen BP, Shi J, Salt D, Senecoff JF, Sashti NA, Meagher RB (2002) Engineering tolerance and hyperaccumulation of arsenic in plants by combining arsenate reductase and gamma-glutamylcysteine synthetase expression. Nat Biotechnol 20:1140–1145

Dhankher OP, Rosen BP, McKinney EC, Meagher RB (2006) Hyperaccumulation of arsenic in the shoots of *Arabidopsis* silenced for arsenate reductase (ACR2). Proc Natl Acad Sci USA 103:5413–5418

Dickinson NM, Pulford ID (2005) Cadmium phytoextraction using short-rotation coppice *Salix*: the evidence trail. Environ Int 31:609–613

Domínguez-Solís JR, López-Martín MC, Ager FJ, Ynsa MD, Romero LC, Gotor C (2004) Increased cysteine availability is essential for cadmium tolerance and accumulation in *Arabidopsis thaliana*. Plant Biotechnol J 2:469–476

Doty SL (2008) Enhancing phytoremediation through the use of transgenics and endophytes. New Phytol 179:318–333

Eapen S, D'Souza S (2005) Prospects of genetic engineering of plants for phytoremediation of toxic metals. Biotechnol Adv 23:97–114

Eapen S, Singh S, D'Souza S (2007) Advances in development of transgenic plants for remediation of xenobiotic pollutants. Biotechnol Adv 25:442–451

Evans K, Gatehouse J, Lindsay W, Shi J, Tommey A, Robinson N (1992) Expression of the pea metallothionein-like gene *PsMTa* in *Escherichia coli* and *Arabidopsis thaliana* and analysis of trace metal ion accumulation: implications for PsMTa function. Plant Mol Biol 20:1019–1028

Foyer CH, Noctor G (2005) Redox homeostasis and antioxidant signaling: a metabolic interface between stress perception and physiological responses. Plant Cell 17:1866–1875

Freeman JL, Salt DE (2007) The metal tolerance profile of *Thlaspi goesingense* is mimicked in *Arabidopsis thaliana* heterologously expressing serine acetyl-transferase. BMC Plant Biol 7:63

Freisinger E (2008) Plant MTs-long neglected members of the metallothionein superfamily. Dalton Trans 47:6663–6675

Gadd GM (2007) Geomycology: biogeochemical transformations of rocks, minerals, metals and radionuclides by fungi, bioweathering and bioremediation. Mycol Res 111:3–49

Gadd GM (2010) Metals, minerals and microbes: geo- microbiology and bioremediation. Microbio 156:609–643

Gasic K, Korban SS (2007) Transgenic Indian mustard (*Brassica juncea*) plants expressing an *Arabidopsis* phytochelatin synthase (*AtPCS1*) exhibit enhanced As and Cd tolerance. Plant Mol Biol 64:361–369

Gilbertson L (2003) Cre-*lox* recombination: Cre-ative tools for plant biotechnology. Trends Biotechnol 21:550–555

Gisbert C, Ros R, De Haro A, Walker DJ, Pilar Bernal M, Serrano R, Navarro-Aviñó J (2003) A plant genetically modified that accumulates Pb is especially promising for phytoremediation. Biochem Biophys Res Comm 303:440–445

Guo J, Dai X, Xu W, Ma M (2008) Overexpressing *gsh1* and *AsPCS1* simultaneously increases the tolerance and accumulation of cadmium and arsenic in *Arabidopsis thaliana*. Chemosphere 72:1020–1026

Hasegawa I, Terada E, Sunairi M, Wakita H, Shinmachi F, Noguchi A, Nakajima M, Yazaki J (1997) Genetic improvement of heavy metal tolerance in plants by transfer of the yeast metal-lothionein gene (CUP1). Plant Soil 196:277–281

He YK, Sun JG, Feng XZ, Czakó M, Márton L (2001) Differential mercury volatilization by tobacco organs expressing a modified bacterial *merA* gene. Cell Res 11:231–236

Heaton ACP, Rugh CL, Kim T, Wang NJ, Meagher RB (2003) Toward detoxifying mercury-polluted aquatic sediments with rice genetically engineered for mercury resistance. Environ Toxicol Chem 22:2940–2947

Hirschi KD, Korenkov VD, Wilganowski NL, Wagner GJ (2000) Expression of Arabidopsis CAX2 in tobacco. Altered metal accumulation and increased manganese tolerance. Plant Physiol 124:125–133

Hsieh J, Chen C, Chiu M, Chein M, Chang J, Endo G, Huang GG (2009) Expressing a bacterial mercuric ion binding protein in plant for phytoremediation of heavy metals. J Hazard Mater 161:920–925

Ike A, Sriprang R, Ono H, Murooka Y, Yamashita M (2007) Bioremediation of cadmium contaminated soil using symbiosis between leguminous plant and recombinant rhizobia with the *MTL4* and the *PCS* genes. Chemosphere 66:1670–1676

Iskandar IK, Adriano DC (1997) Remediation of soils contaminated with metals – a review of current practices in the USA. In: Iskandar IK, Adriano DC (eds.) Remediation of soils contaminated with metals. Science Reviews, Northwood, pp 1–16

Jaffre T, Brooks RR, Lee J, Reeves RD (1976) *Sebertia acuminata*, a hyperaccumulator of nickel from New Caledonia. Science 193:579–580

Kawashima CG, Noji M, Nakamura M, Ogra Y, Suzuki KT, Saito K (2004) Heavy metal tolerance of transgenic tobacco plants over-expressing cysteine synthase. Biotechnol Lett 26:153–157

Kim D, Bovet L, Kushnir S, Noh EW, Martinoia E, Lee Y (2006) AtATM3 is involved in heavy metal resistance in Arabidopsis. Plant Physiol 140:922–932

Kok EJ, Keijer J, Kleter GA, Kuiper HA (2008) Comparative safety assessment of plant-derived foods. Regul Toxicol Pharmacol 50:98–113

Koprivova A, Kopriva S, Jäger D, Will B, Jouanin L, Rennenberg H (2002) Evaluation of transgenic poplars over-expressing enzymes of glutathione synthesis for phytoremediation of cadmium. Plant Biol 4:664–670

Korenkov V, Park S, Cheng N, Sreevidya C, Lachmansingh J, Morris J, Hirschi K, Wagner GJ (2007a) Enhanced Cd^{2+} -selective root-tonoplast-transport in tobaccos expressing Arabidopsis cation exchangers. Planta 225:403–411

Korenkov V, Hirschi K, Crutchfield JD, Wagner GJ (2007b) Enhancing tonoplast Cd/H antiport activity increases Cd, Zn, and Mn tolerance, and impacts root/shoot Cd partitioning in *Nicotiana tabacum* L. Planta 226:1379–1387

Kotrba P, Macek T, Ruml T (1999) Heavy metal-binding peptides and proteins in plants. a review. Collect Czech Chem C 64:1057–1086

Kotrba P, Najmanova J, Macek T, Ruml T, Mackova M (2009) Genetically modified plants in phytoremediation of heavy metal and metalloid soil and sediment pollution. Biotechnol Adv 27:799–810

Krämer U (2010) Metal hyperaccumulation in plants. Ann Rev Plant Biol 61:517–534

Krämer U, Talke I, Hanikenne M (2007) Transition metal transport. FEBS Lett 581:2263–2272

Křížková S, Diopan V, Baloun J, Šupálková V, Shestisvka V, Kotrba P, Mackova M, Macek T, Kizek R (2007) Electrochemical determination of metalothionein in transgenic tobacco plants. In: Book of proceedings of 4th symposium on biosorption and bioremediation, ICT Prague, Prague, pp.21–23

Kuvshinov V, Koivu K, Kanerva A, Pehu E (2001) Molecular control of transgene escape from genetically modified plants. Plant Sci 160:517–522

LeCooper EM, Sims JT, Cunningham SD, Huang JW, Berti WR (1999) Chelate-assisted phytoextraction of lead from contaminated soils. J Environ Qual 28:1709–1719

LeDuc DL, Tarun AS, Montes-Bayon M, Meija J, Malit MF, Wu CP, AbdelSamie M, Chiang CY, Tagmount A, deSouza M, Neuhierl B, Böck A, Caruso J, Terry N (2004) Overexpression of selenocysteine methyltransferase in Arabidopsis and Indian mustard increases selenium tolerance and accumulation. Plant Physiol 135:377–383

LeDuc DL, AbdelSamie M, Móntes-Bayon M, Wu CP, Reisinger SJ, Terry N (2006) Overexpressing both ATP sulfurylase and selenocysteine methyltransferase enhances selenium phytoremediation traits in Indian mustard. Environ Pollut 144:70–76

Lee S, Moon JS, Ko T, Petros D, Goldsbrough PB, Korban SS (2003) Overexpression of Arabidopsis phytochelatin synthase paradoxically leads to hypersensitivity to cadmium stress. Plant Physiol 131:656–663

Leustek T (1996) Molecular genetics of sulfate assimilation in plants. Physiol Plant 97:411–419

Li ZS, Lu YP, Zhen RG, Szczypka M, Thiele DJ, Rea PA (1997) A new pathway for vacuolar cadmium sequestration in *Saccharomyces cerevisiae*: YCF1-catalyzed transport of bis(glutathionato)cadmium. Proc Natl Acad Sci USA 94:42–47

Li Y, Dhankher OP, Carreira L, Lee D, Chen A, Schroeder JI, Balish RS, Meagher RB (2004) Overexpression of phytochelatin synthase in *Arabidopsis* leads to enhanced arsenic tolerance and cadmium hypersensitivity. Plant Cell Physiol 45:1787–1797

Li Y, Dankher OP, Carreira L, Smith AP, Meagher RB (2006) The shoot-specific expression of γ-glutamylcysteine synthetase directs the long-distance transport of thiol-peptides to roots conferring tolerance to mercury and arsenic. Plant Physiol 141:288–298

Lin Z, Schemenauer R, Cervinka V, Zayed A, Lee A, Terry N (2000) Selenium volatilization from a soil-plant system for the remediation of contaminated water and soil in the San Joaquin valley. J Environ Qual 29:1048–1056

Linacre NA, Whiting SN, Baker AJM, Angle S, Ades PK (2003) Transgenics and phytoremediation: the need for an integrated risk assessment, management, and communication strategy. Int J Phytorem 3:181–185

Lyyra S, Meagher RB, Kim T, Heaton A, Montello P, Balish RS, Merkle SA (2007) Coupling two mercury resistance genes in eastern cottonwood enhances the processing of organomercury. Plant Biotechnol J 5:254–262

Macek T, Macková M, Pavlíková D, Száková J, Truksa M, Cundy A, Kotrba P, Yancey N, Scouten WH (2002) Accumulation of cadmium by transgenic tobacco. Acta Biotechnol 22:101–106

Macek T, Kotrba P, Svatos A, Novakova M, Demnerova K, Mackova M (2008) Novel roles for genetically modified plants in environmental protection. Trends Biotechnol 26:146–152

Martínez M, Bernal P, Almela C, Vélez D, García-Agustín P, Serrano R, Navarro-Aviñó J (2006) An engineered plant that accumulates higher levels of heavy metals than *Thlaspi caerulescens*, with yields of 100 times more biomass in mine soils. Chemosphere 64:478–485

Mendoza-Cózatl DG, Butko E, Springer F, Torpey JW, Komives EA, Kehr J, Schroeder JI (2008) Identification of high levels of phytochelatins, glutathione and cadmium in the phloem sap of Brassica napus. A role for thiol-peptides in the long-distance transport of cadmium and the effect of cadmium on iron translocation. Plant J 54:249–259

Meyer A, Fricker M (2002) Control of demand-driven biosynthesis of glutathione in green Arabidopsis suspension culture cells. Plant Physiol 130:1927–1937

Misra S, Gedamu L (1989) Heavy metal tolerant transgenic *Brassica napus* L. and *Nicotiana tabaccum* L. plants. Theor Appl Genet 78:161–168

Moreno FN, Anderson CWN, Stewart RB, Robinson BH (2005) Mercury volatilisation and phytoextraction from base-metal mine tailings. Environ Pollut 136:341–352

Neuhierl B, Thanbichler M, Lottspeich F, Böck A (1999) A family of S-methylmethionine-dependent thiol/selenol methyltransferases. Role in selenium tolerance and evolutionary relation. J Biol Chem 274:5407–5414

Norton GJ, Lou-Hing DE, Meharg AA, Price AH (2008) Rice-arsenate interactions in hydroponics: whole genome transcriptional analysis. J Exp Bot 59:2267–2276

Page M, Page C (2002) Electroremediation of contaminated soils. J Environ Eng 128:208–219

Pan A, Yang M, Tie F, Li L, Chen Z, Ru B (1994) Expression of mouse metallothionein-I gene confers cadmium resistance in transgenic tobacco plants. Plant Mol Biol 24:341–351

Pavlíková D, Macek T, MacKová M, Száková J, Balík J (2004) Cadmium tolerance and accumulation in transgenic tobacco plants with a yeast metallothionein combined with a polyhistidine tail. Int Biodeter Biodegrad 54:233–237

Pickering IJ, Prince RC, George MJ, Smith RD, George GN, Salt DE (2000) Reduction and coordination of arsenic in Indian mustard. Plant Physiol 122:1171–1177

Pilon-Smits H, Mel Lytle C, Zhu T, Bravo CY, Leustek T, Terry N (1999) Overexpression of ATP sulfurylase in Indian mustard leads to increased selenate uptake, reduction, and tolerance. Plant Physiol 119:123–132

Pilon-Smits EAH, Zhu YL, Sears T, Terry N (2000) Overexpression of glutathione reductase in *Brassica juncea*: effects on cadmium accumulation and tolerance. Physiol Plant 110:455–460

Qin J, Lehrb CR, Yuan C, Le XC, McDermott TR, Rosen BP (2009) Biotransformation of arsenic by Yellowstone thermoacidophillic eukaryotic alga. Proc Natl Acad Sci USA 106:5213–5217

Quin J, Rosen BP, Zhang Y, Wang G, Franke S, Rensing C (2006) Arsenic detoxification and evolution of trimethylarsine gas by a microbial arsenite S-adenosylmethionine methyltransferase. Proc Natl Acad Sci USA 103:2075–2080

Ramírez-Solís A, Mukopadhyay R, Rosen BP, Stemmler TL (2004) Experimental and theoretical characterization of arsenite in water: insights into the coordination environment of As-O. Inorg Chem 43:2954–2959

Reeves RD (2006) Hyperaccumulation of trace elements by plants. In: Morel JL, Echevarria G, Goncharova N (eds.) Phytoremediation of metal contaminated soils, vol 68, IVth edn, Earth and Environmental Sciences. NATO Science Series, Springer, Berlin/Heidelberg/New York, pp 25–52

Rugh CL, Wilde HD, Stack NM, Thompson DM, Summers AO, Meagher RB (1996) Mercuric ion reduction and resistance in transgenic *Arabidopsis thaliana* plants expressing a modified bacterial *merA* gene. Proc Natl Acad Sci USA 93:3182–3187

Rugh CL, Senecoff JF, Meagher RB, Merkle SA (1998) Development of transgenic yellow poplar for mercury phytoremediation. Nat Biotechnol 16:925–928

Ruiz ON, Hussein HS, Terry N, Daniell H (2003) Phytoremediation of organomercurial compounds via chloroplast genetic engineering. Plant Physiol 132:1344–1352

Sasaki Y, Hayakawa T, Inoue C, Miyazaki A, Silver S, Kusano T (2006) Generation of mercury-hyperaccumulating plants through transgenic expression of the bacterial mercury membrane transport protein MerC. Transgenic Res 15:615–625

Schützendübel A, Polle A (2002) Plant responses to abiotic stresses: heavy metal-induced oxidative stress and protection by mycorrhization. J Exp Bot 53:1351–1365

Silver S, Phung L (2005) A bacterial view of the periodic table: genes and proteins for toxic inorganic ions. J Ind Microbiol Biotechnol 32:587–605

Song W, Sohn EJ, Martinoia E, Lee YJ, Yang Y, Jasinski M, Forestier C, Hwang I, Lee Y (2003) Engineering tolerance and accumulation of lead and cadmium in transgenic plants. Nat Biotechnol 21:914–919

Sunkar R, Kaplan B, Bouché N, Arazi T, Dolev D, Talke IN, Maathuis FJ, Sanders D, Bouchez D, Fromm H (2000) Expression of a truncated tobacco NtCBP4 channel in transgenic plants and disruption of the homologous *Arabidopsis* CNGC1 gene confer Pb^{2+} tolerance. Plant J 24:533–542

Terry N, Zayed AM, De Souza MP, Tarun AS (2000) Selenium in higher plants. Annu Rev Plant Phys Plant Mol Biol 51:401–432

Thomas JC, Davies EC, Malick FK, Endreszl C, Williams CR, Abbas M, Petrella S, Swisher K, Perron M, Edwards R, Osenkowski P, Urbanczyk N, Wiesend WN, Murray KS (2003) Yeast metallothionein in transgenic tobacco promotes copper uptake from contaminated soils. Biotechnol Prog 19:273–280

Thomine S, Wang R, Ward JM, Crawford NM, Schroeder JI (2000) Cadmium and iron transport by members of a plant metal transporter family in *Arabidopsis* with homology to *Nramp* genes. Proc Natl Acad Sci USA 97:4991–4996

Ullrich-Eberius CI, Sanz A, Novacky AJ (1989) Evaluation of arsenate- and vanadate-associated changes of electrical membrane potential and phosphate transport in *Lemna gibba* G1. J Exp Bot 40:119–128

Van der Zaal BJ, Neuteboom LW, Pinas JE, Chardonnens AN, Schat H, Verkleij JA, Hooykaas PJ (1999) Over-expression of a novel Arabidopsis gene related to putative zinc-transporter genes from animals can lead to enhanced zinc resistance and accumulation. Plant Physiol 119:1047–1055

Van Huysen T, Abdel-Ghany S, Hale KL, LeDuc D, Terry N, Pilon-Smits EAH (2003) Overexpression of cystathionine-γ-synthase enhances selenium volatilization in *Brassica juncea*. Planta 218:71–78

Van Huysen T, Terry N, Pilon-Smits EAH (2004) Exploring the selenium phytoremediation potential of transgenic Indian mustard overexpressing ATP sulfurylase or cystathionine-gamma-synthase. Int J Phytoremediation 6:111–118

Vangronsveld J, Herzig R, Weyens N, Boulet J, Adriaensen K, Ruttens A, Thewys T, Vassilev A, Meers E, Nehnevajova E, van der Lelie D, Mench M (2009) Phytoremediation of contaminated soils and groundwater: lessons from the field. Environ Sci Pollut Res Int 16:765–794

Vasák M (2005) Advances in metallothionein structure and functions. J Trace Elem Med Biol 19:13–17

Wangeline AL, Burkhead JL, Hale KL, Lindblom SD, Terry N, Pilon M, Pilon-Smits EA (2004) Overexpression of ATP sulfurylase in Indian mustard: effects on tolerance and accumulation of twelve metals. J Environ Qual 33:54–60

Wilber CG (1980) Toxicology of selenium: a review. Clin Toxicol 17:171–230

Yang H, Nairn J, Ozias-Akins P (2003) Transformation of peanut using a modified bacterial mercuric ion reductase gene driven by an actin promoter from *Arabidopsis thaliana*. J Plant Physiol 160:945–952

Yazaki K, Yamanaka N, Masuno T, Konagai S, Shitan N, Kaneko S, Ueda K, Sato F (2006) Heterologous expression of a mammalian ABC transporter in plant and its application to phytoremediation. Plant Mol Biol 61:491–503

Zhou SF, Wang LL, Di YM, Xue CC, Duan W, Li CG, Li Y (2008) Substrates and inhibitors of human multidrug resistance associated proteins and the implications in drug development. Curr Med Chem 15:1981–2039

Zhu YL, Pilon-Smits EA, Tarun AS, Weber SU, Jouanin L, Terry N (1999a) Cadmium tolerance and accumulation in Indian mustard is enhanced by over-expressing gamma glutamylcysteine synthetase. Plant Physiol 121:1169–1178

Zhu YL, Pilon-Smits EA, Jouanin L, Terry N (1999b) Overexpression of glutathione synthetase in Indian mustard enhances cadmium accumulation and tolerance. Plant Physiol 119:73–80

Zhu YG, Sun GX, Lei M, Teng M, Liu YX, Chen NC, Wang LH, Carey AM, Deacon C, Raab A, Meharg AA, Williams PN (2008) High percentage inorganic arsenic content of mining impacted and non-impacted Chinese rice. Environ Sci Technol 42:5008–5013

Zuo J, Niu Q, Ikeda Y, Chua N (2002) Marker-free transformation: increasing transformation frequency by the use of regeneration-promoting genes. Curr Opin Biotechnol 13:173–180

Chapter 19
Use of Crop Plants for Removal of Toxic Metals

K.K.I.U. Aruna Kumara

Abstract Phytoextraction is an environmentally sound and cost-effective technology for cleaning up soils contaminated with toxic metals. The success of phytoextraction depends on the ability of plants to produce large amounts of biomass. In addition, plants must be tolerant to the target metals and be efficient to translocate metals from roots to the aboveground organs. The effectiveness of phytoextraction also depends upon site and metal species. However, the amount of metals extracted by plants is basically decided by (1) the metal concentration in dry plant tissues and (2) the total biomass of the plant. Certain varieties of high-biomass crops have been found to have the ability to clean up the contaminated soils. The major advantage of using crop plants for phytoextraction is the known growth requirements and well-established cultural practices. One of the most promising, and perhaps widely studied crop plant for the extraction of heavy metals is Indian mustard. Other crops like sweet sorghum, oat, barley, maize, and sunflower are also reported to accumulate toxic metals. As established cultural practices may not elicit the same plant response as observed under non-contaminated conditions, attention must be paid on developing suitable agronomic practices to optimize the growth of plants even under contaminated conditions. Further, a coordinated effort is required to collect and preserve germplasm of accumulator species where molecular engineering can play a key role in developing engineered plants capable of cleaning up contaminated soils and commercializing phytoextraction strategies.

Keywords Phytoextraction • Toxic metals • Contaminated soils • Crop plants

K.K.I.U. Aruna Kumara (✉)
Department of Crop Science, Faculty of Agriculture, University of Ruhuna,
Mapalana, Kamburupitiya, Sri Lanka
e-mail: kkiuaruna@crop.ruh.ac.lk

M.S. Khan et al. (eds.), *Biomanagement of Metal-Contaminated Soils*,
Environmental Pollution 20, DOI 10.1007/978-94-007-1914-9_19,

19.1 Introduction

Since the Industrial Revolution, pollution of the biosphere with trace elements (heavy metals and metalloids) has accelerated dramatically. Many of these trace elements are toxic even at very low concentrations because of their nonbiodegradable nature, long biological half-life, and potential to accumulate inside the living bodies (Behbahaninia et al. 2009). Excessive deposits of heavy metals in agricultural soils may not only result in soil contamination but also lead to elevated heavy metal uptake by crop plants affecting quality and safety of foods (Muchuweti et al. 2006). Therefore, cleaning up of polluted soils is a subject of utmost concern to human beings. Most of the currently practiced remediation methods are primarily based upon civil engineering techniques whose cost is highly variable and depends on the contaminants of concern, soil properties, and site conditions (Lasat 2002). They are not only expensive but environmentally invasive, too. The search for an alternative remediation technique that is economically viable, environmentally sound, and equally protective of human health is thus urgently required. Strategies of this nature are classified under the generic heading of phytoremediation (Iskandar 2000; Iskandar and Kirtham 2001; Kabata-Pendias 2001), which is an emerging biotechnological application based on "green liver concept" and operates on the principles of biogeochemical cycling (Prasad 2004).

Phytoremediation consists of different plant-based technologies (Table 19.1), each having a different mechanism of action for the remediation of metal-polluted soils, sediment, or water. However, the terms phytoremediation and phytoextraction are often incorrectly used as synonyms, though phytoremediation is a concept, while phytoextraction is a specific cleanup technology (Prasad and Freitas 2003). Phytoextraction is in fact the most commonly recognized of all phytoremediation technologies and is the focus of the present review. Phytoextraction actually refers to a diverse collection of plant-based technologies that use either naturally occurring or genetically engineered plants for cleaning contaminated environments (Flathman and Lanza 1998).

While many plant species avoid uptake of heavy metals from contaminated soils, some characteristic plant species thriving in metal-enriched environments can accumulate significantly high concentrations of toxic metals, to levels that by far exceed the soil levels. These species are generally called hyperaccumulators and, among them, some crop plant species are also found. When phytoextraction is practiced, metal-accumulating plants are seeded or transplanted into metal polluted soil and are cultivated according to the established agricultural practices. The roots of established plants absorb metal elements from the soil and translocate them to the aboveground shoots where they accumulate. If metal availability in the soil is not adequate for sufficient plant uptake, chelates or acidifying agents may be used to liberate them into the soil solution (Huang and Cunningham 1996; Huang et al. 1997; Lasat et al. 1998). After sufficient plant growth and metal accumulation, the aboveground parts of the crop are harvested and removed from the contaminated site.

Table 19.1 Types of phytoremediation techniques

Technique	Process	Medium
Phytoextraction	Accumulation of contaminants in shoots and subsequent shoot harvest	Soil
Rhizofiltration	Absorption/adsorption of contaminants in/on roots	Surface water
Phytostabilization	Root and root exudates reduce bioavailability of contaminant	Soil, groundwater
Phytovolatilization	Evaporation of contaminants through plant transpiration	Soil, groundwater
Phytodegradation	Plant-assisted microbial degradation of contaminants in rhizosphere	Soil, groundwater
Phytotransformation	Plant uptake and degradation of contaminants	Soil, groundwater, surface water
Removal of Aerial	Uptake of volatile contaminants by leaves	Air

(Compiled from Yang et al. 2005; Arthur et al. 2005; Solheim 2008)

19.2   What Merits Does It Have?

The phytoextraction is an environmental friendly green technology involving living plants. These plants act as solar-driven pumps, which can extract and concentrate particular elements from the environment (Raskin et al. 1997). Therefore, phytoextraction offers a cost-effective means for cleaning of metal-contaminated soils, because the cost of metal phytoextraction is only a fraction of that associated with conventional engineering technologies (Zhuang et al. 2009). This technology avoids dramatic landscape disruption as it remediates the soil *in situ*. Furthermore, no artificial materials are used, hence, preserving the ecosystem. In contaminated agricultural lands, metal removal and getting a harvest synchronously can be a key element of a new strategy for land management (Zhuang et al. 2009). However, some limitations avoid the wide application of this technology. The success of phytoextraction is primarily dependent upon the bioavailability of the contaminants of concern for plant uptake. Usually readily available metals in soil solution are free metal ions and soluble metal complexes and metals adsorbed to inorganic soil constitutes at ion exchange site. Therefore, phytoextraction is better suited for metals such as Zn and Cd, which occur primarily in exchangeable and readily bioavailable form, while the others need to be treated separately for making them bioavailable. Selection of plant species is of particular importance as most of accumulator species are slowly growing and produce little biomass over period of time. In addition, slow transport of metals from soil particles to root surface is another major factor limiting metal uptake into roots (Claus et al. 2007). Even after entering to the roots, many heavy metals form sulfate, carbonate, or phosphate precipitates and immobilize these metals in apoplastic (extracellular) and symplastic (intracellular) compartments. Apoplastic transport of metals is further limited by

Table 19.2 Advantages and limitations of phytoextraction with crop plants

Advantages	Limitations
Eco-friendly green technology involving living plants	Better suited for metals that are readily bioavailable
Low cost of implementation as compared to conventional means	Some metals need to be treated separately for making them bioavailable
Aesthetically pleasing and avoids dramatic landscape disruptions	Most of the identified species are slowly growing and produce little biomass over a period of time
No artificial materials are generally used	Long-term remediation effort, requiring many cropping cycles to decontaminate metal pollutants to acceptable levels
Applicable to a range of toxic metals and radionuclides	
Eliminate secondary air- or waterborne wastes.	Depth of soil that can be cleaned or stabilized is restricted to the root zone of the plants being used
Enhance regulatory and public acceptance	
Can get a harvest synchronously with metal removal	Applicable only to sites that contain low to moderate levels of metal pollution
Known agronomic and crop management practices can be used	Potential contamination to food chain
Life cycle and biology of crop are well understood	Results are variable
Easily implemented and maintained	Climate dependent

the high cation-exchange capacity of cell walls (Raskin et al. 1997). The highly insoluble nature of most of the hazardous metals interrupts their free movement in the vascular system of the plant. Therefore, translocating them to the aboveground shoots where their accumulation has taken place is also restricted. Phytoextraction is obviously a long-term remediation effort, requiring many cropping cycles to decontaminate metal pollutants to acceptable levels (Zhuang et al. 2009; Shukla et al. 2010). The depth of soil which can be cleaned or stabilized is restricted to the root zone of the plants being used. Depending on the plant, this depth can range from a few inches to several meters (Schnoor et al. 1995; Chen et al. 2000, 2003). This technology is applicable only to sites that contain low to moderate levels of metal pollution, because plant growth is not sustained in heavily polluted soils. The advantages and limitations of using crop plants for cleaning up contaminated soils are summarized in Table 19.2.

19.3   What Factors Decide the Success of Phytoextraction?

The effectiveness of phytoextraction is dependent upon many factors of which some are plant-, site-, or metal-specific characteristics. However, the amount of metals extracted by plants is basically decided by (1) the metal concentration in dry plant tissues and (2) the total biomass of the plant. Therefore, the product of these factors estimates the total amount of metal extracted from the contaminated soil

(Claus et al. 2007). The time required for remediation is dependent upon the type and extent of metal contamination, the length of the growing season, and the efficiency of metal removal by plants (Blaylock and Huang 2000). In addition, as this is essentially an agronomic approach, some agronomic practices, such as, plant selection, possibility of cultivation, fertilization and irrigation, etc., could also play a crucial role in successful cleaning of a contaminated site (Claus et al. 2007).

As a plant-based technology, the success of phytoextraction inherently depends upon several plant characteristics. The plant should have the ability to produce large amounts of biomass rapidly using standard crop production and management practices (Das and Maiti 2007) together with high efficiency of metal accumulation in shoot biomass (Blaylock et al. 1997; McGrath 1998; Shah and Nongkynrih 2007). Plants considered for use must also be tolerant to the targeted metal, or metals, and be efficient at translocating them after uptake by roots to the harvestable aboveground portions (Blaylock and Huang 2000). In addition to the high shoot biomass, a dense root system is important while growing under hardy conditions. Among the site-specific characteristics, the topography of the land should be acceptable and free from physical barriers, which otherwise could prevent the use of agricultural equipment and machineries. The distribution of metals in soil profiles and their movement in soils, which are primarily determined by many soil related factors, also contribute to the efficiency of metal removal by plants. In fact, a major factor limiting metal uptake into roots is the slow transport from soil particles to root surfaces (Claus et al. 2007). The accumulation of the metals in the surface layer of the soil seems to be related to the properties associated with high adsorption rate of the metals by soil solid phases (Behbahaninia et al. 2009). In this context, soil acidity, light texture, and structural features, such as soil cracks, can be considered as important factors (Smith 1996). Soil pH plays a key role in making the availability of elements in the soil for plant uptake (De Matos et al. 2001; Bambara and Ndakidemi 2010; Yobouet et al. 2010). According to Anton and Mathe-Gaspar (2005), higher temperature and lowering soil pH have resulted in increased cadmium and zinc contents of sorrel and maize shoots. Under acidic conditions, H^+ ions displace metal cations from the cation exchange complex (CEC) of soil components and cause metals to be released from sesquioxides and variable-charged clays to which they have been chemisorbed (McBride 1994).

19.4   Mechanisms of Phytoextraction

Proper understanding of the biological processes associated with metal acquisition, transport, and shoot accumulation is the key to formulate sound strategies for improving phytoextraction. In this context, why do plants absorb metals is the fundamental question to be answered. Plants need nutrients as they are among the key requirements for the growth and development of a plant. Some metals, such as Co, Cr, Cu, Fe, K, Mg, Mn, Na, Ni, and Zn, are essential, serve as micronutrients, and are used for redox processes, to stabilize molecules through electrostatic interactions, as components of various enzymes, and for regulation of osmotic pressure (Bruins et al. 2000; Odjegba and Fasidi 2004). Many other metals have no biological role (e.g., Ag, Al, Cd, Pb, and Hg), and are nonessential (Bruins et al. 2000; Kamal et al. 2004) and potentially toxic to microorganisms. Therefore, it is understood that plants take some metals as they are essential nutrients. The literature on the mechanisms of root and plant cell uptake of elements like N, P, S, Fe, Ca, K, and possibly Cl is reported (Marschner 1995). However, little is known about how plants mobilize, uptake, and transport of most environmentally hazardous heavy metals, such as, Pb, Cd, Cu, Zn, U, Sr, and Cs. Nonessential metals, however, may effectively compete for the same transmembrane carriers used by essential metals (Thangavel and Subbhuraam 2004). Nutrient uptake pathways can also take up heavy metals that are similar in chemical form or behavior to the nutrients (Pivetz 2001). However, even for essential elements, plants keep maintaining the accumulation below their metabolic needs (<10 ppm) (Oyelola et al. 2009). Hyperaccumulator plants, however, can accumulate exceptionally high amounts of micronutrients. They not only accumulate excessively high levels of essential micronutrients, but can also absorb significant quantities of nonessential metals. Hyperaccumulators are capable of accumulating metals 100-fold higher (2% on the dry weight basis) than those typically measured in shoots of the common non-accumulator plants (Claus et al. 2007), and their metal tolerance has enhanced the interest of ecologists, plant physiologists, plant biologists and environmentalists to investigate the physiological and genetical factors responsible for metal uptake and tolerance in plants. Accumulator species have evolved specific mechanisms for detoxifying high metal levels accumulated in the cells, which allow bioaccumulation of extremely high concentration of metals (Yang et al. 2005). In fact, they do have their own mechanisms to absorb, translocate, and store the metals they need. In this regards, the structure and properties of cell membranes play a crucial role in metal absorption process. Because of their charge, metal ions cannot move freely across the cellular membranes and taking up metals into cells are mediated by membrane proteins with transport functions (Hooda 2007).

In soil, metals are found in different forms: (1) in solution as free metal ions and soluble metal complexes; (2) adsorbed to inorganic soil constituents on ion exchange sites; (3) precipitated such as oxides, hydroxides, and carbonates; (4) bound to soil organic matter; and (5) embedded in structures of silicate minerals. Plants do have several mechanisms to solubilize "soil-bound" metals and subsequent uptake (Raskin et al. 1997). Plant roots can solubilize soil-bound metals by acidifying their soil environment with protons extruded from the roots (Thangavel and Subbhuraam 2004). In the rizhosphere, root and microbial activities can influence the chemical mobility of metal ions and ultimately their uptake by plants as consequence of alterations of soil pH or dissolved organic carbon (Hinsinger and Courchesne 2007). Metal-chelating molecules can also be secreted into the rhizosphere to chelate and solubilize "soil-bound" metal (Yang et al. 2005; Hooda 2007). Some rhizosphere microorganisms also secrete plant hormones that increase root growth and thereby the secretion of root exudates (Hooda 2007). In this context, chelating compounds, termed phytosiderophores, have been studied in plants (Higuchi et al. 1999). Some plant roots are capable of reducing "soil-bound" metal ions by specific plasma membrane-bound metal reductases, which may increase metal availability (Thangavel and Subbhuraam 2004). For example, in response to iron deficiency, plants develop several biochemical and morphological reactions to ameliorate iron solubilization and uptake from the soil solution (Hell and Stephan 2003). The biochemical and physiological mechanisms induced in dicotyledonous plants under conditions of iron deficiency comprise three main processes (Babalakova et al. 2005). The first one includes an increased release of protons through the activation of plasmalemma P-type ATPase proton pump to acidify the surrounding solution, thus enhancing Fe(III)-containing compounds solubility (Espen et al. 2000). The second process is an obligatory reduction of ferric-chelates by a membrane-associated Fe(III)-chelate reductase to the more soluble ferro-complexes (Robinson et al. 1999). The third effect of short-term treatment with ionic and chelated copper on membrane adaptive biochemical response is an induction of the synthesis of a specific transporter for ferro-ions in plasmalemma of root cells (Hell and Stephan 2003). In addition, mycorrhizal fungi or root-colonizing bacteria can also be used in increasing the bioavailability of metals (Frey et al. 2000; Khan et al. 2000; Hooda 2007). Mobilized metals then enter the root cells by symplastic or apoplastic pathways (Solheim 2008). Most likely, entrance is via metal ion carriers or channels; however, specialized carriers could also exist for the transport of metal–chelate complexes (Solheim 2008).

The transmembrane structure facilitates the transfer of bound ions from extracellular space through the hydrophobic environment of the membrane into the cell (Lasat 2002). However, of all the adsorbed metals physically at the extracellular negatively charged sites of the root cell walls, only a part enters inside the cells. For success of phytoextraction, absorbed metals, however, should also be transported from roots to shoot, which is primarily controlled by how much water is released from leaves during transpiration and the pressure created by the roots (Welch 1995). Therefore, as the rate of transpiration increases, the internal movement of metal-containing sap from the root to the shoot also increases, allowing roots to absorb more moisture from the soil. Generally, a significant fraction of cell wall-bound metals cannot be translocated to the shoots and, thus, cannot be removed by harvesting shoot biomass (Lasat 2002). Apart from binding onto the cell wall, there are some other means also that determine metal immobilization into roots and subsequent inhibition of ion translocation to the shoot. Complexation in cellular structures of roots could also prevent translocation of metals to the aboveground parts (Lasat et al. 1998). In addition, some plants, coined excluders, possess specialized mechanisms to restrict metal uptake into roots (Lasat 2002). The excluders prevent metal uptake into roots avoiding translocation and accumulation in shoots. Though excluders have a low potential for metal extraction, they can be used to stabilize the soil, and avoid further contamination spread due to erosion (Dahmani-Muller et al. 2000). Most environmentally hazardous metals are too insoluble to move freely in the vascular system of the plant. Many forms like sulfate, carbonate, or phosphate precipitate by immobilizing these metals in apoplastic and symplastic compartments (Raskin et al. 1997; Ghosh and Singh 2005). However, plant species have unique abilities to tolerate, accumulate, and detoxify metals and metalloids (Danika and LeDuc Norman 2005). Several hundred plant species have so far been identified as hyperaccumulators of different metals (McGrath and Zhao 2003; McIntyre 2003; Ghosh and Singh 2005). Hyperaccumulators are found from a wide range of taxonomic groups (45 different families) (Baker et al. 2000) and geographic areas and possess a wide variety of morphologies, physiologies, and ecological characteristics (Pollard et al. 2002). The majority of them accumulate only one metal (Pollard et al. 2002) although a significant number show the ability to accumulate more than one (He et al. 2002; Yang et al. 2004; McIntyre 2003).

19.5   How to Enhance the Efficiency of Phytoextraction?

As many factors either directly or indirectly affect the efficacy of phytoextraction, it is important to employ an integrated approach in order to remove heavy metals from contaminated sites. Such integrated strategy may include selection of high-biomass-producing crops, identify plants that could grow in varying environmental conditions, selection of improved crop husbandry, innovative soil management practices, etc., to ensure high metal removal rates from contaminated soils (Nowack et al. 2006; Evangelou et al. 2007). Therefore, selection, breeding, and genetic engineering of metal accumulators can be considered as the key areas of practical significance. The bioavailability of metals for plant uptake can be altered in several means. For example, if the soil contains chelating agents, they can form soluble complexes with metals, thereby enhancing movement of metals in soil profile (Behbahaninia et al. 2009). To achieve this, use of different chelators has shown a dramatic increase in the metal mobility in soil substrate keeping metals as soluble chelate–metal complexes which become available for uptake by roots and are later on transported within the plants. Many chemical amendments, such as ethylene diamine tetra acetic acid (EDTA), diethylene triamine penta acetic acid (DTPA), nitrilotri acetic acid (NTA), and organic acids, have been used in pot and field experiments to enhance extraction rates of heavy metals and to achieve higher phytoextraction efficiency (Kayser et al. 2000; Thaylakumaran et al. 2003; Tandy et al. 2004; Ke et al. 2006; Wang et al. 2007; Wu et al. 2006; Zhuang et al. 2009). However, the effectiveness of different chelating agents is highly variable with the plant species and metal involved.

Though EDTA has been proved as one of the most efficient chelating agents in enhancing Pb phytoavailability in soil and subsequent uptake and translocation to shoots (Chen and Cutright 2001; Shen et al. 2002; Claus et al. 2007; Zhuang et al. 2009), it has failed, however, in enhancing some other metals such as Cd, Zn, and Cu accumulation in plants (Lai and Chen 2004; McGrath et al. 2006; Zhuang et al. 2009). Furthermore, there is enough evidence that suggest that some plant species had no remarkable response to the application of EDTA (Zhuang et al. 2005, 2007). When several heavy metals are present in the soil, interactions and subsequent inhibitory effects can play a role in responding to the added EDTA. Another key area to be considered is the physical features of the soil, because if the soil allows leaching of metal-chelating agents, it might possibly be a threat to groundwater contamination

(Nowack et al. 2006). Therefore, use of EDTA to enhance phytoextraction requires a critical assessment. Diethylene triamine penta acetic acid is another superior reagent used in extraction of metals, such as Cd, Pb, Zn, and Ni from contaminated soils (Behbahaninia et al. 2009). The DTPA extraction has frequently been found to correlate with amounts of metals taken up by the plants (Nouri et al. 2001). In a similar study, addition of thiosulfate and thiocynate salts to mine spoil has reportedly induced plants to accumulate Hg (Moreno et al. 2005) while chloride anions are shown to increase the Cd solubility in soils by forming relatively stable chloride ion complexes, for example, $CdCl^+$ and $CdCl_2$ (Weggler et al. 2004). According to Zhuang et al. (2005), inorganic agents like elemental sulfur or ammonium sulfate could also enhance metal accumulation. It has repeatedly been reported that the application of ammonium to soil could promote the phytoavailability of heavy metals from the contaminated soil (Xiong and Lu 2002; Zaccheo et al. 2006).

It seems that some soil applications (such as sludge) can produce soluble organic complexes with the heavy metals. These complexes are more mobile and possibly more readily taken up by plants than free metal ions (Shuman 2005; Senesi and Loffrdo 2005; Nouri et al. 2006). However, due to changing of their available forms to some unavailable forms such as fractions associated with organic materials, carbonates, or metal oxides (Walker et al. 2004), bioavailability of metals sometimes can be decreased by the organic amendments (Wei et al. 2010). Due to continuous loading of pollutants, heavy metals can be released into groundwater or soil solution, which are then available for plant uptake (Mapanda et al. 2004). Lowering in soil pH can weaken the retention ability of toxic metals to soil organic matter resulting in more available metal in soil solution for root absorption. In fact, many metal cations (e.g., Cd, Cu, Hg, Ni, Pb, and Zn) are more soluble and available in the soil solution at low pH (below 5.5) (Blaylock and Huang 2000). It could, therefore, be suggested that the phytoextraction process is enhanced when metal availability to plant roots is facilitated through the addition of acidifying agents to the soil (Brown et al. 1994; Salt et al. 1995). Possible amendments of acidification include NH_4-containing fertilizers, organic and inorganic acids, and elemental S.

Fertilization, on the other hand, can enhance the growth of the plants resulting in high biomass, which has also been used in increasing the efficiency of phytoextraction (Wei et al. 2010). For example, Wei et al. (2010), in a study with *Solanum nigrum*, reported that the application of urea has enhanced the efficiency of phytoextraction. After application of natural N-P-K fertilizer, particularly at the early stage of growth, the biomass of common reed *(Phragmites australis)* was increased by twofold compared to control plants that subsequently improved phytoextraction of Ni and Zn by 2–3-folds (Claus et al. 2007). In addition, fertilizers with high content of NH_4^+ have the additional benefit of lowering the soil pH, leading to an increase in plant uptake of metals. According to Zaccheo et al. (2006), soils amended with $(NH_4)_2SO_4$ and $(NH_4)_2S_2O_3$ led to an increase in metal availability due to decreased soil pH. The addition of NH_4NO_3 and $(NH_4)_2SO_4$ to soil, however, did not increase Zn and Cu accumulation in three sorghum varieties (Zhuang et al. 2009). The contradictory reports on the effect of ammonium fertilization on phytoextraction are basically due to the degree of solubilization of metals under different soil pH levels. Generally, Zn and Cd can easily be solubilized at pH values of conventional soils, whereas the solubilization of Pb and Cu occurs at lower pH (Schmidt 2003). Therefore, metal availability in soil can be manipulated by the proper ratio of NO_3 to NH_4 used for plant fertilization.

19.6 Promising Crop Plants

Many studies have indicated that certain varieties of high-biomass crops display heavy metal tolerance and/or ability to cleaning up the contaminated soils. In this regard, Kumar et al. (1995) evaluated several fast-growing Brassicas such as Indian mustard (*Brassica juncea* L. Czern), black mustard (*Brassica nigra* Koch), turnip (*Brassica campestris* L.), rape (*Brassica napus* L.), and kale (*Brassica oleracea* L) for their ability to tolerate and accumulate metals. Indeed, Indian mustard is one of the most promising, and perhaps most studied, non-hyperaccumulator plant for the extraction of heavy metals from contaminated sites (Prasad and Freitas 2003). Upon further screening, it was found effective in sorbing particularly divalent cations of toxic metals (Salt and Kramer 2000). In a similar study, Dushenkov et al. (1995) observed that the roots of Indian mustard are effective in the removal of Cd, Cr, Cu, Ni, Pb, and Zn as also reported by others (Ebbs and Kochian 1998; Prasad and Freitas 2003). In a recent investigation, the leaves of sorghum plants have been found very effective in the removal of Pb, while the removal of Cd, Zn, and Cu was maximum by stems (Zhuang et al. 2009). Sweet sorghum (*Sorghum bicolor* L.) a hardy, C4 grass widely used as a forage crop (Buxton et al. 1998; Unger 2001) and as a great promising energy plant, has also shown to display a potential removing ability also due to its fast-growing and high-biomass production capacity. Zhuang et al. (2009) have used three varieties of sweet sorghum to evaluate the phytoextraction efficiency of heavy metals. Their results revealed that even when grown in the contaminated soil, sorghum plants can extract more than 0.05 kg/ha of Cd in a single crop and the removal of Pb and Zn was 0.35 and 1.44 kg/ha, respectively. Similar findings for sorghum plant were also reported by Marchiol et al. (2007) who calculated the values of 0.38 kg/ha for Pb and 1.22 kg/ha for Zn in an alkaline, industrial-polluted soil. These reports confirmed the findings of An (2004) who also reported the ability of sweet sorghum to accumulate metal elements. According to Madejón et al. (2003), compared to sorghum plant, sunflower (*Helianthus annuus* L.) could extract significantly greater amount of Zn (2.14 kg/ha), when the roots were also considered in calculations. Studies conducted with hydroponic solutions revealed that sunflower can remove Pb (Dushenkov et al. 1995), U (Dushenkov et al. 1997a), ^{137}Cs, and [90]Sr (Dushenkov et al. 1997b). Claus et al. (2007) have used sunflower, maize (*Zea mays* L.), and rape (*Brassica napus*) to assess the removal of Cd, Cu, Ni, Zn, Cr, and Pb from a contaminated site. According to their findings, rape plants bioconcentrated up to 40 ppm Cr and Pb. Even though maize produced the largest biomass, the total amount of metals taken up by this plant was lower than sunflower and rape plants. Metal removal capacity of different plants has also been studied in various cultural practices by Keller et al. (2003) and Ciura et al. (2005) using maize as the test plant, while Madejón et al. (2003) and Soriano and Fereres (2003) tested sunflower and barley respectively for assessing their metal-removing potential.

In addition to Indian mustard, Zn has also been removed successfully by oat (*Avena sativa* L.) and barley (*Hordium vulgare* L.) with the established cultural practices (Ebbs and Kochian 1998). Some more reports are also available on Indian mustard, oat, maize, barley, sunflower, and ryegrass (Salt et al. 1998; Shen et al. 2002; Meers et al. 2005; Komárek et al. 2007). Moreover, fast-growing willows (*Salix viminalis*) and poplars (*Populus* sp.) are excellent producers of biomass and have characteristics that make these species promising for phytoremediation application (Vervaeke et al. 2003). Keller et al. (2003) reported that *Nicotiana tabacum* L. has the ability to produce 12.6 t/ha of biomass, which could extract 1.83 kg/ha of Zn, 0.47 kg/ha of Cu and 0.042 kg/ha of Cd. Potentially promising crop plants with respective metals are given in Table 19.3.

Table 19.3 Potentially promising crop plants for phytoextraction

Metal	Species	Reference
Pb	*Lycopersicon esculentum*	Cornu et al. (2007) and Oyelola et al. (2009)
	Sorghum bicolor	Marchiol et al. (2007) and Zhuang et al. (2009)
	Helianthus annuus	Madejón et al. (2003), Marchiol et al. (2007), and Claus et al. (2007)
	Zea mays	Ciura et al. (2005) and Claus et al. (2007)
	Hordeum vulgare	Soriano and Fereres (2003)
	Brassica juncea	Ebbs and Kochian (1997) and Prasad and Freitas (2003)
	Brassica napus	Claus et al. (2007)
	Pisum sativum	Huang et al. (1997)
	Amaranthus cruentus	Oyelola et al. (2009)
Cd	*Sorghum bicolor*	Zhuang et al. (2009)
	Helianthus annuus	Turgut et al. (2004), Claus et al. (2007), and Marchiol et al. (2007)
	Zea mays	Ciura et al. (2005) and Claus et al. (2007)
	Hordeum vulgare	Soriano and Fereres (2003)
	Brassica juncea	Zavoda et al. (2001), Keller et al. (2003), and Prasad and Freitas (2003)
	Nicotiana tabacum	Keller et al. (2003)
	Brassica napus	Claus et al. (2007)
Zn	*Sorghum bicolor*	Madejón et al. (2003), Marchiol et al. (2007), and Zhuang et al. (2009)
	Helianthus annuus	Madejón et al. (2003), Marchiol et al. (2007), and Claus et al. (2007)
	Zea mays	Ciura et al. (2005) and Claus et al. (2007)
	Hordeum vulgare	Ebbs and Kochian (1998) and Soriano and Fereres (2003)

(continued)

Table 19.3 (continued)

Metal	Species	Reference
	Brassica juncea	Kumar et al. (1995), Keller et al. (2003), and Prasad and Freitas (2003)
	Nicotiana tabacum	Keller et al. (2003)
	Brassica napus	Claus et al. (2007)
	Avena sativa	Ebbs and Kochian (1998)
Cr	Helianthus annuus	Zavoda et al. (2001), Turgut et al. (2004), and Claus et al. (2007)
	Brassica juncea	Kumar et al. (1995) and Zavoda et al. (2001)
	Zea mays	Claus et al. (2007)
	Brassica napus	Claus et al. (2007)
Cu	Sorghum bicolor	Zhuang et al. (2009)
	Helianthus annuus	Madejón et al. (2003), Marchiol et al. (2007), and Claus et al. (2007)
	Zea mays	Brun et al. (2001), Ciura et al. (2005), and Claus et al. (2007)
	Hordeum vulgare	Soriano and Fereres (2003)
	Brassica juncea	Prasad and Freitas (2003)
	Nicotiana tabacum	Keller et al. (2003)
	Brassica napus	Claus et al. (2007)
	Lycopersicon esculentum	Cornu et al. (2007) and Oyelola et al. (2009)
	Amaranthus cruentus	Oyelola et al. (2009)
Ni	Helianthus annuus	Zavoda et al. (2001), Turgut et al. (2004), and Claus et al. (2007)
	Brassica juncea	Kumar et al. (1995) and Zavoda et al. (2001)
	Zea mays	Claus et al. (2007)
	Brassica napus	Claus et al. (2007)
Cs	Brassica oleracea	Lasat et al. (1997)
	Phaseolus acutifolius	Lasat et al. (1997)
	Brassica juncea	Lasat et al. (1997)

19.7   What Aspects Need More Investigations?

Though, phytoextraction has been intensively investigated over the years, only a scanty of information is available on the usage of crop plants for the metal removal from contaminated sites. The prime advantage of using common crop species for phytoextraction is the known growth requirements and well-established cultural practices. Although some crop species were found to accumulate heavy metals while producing high biomass in response to established agricultural management (Ebbs and Kochian 1998), growth and yield performances may vary widely under contaminated conditions (Blaylock et al. 1997), and even established cultural practices sometimes may not elicit the same plant response as observed under non-contaminated environment. The fundamental aim of the agronomic research is to enhance the growth and yield performance. But in general, no attention is paid on how to enhance metal accumulation in the tissues of crop species. However, with the merits of phytoextraction, it is necessary to develop suitable agronomic practices to optimize the growth of crop plants even under contaminated conditions. In this context, research must be focused on agronomic practices such as crop establishment (planting season, spacing, establishment method), irrigation (frequency, amount, method), fertilization, weeding (method and frequency), and other cultural practices including mulching, pruning, pest and disease control, and harvesting (method and time) to increase the efficiency of phytoextraction. Among the different agronomic practices, the composition, frequency, and method of application of fertilizers need to be assessed thoroughly in order to find potential crop species. Furthermore, over dosage and/or frequent application of certain plant nutrients can limit/suppress the absorption of the target element. To make phytoextraction economically viable, the cost of fertilization should also be considered while formulating fertilizer mixtures.

Another factor that makes phytoextraction successful is the biomass and ability of plants to accumulate metals within the tissues (Blaylock et al. 1997; McGrath 1998). Increased plant biomass can obviously take up and store more metals. Well-developed root system can provide more surface area to take up metals and the aboveground components should be ready to store them. However, increase in aerial and belowground biomass cannot be achieved simultaneously, because plants generally tend to develop more roots under stressed conditions, which negatively affect the aboveground biomass. Since conclusive reports on these aspects are still lacking, scientists need to address these issues seriously. The majority of phytoextraction research has focused on finding the ideal metal-accumulating plant species and the means by which metals can be removed from soils. Once any promising crop species is identified, genetic factors responsible for their hyperaccumulating nature should be investigated. Despite recent advances in biotechnology, little is known about the genetics of metal hyperaccumulators. Particularly, the heredity of relevant plant mechanisms, such as metal transport and storage (Lasa et al. 2000) and metal tolerance (Ortiz et al. 1992, Ortiz et al. 1995), must be better understood. Bioengineering of plants capable of cleaning up contaminated soils could be the next step that has been successfully performed for several species. Manipulation of genes involved in the biosynthesis of metal sequestering compounds and subsequent introduction and expression of the engineered genes into desirable plant species might attract plant growers to adopt phytoremediation strategies (Prasad and Strzalka 2002). Meanwhile, Chaney et al. (1999) proposed the use of traditional breeding approaches for improving metal hyperaccumulator species and possibly incorporating significant traits, such as metal tolerance and uptake characteristics, into high-biomass-producing plants. Further, it is important to collect and preserve germplasm of accumulator species. The USDA-ARS Plant Introduction Station maintains a worldwide collection of *B. juncea* accessions that are known metal accumulators, and the seeds are distributed to public and private research institutions at no cost (Prasad and Freitas 2003).

## 19.8	Conclusion

Since it evidently does indicate several benefits, phytoextraction can be considered as one of the most preferred methods for restoring metal contaminated environments. In order to exploit the full potential of phytoextraction, a comprehensive understanding is needed on as to how metal uptake, transport, and trafficking across plant membranes and distribution, tolerance, sensitivity, etc., take place under different cultural practices. Furthermore, phytoextraction should be viewed as a long-term remediation solution because many cropping cycles may be needed over several years to reduce metals to acceptable regulatory levels. Taking all these into consideration, it could be concluded that phytoextraction with crop plants is still in the research and developmental phase, which requires further attention.

References

An YJ (2004) Soil ecotoxicity assessment using cadmium sensitive plants. Environ Poll 127:21–26

Anton A, Mathe-Gaspar G (2005) Factors affecting heavy metal uptake; plant selection for phytoremediation. Z Naturforsch 60:244–246

Arthur E, Rice P, Rice P, Anderson T, Baladi S, Henderson K, Coats J (2005) Phytoremediation–an overview. Crit Rev Plant Sci 24:109–122

Babalakova N, Boycheva S, Rocheva S (2005) Effects of short-term treatment with ionic and chelated copper on membrane redox-activity induction in roots of iron – deficient cucumber plants. Gen Appl Plant Physiol 31:143–155

Baker A, McGrath S, Reeves R, Smith J (2000) Metal hyperaccumulator plants: a review of the ecology and physiology of a biological resource for phytoremediation of metal-polluted soils. In: Terry N, Bañuelos G (eds.) Phytoremediation of contaminated soil and water. Lewis Publishers, Boca Raton

Bambara S, Ndakidemi PA (2010) Changes in selected soil chemical properties in the rhizosphere of *Phaseolus vulgaris* L. supplied with *Rhizobium* inoculants, molybdenum and lime. Sci Res Ess 5:679–684

Behbahaninia A, Mirbagheri SA, Khorasani N, Nouri J, Javid AH (2009) Heavy metal contamination of municipal effluent in soil and plants. J Food Agric Environ 7:852–856

Blaylock MJ, Huang JW (2000) Phytoextraction of metals. In: Rakshin I, Ensley BD (eds.) Phytoremediation of toxic metals: using plants to clean up the environment. Wiley, New York, p 314

Blaylock MJ, Salt DE, Dushenkov S, Zakharova O, Gussman C, Kapulnik Y (1997) Enhanced accumulation of Pb in Indian mustard by soil-applied chelation agents. Environ Sci Technol 31:860–865

Brown SL, Chaney RL, Angle JS, Baker AJM (1994) Phytoremediation potential of *Thlaspi caerulescens* and bladder campion for zinc and cadmium contaminated soil. J Environ Qual 23:1151–1157

Bruins MR, Kapil S, Oehme FW (2000) Microbial resistance to metals in the environment. Ecotoxicol Environ Saf 45:198–207

Brun LA, Maillet J, Hinsinger P, Pépin M (2001) Evaluation of copper availability to plants in copper-contaminated vineyard soils. Environ Poll 111:293–302

Buxton DR, Anderson IC, Hallam A (1998) Intercropping sweet sorghum into alfalfa and reed canarygrass to increase biomass yield. J Pro Agric 11:481–486

Chaney RL, Li YM, Angle JS, Baker AJM, Reeves RD, Brown SL, Homer FA, Malik M, Chin M (1999) Improving metal-hyperaccumulators wild plants to develop commercial phytoextraction systems: approaches and progress. In: Terry N, Bañuelos GS (eds.) Phytoremediation of contaminated soil and water. CRC Press, Boca Raton

Chen H, Cutright T (2001) EDTA and HEDTA effects on Cd, Cr, and Ni uptake by *Helianthus annuus*. Chemosphere 45:21–28

Chen HM, Zheng CR, Tu C, Shen ZJ (2000) Chemical methods and phytoremediation of soil contaminated with heavy metals. Chemosphere 41:229–234

Chen YX, Lin Q, Luo YM, He YF, Zhen SJ, Yu YL, Tian GM, Wong MH (2003) The role of citric acid on phytoremediation of heavy metal contaminated soils. Chemosphere 50:807–811

Ciura J, Poniedzialek M, Sekara A, Je drszczyk E (2005) The possibility of using crops as metal phytoremediation. Pol J Environ Stu 14:17–22

Claus D, Dietze H, Gerth A, Grosser W, Hebner A (2007) Application of agronomic practice improves phytoextraction on a multipolluted site. J Environ Eng Lands Manage 15:208–212

Cornu JY, Staunton S, Hinsinger P (2007) Copper concentration in plants and in the rizhosphere as influenced by the iron status of tomato (*Lycopersicon esculentum* L.). Plant Soil 292:63–77

Dahmani-Muller H, van Oort F, Ge lie B, Balabane M (2000) Strategies of heavy metal uptake by three plant species growing near a metal smelter. Environ Poll 109:231–238

Danika L, LeDuc Norman T (2005) Phytoremediation of toxic trace elements in soil and water. J Ind Microbiol Biotechnol 32:514–520

Das M, Maiti SK (2007) Metal accumulation in 5 native plants growing on abandoned CU-tailings ponds. Appl Ecol Environ Res 5:27–35

De Matos AT, Fontes MPF, Da Costa LM, Martinez MA (2001) Mobility of heavy metals as related to soil chemical and mineralogical characteristics of Brazilian soils. Environ Poll 111:429–435

Dushenkov V, Kumar PBAN, Motto H, Raskin I (1995) Rhizofiltration: the use of plants to remove heavy metals from aqueous streams. Environ Sci Technol 29:1239–1245

Dushenkov S, Vasudev D, Kapulnik Y, Gleba D, Fleisher D, Ting KC, Ensley B (1997a) Removal of uranium from water using terrestrial plants. Environ Sci Technol 31:3468–3474

Dushenkov S, Vasudev D, Kapulnik Y, Gleba D, Fleisher D, Ting KC, Ensley B (1997b) Phytoremediation: a novel approach to an old problem. In: Wise DL (ed.) Global environmental biotechnology. Else Sci BV, Amsterdam, pp 563–572

Ebbs SD, Kochian LV (1997) Toxicity of zinc and copper to *Brassica* species: implications for phytoremediation. J Environ Qual 26:776–781

Ebbs SD, Kochian LV (1998) Phytoextraction of Zn by oat (*Avena sativa*), barley (*Hordium vulgare*) and Indian mustard (*Brassica juncea*). Sci Total Environ 32:802–806

Espen L, Dell'Orto M, De Nisi P, Zocchi G (2000) Metabolic responses in cucumber (*Cucumis sativus* L.) roots under Fe-deficiency: a ^{31}P-nuclear magnetic resonance in-vivo study. Planta 210:985–992

Evangelou MWH, Ebel M, Schaefer A (2007) Chelate assisted phytoextraction of heavy metals from soil. Effect, mechanism, toxicity and fate of chelating agents. Chemosphere 68:989–1003

Flathman PE, Lanza GR (1998) Phytoremediation: current views on an emerging green technology. J Soil Contam 7:415–432

Frey B, Zierold K, Brunner I (2000) Extracellular complexation of Cd in the Hartig net and cytosolic Zn sequestration in the fungal mantle of *Picea abies–Hebeloma crustuliniforme* ectomycorrhizas. Plant Cell Environ 23:1257–1265

Ghosh M, Singh SP (2005) A review on phytoremediation of heavy metals and utilization of its byproducts. Appl Ecol Environ Res 3:1–18

He B, Yang X, Wei Y, Ye Z, Ni W (2002) A new lead resistant and accumulating ecotype – *Sedum alfredii* H. Acta Bot Sinica 44:1365–1370

Hell R, Stephan UW (2003) Iron uptake and homeostasis in plants. Planta 216:541–551

Higuchi K, Suzuki K, Nakanishi H, Yamaguchi H, Nishizawa NK, Mori S (1999) Cloning of nicotianamine synthase genes, novel genes involved in the biosynthesis of phytosiderophores. Plant Physiol 119:471–479

Hinsinger P, Courchesne F (2007) Mobility and bioavailability of heavy metals and metalloids at soil-root interface. In: Violante A, Huang PM, Gadd GM (eds.) Biophysico-chemical processes of heavy metals and metalloids in soil environments, vol 1. Wiley-IUPAC Series Biophisico-Chemical processes in Environmental Systems, Chichester

Hooda V (2007) Phytoremediation of toxic metals from soil and waste water. J Environ Biol 28:367–376

Huang JW, Cunningham SD (1996) Lead phytoextraction: species variation in lead uptake and translocation. New Phytol 134:75–84

Huang JW, Chen J, Berti WB, Cunningham SD (1997) Phytoremediation of lead-contaminated soils: role of synthetic chelates in lead phytoextraction. Sci Total Environ 31:800–805

Iskandar IK (2000) Environmental restoration of metal contaminated soils. CRC Press, Boca Raton, pp 320

Iskandar IK, Kirtham MB (2001) Trace elements in soil; bioavailability, flux and transfer. CRC Press, Boca Raton, pp 304

Kabata-Pendias A (2001) Trace elements in soils and plants. CRC Press, Boca Raton, pp 432

Kamal M, Ghaly AE, Mahamoud N, Cote R (2004) Phytoaccumulation of heavy metals by aquatic plants. Environ Int 29:1029–1039

Kayser A, Wenger K, Keller A, Attinger W, Felix HR, Gupta SK (2000) Enhancement of phytoextraction of Zn, Cd and Cu from calcareous soil: the use of NTA and sulfur amendments. Sci Total Environ 34:1778–1783

Ke X, Li PJ, Zhou QX, Zhang Y, Sun TH (2006) Removal of heavy metals from a contaminated soil using tartaric acid. J Environ Sci 18:727–733

Keller C, Hammer D, Kayser A, Richner W, Brodbeck M, Sennhauser M (2003) Root development and heavy metal phytoextraction efficiency: comparison of different plant species in the field. Plant Soil 249:67–81

Khan AG, Keuk C, Chaudhry TM, Khoo CS, Hayes WJ (2000) Role of plants, mycorrhizae and phytochelators in heavy metal contaminated land remediation. Chemosphere 41:197–207

Komárek M, Tlustoš P, Szákova J, Richner W, Brodbeck M, Sennhauser M (2007) The use of maize and poplar in chelant-enhanced phytoextraction of lead from contaminated agricultural soils. Chemosphere 67:640–651

Kumar PBAN, Dushenkov V, Motto H, Raskin I (1995) Phytoextraction: the use of plants to remove heavy metals from soils. Environ Sci Technol 29:1232–1238

Lai HY, Chen ZS (2004) Effects of EDTA on solubility of cadmium, zinc, and lead and their uptake by rainbow pink and vetiver grass. Chemosphere 55:421–430

Lasa B, Frechilla S, Lamsfus C, Aparicio-Tejo PM (2000) Effects of low and high levels of magnesium on the response of sunflower plants grown with ammonium and nitrate. Plant Soil 225:167–174

Lasat MM (2002) Phytoextraction of metals from contaminated soil: a review of plant/soil/metal interaction and assessment of pertinent agronomic issues. J Hazard Subs Res 5:1–25

Lasat MM, Norvell WA, Kochian LV (1997) Potential for phytoextraction of ^{137}Cs from a contaminated soil. Plant Soil 195:99–106

Lasat MM, Fuhrmann M, Ebbs SD, Cornish JE, Kochian LV (1998) Phytoremediation of a radiocesium-contaminated soil: evaluation of cesium-137 bioaccumulation in the shoots of three plant species. J Environl Qual 27:165–169

Madejón P, Murillo JM, Marañón T, Cabrera F, Soriano MA (2003) Trace element and nutrient accumulation in sunflower plants two years after the Aznalcóllar spill. Sci Total Environ 307:239–257

Mapanda F, Mangwayana EN, Nyamangara J, Giller KE (2004) The effects of long-term irrigation using wastewater on heavy metal contents of soils under vegetables in Harare, Zimbabwe. Agric Eco Environ 107:151–156

Marchiol L, Fellet G, Perosa D, Zerbi G (2007) Removal of trace metals by *Sorghum bicolor* and *Helianthus annuus* in a site polluted by industrial wastes: a field experience. Plant Physiol Biochem 45:379–387

Marschner H (1995) Mineral nutrition of higher plants. 2nd ed. Academic Press, New York

McBride MB (1994) Environmental chemistry of soils. Oxford University Press, New York

McGrath SP (1998) Phytoextraction for soil remediation. In: Brooks RR (ed.) Plants that hyperaccumulate heavy metals: their role in phytoremediation, microbiology, archaeology, mineral exploration and phytomining. CAB International, New York, pp 261–288

McGrath SP, Zhao F (2003) Phytoextraction of metals and metalloids from contaminated soils. Curr Opin Biotechnol 14:277–282

McGrath SP, Lombi E, Gray CW, Caille N, Dunham SJ, Zhao FJ (2006) Field evaluation of Cd and Zn phytoextraction potential by the hyperaccumulators *Thlaspi caerulescens* and *Arabidopsis halleri*. Environ Poll 141:115–125

McIntyre T (2003) Phytoremediation of heavy metals from soils. Adv Biochem Engg Biotechnol 78:97–123

Meers E, Ruttens A, Hopgood M, Lesage E, Tack FMG (2005) Potential of *Brassic rapa*, *Cannabis sativa*, *Helianthus annuus* and *Zea mays* for phytoextraction of heavy metals from calcareous dredged sediment derived soils. Chemosphere 61:561–572

Moreno FN, Anderson CWN, Stewart RB, Robinson BH, Ghoshei M, Meech JA (2005) Induced plant uptake and transport of mercury in the presence of sulphur-containing ligands and humic acid. New Phytol 166:445–454

Muchuweti M, Birkett JW, Chinyanga E, Zvauya R, Scrimshaw MD, Lester JN (2006) Heavy metal content of vegetables irrigated with mixture of wastewater and sewage sludge in Zimbabwe: implications for human health. Agric Eco Environ 112:41–48

Nouri J, Alloway BJ, Peterson PJ (2001) Forms of heavy metals in sewage sludge and soil amended with sludge. Pak J Biol Sci 4:1460–1465

Nouri J, Mahvi AH, Babaei AA, Ahmadpour E (2006) Regional pattern distribution of groundwater fluoride in the Shush aquifer of Khuzestan county. Fluoride 39:321–325

Nowack B, Schulin R, Robinson B (2006) Critical assessment of chelant enhanced metal phytoextraction. Sci Total Environ 40:5225–5232

Odjegba VJ, Fasidi IO (2004) Accumulation of trace elements by *Pistia stratiotes*: implications for phytoremediation. Ecotoxicol 13:637–646

Ortiz DF, Kreppel L, Speiser DM, Scheel G, McDonald G, Ow DV (1992) Heavy metal tolerance in the fission yeast requires an ATP-binding cassette-type vacuolar membrane transporter. EMBO J 11:3491–3499

Ortiz DF, Ruscitti T, McCue KF, Ow DV (1995) Transport of metal-binding peptides by HMT1, a fission yeast ABC-type B vacuolar membrane protein. J Biol Chem 270:4721–4728

Oyelola OT, Babatunde AI, Odunlade AK (2009) Phytoremediation of Metals from Contaminated Soil using *Lycopercium Esculentum* (Tomato) Plant. Int J Pure Appl Sci 3:44–48

Pivetz BE (2001) Phytoremediation of contaminated soil and groundwater at hazardous waste sites. Ground Water Issue, United States Environmental Protection Agency, EPA/540/S-01/500

Pollard A, Powell K, Harper F, Smith J (2002) The genetic basis of metal hyperaccumulation in plants. Crit Rev Plant Sci 21:539–566

Prasad MNV (2003) Phytoremediation of metal polluted ecosystems – Hype for commercialization. Russ J Plant Physiol 50:686–701

Prasad MNV (2004) Heavy metals stress in plants: from biomolecules to ecosystem. Springer-Verlag/Narosa, Heidelberg/New Delhi, p 1462

Prasad MNV, Freitas HM (2003) Metal hyperaccumulation in plants-Biodiversity prospecting for phytoremediation technology. Elect J Biotechnol 16:285–321

Prasad MNV, Strzalka K (2002) Physiology and biochemistry of metal toxicity and tolerance in plants. Kluwer Academic Publishers, Dordrecht, p 432, ISBN 1-40-200468-0

Raskin I, Smith RD, Salt DE (1997) Phytoremediation of metals: using plants to remove pollutants from the environment. Curr Opin Biotechnol 8:221–226

Robinson NJ, Proctor CM, Connolly EL, Guerinot ML (1999) A ferric chelate reductase for iron uptake from soils. Nature 397:694–697

Salt DE, Kramer U (2000) Mechanisms of metal hyperaccumulation in plants. In: Raskin I, Ensley BD (eds.) Phytoremediation of toxic metals using plants to clean-up the environment. Wiley, New York, pp 231–246

Salt DE, Prince RC, Pickering IJ, Raskin I (1995) Mechanisms of cadmium mobility and accumulation in Indian mustard. Plant Physiol 109:1427–1433

Salt DE, Smith RD, Raskin I (1998) Phytoremediation. Annu Rev Plant Physiol Plant Mol Biol 49:643–668

Schmidt U (2003) Enhancing phytoextraction: the effect of chemical soil manipulation on mobility, plant accumulation, and leaching of heavy metals. J Environ Qual 32:1939–1954

Schnoor JL, Light LA, McCutcheon SC, Wolfe NL, Carreira LH (1995) Phytoremediation of organic and nutrient contaminants. Environ Sci Technol 29:318–323

Senesi N, Loffrdo E (2005) Metal ion complexation by soil humic substances. In: Tabatabai MA, Sparks DL (eds.) Chemical processes in soils. SSSA, Madison

Shah K, Nongkynrih J (2007) Metal hyperaccumulation and bioremediation. Biol Plant 51:618–634

Shen ZG, Li XD, Wang CC, Chen HM, Chua H (2002) Lead phytoextraction from contaminated soil with high-biomass plant species. J Environ Qual 31:1893–1900

Shukla KP, Singh NK, Sharma S (2010) Bioremediation: developments, current practices and perspectives. Genet Engg Biotechnol J 3:1–20

Shuman LM (2005) Chemistry of micronutrients in soils. In: Tabatabai MA, Sparks DL (eds.) Chemical processes in soils. SSSA, Madison

Smith SR (1996) Agricultural recycling of sewage sludge and the environment. CAB International, Wallingford

Solheim C (2008) Identification and characterization of copper responsive proteins in *Arabidopsis*. Ph.D. thesis, Department of Plant Sciences, University of Saskatchewan

Soriano MA, Fereres E (2003) Use of crops for *in situ* phytoremediation of polluted soils following a toxic flood from a mine spill. Plant Soil 256:253–264

Tandy S, Bossart K, Mueller R, Ritschel J, Hausar L, Schulin R, Nowack B (2004) Extraction of heavy metals from soils using biodegradable chelating agents. Environ Sci Technol 40:2753–2758

Thangavel P, Subbhuraam CV (2004) Phytoextraction: role of hyperaccumulators in metal contaminated soils. Proc Ind Natl Sci Acad 70:109–130

Thaylakumaran T, Robinson BH, Vogeler I, Scotter DR, Clothier BE, Percivel HJ (2003) Plant uptake and leaching of copper during EDTA-enhanced phytoremediation of repacked and undisturbed soil. Plant Soil 254:415–423

Turgut C, Katie Pepe M, Cutright TJ (2004) The effect of EDTA and citric acid on phytoremediation of Cd, Cr and Ni from soil using *Helianthus annuus*. Environ Poll 131:147–154

Unger PW (2001) Alternative and opportunity dry land crops and related soil conditions in the Southern Great Plains. Agron J 93:216–226

Vervaeke P, Luyssaert S, Mertens J, Meers E, Tack FMG, Lust N (2003) Phytoremediation prospects of willow stands on contaminated sediment: a field trial. Environ Poll 126:275–282

Walker DJ, Clemente R, Bernal MP (2004) Contrasting effects of manure and compost on soil pH, heavy metal availability and growth of *Chenopodium album* L. in a soil contaminated by pyritic mine waste. Chemosphere 57:215–224

Wang HQ, Lu SJ, Li H, Yao ZH (2007) EDTA-enhanced phytoremediation of lead contaminated soil by *Bidens maximowicziana*. J Environ Sci 19:1496–1499

Weggler K, Mclaqhlin MJ, Graham RD (2004) Effect of chloride in soil solution on the plant availability of biosolid-borne cadmium. J Environ Qual 33:496

Wei S, Li Y, Zhou Q, Srivastava M, Chiu S, Zhan J, Wu Z, Sun T (2010) Effect of fertilizer amendments on phytoremediation of Cd contaminated soil by a newly discovered hyperaccumulator *Solanum nigrum* L. J Hazard Mat 176:269–273

Welch RM (1995) Micronutrient nutrition of plants. Crit Rev Plant Sci 14:49–82

Wu QT, Deng JC, Long XX, Morel JL, Schwartz C (2006) Selection of appropriate organic additives for enhancing Zn and Cd phytoextraction by hyperaccumulators. J Environ Sci 18:1113–1118

Xiong ZT, Lu P (2002) Joint enhancement of lead accumulation in *Brassica* plants by EDTA and ammonium sulfate in sand culture. J Environ Sci 14:216–220

Yang X, Long X, Ye H, He Z, Stofella P, Calvert D (2004) Cadmium tolerance and hyperaccumulation in a new Zn hyperaccumulating plant species (*Sedum alfredii* Hance). Plant Soil 259:181–189

Yang X, Feng Y, He Z, Stofella P (2005) Molecular mechanisms of heavy metal hyperaccumulation and phytoremediation. J Trace Ele Med Biol 18:339–353

Yobouet YA, Adouby K, Trokourey A, Yao B (2010) Cadmium, Copper, Lead and Zinc speciation in contaminated soils. Int J Engg Sci Technol 2:802–812

Zaccheo P, Crippa L, Pasta VDM (2006) Ammonium nutrition as a strategy for cadmium mobilisation in the rhizosphere of sunflower. Plant Soil 283:43–56

Zavoda J, Cutright T, Szpak J, Fallon E (2001) Uptake, selectivity, and inhibition of hydroponic treatment of contaminants. J Environ Engg 127:502

Zhuang P, Ye ZH, Lan CY, Xie ZW, Shu WS (2005) Chemically assisted phytoextraction of heavy metals contaminated soils using three plant species. Plant Soil 276:153–162

Zhuang P, Yang QW, Wang HB, Shu WS (2007) Phytoextraction of heavy metals by eight plant species in the field. Water Air Soil Poll 184:235–242

Zhuang P, Shu WS, Li Z, Liao B, Li J, Shao J (2009) Removal of metals by sorghum plants from contaminated land. J Environ Sci 21:1432–1437

Chapter 20
Bioremediation Potential of Heavy Metal–Resistant Actinobacteria and Maize Plants in Polluted Soil

Claudia S. Benimeli, Marta A. Polti, Virginia H. Albarracín, Carlos M. Abate, and María J. Amoroso

Abstract The screening and characterization of metal resistant microorganisms and plants are important for developing novel bioremediation processes. Considering these, we assessed the potential of copper- and chromium-resistant actinomycetes for bioremediation activity in polluted soils. Also, we assessed the effects of copper concentrations on roots, shoots, and leaf growth of maize and the copper uptake and accumulation by the maize plants. Four chromium resistant *Streptomyces* strains reduced hexavalent chromium up to 85–95% after 21 days. The novel copper-resistant actinobacterium *Amycolatopsis tucumanensis* efficiently immobilized copper when inoculated into copper-polluted soil microcosms: bioavailable Cu was 31% lower in soil compared to non-bioaugmented soil. Maize plant was found

C.S. Benimeli (✉)
Planta Piloto de Procesos Industriales y Microbiológicos (PROIMI-CONICET),
Tucumán, Argentina

Universidad del Norte Santo Tomás de Aquino, Tucumán, Argentina
e-mail: cbenimeli@yahoo.com.ar

M.A. Polti • V.H. Albarracín • C.M. Abate
Planta Piloto de Procesos Industriales y Microbiológicos (PROIMI-CONICET),
Tucumán, Argentina

Universidad Nacional de Tucumán, Avenida Belgrano y Pasaje Caseros,
4000 Tucumán, Argentina

M.J. Amoroso
Planta Piloto de Procesos Industriales y Microbiológicos (PROIMI-CONICET),
Tucumán, Argentina

Universidad del Norte Santo Tomás de Aquino,
Tucumán, Argentina

Universidad Nacional de Tucumán, Avenida Belgrano y Pasaje Caseros,
4000 Tucumán, Argentina

M.S. Khan et al. (eds.), *Biomanagement of Metal-Contaminated Soils*,
Environmental Pollution 20, DOI 10.1007/978-94-007-1914-9_20,
© Springer Science+Business Media B.V. 2011

interesting both as biomarker and bioremediation tool. The bioremediation activity of *A. tucumanensis* inoculated maize plants grown in polluted soil microcosms correlated well with the values obtained with chemical and physical methods: 20% and 17% lower tissue contents of copper were measured in roots and leaves, respectively. The roots, shoots, and leaves of maize plants also showed a great ability to accumulate copper, which however increased with metal concentration. The metal concentrations were 382 times more in roots, 157 in shoots, and only 16 in leaves, compared to the control (without $CuSO_4$).

Keywords Bioremediation • Phytoremediation • Actinomycetes • *Zea mays* • Heavy metals

20.1   Introduction

Metals are natural components of soil and some metals are required as micronutrients by plants. However, pollution of biosphere by toxic metals has increased alarmingly since the beginning of the industrial revolution. Among heavy metals, chromium is one of the most widely used metals in industrial processes, like steel production, wood preservation, leather tanning, metal corrosion inhibition, paints, and pigments. It is mainly used as chromate or dichromate (Baldi et al. 1990). Industrial effluents containing chromium compounds are released directly or indirectly into natural water resources, mostly without proper effluent treatment, resulting in anthropogenic contamination of non industrial environments (Cefalu and Hu 2004; Cheung and Gu 2007). Of the different forms of chromium, hexavalent chromium Cr (VI) and trivalent chromium Cr (III) are ecologically important as they are most stable in a natural environment (Megharaj et al. 2003). Of these, Cr (III) is an essential micronutrient for proper glucose metabolism, and stimulates the enzyme system and stabilizes nucleic acids (Viti et al. 2003). While Cr (VI) is more mobile and soluble in water than Cr (III), which is relatively inert, chemically more stable, and less bioavailable due to its negligible permeability to biomembranes. Besides, Cr (VI) is approximately 100 times more toxic (Beleza et al. 2001) and 1,000 times more mutagenic than Cr (III) (Czakó-Vér et al. 1999; Ganguli and Tripathi 2002). In view of its alarming effects on human health, Cr (VI) has been listed as a priority pollutant and classified as a class A human carcinogen by the US Environmental Protection Agency (USEPA) (Costa and Klein 2006).

Due to the ubiquity and toxicity of chromium, there is considerable interest in identifying, low-cost methods for the remediation of Cr (VI) from contaminated environments (Smith et al. 2002). Biological transformation of Cr (VI) to Cr (III) by enzymatic reduction has been recognized as a means of chromium decontamination from effluents (Laxman and More 2002). Reduction of Cr (VI) by species of *Bacillus* (Lloyd 2003; Camargo et al. 2004; Liu et al. 2006), *Pseudomonas* (Lloyd 2003; Park et al. 2000), *Escherichia* (Bae et al. 2005), *Desulfovibrio* (Mabbett and Macaskie 2001), *Microbacterium* (Pattanapipitpaisal et al. 2001), *Shewanella*

(Myers et al. 2000; Vaimajala et al. 2002), and *Arthrobacter* (Horton et al. 2006) have been reported. However, there are only a few studies on Cr (VI) reduction by actinomycetes and their possible role in bioremediation processes. The Cr (VI) reduction ability of *Streptomyces* was reported first time by Das and Chandra (1990), which was followed by Amoroso et al. (2001). Recently, Polti et al. (2010a) did find chromate reductase activity in *Streptomyces* sp. MC1, a strain able to remove and accumulate chromium from soil samples (Polti et al. 2009, 2010b). These and other studies have generated interest in the use of microorganisms, and biological methods in general, for metal decontamination as alternatives to the conventional methods.

Copper (Cu) is another essential and versatile heavy metal which has many known functions in biological systems. However, at elevated concentration, Cu becomes toxic. Copper cannot be destroyed and accumulates in soils, plants, and animals (Georgopoulus et al. 2002). In Argentina, for instance, the legal limit permissible for Cu in drinking water is 1 mg L^{-1}, whereas in European Union, it is 3 mg L^{-1}. Plants require approximately 5–30 mg Cu kg^{-1} dry weight for normal growth (Kabata-Pendias and Pendias 1992) while Cu deficiency usually occurs when plant Cu concentration is less than 5 mg kg^{-1} dry weight (Marschner 1995). When it is absorbed in excess, Cu can alter mitosis, inhibit root elongation, photosynthesis, pigment synthesis, nitrogen and protein metabolism, membrane integrity, mineral uptake and consequently cause total inhibition of plant growth (Luna et al. 1994; Ouzounidou et al. 1995; Shen et al. 1998; Nielsen et al. 2003; Demirevska-Kepova et al. 2004). However, many uses of copper in several applications lead to their wide distribution in soil, silt, waste, and wastewater resulting in significant environmental problems; that require attention of the scientists (Lloyd and Lovley 2001). According to Kabata-Pendias and Pendias (1984), 60–125 mg/kg Cu, based on total fractions in soil, would be considered toxic to plants. In particular, mining and industrial activities in the province of Tucumán, Argentina have led to large-scale contamination of the environment with Cu (Albarracín et al. 2005). To remediate the polluted sites, many conventional approaches like land-filling, recycling, pyrolysis, and incineration are used, which, however, are inefficient and costly. Thus, biological decontamination methods are preferable to conventional systems for their better efficiency and more so they do not produce toxic intermediates (Kothe et al. 2005).

Among microbes, copper-resistant actinobacteria isolated from various polluted areas have been used as potential organisms in bioremediation technologies (Amoroso et al. 1998; Richards et al. 2002). They are metabolically and morphologically versatile, which provide them a great opportunity to accomplish bioremediation processes. As actinomycetes are indigenous soil microorganisms, they have been applied successfully to bioremediate xenobiotics and metal-polluted soil microcosms (Jézéquel and Lebeau 2008; Benimeli et al. 2007, 2008; Albarracín et al. 2010b). Nevertheless, still more research is needed as soil bioremediation constitutes a special challenge because of its heterogeneity. On the other hand, phytoremediation, the use of plants to restore polluted sites, has recently become a tangible alternative to traditional methodologies (Glass 2000; Lasat 2002; Jing et al. 2007).

It has been established that certain wild and crop plant species have the ability to accumulate elevated amounts of toxic heavy metals (Reeves and Baker 2000; Ghosh and Spingh 2005; Brunet et al. 2008). Thus, researchers all over the world are searching new plant species so that they could be used at large scale in removing metals from contaminated sites (Rai et al. 2002; Del Rio et al. 2002; Wang et al. 2007). Maize (*Zea mays*) is one of the most important cereal crops. However, few reports on copper accumulation by maize are available. For example, Liu et al. (2001) studied the uptake and accumulation of metals by roots and shoots of maize. They found that root growth decreased progressively with increased concentrations of Cu^{2+} in solution, but the shoot growth was similar to the control. However, the plants transported and concentrated only a small amount of copper in their roots. The importance of copper- and chromium-resistant actinomycetes strains in bioremediation and how copper affects the overall growth of maize plant is reviewed and discussed.

20.2 Importance of Chromium-Resistant Actinomycetes in Remediation of Metal-Polluted Soils

Streptomyces, commonly found in both conventional (Arifuzzaman et al. 2010) and metal-polluted ecosystems (Guo et al. 2009; Polti et al. 2007), have demonstrated metal-reducing ability. In order to evaluate the Cr (VI)-reducing activity of *Streptomyces* strains in soil, a study was conducted using sterile and non-sterile soil samples, collected from agronomic sites of Aspach-le-Bas (Haut-Rhin, France). Glass pots filled with 200 g soil containing 50 mg Cr (VI) kg^{-1} soil were inoculated with 0.5 mg kg^{-1} soil dry weight by strains R22, MC1, M3, and C55 of *Streptomyces* species. Non-inoculated soil pots were used to determine Cr (VI) reduction by soil and/or autochthonous microflora. Soil pots were incubated at 30°C for 21 days.

All strains were able to grow in soil whether it was contaminated or not with Cr (VI) (Fig. 20.1). Moreover, no significant ($P \leq 0.05$) difference in growth pattern was observed when the strains were developed with or without Cr (VI). *Streptomyces* sp. MC1 and *Streptomyces* sp. C55 reached $2-2.5.10^5$ CFU g^{-1} soil, while *Streptomyces* sp. M3 and *Streptomyces* sp. C55 reached $2-6.10^6$ CFU g^{-1} soil after 21 days of incubation.

The Cr (VI) concentration reduced from 50 to 2 mg kg^{-1} in soils inoculated with the actinomycetes strains, whether the soil was sterilized or not (Fig. 20.2). However, the Cr (VI) reduction was significantly ($P < 0.05$), higher in sterilized soil, where the Cr (VI) removal was $94.26 \pm 1.80\%$, while in non-sterilized soil it was $86.51 \pm 1.01\%$. Non-inoculated samples showed a slight decrease in Cr (VI) concentration (2%), indicating that soils probably does not contain significant amounts of any substance or microorganism able to reduce Cr (VI). The autochthonous culturable microflora in non-sterilized soil was 3.10^5 at the beginning and 2.10^5 CFU g^{-1} soil at the end of the assay, showing the microflora viability.

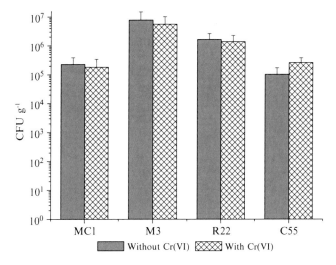

Fig. 20.1 Growth of *Streptomyces* in the absence or presence of 50 mg Cr (VI) kg⁻¹, after 21 days of incubation

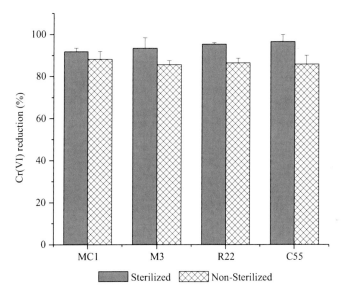

Fig. 20.2 Percentage of Cr (VI) reduction by *Streptomyces* after 21 days of growth in soil

Furthermore, the Cr-reducing *Streptomyces* sp. strains used showed bioremediation ability and reduced up to 85–95% of Cr (VI) (50 mg kg⁻¹) after 21 days in soils, without any prior treatment, or addition of any substrate at a normal soil humidity level. Other studies have in contrast demonstrated that the addition of organic substrate to soil samples is needed to obtain higher Cr (VI) reduction levels

(Turick et al. 1998; Vainshtein et al. 2003). However, Polti et al. (2009) found a total Cr (VI) removal without amending any substrate with a normal humidity level for a soil without previous treatment, and using a higher soil quantity.

20.3   Importance of Copper-Resistant Actinomycetes in Bioremediation

In a previous screening program, 50 copper-resistant actinomycetes were isolated from copper-polluted (DP2 Channel, Ranchillos; 600 mg l^{-1} Cu^{2+}) and non-copper polluted sediments (El Cadillal Dam; 20 mg l^{-1} Cu^{2+}) at Tucumán, Argentina, suggesting that copper-resistant phenotypes were widespread among indigenous actinomycetes (Albarracín et al. 2005). Living organisms have been exposed to heavy metals released into the environment by geochemical processes (Brown et al. 1998). Since the age of industrialization and enhanced mining activities, this exposure has been dramatically increased by human activities. Therefore, it is not surprising to find that copper resistance ability is common among actinomycetes growing both in contaminated and non-contaminated soils.

Using qualitative and semi-quantitative screening assays (Fig. 20.3) using minimal agar medium supplemented with CuSO$_4$ at different concentrations up to

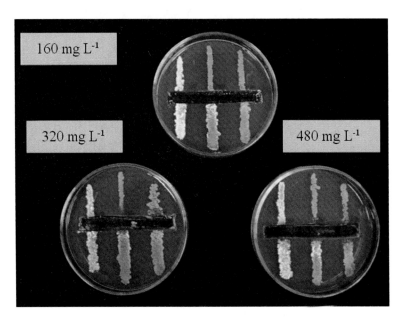

Fig. 20.3 Qualitative assay for assessing copper resistance in actinobacterial strains: *Streptomyces* spp. AB2A, AB2B, and AB2C (*left* to *right*, in the *upper row*) and *Amycolatopsis tucumanensis*, *Streptomyces* sp. C16, and C39 (*left* to *right*, in the *lower row*) tested at three copper concentrations (160, 320, and 480 mg L^{-1} Cu^{2+}). Wells of 5 mm were made in the center of the plate and filled with the copper solutions. Microbial growth around the well was used as the qualitative parameter of metal resistance

1,000 mg L^{-1} Cu^{2+}, it was determined that isolates from the polluted area displayed copper resistance to a level of 1,000 mg L^{-1} Cu^{2+} than cultures isolated from the non-polluted area (only up to 200 and 400 mg L^{-1} Cu^{2+}) (Albarracín et al. 2005). This variation in copper tolerance among actinomycetes suggests that the actinomycete strains recovered from copper-polluted soils might have evolved physiological and genetic mechanisms that allowed them to survive in adverse environments, which in turn might have give them a competitive advantage when growing in polluted environment (Boopathy 2000).

Following the same procedure, we selected the most copper-resistant actinobacterial strains (N = 11) in order to characterize them morphologically, physiologically, and at molecular level (Table 20.1). Morphological studies were performed by growing the strains in minimal agar medium and by observing the macro- and microstructures of the colonies and mycelia (Fig. 20.4) after 10 days of incubation at 30°C by both optical and electron microscopy (Albarracín et al. 2008b; Albarracín et al. 2010a).

Physiological studies were performed to assess the toxic effect of Cu^{2+} by growing the strains in liquid medium amended with copper. Highly dissimilar growth patterns and copper removal efficiency were observed for the selected strains grown in copper-treated medium. Among them, ABO strain displayed the higher copper-specific biopsorption ability (25 mg g^{-1}) (Table 20.1). For this reason, this strain was applied to polluted soil microcosms to assess its bioremediation ability.

The isolated actinomycetes strains were later subjected to 16s rDNA sequencing. Of the total, 11 strains belonged to the genus *Streptomyces* and only one (ABO/DSM 45259^{T}) to the genus *Amycolatopsis* (Table 20.1). Since *Amycolatopsis* sp. ABO was found as the most resistant strain, it was further identified to species level employing a polyphasic taxonomical approach (Colwell 1970; Albarracín et al. 2010a). Strain ABO (Fig. 20.5) was distinguished from its closest phylogenetic neighbors including *Amycolatopsis eurytherma*, using a combination of phenotypic and molecular tests. The strain ABO was identified as *Amycolatopsis tucumanensis* sp. nov by genotypic and phenotypic characteristics (Albarracín et al. 2010a).

20.3.1 Amycolatopsis tucumanensis: A Novel Copper-Resistant Actinobacterium Able to Colonize and Bioremediate Polluted Soil Microcosms

Bioremediation technologies can be broadly classified as ex situ or in situ (Iwamoto and Nasu 2001). Ex situ technologies are the treatments that remove contaminants at a separate treatment site. In situ bioremediation technologies involves the treatment of the contaminants at the contaminated sites and are currently classified into the following three categories: (1) bioattenuation – involves monitoring of the natural progress of degradation to ensure that contaminant concentration decreases with time; (2) biostimulation – when natural biodegradation or biotransformation is

Table 20.1 Morphological characteristics, copper biosorption (µg Cu/mg cells), and taxonomic affiliation of the copper-resistant strains of *Amycolatopsis tucumanensis* (ABO) and *Streptomyces* spp.

Strain	Morphological characteristics	Taxonomic affiliation according to the 16s rDNA gene	Genebank accession number	Copper specific biosorption (µg Cu/mg cells)[a]	References
ABO/DSM 45259[T]	White aerial mycelium, rectuflexibilis spore chains, smooth spores	*Amycolatopsis tucumanensis*	DQ886938	25 ± 1.04	Albarracín et al. (2005, 2008b, 2010a)
AB2A	White, gray, and black aerial mycelium, spiral spore chains, smooth spores	*Streptomyces* sp.	AY741363	23 ± 1.13	Albarracín et al. (2005, 2008a)
AB2B	White aerial mycelium, retinaculum spore chains, smooth spores	*Streptomyces* sp.	EF527809	8 ± 0.82	Albarracín et al. (2005)
AB2C	White-brown aerial mycelium, retinaculum spore chains, smooth spores	*Streptomyces* sp.	EF493850	9.5 ± 0.71	Albarracín et al. (2005)
AB3	White-gray aerial mycelium, spiral spore chains, smooth spores	*Streptomyces* sp.	AY741364	5.5 ± 0.71	Albarracín et al. (2005, 2008a)
AB5A	White aerial mycelium, spiral spore chains, smooth spores	*Streptomyces* sp.	EF527810	20 ± 0.36	Albarracín et al. (2005, 2008a)
AB5B	White aerial mycelium, retinaculum spore chains, smooth spores	*Streptomyces* sp.	EF527811	6.5 ± 0.70	Albarracín et al. (2005)
AB5C	White, gray, and black aerial mycelium, retinaculum spore chains, hairy spores	*Streptomyces* sp.	AY741365	7 ± 1.41	Albarracín et al. (2005)
AB5D	White aerial mycelium, retinaculum spore chains, smooth spores	*Streptomyces* sp.	EF527812	11.5 ± 3.53	Albarracín et al. (2005)
AB5E	White aerial mycelium, retinaculum spore chains, smooth spores	*Streptomyces* sp.	EF527813	6 ± 1.41	Albarracín et al. (2005)
AB5F	Brown and white aerial mycelium, retinaculum spore chains, smooth spores	*Streptomyces* sp.	EF527814	2.5 ± 0.69	Albarracín et al. (2005)

[a]Mean values and standard deviations are indicated

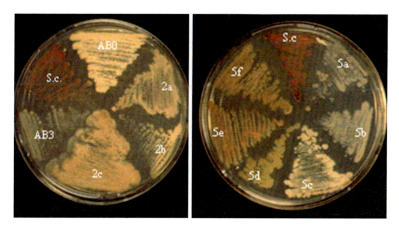

Fig. 20.4 Macroscopic observation of copper-resistant strains: *Amycolatopsis tucumanensis* (*ABO*) and *Streptomyces* spp. AB2A, AB2B, AB2C, AB3, AB5A, AB5B, AB5C, AB5D, AB5E, AB5F on MM agar. *Streptomyces coelicolor* (*Sc*) was included as a reference strain

Fig. 20.5 Magnified morphology of the novel copper-resistant strain *Amycolatopsis tucumanensis* (*ABO^T*) developed on ISP2 agar. (**a** and **b**) Stereoscopic lamp (Nikon), 4× and 8× respectively; (**c**) optic microscope, 400x (Nikon); (**d**) scanning electron microscopy, 14,810× (Zeiss Supra 55VP, Carl Zeiss NTS GmbH, Germany)

stimulated with nutrients, electron acceptors, or substrates; and (3) bioaugmentation – a way to enhance the biodegradability or bio-transforming capacity of contaminated sites by inoculation of bacteria with the desired catalytic capabilities (Iwamoto and Nasu 2001).

Soil bioremediation is a major challenge because of its heterogeneity and due to requirement of well-adapted microorganisms to remediate contaminated environment (Tabak et al. 2005). Hence, it is essential to identify microorganisms capable of cleaning up heavy metal–polluted soils. In this context, a novel *A. tucumanensis* was used to remediate copper-polluted soil microcosms (SM).

A. tucumanensis displayed high colonization ability when inoculated in SM with 20, 80, or 300 mg of copper kg⁻¹ soil. Interestingly, growth of *A. tucumanensis* was not inhibited when *A. tucumanensis* was grown in SM even with higher copper concentrations (Fig. 20.6). Instead, the maximum growth was obtained at the maximum copper concentration (300 mg Cu kg⁻¹ soil) tested. This result may be surprising but

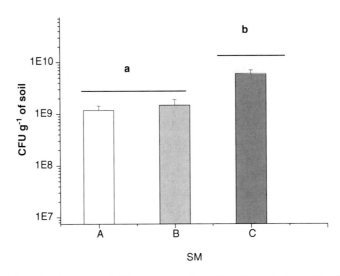

Fig. 20.6 Colony-forming units g^{-1} of *A. tucumanensis* developed in soil microcosms after 28 days of incubation. (**a**) $SM20_b$, (**b**) $SM80_b$, and (**c**) $SM300_b$. The values of the group a and b are significantly ($P \leq 0.05$) different

can be explained because the original sediment environment where from the strain was isolated had 600 mg Cu kg^{-1} soil (Albarracín et al. 2005). In a follow-up study, *A. tucumanensis* trapping ability of copper from the soil solution was tested by chemical and physical methods (Albarracín et al. 2010b). For $SM20_{nb}$, the bioavailable copper measured in the soil solution was approximately 30% with respect to the one recorded for $SM80_{nb}$. After application to soil, the strain did not show any, significant difference between the bioavailable copper from $SM20_{nb}$ and $SM20_b$. On the contrary, a significant depletion in the bioavailable copper (31%) in $SM80_b$ was observed with respect to the total bioavailable copper present in $SM80_{nb}$, demonstrating *A. tucumanensis* copper biosorption ability in a polluted soil (Albarracín et al. 2010b). The biotrapping copper ability of *A. tucumanensis* could be used to enhance bioremediation process of polluted soils, as proposed for other microorganisms (Gadd 2004; Jézéquel et al. 2005). In other study, Roane et al. (2001) used *Arthrobacter* sp. D9 to diminish the bioavailable Cd fraction in soils co-contaminated with pyrene and Cd while Groudev et al. (2001) used a bacterial consortium including *Streptomyces* representatives for the successful in situ bioremediation of soil highly polluted with radionucleids and heavy metals. Jézéquel and Lebeau (2008) found between 26 and 50% reduction in the bioavailable Cd when *Streptomyces* sp. R25 was applied to polluted soil while Polti et al. (2009) achieved a 90% reduction of Cr (VI) for soil bioaugmented with *Streptomyces* sp. MC1.

20.4   Effects of Copper on Maize: Copper Uptake and Accumulation by the Plants

Benimeli et al. (2010) studied the growth of *Zea mays* seedlings in vermiculite, an inert material, with the addition of Hoagland's nutrient solution supplemented with different concentrations of Cu (10^{-2}, 10^{-3}, and 10^{-4} M Cu^{2+}). Plants from all populations of *Zea mays* grew well in the presence of 10^{-4}–10^{-2} M Cu^{2+} with a similar leaf color to those grown under the control conditions; however, the effects of Cu^{2+} on roots of *Z. mays* varied with concentrations (Fig. 20.7). The 10^{-4} M or 10^{-3} M Cu^{2+} did not cause any significant changes in roots length compared to the control roots. Seedlings exposed to 10^{-2} M Cu^{2+} solution, however, reduced the root growth by 56% compared to untreated plants. The roots appeared thinner and the root tips were slightly blue.

The effect of Cu^{2+} on shoots and leaves length varied considerably (Fig. 20.7). When 10^{-4} M Cu^{2+} was applied to seedlings, it increased the shoots and leaves by 16% and 42%, respectively, relative to the control seedlings. The seedlings treated with 10^{-3} and 10^{-2} M Cu^{2}+ had, however, poor shoot and leaves growth and were smaller and appeared slightly yellow. Excess concentration of Cu is reported to produce toxic effects on plants. The toxicity of Cu inhibits plant growth, causing chlorosis of leaves and increasing leakage of solutes from root cell membranes (Shen et al. 1998; Murphy et al. 1999). For example, Ali et al. (2002) found that

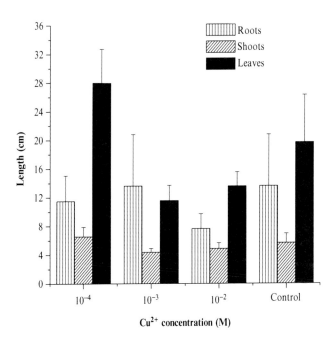

Fig. 20.7 Effect of different concentrations of Cu^{2+} on roots, shoots, and leaves growth of *Z. mays*. *Vertical bars* denote SE (*N* = 15) (Adapted from Benimeli et al. 2010)

Table 20.2 Effects of Cu^{2+} on fresh weight of roots, shoots, and leaves of Z. *mays*

Treatment (M)[a]	Roots (g)	Shoots (g)	Leaves (g)
Control	0.50 ± 0.10	0.21 ± 0.10	0.48 ± 0.20
10^{-4}	0.51 ± 0.10	0.25 ± 0.06	0.66 ± 0.20
10^{-3}	0.60 ± 0.10	0.14 ± 0.03	0.25 ± 0.10
10^{-2}	0.35 ± 0.08	0.18 ± 0.06	0.29 ± 0.10

Adapted from Benimeli et al. (2010)
Values indicate means $\pm$ SE ($N = 15$)
[a]M: mol L^{-1}

Table 20.3 Tolerance index (TI) of roots and total metal accumulation rate

Treatment (M)[a]	TI (%)	Accumulation rate (μg g^{-1}DW day^{-1})
10^{-4}	82.6	2.71
10^{-3}	91.3	4.00
10^{-2}	82.6	156.41

DW dry weight
[a]M: mol L^{-1}
Adapted from Benimeli et al. (2010)

$$TI(\%) = \frac{\text{Mean length of longest root in presence of added Cu}}{\text{Mean length of longest root in unamended control}} \times 100$$

$$\text{Accumulation rate (mg/g} \times \text{DW/day)}$$
$$= \frac{([\text{metal}]\text{root} \times \text{DWroot} + [\text{metal}]\text{shoot} \times \text{DWshoot} + [\text{metal}]\text{leave} \times \text{DWleave})}{6 \times (\text{DWroot} + \text{DWshoot} + \text{DWleave})}$$

root length of reed and maize seedlings was more sensitive than other measured growth parameters. The results observed here showed that seedlings treated with 10^{-3} and 10^{-2} M Cu^{2+} resulted in inhibition of shoot and leaves growth but not roots. In a similar report, Meng et al. (2007) also observed reduction in garlic (*Allium sativum*) seedlings following exposure to 10^{-4} M and 10^{-3} M Cu^{2+}. Furthermore, Cu^{2+} can, to some degree, cause partial improvement in fresh biomass of the roots, shoots, and leaves of Z. *mays* (Table 20.2). The fresh weights of roots, shoots and leaves were slightly increased or decreased in the presence of 10^{-4} and 10^{-3} M Cu^{2+}; however, the decrease was significant at 10^{-2} M Cu^{2+}.

The tolerance index (TI), based on root length, was, however, not significantly different for the three Cu treatments indicating that the sensitivity of the plant was similar in all studied cases. The total accumulation rate of Cu was very low (almost ten times less) at 10^{-4} and 10^{-3} M Cu^{2+} treatments compared to 10^{-2} (Table 20.3). The root had similar TI at the three Cu concentrations tested; the total metal accumulation rate by the seedlings was, however, increased more than 30 times at 10^{-2} M Cu^{2+}. These observations indicate that maize plants can tolerate and accumulate high Cu concentrations without visible morphological changes.

Accumulation of Cu in roots, shoots, and leaves of maize considerably varied, depending on Cu concentration used (Table 20.4). Interestingly, the Cu accumulation in maize plants increased with increase in Cu concentration, which was 382 times higher in roots, 157 in shoots, and only 16 in leaves compared to the respective controls.

Table 20.4 Copper accumulation by roots, shoots, and leaves of Z. *mays* after 6-day treatment

Treatment (M)[a]	Roots (µg/g DW)	Shoots (µg/g DW)	Leaves (µg/g DW)
Control	4.37 ± 1.70	3.78 ± 2.68	10.21 ± 5.76
10^{-4}	5.92 ± 0.70	5.83 ± 1.68	13.57 ± 1.76
10^{-3}	8.34 ± 0.10	6.52 ± 0.16	22.06 ± 3.49
10^{-2}	1668.25 ± 23.28	594.82 ± 2.73	160.97 ± 31.71

Adapted from Benimeli et al. (2010)
Values indicate means ± SE (N = 15)
[a]M: mol L^{-1}

Table 20.5 Distribution of copper in roots, shoots, and leaves of maize grown in soils treated with different concentrations of copper

Treatment (M)[a]	Total amount (µg/g DW)	Roots (%)	Shoots (%)	Leaves (%)
Control	18.36	23.8	20.6	55.6
10^{-4}	25.32	23.4	23	53.6
10^{-3}	36.92	22.6	17.6	59.8
10^{-2}	2424.04	68.8	24.5	6.7

Adapted from Benimeli et al. (2010)
[a]M: mol L^{-1}

When the average copper accumulation was compared to Cu added to the plant (Table 20.5), it was found that this metal could be accumulated by roots, shoots, and leaves, when the initial concentrations were 10^{-3} and 10^{-4} M. However, at 10^{-2} M Cu^{2+} concentration, the metal could not be accumulated by leaves and shoots, but the roots could increase their Cu accumulation capacity three times compared to the control, probably due to the interference of the high Cu concentration in the nutrient solution. Even, after 6 days of treatment, the seedling increased the capacity of Cu accumulation in roots, shoots, and leaves by 68.8%, 24.5%, and 6.7%, respectively, when 10^{-2} M Cu^{2+} was added with the nutrient solution. In a similar study, Liu et al. (2001) found that the Cu content in roots of Z. *mays* increased with increasing concentration of Cu^{2+}; however, they could not find significant Cu accumulation in shoots and leaves.

Copper is required by biological systems as a structural and catalytic enzyme component. When present in excess in soil, Cu^{2+} can be a stress factor and may alter physiological responses that can decrease the vigor of the plants and inhibit plant growth (Ouzounidou et al. 1995). Copper pollution has, therefore, become a major environmental problem due to the long-term use of copper-containing fungicides, industrial and urban activities (e.g., air pollution, city waste, and sewage sludge), and the application of pig and poultry slurries that contain significantly higher amounts of copper (Marschner 1995). To offset Cu toxicity problems, phytoremediation has been considered as an emerging technology that involves the use of selected and engineered metal-accumulating plants for environmental clean-up. In this context, many studies on uptake and accumulation of heavy metals by plants have been reported recently. For example, Wei et al. (2008) found that concentrations of Cu accumulated in plants of *Chrysanthemum coronarium* L. and *Sorghum sudanense* L. increased greatly with the increasing Cu level. To validate this further, Ucun et al. (2009) proposed the use of *Pinus sylvestris* L. biomass as biosorbent for removing Zn(II) and Cu(II) with the maximum biosorption efficiency (67%) observed for Cu(II). Maize plants can also be used for phytoremediation because of its high biomass yields and heavy metal tolerance. Ali et al. (2002), in their study, proposed that maize plant could provide a possible solution for the stabilization and restoration of Cu-polluted soils besides creating suitable environmental conditions for soil microorganisms and microfauna (Lin et al. 2008). However, few reports on copper accumulation by wild *Zea mays* from Argentina are available. The results of this study indicated that *Z. mays* plants have the potential ability to remove and accumulate Cu^{2+} from aqueous solutions. Yet, to our knowledge, no study has demonstrated copper accumulation in maize grain, even in Argentina.

20.4.1   Maize Plants as a Bioremediation Marker

Traditionally, the physical and chemical techniques are quite often used to detect environmental pollutants (Atlas and Bartha 2002). Spectroscopy (ultraviolet radiation, visible light, and infrared radiation) and chromatography (gas, liquid, and thin-layer high resolution) have been found as powerful tools since they can detect, quantify, and identify xenobiotics at parts per million (ppm) or even per billion (ppb) level. With respect to heavy metals, such techniques are very useful and accurate to detect the total content of metals in any environment. However, for practical purposes, the effective menace of heavy metals on organisms living in a particular environment is not reflected by the total concentration of metal present but depends on its bioavailable fraction. For this purpose, it is convenient to use physical and/or chemical extraction (Csillag et al. 1999; Gray et al. 1999), although the use of biosensors and biomarkers is a much more attractive approach (Atlas and Bartha 2002). The biomarkers are biological systems that modify their response upon changes in the environment and, thus, are extremely useful in monitoring bioremediation processes.

Ogboghodo et al. (2004) monitored the effect of manure on soil contaminated with oil using maize as biomarker. Reduction of soil contamination after fertilizer application improved growth and yield of plants compared to those grown in contaminated soil but not fertilized. In this case, maize was used not only as a biological indicator of oil pollution, but also as an indicator of the ability of manure to achieve soil remediation. In other study, Benimeli et al. (2008) monitored the efficiency of degradation of lindane in soil by the strain *Streptomyces* sp. M7 using maize plants. Lindane concentrations of 100, 200, and 400 g kg^{-1} soil did not affect the germination and vigor index of maize plants seeded in contaminated soils without *Streptomyces* sp. M7. When this microorganism was inoculated under the identical conditions, a better vigor index was observed with 68% of lindane removal.

Similarly, Albarracín et al. (2010b) confirmed the ability of *A. tucumanensis* to effectively bioremediate copper-polluted soil microcosms by using *Zea mays*; the plants were seeded in non-bioremediated soils ($SM20_{nb}$ and $SM80_{nb}$) and bioremediated soils ($SM20_b$ and $SM80_b$). The plants grown in $SM80_{nb}$ took up eightfold more copper into their roots than the ones grown in $SM20_{nb}$ without phenotypical modification. Neither was it observed a significant reduction in biomass and length of the plants seeded in $SM80_{nb}$ with respect to the control. Similar observations were made by Lin et al. (2008), testing different copper concentrations (200 and 400 mg kg^{-1}) on soil seeded with *Z. mays*. The plants were able to take great quantities of copper without displaying morphological modification.

20.5 Conclusion

Since heavy metal pollution is a worldwide problem, new strategies must be developed to overcome this situation. Among the diverse methods of remediation, technologies that use living organisms and/or their biomolecules for this purpose are preferable to conventional systems due to their better efficiency. In particular, the application of microorganisms and/or plants to bioremediate heavy metal–polluted soil has become important. The results achieved so far from different studies indicate the feasibility of using copper- and chromium-resistant actinobacteria to efficiently bioremediate polluted soils. Some of the actinobacteria, like *Streptomyces* and *Amycolatopsis* strains, isolated from polluted environments have shown efficient bioreduction or bioimmobilization abilities when inoculated onto heavy metal–polluted soil microcosms. On the other hand, *Zea mays* plants have also been found as interesting both as biomarker and bioremediation tool. Current research is, however, directed toward employing composite application of both actinobacteria and maize plants in order to achieve a more effective, viable, and ecologically balanced bioremediation strategy to clean up polluted soils at a larger scale.

References

Ait Ali N, Bernal MP, Ater M (2002) Tolerance and bioaccumulation of copper in *Phragmites australis* and *Zea mays*. Plant Soil 239:103–111

Albarracín VH, Amoroso MJ, Abate CM (2005) Isolation and characterization of indigenous copper resistant actinomycete strains. Chem Erde-Geochem 65(S1):145–156

Albarracín VH, Avila AL, Amoroso MJ, Abate CM (2008a) Copper removal ability by *Streptomyces* strains with dissimilar growth patterns and endowed with cupric reductase activity. FEMS Microbiol Lett 288:141–148

Albarracín VH, Winik B, Kothe E, Amoroso MJ, Abate CM (2008b) Copper bioaccumulation by the actinobacterium *Amycolatopsis* sp. AB0. J Basic Microbiol 48:323–330

Albarracín VH, Alonso-Vega P, Trujillo ME, Amoroso MJ, Abate CM (2010a) *Amycolatopsis tucumanensis* sp. nov., a novel copper resistant actinobacterium isolated from polluted sediments in Tucumán, Argentina. Int J Syst Evol Microbiol 60:397–401

Albarracín VH, Amoroso MJ, Abate CM (2010b) Bioaugmentation of copper polluted soil by *Amycolatopsis tucumanensis* to diminish phytoavailable copper for *Zea mays* plants. Chemosphere 79:131–137

Amoroso MJ, Castro GR, Carlino FJ, Romero NC, Hill RT et al (1998) Screening of heavy metal-tolerant actinomycetes isolated from the Salí River. J Gen Appl Microbiol 44:129–132

Amoroso MJ, Castro GR, Durán A, Peraud O, Oliver G, Hill RT (2001) Chromium accumulation by two *Streptomyces* spp. isolated from riverine sediments. J Indian Microbiol Biotechnol 26:210–215

Arifuzzaman M, Khatun MR, Rahman H (2010) Isolation and screening of actinomycetes from Sundarbans soil for antibacterial activity. Afr J Biotechnol 9:4615–4619

Atlas RM, Bartha R (2002) Ecología microbiana y microbiología ambiental. Pearson Educación, Madrid

Bae WC, Lee HK, Choe YC, Jahng DJ, Lee SH, Kim SJ, Lee JH, Jeong BC (2005) Purification and characterization of NADPH-dependent Cr (VI) reductase from *Escherichia coli* ATCC 33456. J Microbiol 43:21–27

Baldi F, Vaughan AM, Olson GJ (1990) Chromium (VI) resistant yeast isolated from a sewage treatment plant receiving tannery wastes. Appl Environ Microbiol 56:913–918

Beleza VM, Boaventura RA, Almeida MF (2001) Kinetics of chromium removal from spent tanning liquors using acetylene production sludge. Environ Sci Technol 35:4379–4383

Benimeli CS, González AJ, Chaile AP, Amoroso MJ (2007) Temperature and pH effect on lindane removal by *Streptomyces* sp. M7 in soil extract. J Basic Microbiol 47:468–473

Benimeli CS, Fuentes MS, Abate CM, Amoroso MJ (2008) Bioremediation of lindane-contaminated soil by *Streptomyces* sp. M7 and its effects on *Zea mays* growth. Int Biodeterior Biodegr 61: 233–239

Benimeli CS, Medina A, Navarro CM, Medina RB, Amoroso MJ, Gómez MI (2010) Bioaccumulation of copper by *Zea mays:* impact on roots, shoots and leaves growth. Water Air Soil Poll 210:365–370

Boopathy R (2000) Factors limiting bioremediation technologies. Biores Technol 74:63–67

Brown NL, Lloyd JR, Jakeman K, Hobman JL, Bontidean I, Mattiasson B, Csöregi E (1998) Heavy metal resistance genes and proteins in bacteria and their application. Biochem Soc Trans 26:662–664

Brunet J, Repellin A, Varrault G, Terryn N, Zuily-Fodil Y (2008) Lead accumulation in the roots of grass pea (Lathyrus sativus L.): a novel plant for phytoremediation systems? C R Biol 331:859–864

Camargo FAO, Bento FM, Okeke BC, Frankenberger WT (2004) Hexavalent chromium reduction by an actinomycete, *Arthrobacter crystallopoietes* ES 32. Biol Trace Elem Res 97:183–194

Cefalu WT, Hu FB (2004) Role of chromium in human health and in diabetes. Diabetes Care 27:2741–2751

Cheung KH, Gu JD (2007) Mechanism of hexavalent chromium detoxification by microorganisms and bioremediation application potential: a review. Int Biodeter Biodegr 59:8–15

Colwell RR (1970) Polyphasic taxonomy of the genus *Vibrio*: numerical taxonomy of *Vibrio cholerae*, *Vibrio parahaemolyticus*, and related *Vibrio* species. J Bacteriol 104:410–433

Costa M, Klein CB (2006) Toxicity and carcinogenicity of chromium compounds in humans. Crit Rev Toxicol 36:155–163

Csillag J, Pártay G, Lukács A, Bujtás K, Németh T (1999) Extraction of soil solution for environmental analysis. Int J Environ Anal Chem 74:305–324

Czakó-Vér K, Batic M, Raspor P, Sipicki M, Pesti M (1999) Hexavalent chromium uptake by sensitive and tolerant mutants of *Schizosacchoromyces pombe*. FEMS Microbiol Lett 178:109–115

Das S, Chandra AL (1990) Chromate reduction in *Streptomyces*. Experientia 46:731–733

Del Rio M, Font R, Almela C, Velez D, Montoro R, De Haro A (2002) Heavy metals and arsenic uptake by wild vegetation in the Guadiamar river area after the toxic spill of the Aznalcollar mine. J Biotechnol 98:125–137

Demirevska-Kepova K, Simova-Stoilova L, Stoyanova Z, Holzer R, Feller U (2004) Biochemical changes in barley plants after excessive supply of copper and manganese. Environ Exp Bot 52:253–266

Gadd GM (2004) Microbial influence on metal mobility and application for bioremediation. Geoderma 122:109–119

Ganguli A, Tripathi AK (2002) Bioremediation of toxic chromium from electroplating effluents by chromate-reducing *Pseudomonas aeruginosa* A2Chr in two bioreactors. Appl Microbiol Biotech 58:416–420

Georgopoulus PG, Roy A, Opiekun RE, Yonone-Lioyand MJ, Lioy PJ (2002) Introduction: copper and man. In: Georgopoulus PG, Roy A, Opiekun RE, Yonone-Lioyand MJ, Lioy PJ (eds.) Environmental dynamics and human exposure to copper, vol 1, Environmental dynamics and human exposure issues3. International Copper Association Ltd, New York, pp 15–26

Ghosh M, Spingh SP (2005) A review on phytoremediation of heavy metals and utilization of its by products. Appl Ecol Environ Res 3:1–18

Glass DJ (2000) Economical potential of phytoremediation. In: Raskin I, Ensley BD (eds.) Phytoremediation of toxic metals: using pants to clean up the environment. Wiley, New York, pp 15–31

Gray CW, McLaren RG, Roberts AHC, Condron LM (1999) Cadmium phytoavailability in some New Zealand soils. Aust J Soil Res 37:461–477

Groudev SN, Spasova II, Georgiev PS (2001) *In situ* bioremediation of soils contaminated with radioactive elements and toxic heavy metals. Int J Miner Process 62:301–308

Guo JK, Lin YB, Zhao ML, Sun R, Wang TT, Tang M, Wei GH (2009) *Streptomyces plumbiresistens* sp. nov., a lead-resistant actinomycete isolated from lead-polluted soil in north-west China. Int J Syst Evol Microbiol 59:1326–1330

Horton RN, Apel WA, Thompson VS, Sheridan PP (2006) Low temperature reduction of hexavalent chromium by a microbial enrichment consortium and a novel strain of *Arthrobacter aurescens*. BMC Microbiol 6:5

Iwamoto T, Nasu M (2001) Current bioremediation practice and perspective. J Biosci Bioeng 92:1–8

Jézéquel K, Lebeau T (2008) Soil bioaugmentation by free and immobilized bacteria to reduce potentially phytoavailable cadmium. Biores Technol 99:690–698

Jézéquel K, Perrin J, Lebeau T (2005) Bioaugmentation with a *Bacillus* sp. to reduce the phytoavailable Cd of an agricultural soil: comparison of free and immobilized microbial inocula. Chemosphere 59:1323–1331

Jing Y, Zhen-Li HE, Yang X (2007) Role of soil rhizobacteria in phytoremediation of heavy metal contaminated soils. J Zhejiang Univ Sci B 8:192–207

Kabata-Pendias A, Pendias H (1984) Trace elements in soils and plants. CRC Press, Boca Raton

Kabata-Pendias A, Pendias H (1992) Trace elements in soil and plants. CRC Press, Boca Raton

Kothe E, Bergmann H, Büchel G (2005) Molecular mechanisms in bio-geo-interactions: from a case study to general mechanisms. Chem Erde-Geochem 65(S1):7–27

Lasat HA (2002) Phytoextraction of toxic metals: a review of biological mechanisms. J Environ Qual 31:109–120

Laxman SR, More S (2002) Reduction of hexavalent chromium by *Streptomyces griseus*. Miner Eng 15:831–837

Lin Q, Shen KL, Zhao HM, Li WH (2008) Growth response of Zea mays L. in pyrene-copper co-contaminated soil and the fate of pollutants. J Hazard Mat 150:515–521

Liu DH, Jiang WS, Hou WQ (2001) Uptake and accumulation of copper by roots and shoots of maize (*Zea mays* L.). J Environ Sci 13:228–232

Liu YG, Xu WH, Zeng GM, Gao H (2006) Cr (VI) reduction by Bacillus sp. isolated from chromium landfill. Process Biochem 41:1981–1986

Lloyd JR (2003) Microbial reduction of metals and radionuclides. FEMS Microbiol Rev 777:1–15

Lloyd JR, Lovley DR (2001) Microbial detoxification of metals and radionuclides. Curr Opin Biotechnol 12:248–253

Luna CM, Gonzalez CA, Trippi VS (1994) Oxidative damage caused by an excess of copper in oat leaves. Plant Cell Physiol 35:11–15

Mabbett AN, Macaskie LE (2001) A novel isolate of *Desulfovibrio* sp. with enhanced ability to reduce Cr (VI). Biotechnol Lett 23:683–687

Marschner H (1995) Mineral nutrition of higher plants. Academic, London

Megharaj M, Avudainayagam S, Naidu R (2003) Toxicity of hexavalent chromium and its reduction by bacteria isolated from soil contaminated with tannery waste. Curr Microbiol 47:51–54

Meng Q, Zou J, Zou J, Jiang W, Liu D (2007) Effect of Cu^{2+} concentration on growth, antioxidant enzyme activity and malondialdehyde content in garlic (*allium sativum* L.). Acta Biol Cracoviensia Serie Bot 49:95–101

Myers CR, Carstens BP, Antholine WE, Myers JM (2000) Chromium(VI) reductase activity is associated with the cytoplasmic membrane of anaerobically grown *Shewanella putrefaciens* MR-1. J Appl Microbiol 88:98–106

Murphy AS, Eisinger WR, Shaff JE, Kochian LV, Taiz L (1999) Early copper-induced leakage of K^+ from Arabidopsis seedlings is mediated by ion channels and coupled to citrate efflux. Plant Physiol 121:1375–1382

Nielsen HD, Brownlee C, Coelho SM, Brown M (2003) Inter-population differences in inherited copper tolerance involve photosynthetic adaptation and exclusion mechanisms in *Fucus serratus*. New Phytol 160:157–165

Ogboghodo IA, Erebor EB, Osemwota IO, Isitekhale HH (2004) The effects of application of poultry manure to crude oil polluted soils on maize (Zea mays) growth and soil properties. Environ Monit Assess 96:153–161

Ouzounidou G, Ciamporova M, Moustakas M (1995) Responses of maize (*Zea mays* LR) plants to copper stress: IR Growth, mineral content and ultrastructure of roots. Environ Exp Bot 35:167–176

Park CH, Keyhan M, Wielinga B, Fendorf S, Matin A (2000) Purification to homogeneity and characterization of a novel *Pseudomonas putida* chromate reductase. Appl Environ Microbiol 66:1788–1795

Pattanapipitpaisal P, Brown NL, Macaskie LE (2001) Chromate reduction by *Microbacterium liquefaciens* immobilised in polyvinyl alcohol. Biotechnol Lett 23:61–65

Polti MA, Amoroso MJ, Abate CM (2007) Chromium (VI) resistance and removal by actinomycete strains isolated from sediments. Chemosphere 67:660–667

Polti MA, García RO, Amoroso MJ, Abate CM (2009) Bioremediation of Chromium(VI) contaminated soil by *Streptomyces* sp. MC1. J Basic Microbiol 49:285–292

Polti MA, Amoroso MJ, Abate CM (2010a) Chromate reductase activity in *Streptomyces* sp. MC1. J Gen App Microbiol 56:11–18

Polti MA, Amoroso MJ, Abate CM (2010b) Intracellular chromium accumulation by *Streptomyces* sp. MC1. Water Air Soil Poll 24:49–57

Rai UN, Tripathi RD, Vajpayee P, Jha V, Ali MB (2002) Bioaccumulation of toxic metals (Cr, Cd, Pb, and Cu) by seeds of *Euryale ferox* Salisb. (Makhana). Chemosphere 46:267–272

Reeves RD, Baker AJM (2000) Metal accumulating plants. In: Raskin I, Ensley BD (eds.) Phytoremediation of toxic metals: using plants to clean up the environment. Wiley, New York, pp 193–229

Richards JW, Krumholz GD, Chval MS, Tisa LS (2002) Heavy metal resistance patterns of *Frankia* strains. Appl Environ Microbiol 68:923–927

Roane TM, Josephson KL, Pepper IL (2001) Dual-bioaugmentation strategy to enhance remediation of co-contaminated soil. Appl Environ Microbiol 67:3208–3215

Shen ZG, Zhang FQ, Zhang FS (1998) Toxicity of copper and zinc in seedings of Mung Bean and inducing accumulation of polyamine. J Plant Nutr 21:1153–1162

Smith WA, Apel WA, Petersen JN, Peyton BM (2002) Effect of carbon and energy source on bacterial chromate reduction. Bioremediation J 6:205–215

Tabak HH, Lens P, Van Hullebusch ED, Dejonghe W (2005) Developments in bioremediation of soils and sediments polluted with metals and radionuclides – 1. Microbial processes and mechanisms affecting bioremediation of metal contamination and influencing metal toxicity and transport. Rev Environ Sci Biotechnol 4:115–156

Turick CE, Graves C, Appel WA (1998) Bioremediation potential of Cr (VI)-contaminated soil using indigenous microorganisms. Bioremediat J 2:1–6

Ucun H, Aksakal O, Yildiz E (2009) Copper(II) and zinc(II) biosorption on *Pinus sylvestris* L. J Hazard Mater 161:1040–1045

Vaimajala S, Peyton BM, Apel WA, Peterson JN (2002) Chromate reduction in *Shewanella oneidensis* MR-1 is an inducible process associated with anaerobic growth. Biotechnol Prog 18:290–296

Vainshtein M, Kuschk P, Mattusch J, Vatsourina A, Wiessner A (2003) Model experiments on the microbial removal of chromium from contaminated groundwater. Water Res 37:1401–1405

Viti C, Pace A, Giovannetti L (2003) Characterization of Cr (VI)-resistant bacteria isolated from chromium-contaminated soil by tannery activity. Curr Microbiol 46:1–5

Wang M, Zou J, Duan X, Jiang W, Liu D (2007) Cadmium accumulation and its effects on metal uptake in maize (*Zea mays* L.). Biores Technol 98:82–88

Wei L, Luo C, Li X, Shen Z (2008) Copper accumulation and tolerance in *Chrysanthemum coronarium* L. and *Sorghum sudanense* L. Arch Environ Contam Toxicol 55:238–246

Chapter 21
Importance of Free-Living Fungi in Heavy Metal Remediation

Almas Zaidi, Mohammad Oves, Ees Ahmad, and Mohammad Saghir Khan

Abstract Discharge of heavy metals from various human activities including agricultural practices and metal processing industries is known to cause adverse effects on the environment. Even though conventional technologies adopted for removal of heavy metals from polluted environment tend to be efficient, they are generally expensive and produce huge quantity of toxic chemical products. The use of biological materials including fungal biomass offers an economical, effective, and safe option for removing heavy metals and, therefore, has emerged as a potential alternative method to conventional treatment techniques. Among the various remediation strategies, biosorption of heavy metals by metabolically active or inactive nonliving (dead) biomass of fungal origin is an innovative and alternative technology for removal of metals from contaminated sites. Due to unique chemical composition, fungal biomass sequesters metal ions by forming metal complexes with certain reactive groups on their cell surface and does not require growth-supporting conditions. Biomass of numerous fungi like *Aspergillus, Penicillium, Mucor, Rhizopus*, etc., has been found to have highest metal adsorption capacities. Biomass generated as a by-product of fermentative processes offers great potential for adopting an economical metal-recovery system. The purpose of this chapter is to gather state of the art information on the use of fungal biomass and explores the possibility of exploiting them for heavy metal remediation.

Keywords Biosorption • Fungal biomass • Metal uptake • Extracellular polysaccharides

A. Zaidi (✉) • M. Oves • E. Ahmad • M.S. Khan
Department of Agricultural Microbiology, Faculty of Agricultural Sciences,
Aligarh Muslim University, Aligarh 202002, Uttar Pradesh, India
e-mail: alma29@rediffmail.com

M.S. Khan et al. (eds.), *Biomanagement of Metal-Contaminated Soils*, 479
Environmental Pollution 20, DOI 10.1007/978-94-007-1914-9_21,
© Springer Science+Business Media B.V. 2011

21.1 Introduction

When the amount of heavy metals exceeds a certain level due to pollutants emanating from various anthropogenic sources, it causes soil contamination and adversely affects agricultural produce (Gupta et al. 2008; Bhattacharyya et al. 2008). The primary sources of heavy metal pollution include the burning of fossil fuels, mining and smelting of metalliferous ores, municipal wastes, fertilizers, pesticides, and sewage (Marcovecchio et al. 2007; Wei and Zhou 2008; Adepoju-Bello et al. 2009). In some areas, sewage when used for irrigation is known to contribute significantly to the heavy metal content of soils (Singh et al. 2004; Mapanda et al. 2005; Wu and Cao 2010). Soil metal content in general is, however, significantly higher in industrial area where accumulation may be several times higher than the average content in non-contaminated areas. The distribution of metals is influenced by the nature of parent materials and climate while their relative mobility depends on soil characteristics (Krishna and Govil 2007). Additionally, areas distant from industrial centers also show increased metal concentrations due to long-range atmospheric transport as reported by numerous authors (Jonathan et al. 2004; Wilson et al. 2005).

To overcome heavy metal toxicity to living organisms or to make metal-contaminated soil suitable for cultivation, various approaches have been applied. The conventional treatment processes for example have been found neither effective nor economical (Amini et al. 2008). Moreover, chemical precipitation of heavy metals produces large amounts of sludge and is ineffective when metal ion concentrations are lower than 100 mg l^{-1} (Wang and Chen 2006). In addition, solvent extraction techniques are not suitable for effluents with low heavy metal concentrations (Mameri et al. 1999) while multi-metal contamination is a common problem in the industrial effluents (Gikas 2008). In contrast, the biological approaches that may involve the use of stress-tolerant organisms like fungi for example *Fusarium, Gliocladium, Penicillium,* and *Trichoderma* have been found effective and inexpensive in metal decontamination/removal from polluted environment. Among microorganisms, fungi, which adopt various strategies for metal removal (Fig. 21.1), display a high ability to immobilize toxic metals by insoluble metal oxalate formation, biosorption, or chelation onto melanin-like polymers (Baldrian 2003; Pal et al. 2006). Fungal biomass have been found to accumulate heavy metals such as cadmium, copper, mercury, lead, and zinc very efficiently and systems using *Rhizopus arrhizus* have been developed for treating uranium and thorium (Gavrilesca 2004; Li and Yuan 2006; Javaid et al. 2010). In a recent study, Vala et al. (2010) have found *Aspergillus flavus* as a promising candidate for environmental bioremediation. And hence, the ability of different mesophilic, psychrophilic, or thermophilic fungi to transform a wide range of hazardous chemicals to non-toxic forms has generated interest in using them in bioremediation (Alexander 1994). In other study, Rehman et al. (2007) reported that *Candida tropicalis* removed 64% copper from the industrial wastewater after 4 days and 74% after 8 days. A study by Kahraman et al. (2005) demonstrated that the live biomass of two white rot fungi had a higher copper adsorption capacity when compared with dried biomass. Pan et al. (2009) analyzed the effects of single and multiple heavy metals on the growth and uptake of

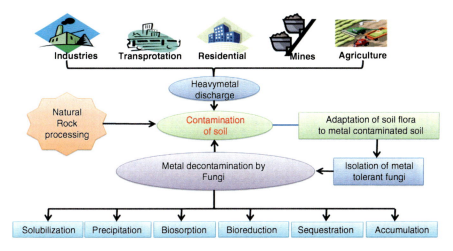

Fig. 21.1 Source of heavy metal pollution and strategies adopted by fungi for metal decontamination

consortium of two types of fungal strains, *Penicillium* sp. A1 and *Fusarium* sp. A19. These fungal strains were tested to be tolerant to several heavy metals. Combined inoculation of A1 and A19 had profound effects on the growth of the two fungi in potato dextrose agar (PDA) and Czapex Dox agar (CDA) under the treatments with Cu^{2+} and mixed $Cd^{2+}+Zn^{2+}$. The amount of metals through bioaccumulation by A1, A19, and A1+A19 was significantly higher than that through biosorption by these fungi. Similarly, El-Morsy (2004) studied 32 fungal species isolated from polluted water in Egypt for their resistance to metals and found that *Cunninghamela echinulata* biomass could be employed as a biosorbent of metal ions in wastewater. In other studies (Svoboda et al. 2006; Villegas et al. 2008; Antonijevic and Maric 2008), the concentrations of heavy metals have also been observed in the fruiting bodies (Courtecuisse 1999) of different mushrooms collected from sites adjacent to heavy metal smelters, landfills of sewage sludge, emission area. Mushrooms are generally capable of accumulating heavy metals, which subsequently become the source in food chain (Kalac 2009) as reported by Xiangliang et al. (2005). Ayodele and Odogbili (2010) reported heavy metals in three edible mushrooms, like *Lentinus squarrosulus*, *Pleurotus tuberregium*, and *Psathyrella atroumbonata*, growing in Abraka, Delta State, Nigeria. Likewise, filamentous fungi have been revealed as promising candidates for Cr(VI) bioremediation (Morales-Barrera and Cristiani-Urbina 2008; Morales-Barrera et al. 2008).

21.2 Heavy Metal Toxicity and Tolerance in Fungi

Some of the metals like magnesium, potassium, calcium, and sodium must be present for normal body functions. Others like copper, iron, cobolt, manganese, molybdenum, and zinc are required at low levels as catalyst for enzyme activities

(Adepoju-Bello et al. 2009). Of these micronutrients, Zn, Cu, Mn, Ni, and Co are important for plant growth (Marschner 1995). Some of the other metals like Cd, Pb, and Hg have no known biological function. However, excess exposure to heavy metals can result in toxicity to both microbes like fungi and crop plants (Alkorta et al. 2004; Van-der-Heggen et al. 2010; Chatterjee and Luo 2010). Heavy metal can cause toxicity by forming complexes with protein or inactivate important enzyme systems. The modified biological molecules lose their ability to function properly and result in the malfunction or death of the cells. The toxicity can last longer, but some heavy metals could even be transformed from relatively low toxic species into more toxic forms. The bioaccumulation and bioaugmentation of heavy metal through food chain could damage normal physiological activity and endanger human life. Therefore, once the agricultural land is contaminated, it becomes important to solve this problem. To combat metal toxicity, fungi have evolved mechanisms. For example, tolerance of a facultative marine fungus *Aspergillus flavus* toward As (V) was tested by Vala et al. (2010). The tolerance of fungi strains including *Penicillium funiculosum, Aspergillus foetidus, Penicillium simplicissimum* for different heavy metals, which could be leached, from nickel laterite ores (Ni, Co, Fe, Mg, and Mn) was studied. These strains were exposed to heavy metals up to 2,000 ppm. The tolerant strains were selected by repeated subculturing in petri dishes with increasing metal concentration in the medium. The degree of tolerance was measured from the growth rate in the presence of the various heavy metals and compared to a control, which contained no heavy metals. Rehman and Anjum (2010) isolated multiple metal-tolerant fungi (*Candida tropicalis*) from industrial effluents. It appears that *Penicillium funiculosum* and *Aspergillus foetidus* were the most tolerant to the heavy metals and exhibited strong growth even exceeding the control. *Penicillium simplicissimum* showed the least tolerance particularly for Ni and Co. A growth pattern, which was consistent for each strain under various heavy metals, was observed as a function of time. The growth pattern of the fungi exhibited a lag, retarded, similar, and enhanced rate of growth in the presence of heavy metal relative to the control. The similarity in the pattern appears to suggest the tolerance development or adaptation of the fungi for heavy metals (Valix et al. 2001). The role of vacuole in the detoxification of metal ions was investigated, and the results showed that vacuole-deficient strain displayed much higher sensitivity and the biosorption capacity for Zn, Mn, Co, and Ni decreased (Ramsay and Gadd 1997). However, no significant difference for Cd and Cu biosorption or sensitivity to both the metal ions was observed between wild type and mutant of *S. cerevisiae*. Gharieb and Gadd (1998) found that the vacuolar-lacking strains and the defective mutants of *S. cerevisiae* display higher sensitivity to chromate and tellurite with a decrease in the cellular content of each metal, whereas the tolerance to selenite increased with the cellular content of Se. Many genes involved in the uptake or detoxification or tolerance to metal ions have been identified (Rosen 2002). For example, the *S. cerevisiae* Arr4p plays an important role in the tolerance to metal ions like As^{3+}, As^{5+}, Co^{2+}, Cr^{3+}, Cu^{2+}, VO_4^{3-} (Shen et al. 2003).

21.3   Metal Ion Uptake by Fungi

The metal uptake by living and dead cells can occur by (1) surface binding of metal ions to cell wall and extracellular material and (2) intracellular uptake or bioaccumulation – uptake into the cell across the cell membrane, which is dependent on the cell metabolism (Volesky 1990). The first mode of metal uptake is commonly employed by both living and dead cells while the intracellular uptake occurs only in living cells. Among living cells, metal uptake is also facilitated by the production of metal-binding proteins. However, whatever may be the mode of metal uptake, both living and dead cells of fungi are capable of metal adsorption.

21.3.1   Metal Uptake by Living Cells

Fungi can adapt and grow under various extreme conditions of pH, temperature, and nutrient availability, as well as high metal concentrations (Anand et al. 2006). The cell wall material of fungi shows excellent metal-binding properties (Gupta et al. 2000). Generally, microbial biomasses including those of fungi have evolved various measures to respond to heavy metals stress. Such processes include transport across the cell membrane, biosorption to cell walls, entrapment in extracellular capsules, and precipitation and transformation of metals. The living cells of *Penicillium, Aspergillus, Rizopus, Mucor, Saccharomyces,* and *Fusarium* have been shown to biosorb metal ions (Volesky et al. 1993; Tan and Cheng 2003). The metal uptake by living fungal cells, however, depends on the composition of media and growth environment, contact time, age of cells, and biomass-producing ability. Volesky (1994) for example showed that *R. nigricans* when grown in potato-dextrose medium supplemented with different sugars like glucose and sucrose showed a variable uranium uptake capacity. Similarly, the amount of chromium biosorbed per unit weight of biomass decreased with an increase in concentration of *R. arrhizus, R. nigricans, A. oryzae,* and *A. Niger* (Niyogi et al. 1998).

21.3.2   Cell Surface Precipitation of Metals

The cell wall is the first cellular structure to come in contact with metal ions. After contact, the heavy metals interact stoichiometrically with functional groups of cell wall including phosphate, carboxyl, amine, and phosphodiesters. To consolidate these facts, several studies have been conducted (Simmons and Singleton 1996; Machado et al. 2009). For example, Brady and Duncan (1994a) observed that the uptake capacity of metals can be reduced by blocking the functional groups (amino, carboxyl, or hydroxyl) of fungal/actinomycetal cell walls suggesting that the cell wall components do play a major role in metal binding. Similarly, the characterization of biosorbents surface by infrared spectroscopy has also suggested the involvement of carboxyl and amino groups in the metals removal (Machado et al. 2009). At very low pH values, these groups are protonated and as a result, the surface of biosorbent is surrounded by H^+ ions (Parvathi and Nagendran 2007), which enhance the metal interaction with binding sites of the biosorbent due to electrostatic forces (Özer and Özer 2003).

The synthesis of exracellular polymeric substances (EPS), such as polysaccharides, glucoprotein, lipopolysaccharide, and soluble peptide, also possesses functional groups, which can adsorb metal ions. Generally, complexation, ion exchange, adsorption (by electrostatic interaction or van der Waals force), inorganic microprecipitation, oxidation, and/or reduction have been proposed to explain metal uptake by fungi (Jung et al. 1998). The roles of EPS in metal removal in a biosorption system are usually neglected or ignored, especially in the case of fungi and yeast. Among the limited studies on metal removal by EPS, most of them are related to the EPS extracted from intact organism cells, but not the EPS in living cells. However, Suh et al. (1999b), for example, investigated the effect of EPS on Pb^{2+} removal by a polymorphic fungus *Aureobasidium pullulans* and observed that Pb^{2+} accumulated only on the surface of the intact cells of *A. pullulans* due to the presence of EPS. Lead also penetrated into the inner parts of the EPS-extracted cells of *A. pullulans*. The uptake of Pb^{2+} increased with storage period of cells and more than 90% of the Pb^{2+} was removed due to excreted EPS. However, the ability of EPS-extracted cells to biosorb Pb^{2+} was significantly lower compared to the intact cells and remained constant, irrespective of the storage time. Suh et al. (1998) also discovered that the initial rate of Pb^{2+} uptake by live cells of *S. cerevisiae* is lower than that of dead cells, while in the case of *A. pullulans*, both the capacity and the initial rate of Pb^{2+} accumulation in the live cells are higher than those in the dead cells, due to the presence of EPS for live *A. pullulans*.

21.3.3   Intracellular Accumulation

Metal ions can also enter the cell provided the cell wall is disrupted naturally (e.g., autolysis) or artificially by mechanical forces or alkali treatment. The intracellular accumulation of metal is an energy-driven process and depends on functional metabolism of organisms. Once inside, metal ions are transformed into species other than parent ones or could be precipitated within the cell. After entering into the cell, the metal ions are compartmentalized into different subcellular organelles (Vijver et al. 2004). Metal accumulation strategies for essential and non-essential metal ions may, however, be different. Limiting metal uptake or active excretion, storage in an inert form, and/or excretion of stored metal are the main strategies used in removal of essential metals. For non-essential metals, excretion from the metal excess pool and internal storage are the major strategies. In general, the cellular sequestration mechanism involves the formation of distinct inclusion bodies and bind metals to heat-stable proteins. The former includes three types of granules: (1) type A: amorphous deposits of calcium phosphates; (2) type B: mainly containing acid phosphatase, accumulating Cd, Cu, Hg, and Ag; and (3) type C: excess iron stored in granules as haemosiderin. The latter mechanism involves metal binding protein, metallothioneins (MT), which can be induced by many substances, including heavy metal (for details, see Chap. 9).

21.4 Metal Uptake by Dead Cells

The application of dead biomass offers certain advantages over living cells. For example, living cells are more likely to be sensitive to metal ion concentration, environmental variables, and operating conditions. Furthermore, a consistent nutrient supply is required for systems using living cells besides the recovery of metals and regeneration of biosorbent is more complicated. On the other hand, dead biomass can easily and inexpensively be procured from industrial sources as a waste product. The use of dead cells in the biosorbent studies is receiving acceptance due to the absence of toxicity and it does not require growth media and nutrients. Moreover, the biosorbed metals can be adsorbed and recovered easily, the regenerated biomass can be reused, and the metal uptake reactors can be easily modeled mathematically. The dead mass of various fungal cells has shown the metal binding ability even at level greater than live cells (Merrin et al. 1998; Kogej and Pavko 2001).

21.5 Biosorption of Heavy Metal with Fungi

Today, biosorption is one of the main components of the environmental and bioresource technology (Park et al. 2010), which is considered as an alternative sustainable strategy for cleaning up the contaminated sites (Ngwenya et al. 2009). Fruiting bodies of macrofungi are considered to be ideal materials as biosorbents. It has been demonstrated that many fungal species exhibit high biosorptive potentials (Collin-Hansen et al. 2007; García et al. 2009) as listed in Table 21.1.

21.5.1 Factors Affecting Metal Sorption by Fungi

21.5.1.1 Pretreatment Effect

Pretreatment methods have usually shown an increase in the metal sorption capacity for a variety of fungal species. For example, alkali treatment (usually with NaOH) of fungal biomass for 4–6 h at 95–100°C deacetylates chitin present in the cell wall to form chitosan–glucan complexes with higher affinity for metal ions. It is reported

Table 21.1 Biosorption of metal by various fungus species

Fungi	Metals studied	Biosorption capacity	References
Saccharomyces cerevisiae	Cr^{6+}	93%	Peng et al. (2010)
Trichoderma harzianum	Cu^{2+}; Pb^{2+} Zn^{2+}	97%	Akhtar et al. (2007)
Trametes versicolor	Cd^{2+}	80%	Arıca et al. (2001)
Botrytis cinerea	Pb^{2+}	97%	Akar et al. (2005)
Inonotus hispidus	Zn^{2+}	30–60%	Sari and Tuzen (2009)
Aspergillus niger	Ni^{2+}	96%	Amini et al. (2008, 2009)
Saccharomyces cerevisiae	Ni^{2+}	89%	Machado et al. (2010)
Tremella fuciformis	Pb^{2+}	97%	Pan et al. (2010)
Auricularia polytricha	Pb^{2+}	91%	Pan et al. (2010)
Roccella phycopsis	Zn^{2+}, Cu^{2+}	37.8, 22.79 mg/g	Yalçln et al. (2010)
Ganoderma carnosum	Pb^{2+}	38.40 mg/g	Akar et al. (2006)
Amanita rubescens	Pb^{2+}	27.30 mg/g	Sari and Tuzen (2009)
Amanita rubescens	$Cd^{2+}As^{3+}$	59.6 mg/g	Sari and Tuzen (2009)
Fusarium spp.	Zn^{2+}	42.75 mg/g	Velmurugan et al. (2010)
Streptomyces ciscaucasicus	Cr^{6+}	50 mg/l	Li et al. (2010)
Agaricus bisporus	Pb^{2+}, Hg^{2+}, Cd^{2+}	247.2, 37.7, 23.8 mg/g	Ertugay and Bayhan (2007)
Aspergillus terreus	U(VI)	60 mg/l	Sun et al. (2010)
Aspergillus fumigatus	Cr(VI)	78 mg/g	Wang et al. (2010)
Rhizopus arrhizus	Cu^{2+}	79.37 mg/g	Aksu and Balibek (2007)
Candida lipolytica	Cu^{2+}	60 mg/l	Ye et al. (2010)
Rhodotorula glutinis	U	612 mg/g	Bai et al. (2009)
Trametes versicolor	Pb^{2+}	57.5 mg/g	Bayramoglu et al. (2003)
Polyporous versicolor	Pb^{2+}	110 mg/g	Yetis et al. (1998)
Phanerochaete chryosporium	Cd^{2+}	120.6 mg/g	Say et al. (2001)
Rhizopus cohnii	Hg^{2+}, Cd^{2+}, Zn^{2+}	403.2, 191.6, 4 mg/g	Jin-ming et al. (2010)
Funalia trogii	Cu^{2+}, Zn^{2+}, Cr^{6+}		Arıca et al. (2004)
Streptomyces rimosus	Cr^{6+}		Chergui et al. (2007)
Streptomyces rimosus	Pb^{2+}		Ammar (2009)
Rhizopus oligosporus	Cr^{6+}, Cu, Ni^{2+}, Zn		Ozsoy et al. (2008)
Phanarochaete chrysosporium	Cd^{2+}, Pb^{2+}, Cu^{2+}		Yetis et al. (1998)
Penicillium sp.	Cr^{6+}		Fukuda et al. (2008)
Aspergillus tubingensis	U		Coreño-Alonso et al. (2009)

that NaOH removes protein content of the cell wall, exposes more available metal binding sites, and increases the negative charge, thereby increasing the biosorption (Fourest and Roux 1992; Göksungur et al. 2005). On the contrary, biosorption is reduced in the presence of ethylenediamine tetra acetate (EDTA), sulfate, chloride, phosphate, carbonate, glutamate, citrate, and pyrophosphate. The presence of EDTA has been found to severely affect the biosorption of Cu, La, U, Ag, Cd, and Pb. In other study, acetone-pretreated *R. glutinis* cells showed higher Ni(II) biosorption capacity than untreated cells at pH values ranging from 3 to 7.5, with an optimum pH of 7.5. The effects of other relevant environmental parameters, such as initial Ni(II) concentration, shaking contact time, and temperature on Ni(II) biosorption onto acetone-pretreated *R. glutinis,* were also evaluated. Significant enhancement of Ni(II) biosorption capacity was observed by increasing initial metal concentration and temperature (Suazo-Madrid et al. 2010).

21.5.1.2 pH Effect

Hydrogen ion concentration is other factor that strongly affects the biosorptive ability of fungal species. For example, the biosorption of Cr, Ni, Zn, and Pb by *P. chrysogenum* was inhibited below pH 3 while it increased at acidic to basic range (Tan and Cheng 2003). The biosorption of Pb, Cd, Ni, and Zn was severely inhibited at pH below 4 (Brady et al. 1994). Fourest et al. (1994) observed that Zn biosorption on *M. miehei* and *P. chrysogenum* occurred at pH less than 4 and for *R. arrhizus,* which exhibited a higher Zn uptake, it was 5.8. The metal uptake for *R. arrhizus, M. miehei,* and *P. chrysogenum* increased from 16 to 35, 3 to 32, and 4.5 to 22 mg/g, respectively, when the pH of the reaction mixture was controlled at 7. Similarly, cadmium biosorption by fungal strains was pH sensitive. *Aspergillus oryzae, A. niger, F. solani,* and *Candida utilis* were found to perform better in the acidic range. The variation in the sorption capacity following change in pH range could be due to proton-competitive adsorption reaction (Huang 1986). Under uncontrolled conditions of pH, the drop in pH may create an undesired competition for metal ions from protons, thus lowering the metal uptake capacity. The protonation or poor ionization of acidic functional group of cell wall at low pH induces a weak complexation affinity between the cell wall and the metal ions. The reduction in metal ions uptake displayed by fungus at pH > 5.5 can be explained on the basis that at higher pH values, the metal ions may accumulate inside the cells, and/or the intra-fibular capillarities of the cell walls by a combined sorption microprecipitation mechanism; therefore, biosorption experiments are meaningless at higher pH.

21.5.1.3 Multi-metals Effect

Yan and Viraraghavan (2001) observed that the biosorption column of *Mucor rouxii* biomass was able to remove metal ions like Pb, Cd, Ni, and Zn not only from single component metal solutions but also from multi-component metal solutions. The metal adsorption rates and amount by the different fungal fruiting bodies in the multi-metal solutions are, however, generally lower than those in the single-metal solutions under the same experimental conditions. With more metal types involved, the metal rates and amount adsorbed by the fungal biomass decrease. The interactions among the different metals may influence the binding capacity of metals to the adsorption sites. Therefore, the uptake of metal ions in a competitive adsorption process would be lower than that for individual adsorption (Arief et al. 2008). In other study, Yakubu and Dudeney (1986) showed that biosorption of uranium on *A. niger* was substantially reduced in the presence of Cu, Zn, and Fe and the preferential order for biosorption was: Fe > U>Cu > Zn. Zhou and Kiff (1991) indicated that Mn, Zn, Cd, Mg, and Ca inhibited Cu biosorption by *R. arrhizus*. The metal uptake followed the order: Cu>Cr>Cd and Cu>Pb>Ni. The presence of anions also affects the biosorption of metal ions.

21.5.1.4   Cell Age and Contact Time

The age of cell also affects the biosorption of metal ions. Increased biosorption has been observed during the lag period or early stages of growth while it declines as cultures reaches stationary phase, as observed for *A. niger, P. spinulosum,* and *T. viride.* Volesky and May Phillips (1995) observed that 12-hour-old cultures of baker's yeast were able to biosorb 2.6 times more uranium than 24-hour-grown cultures. Biosorption of Cu, Zn, Cd, Pb, and U by non growing cells of *Penicillium, Aspergillus, Saccharomyces, Rhizopus,* and *Mucor* attained equilibrium in 1–4 h (Gadd et al. 1988; Mullen et al. 1992). Biosorption kinetics of metals is usually biphasic in nature, consisting of an initial rapid phase, contributing up to 90% biosorption, and lasting for 10 min. Second phase is slower and lasts up to 4 h (Huang et al. 1990). According to Kinetic studies, a contact time of 30 min was found enough to reach the equilibrium between cells and metals solution (Machado et al. 2009).

21.6   Biosorption Equilibrium Modeling

The kinetic mechanism that controls the metal biosorption process involves the pseudo-first-order and pseudo-second order kinetic models to interpret the experimental data (Ho and McKay 1998; Malkoc 2006). Generally, the pseudo-first-order kinetic model does not fit well to the whole range of an adsorption process and is usually applicable over the initial stage of the process, whereas the pseudo-second-order model fits experimental results better (Bulut et al. 2008; Gupta and Rastogi 2008; Kílíc et al. 2009). The pseudo-second-order model has been successfully used to describe chemisorptions involving valency forces through sharing or exchanging electrons between the adsorbent and adsorbate and through exchanging electrons among the particles involved (Kílíc et al. 2009). Several two-parameter (Langmuir, Freundlich, Temkin and Dubinin-Radushkevich) (Ho and McKay 1998; Özer and Özer 2003; Febrianto et al. 2009), three-parameter (Sips-Toth, Redlich-Peterson and Radke-Prausnitz) (Febrianto et al. 2009; Cayllahua et al. 2009; Abdel-Salam and Burk 2010), and four-parameter (Fritz-Schluender) (Abdel-Salam and Burk 2010) sorption isotherm models have been proposed, which are used to fit the experimental equilibrium data obtained at different initial metal concentrations (For details, see Chap. 8).

21.7 Conclusion

Fungi are known to tolerate and detoxify metals by several mechanisms including transformation, extra and intracellular precipitation, and active uptake. The ability of fungi to detoxify metals is the reasons that they are considered as potential alternative to chemical means of remediation of metals. Considering this, it is expected that identifying metal tolerant/metal removing fungi may help to clean up the contaminated environment. Biosorption as metal removal strategy can be useful in the decontamination of heavy metal–contaminated soils. More information is, however, required to understand the mechanistic basis of biosorption process. The methods to harvest more and more fungal biomass need to be developed. As biosorption technology decreases the costs of metal removal due to the usage of natural biological materials, it might be considered as an additional process for the decontamination of lands.

References

Abdel-Salam M, Burk RC (2010) Thermodynamics and kinetic studies of pentachlorophenol adsorption from aqueous solutions by multi-walled carbon nanotubes. Water Air Soil Pollut 210:101–111

Adepoju-Bello AA, Ojomolade OO, Ayoola GA, Coker HAB (2009) Quantitative analysis of some toxic metals in domestic water obtained from Lagos metropolis. Niger J Pharm 42:57–60

Akar T, Tunali S, Kiran I (2005) *Botrytis cinerea* as a new fungal biosorbent for removal of Pb(II) from aqueous solutions. Biochem Eng J 25:227–235

Akar T, Cabuk A, Tunali S, Yamac M (2006) Biosorption potential of the macrofungus Ganoderma carnosum for removal of lead(II) ions from aqueous solutions. J Environ Sci Health A Tox Hazard Subst Environ Eng 41:2587–2606

Akhtar K, Akhtar MW, Khalid AM (2007) Removal and recovery of uranium from aqueous solutions by *Trichoderma harzianum*. Water Res 41:1366–1378

Aksu Z, Balibek E (2007) Chromium (VI) biosorption by dried Rhizopus arrhizus: effect of salt (NaCl) concentration on equilibrium and kinetic parameters. J Hazard Mater 25:210–220

Alexander M (1994) Biodegradation and bioremediation. Academic, San Diego

Alkorta I, Hernández-Allica Becerril JM, Amezaga I, Albizu I, Garbisu C (2004) Recent findings on the phytoremediation of soils contaminated with environmentally toxic heavy metals and metalloids such as zinc, cadmium, lead, and arsenic. Rev Environ Sci Biotechnol 3:71–90

Amini M, Younesi H, Bahramifar N, Lorestani AA, Ghorbani F, Daneshi A, Sharifzadeh M (2008) Application of response surface methodology for optimization of lead biosorption in an aqueous solution by *Aspergillus niger*. J Hazard Mater 154:694–702

Amini M, Younesi H, Bahramifar N (2009) Biosorption of nickel(II) from aqueous solution by *Aspergillus niger*: response surface methodology and isotherm study. Chemosphere 75:1483–1491

Ammar S (2009) Biosorption of Cr^{6+} ions from aqueous solution by a *Streptomyces rimosus* biomass. New Biotechnol 25:S274

Anand P, Isar J, Saran S, Saxena RK (2006) Bioaccumulation of copper by *Trichoderma viride*. Bioresour Technol 97:1018–1025

Antonijevic MM, Maric M (2008) Determination of the content of heavy metals in pyrite contaminated soil and plants. Sensors 8:5857–5865

Arıca MY, Kaçar Y, Genç O (2001) Entrapment of white-rot fungus Trametes versicolor in Ca-alginate beads: preparation and biosorption kinetic analysis for cadmium removal from an aqueous solution. Bioresour Technol 80:121–129

Arıca MY, Lu GB, Yılmaz M, Bekta S, Genç O (2004) Biosorption of Hg2+, Cd2+, and Zn2+ by Ca-alginate and immobilized wood-rotting fungus *Funalia trogii*. J Hazard Mater 109:191–199

Arief VO, Trilestari K, Sunarso J, Indraswati N, Ismadji S (2008) Recent progress on biosorption of heavy metals from liquids using low cost biosorbents: characterization, biosorption parameters and mechanism studies. Clean 36:937–962

Ayodele SM, Odogbili OD (2010) Metal impurities in three edible mushrooms collected in Abraka, Delta State, Nigeria. Micol Apl Int 22:27–30

Bai J, Qin Z, Wang JF, Guo JS, Zhang LN, Fan FL, Lin MS, Ding HJ, Lei FA, Wu XL, Li XF (2009) Study on biosorption of uranium by *Rhodotorula glutinis*. Guang Pu Xue Yu Gang Pu Fen Xi 29:1218–1221

Baldrian P (2003) Interactions of heavy metals with white-rot fungi. Enzyme Microb Technol 32:78–91

Bayramoglu G, Baktas S, Arica MY (2003) Biosorption of heavy metal ions on immobilized white-rot fungus *Trametes versicolor*. J Hazard Mater 101:285–300

Bhattacharyya P, Tripathy S, Chakrabarti K, Chakraborty A, Banik P (2008) Fractionation and bioavailability of metals and their impacts on microbial properties in sewage irrigated soil. Chemosphere 72:543–550

Brady D, Stoll A, Ducan JR (1994) Biosorption of heavy metal cations by non-viable yeast biomass. Environ Technol 15:429–438

Brady D, Duncan JR (1994a) Binding of heavy metals by the cell walls of *Saccharomyces cerevisiae*. Enzyme Microb Technol 16:633–638

Bulut E, Ozacara M, Sengil IA (2008) Equilibrium and kinetic data and process design for adsorption of Congo Red onto bentonite. J Hazard Mater 154:613–622

Cayllahua JEB, de Carvalho RJ, Torem ML (2009) Evaluation of equilibrium, kinetic and thermodynamic parameters for biosorption of nickel(II) ions onto bacteria strain, *Rhodococcus opacus*. Miner Eng 22:1318–1325

Chatterjee N, luo Z (2010) Exposure-response of Cr (iii)-organic complexes to *Saccharomyces cerevisiae*, front. Environ Sci Eng China 4:196–202

Chergui A, Bakhti MZ, Chahboub A, Haddoum S, Selatnia A, Junter GA (2007) Simultaneous biosorption of Cu$^{2+,}$ Zn^{2+} and Cr^{6+} from aqueous solution by *Streptomyces rimosus* biomass. Desalination 206:179–184

Collin-Hansen C, Pedersen SA, Andersen RA, Steinnes E (2007) First report of phytochelatins in a mushroom: induction of phytochelatins metal exposure in Boletus edulis. Mycologia 99:161–174

Coreño-Alonso A, Acevedo-Aguilar FJ, Reyna-López GE, Tomasini A, Fernández FJ, Wrobel K, Wrobel K, Gutiérrez-Corona JF (2009) Cr(VI) reduction by an Aspergillus tubingensis strain: role of carboxylic acids and implications for natural attenuation and biotreatment of Cr(VI) contamination. Chemosphere 76:43–47

Courtecuisse R (1999) Collins guide to the mushrooms of Britain and Europe. Harper Collins Publishers, London

El-Morsy EM (2004) *Cunninghamella echinulata* a new biosorbent of metal ions from polluted water in Egypt. Mycologia 96:1183–1189

Ertugay N, Bayhan YK (2007) Biosorption of Cr (VI) from aqueous solutions by biomass of Agaricus bisporus. J Hazard Mater 154:432–439

Febrianto J, Kosasih AN, Sunarso J, Ju YH, Indraswati N, Ismadju S (2009) Equilibrium and kinetic studies in adsorption of heavy metals using biosorbent: a summary of recent studies. J Hazard Mater 162:616–645

Fourest E, Roux JC (1992) Heavy metal biosorption by fungal mycelial by-products: mechanisms and influence of pH. Appl Microbiol Biotechnol 37:399–403

Fourest E, Canal C, Roux J (1994) Improvement of heavy metal biosorption by mycelial dead biomass (*Rhizopus arrhizus, Mucor miehei, Penicillium chrysogenum*): pH control and cationic activation. FEMS Microbiol Rev 14:325–332

Fukuda T, Ishino Y, Ogawa A, Tsutsumi K, Morita H (2008) Cr(VI) reduction from contaminated soils by *Aspergillus* sp. N2 and *Penicillium* sp. N3 isolated from chromium deposits. J Gen Appl Microbiol 54:295–303

Gadd GM, White C, de Rome L (1988) Heavy metal and radionuclide uptake by by fungi and yeasts. In Norris PR, Kelly DP (eds) *BioHydroMetallurgy: Proceedings of the international symposium*. Sci Tech Letters Kew, pp 421–436

García MA, Alonso J, Melgar MJ (2009) Lead in edible mushrooms levels and bioaccumulation factors. J Hazard Mater. doi:10.1016/ j.jhazmat.2009.09.058

Gavrilesca M (2004) Removal of heavy metals from the environment by biosorption. Eng Life Sci 4:219–232

Gharieb MM, Sayer JA, Gadd GM (1998) Solubilization of natural gypsum (CaSO$_4$.2H$_2$O) and the formation of calcium oxalate by *Aspergillus niger* and *Serpula himantioides*. Mycol Res 102:825–830

Gikas P (2008) Single and combined effects of nickel (Ni(II)) and Cobalt (Co(II)) ions on activated sludge and on aerobic microorganisms: a review. J Hazard Mater 159:187–203

Göksungur Y, Üren S, Güvenc U (2005) Biosorption of cadmium and lead ions by ethanol treated waste baker's yeast biomass. Bioresour Technol 96:103–109

Gupta VK, Rastogi A (2008) Equilibrium and kinetic modelling of cadmium(II) biosorption by nonliving algal biomass *Oedogonium* sp. from aqueous phase. J Hazard Mater 153:759–766

Gupta R, Ahuja P, Khan S, Saxena RK, Mohapatra H (2000) Microbial biosorbents: Meeting challenges of heavy metal pollution in aqueous solutions. Curr Sci 78:967–973

Gupta S, Nayek S, Saha RN, Satpati S (2008) Assessment of heavy metal accumulation in macrophyte, agricultural soil and crop plants adjacent to discharge zone of sponge iron factory. Environ Geol 55:731–739

Ho YS, McKay G (1998) Sorption of dye from aqueous solution by peat. Chem Eng J 70:115–124

Huang CP (1986) Management alternatives for plating waste. In: Proceedings, pollution control engineering, 11th modern engineering technology seminar, IX, pp 20

Huang C, Huang CP, Morehart AL (1990) The removal of Cu (II) from dilute aqueous solutions by *Saccharomyces cerevisiae*. Water Res 24:433–439

Javaid A, Rukhsana B, Arshad J (2010) Biosorption of heavy metals using a dead macro fungus *Schizophyllum commune* fries: evaluation of equilibrium and kinetic models. Pak J Bot 42:2105–2118

Jin-ming LUO, Xiao XIAO, sheng-lian LUO (2010) Biosorption of cadmium(II) from aqueous solutions by industrial fungus *Rhizopus cohnii*. Trans Nonferrous Metals Soc China 20:1104–1111

Jonathan MP, Ram Mohan V, Srinivasulu S (2004) Geochemical variation of major and trace elements in recent sediments of the Gulf of Mannar the southeast coast of India. Environ Geol 45:466–480

Jung SUH, Ho KIM, Seog D, Jong Won YON, Seung Koo SONG (1998) Process of Pb^{2+} accumulation in *Saccharomyces cerevisiae*. Biotechnol Lett 20:153–156

Kahraman S, Asma D, Erdemoglu S, Yesilada O (2005) Biosorption of copper(II) by live and dried biomass of the white rot fungi *Phanerochaete chrysosporium* and *Funalia trogii*. Eng Life Sci 5:72–77

Kalac P (2009) Chemical composition and nutritional value of European species of wild growing mushrooms: a review. Food Chem 113:9–16

Kílíc M, Yazící H, Solak M (2009) A comprehensive study on removal and recovery of copper(II) from aqueous solutions by NaOH pretreated *Marrubium globosum* spp. globosum leaves powder: potential for utilizing the copper(II) condensed desorption solutions in agricultural applications. Bioresour Technol 100:2130–2137

Kogej A, Pavko A (2001) Laboratory experiments of lead biosorption by self-immobilized *Rhizopus nigricans* pellets in the batch stirred tank reactor and the packed bed column. Chem Biochem Eng 15:75–79

Krishna AK, Govil PK (2007) Soil contamination due to heavy metals from an industrial area of surat, Gujarat, Western India. Environ Monit Assess 124:263–275

Li Z, Yuan H (2006) Characterization of cadmium removal by Rhodotorrula sp. Y11. Appl Microbiol Biotechnol 73:458–463

Li H, Lin Y, Guan W, Chang J, Xu L, Guo J, Wei GI (2010) Biosorption of Zn(II) by live and dead cells of *Streptomyces ciscaucasicus* strain CCNWHX 72–14. J Hazard Mater 179:151–159

Machado MD, Janssens S, Soares HMVM, Soares EV (2009) Removal of heavy metals using a brewer's yeast strain of Saccharomyces cerevisiae: advantages of using dead biomass. J Appl Microbiol 106:1792–1804

Machado MD, Soares EV, Soares HM (2010) Removal of heavy metals using a brewer's yeast strain of Saccharomyces cerevisiae: chemical speciation as a tool in the prediction and improving of treatment efficiency of real electroplating effluents. J Hazard Mater 180:347–353

Malkoc E (2006) Ni(II) removal from aqueous solutions using cone biomass of Thuja orientalis. J Hazard Mater B 137:899–908

Mameri N, Boudries N, Addour L, Belhocine D, Lounici H, Grib H, Pauss A (1999) Batch zinc biosorption by a bacterial nonliving Streptomyces rimosus biomass. Water Res 33:1347–1354

Mapanda F, Mangwayana EN, Nyamangara J, Giller KE (2005) The effect of long-term irrigation using wastewater on heavy metal contents of soils under vegetables in Harare, Zimbabwe. Agric Ecosyst Environ 107:151–165

Marcovecchio JE, Botte SE, Freije RH (2007) Heavy metals, major metals, trace elements. In: Nollet LM (ed.) Handbook of water analysis, 2nd edn. CRC Press, London, pp 275–311

Marschner H (1995) Mineral nutrition of higher plants. Academic, London

Merrin JS, Sheela R, Saswathi N, Prakasham RS, Ramakrishna SV (1998) Biosorption of chromium (VI) using *Rhizopus arrhizus*. Indian J Exp Biol 36:1052–1055

Morales-Barrera L, Cristiani-Urbina E (2008) Hexavalent chromium removal by a Trichoderma inhamatum fungal strain isolated from tannery effluent. Water Air Soil Pollut 187:327–336

Morales-Barrera L, Guillén-Jiménez FM, Ortiz-Moreno A, Villegas-Garrido TL, Sandoval-Cabrera A, Hernández-Rodríguez CH (2008) Isolation, identification and characterization of a Hypocrea tawa strain with high Cr(VI) reduction potential. Biochem Eng J 40:284–292

Mullen MD, Wolf Dc, Beveridge TJ, Bailey GW (1992) Sorption of heavy metals by the soil fungi *Aspergillus niger* and *Mucor roxii*. Soil Biol Biochem 24:129–135

Ngwenya BT, Mosselmans WJ, Magennis M, Atkinson KD, Tourney J, Olive V, Ellam RM (2009) Macroscopic and spectroscopic analysis of lanthanide adsorption to bacterial cells. Geochim Cosmochim Acta 73:3134–3147

Niyogi S, Abraham TE, Ramakrishna SV (1998) Removal of chromium (VI) ions from industrial effluents by immobilized biomass of *Rhizopus arrhizus*. J Sci Ind Res 57:809–816

Özer A, Özer D (2003) Comparative Study of the Biosorption Pb(II), Ni(II) and Cr(VI) Ions onto *S. cerevisiae*: Determination of biosorption heats. J Hazard Mater B100:219–229

Ozsoy HD, Kumbur H, Saha B, van Leeuwen JH (2008) Use of Rhizopus oligosporus produced from food processing wastewater as a biosorbent for Cu(II) ions removal from the aqueous solutions. Bioresour Technol 99:4943–4948

Pal A, Ghosh S, Paul AK (2006) Biosorption of cobalt by fungi from serpentine soil of Andaman. Bioresour Technol 97:1253–1258

Pan R, Cao L, Zhang R (2009) Combined effects of Cu, Cd, Pb, and Zn on the growth and uptake of consortium of Cu-resistant *Penicillium* sp. A1 and Cd-resistant *Fusarium* sp. A19. J Hazard Mater 171:761–766

Pan R, Cao L, Huang H, Zhang R, Mo Y (2010) Biosorption of Cd, Cu, Pb, and Zn from aqueous solutions by the fruiting bodies of jelly fungi (*Tremella fuciformis* and *Auricularia polytricha*). Appl Microbiol Biotechnol 88:997–1005

Park JY, Jung JH, Seo DH, Ha SJ, Yoon JW, Kim YC, Shim JH, Park CS (2010) Microbial production of palatinose through extracellular expression of a sucrose isomerase from *Enterobacter sp.* FMB-1 in *Lactococcus lactis* MG1363. Bioresour Technol 101:8828–8833

Parvathi K, Nagendran R (2007) Biosorption of chromium from effluent generated in chrome-electroplating unit using *Saccharomyces cerevisiae*. Sep Sci Technol 42:625–638

Peng Q, Liu Y, Zeng G, Xu W, Yang C, Zhang J (2010) Biosorption of copper(II) by immobilizing *Saccharomyces cerevisiae* on the surface of chitosan-coated magnetic nanoparticles from aqueous solution. J Hazard Mater 177:676–682

Ramsay LM, Gadd GM (1997) Mutants of Saccharomyces cerevisiae defective in vacuolar function confirms a role for the vacuole in toxic metal ion detoxification. FEMS Microbiol Lett 152:293–298

Rehman A, Anjum MS (2010) Cadmium uptake by yeast, *Candida tropicalis*, Isolated from Industrial Effluents and Its Potential Use in Wastewater Clean-Up Operations. Water Air Soil Pollut 205:149–159

Rehman A, Farooq H, Shakoori AR (2007) Copper tolerant yeast, Candida tropicalis, isolated from industrial effluents: Its potential use in wastewater treatment. Pak J Zool 39:405–412

Rosen BP (2002) Transport and detoxification systems for transition metals, heavy metals and metalloids in eukaryotic and prokaryotic microbes. Comp Biochem Physiol A Mol Integr Physiol 133:689–693

Sari A, Tuzen M (2009) Biosorption of As(III) and As(V) from aqueous solution by macrofungus (Inonotus hispidus) biomass: equilibrium and kinetic studies. J Hazard Mater 164:1372–1378

Say R, Denizli A, Arıca MY (2001) Biosorption of cadmium(II), lead(II) and copper(II) with the filamentous fungus *Phanerochaete chrysosporium*. Bioresour Technol 76:67–70

Shen J, Hsu CM, Kang BK, Rosen BP, Bhattacharjee H (2003) The Saccharomyces cerevisiae Arr4p is involved in metal and heat tolerance. Biometals 16:369–378

Simmons P, Singleton I (1996) A method to increase silver biosorption by an industrial strain of *Saccharomyces cerevisiae*. Appl Microbiol Biotechnol 45:278–285

Singh KP, Mohon D, Sinha S, Dalwani R (2004) Impact assessment of treated/untreatedwastewater toxicants discharge by sewage treatment plants on health, agricultural, and environmental quality in wastewater disposal area. Chemosphere 55:227–255

Suazo-Madrid A, Morales-Barrera L, Aranda-García E, Cristiani-Urbina E (2010) Nickel(II) biosorption by Rhodotorula glutinis. J Ind Microbiol Biotechnol 38:51–64

Suh JH, Yun JW, Kim DS (1998) Comparison of Pb2+ accumulation characteristics between live and dead cells of Saccharomyces cerevisiae and *Aureobasidium pullulans*. Biotechnol Lett 20:247–251

Suh SJ, Silo-Suh LA, Woods DE, Hassett DJ, West SEH, Ohman DE (1999) Effect of *rpoS* mutation on the stress response and expression of virulence factors in *Pseudomonas aeruginosa*. J Bacteriol 181:3890–3897

Sun YM, Horng CY, Chang FL, Cheng LC, Tian WX (2010) Biosorption of lead, Mercury, and cadmium ions by *Aspergillus terreus* immobilized in a natural matrix. Pol J Microbiol 59:37–44

Svoboda L, Havlickova B, Kalac P (2006) Contents of cadmium, mercury and lead in edible mushrooms growing in a historical silver-mining area. Food Chem 96:580–585

Tan T, Cheng P (2003) Biosorption of metal ions with *Penicillium chrysogenum*. Appl Biochem Biotechnol 104:119–128

Vala AK, Davariya V, Upadhyay RV (2010) An investigation on tolerance and accumulation of a facultative marine fungus *Aspergillus flavus* to pentavalent arsenic. Ocean Univ China (Oceanic and Coastal Sea Research) 9:65–67

Valix M, Tang JY, Malik R (2001) Heavy metal tolerance of fungi. Miner Eng 14:499–505

Van-der-Heggen M, Martins S, Flores G, Soares EV (2010) Lead toxicity in *Saccharomyces cerevisiae*. Appl Microbiol Biotechnol. doi:10.1007/s00253-010-2799-5

Velmurugan P, Shim J, You Y, Choi S, Kamala-Kannan S, Lee KJ, Kim HJ, Oh BT (2010) Removal of zinc by live, dead, and dried biomass of *Fusarium* spp. isolated from the abandoned-metal mine in South Korea and its perspective of producing nanocrystals. J Hazard Mater 182:317–324

Vijver MG, Van Gestel CA, Lanno RP, Van Straalen NM, Peijnenburg WJ (2004) Internal metal sequestration and its ecotoxicological relevance: a review. Environ Sci Technol 38:4705–4712

Villegas LB, Fernández PM, Amoroso MJ, Figueroa LIC (2008) Chromate removal by yeasts isolated from sediments of a tanning factory and a mine site in Argentina. Biometals 21:591–600

Volesky B (1990) Biosorption and biosorbent. In: Biosorption of heavy metals. CRC Press, Boca Raton, pp 3–43

Volesky B (1994) Advances in biosorption of metals: selection of biomass types. FEMS Microbiol Rev 14:291–302

Volesky B, May Phillips HA (1995) Biosorption of heavy metals by *Saccharomyces cerevisiae*. Appl Microbiol Biotechnol 42:797–806

Volesky B, May H, Holan ZR (1993) Cadmium biosorption by *Saccharomyces cerevisiae*. Biotechnol Bioeng 41:826–829

Wang J, Chen C (2006) Biosorption of heavy metals by Saccharomyces cerevisiae: a Review. Biotechnol Adv 24:427–451

Wang JS, Hu XJ, Liu YG, Xie SB, Bao ZL (2010) Biosorption of uranium (VI) by immobilized *Aspergillus fumigatus* beads. J Environ Radioact 101:504–508

Wei S, Zhou Q (2008) Trace elements in agro-ecosystems. In: Prasad MNV (ed.) Trace elements as contaminants and nutrients consequences in ecosystems and human health. Wiley, Somerset, pp 55–80

Wilson B, Brennan L, Brian PF (2005) The dispersion of heavy metals in the vicinity of Britannia Mine, British Columbia, Canada. Ecotoxicol Environ Saf 60:269–276

Wu GH, Cao SS (2010) Mercury and cadmium contamination of irrigation water, sediment, soil and shallow groundwater in a wastewater- irrigated field in Tianjin, China. Bull Environ Contam Toxicol 84:336–341

Xiangliang P, Jianlong W, Daoyong Z (2005) Biosorption of Pb(II) by *Pleurotus ostreatus* Immobilized in Calcium Alginate Gel [J]. Process Biochem 40:2799–2803

Yalçln E, Cavu oglu K, Kinalioglu K (2010) Biosorption of Cu2+ and Zn2+ by raw and autoclaved Rocella phycopsis. J Environ Sci (China) 22:367–373

Yakubu NA, Dudeney AWL (1986) Biosorption of uranium with *Aspergillus niger*. In: Eccles HH, Hunt S (eds) Immobilisation of ions by biosorption. Ellis Horwood, Chichester, pp 183–200

Yan G, Viraraghavan T (2001) Heavy metal removal in a biosorption column by immobilized *Mucor rouxii* biomass. Bioresour Technol 78:243–249

Ye J, Yin H, Mai B, Peng H, Qin H, He B, Zhang N (2010) Biosorption of chromium from aqueous solution and electroplating wastewater using mixture of Candida lipolytica and dewatered sewage sludge. Bioresour Technol 101:3893–3902

Yetis U, Özcengiz G, Dilek FB, Ergen N, Erbay A, Dölek A (1998) Heavy metal biosorption by white-rot fungi. Water Sci Technol 38:323–330

Zhou JL, Kiff RJ (1991) The uptake of copper from aqueous solution by immobilized fungal biomass. J Chem Technol Biotechnol 52:317–330

Index